Genetics of Sleep and Sleep Disorders

Philip Gehrman • Alex C. Keene
Struan F. Grant
Editors

Genetics of Sleep and Sleep Disorders

Springer

Editors
Philip Gehrman
Department of Psychiatry
Perelman School of Medicine of the
University of Pennsylvania
Philadelphia, PA, United States

Alex C. Keene
Department of Biology
Texas A&M University
College Station, TX, USA

Struan F. Grant
Divisions of Human Genetics and
Endocrinology & Diabetes
CHOP Research Institute
Philadelphia, PA, USA

ISBN 978-3-031-62722-4 ISBN 978-3-031-62723-1 (eBook)
https://doi.org/10.1007/978-3-031-62723-1

This Springer imprint is published by the registered company Springer Nature Switzerland AG
The registered company address is: Gewerbestrasse 11, 6330 Cham, Switzerland

Contents

About the Editors

Philip Gehrman is Professor of Psychology in the Department of Psychiatry of the University of Pennsylvania School of Medicine, and a clinical psychologist at the Philadelphia VA Medical Center. He directs the Sleep, Neurobiology and Psychopathology Lab at Penn. He has an active research program exploring the mechanisms and treatment of sleep and circadian dysregulation in the context of mental health disorders. Dr. Gehrman's clinical specialization is on the delivery of cognitive behavioral and chronotherapeutic interventions for insomnia, circadian rhythm disorders, and other sleep disorders. The overarching goal of his work is to advance the understanding of the links between sleep and mental illness through translational research that spans biology to therapeutics.

Alex C. Keene, PhD, studies genetic regulation of sleep in a variety of organisms including mosquitos, fruit flies, zebrafish, Mexican cavefish, and African cichlids. He has applied genetic screening in fruit flies to identify novel genes that contribute to sleep, particularly focusing on identifying genes and neurons that link sleep and metabolism. He has also led the development of Mexican cavefish as a model for insomnia, identifying multiple, independently evolved populations of cavefish that have lost sleep. His current work, in collaboration with the co-authors of this book, has used fly and fish models to validate sleep genes first identified through large-scale human genome-wide association studies. The use of these integrative approaches has the potential to novel biological processes.

Struan F. Grant, PhD, has been conducting human genomics research for over 20 years. Some of the highlights of his career include the discovery of variation in the TCF7L2 gene playing a key role in conferring type 2 diabetes risk and providing leadership in international genetics efforts to characterize genes influencing birth weight and common childhood obesity risk. His current work continues to primarily investigate disease genomics, with a specific focus on pediatrics. Utilizing high-throughput genotyping and sequencing technologies, combined with statistical and bioinformatic approaches, his goals include unraveling genomic puzzles related to childhood obesity, early onset diabetes, and sleep-related traits.

About the Editors

Introduction

Philip Gehrman, Struan F. Grant, and Alex C. Keene

Abstract As sleep is increasingly recognized as a key pillar of a balanced and healthy lifestyle, this awareness ironically comes at a time when the rates of insufficient quantity and quality of sleep are steadily increasing. There is substantial evidence that there are negative consequences to habitual "short sleep," defined as sleeping <6 h per night [1]. There appears to be a particularly strong relationship between short sleep and the health span, in particular with respect to negative cardiovascular and metabolic outcomes. For example, a number of studies have demonstrated strong associations between sleep duration and obesity across all age groups [2, 3]. There are also a number of common sleep disorders, such as insomnia and sleep apnea, that have major public health consequences [4], including increased motor vehicle accidents [5], higher likelihood of depression/anxiety [6], greater risk for insulin resistance/obesity, altered appetite control [3], and cardiovascular morbidity [7]. The pervasive impacts of sleep on societal health underscore the need for improved understanding and treatments for sleep disorders.

P. Gehrman (✉)
Department of Psychiatry, Perelman School of Medicine of the University of Pennsylvania, Philadelphia, PA, United States
e-mail: philip.gehrman@pennmedicine.upenn.edu

S. F. Grant
Center for Spatial and Functional Genomics, Children's Hospital of Philadelphia, Philadelphia, PA, USA

Department of Genetics, University of Pennsylvania, Philadelphia, PA, USA

Institute for Diabetes, Obesity and Metabolism, Perelman School of Medicine, University of Pennsylvania, Philadelphia, PA, USA

Department of Pediatrics, The University of Pennsylvania Perelman School of Medicine, Philadelphia, PA, USA

Divisions of Human Genetics and Endocrinology & Diabetes, Children's Hospital of Philadelphia, Philadelphia, PA, USA
e-mail: grants@chop.edu

A. C. Keene
Department of Biology, Texas A&M University, College Station, TX, USA

P. Gehrman et al. (eds.), *Genetics of Sleep and Sleep Disorders*,
https://doi.org/10.1007/978-3-031-62723-1_1

As sleep is increasingly recognized as a key pillar of a balanced and healthy lifestyle, this awareness ironically comes at a time when the rates of insufficient quantity and quality of sleep are steadily increasing. There is substantial evidence that there are negative consequences to habitual "short sleep," defined as sleeping <6 h per night [1]. There appears to be a particularly strong relationship between short sleep and the health span, in particular with respect to negative cardiovascular and metabolic outcomes. For example, a number of studies have demonstrated strong associations between sleep duration and obesity across all age groups [2, 3]. There are also a number of common sleep disorders, such as insomnia and sleep apnea, that have major public health consequences [4], including increased motor vehicle accidents [5], higher likelihood of depression/anxiety [6], greater risk for insulin resistance/obesity, altered appetite control [3], and cardiovascular morbidity [7]. The pervasive impacts of sleep on societal health underscore the need for improved understanding and treatments for sleep disorders.

One of the fundamental components of sleep/wake regulation is circadian timing. Virtually all physiologic processes follow predictable circadian rhythms that synchronize most organic processes in response to changes to the external environment, in particular light, temperature, and nutrient availability [8]. While the presence of circadian rhythms is nearly universal in humans, there is considerable interindividual variability in the timing of rhythms, leading to strong preferences for daytime or nighttime activity. This variability can be measured via the "preferred sleep phase" or chronotype. When individuals need to follow a schedule that is inconsistent with their circadian tendency, they often experience circadian rhythm sleep/wake disorders, such as commonly seen in shift workers and during transmeridian travel [9]. Therefore, it is critical to understand how natural variation in circadian systems impact sleep and health outcomes.

It has long been demonstrated that sleep and circadian rhythms are under strong genetic control. There have been tremendous advances in both humans and model organisms to gain critical insight into the genetic factors that regulate these processes and the characterization of genetic variants that contribute to disorders in humans. The role of genetic factors in humans has been studied for decades, progressing from classical heritability and linkage studies, to more focused candidate gene analyses, then on to more hypothesis-free genome-wide approaches. Genome-wide analyses have been made possible by the sequencing of the human genome, which has led on to scientific advances in array-based genotyping technologies and, more recently, whole-exome and whole-genome sequencing analyses, as well as typing of genome-wide epigenetic modifications. However, there remains an opportunity for further discoveries, particularly for sleep disorders. To date, only a relatively small number of genetic risk variants have been discovered for sleep-related traits which only explain a minority of the predicted genetic component of these traits. There are several reasons for this lack of discovery, including inadequate sample sizes, heterogeneity in phenotypes, lack of trans-ethnic analyses, and multifactorial pathways to disease.

Striking advances in sleep and circadian genetics have often resulted from research in model systems. Indeed, the core components of the molecular circadian clock were identified using large-scale genetic screens in fruit flies over dating back nearly half a century. These discoveries led to the 2017 Nobel Prize in Physiology or Medicine to Jeffrey Hall, Michael Rosbash, and Michael Young, and subsequent work has shown the mechanisms they identified to be highly conserved in mammals. More recently, these advances have been extended to the study of sleep. Behavioral methods for defining sleep were first applied to simple genetic models, including flies and worms, in the early 2000s, leading to the identification of many new genes involved in sleep. These studies have been extended to flies and fish to screen for neural circuits and drug screens that regulate sleep. More recently, this work has led to a recognition that both the genes and neural circuits are evolutionary ancient, resulting in the study of sleep in simple animal models including jellyfish. Further, physiological approaches suggest that the evolutionarily conserved mechanisms of sleep extend beyond behavior to the fundamental components of cellular function. This book highlights the growing role that diverse animal models play in the regulation of sleep.

The goals of this book are to review the present science on the genomics of sleep and sleep disorders and to highlight gaps in knowledge to inform the next steps for research in this area. We have invited leaders in the field to describe both their own work and the broader science in each topic area. Together, these chapters highlight the rapidly increasing emphasis of interdisciplinary approaches and diversification of animal models that are being used in the field.

In the first section of the book, we focus on genetic studies of sleep and sleep disorders in model systems, beginning with the pioneering studies in *Drosophila*. Additional chapters focus on genetic studies in *C. elegans*, zebrafish, and mouse models, highlighting the deep evolutionary conservation of sleep regulation. Finally, there will be a discussion of the latest research on jellyfish and cavefish as novel model systems for studying sleep.

The second section of the book will review the genetics of human and sleep and sleep disorders, starting with a methodological primer on approaches to human genetic studies. The availability of large genotyped cohorts such as the UK Biobank has led to a surge in Mendelian randomization studies to establish the cause and effect relationships between sleep/circadian traits and a range of health outcomes. A review of these studies will be followed by chapters focused on individual sleep disorders including insomnia, chronotype/circadian rhythm disorders, sleep apnea, narcolepsy, and restless legs syndrome.

The final section of the book will round out the coverage of this topic by reviewing the latest science in a series of domains. The detailed biology of the circadian clock is one of the most significant success stories in genetics research to date. The extent to which the response to sleep deprivation is affected by genetic factors will be discussed. Additional chapters will focus on the genetics of sleep in the context of neurodevelopmental disorders, and in children. The genetic links between sleep and chronic disease will be reviewed. And the final chapter will demonstrate the role of genetics in understanding the importance of sleep in learning and memory.

This list of topics is far from exhaustive, but this content covers the breadth and depth of research on the genetics of sleep and sleep disorders. This book demonstrates the impressive progress the collective field has made in this area, particularly in the past 10–15 years. We trust that this book provides a foundation for today and points the direction for the next generation of research. A detailed understanding of genetics is critical to understand the importance of sleep and circadian rhythms to organismal biology and to the prevention and treatment of sleep disorders.

References

1. Grandner MA, Patel NP, Gehrman PR, Perlis ML, Pack AI. Problems associated with short sleep: bridging the gap between laboratory and epidemiological studies. Sleep Med Rev. 2010;14(4):239–47.
2. Hall MH, Muldoon MF, Jennings JR, Buysse DJ, Flory JD, Manuck SB. Self-reported sleep duration is associated with the metabolic syndrome in midlife adults. Sleep. 2008;31(5):635–43.
3. Knutson KL, Spiegel K, Penev P, Van Cauter E. The metabolic consequences of sleep deprivation. Sleep Med Rev. 2007;11(3):163–78.
4. Colten HR, Altevogt BM. Sleep Disorders and Sleep Deprivation: An Unmet Public Health Problem. Washington DC: The National Academies Press; 2006.
5. Banks S, Dinges DF. Behavioral and physiological consequences of sleep restriction. J Clin Sleep Med JCSM Off Publ Am Acad Sleep Med. 2007;3(5):519–28.
6. Riemann D, Voderholzer U. Primary insomnia: a risk factor to develop depression? J Affect Disord. 2003;76(1–3):255–9.
7. Grandner MA, Jackson NJ, Pak VM, Gehrman PR. Sleep disturbance is associated with cardiovascular and metabolic disorders. J Sleep Res. 2012;21(4):427–33.
8. Rijo-Ferreira F, Takahashi JS. Genomics of circadian rhythms in health and disease. Genome Med. 2019;11(1):82.
9. Burgess HJ, Emens JS. Circadian-Based Therapies for Circadian Rhythm Sleep-Wake Disorders. Curr Sleep Med Rep. 2016;2(3):158–65.

Part I
Animal Models: Genetic Studies of Sleep In

The Genetics of Sleep in *Drosophila*

Narendra Pratap Singh, Arijit Ghosh, and Susan T. Harbison

Abstract This chapter explores two decades of research using the model organism *Drosophila melanogaster* to decipher the genetic underpinnings of sleep. The approaches used to map sleep phenotypes to genomic variants and genes are discussed. Also presented are genetic approaches to localize sleep circuitry in the brain and peripheral tissues. Additionally, the chapter explores the role of physiological and environmental context on sleep, including aging, immune response, metabolism, and sleep deprivation. The chapter concludes with some emerging trends based on the present understanding in the field.

Keywords Genome-wide association · Selective breeding · Sleep circuitry · Context-dependent

Introduction

Prior to 2000, studies of the cockroach and honey bee suggested that sleep may occur in invertebrate species [1–3]. Cockroaches had periodic bouts of immobility across the day that increased after gentle handling or exposure to a novel environment [2]. Likewise, the locomotor activity of forager bees exhibited daily activity and rest patterns that correlated with the activity of their optomotor interneurons [3]. In their review of these studies, Campbell and Tobler concluded that a complete evaluation of sleep in invertebrates was necessary to demonstrate the evolutionary significance of sleep [1]. In 2000, two landmark papers did just that, describing a sleep-like state in *Drosophila* [4, 5]. These papers provided evidence that flies sleep by demonstrating that their behavior meets the mammalian behavioral criteria for sleep. Like mammals, flies enter a reversible state of quiescence each day. Quiescent

N. P. Singh · A. Ghosh · S. T. Harbison (✉)
Laboratory of Systems Genetics, Systems Biology Center, National Heart Lung and Blood Institute, National Institutes of Health, Bethedsa, MD, USA
e-mail: susan.harbison@nih.gov

P. Gehrman et al. (eds.), *Genetics of Sleep and Sleep Disorders*,
https://doi.org/10.1007/978-3-031-62723-1_2

flies probed with a gentle stimulus do not respond, indicating that their arousal threshold is higher during periods of prolonged quiescence. Flies also compensate for lost sleep by increasing sleep during recovery periods, indicating that they exhibit sleep homeostasis. The realization that flies sleep led to widespread use of this classic model organism to uncover the genetic networks influencing sleep. Since 2000, researchers identified genes that modify 81 unique sleep parameters. Transgenic approaches localized the role of these genes to specific regions of the brain and to other structures in the body. Flies were shaken mechanically, subjected to varying temperatures, fed different diets and drugs, and exposed to varying light/dark cycles in order to understand the relationship between genes and environment. Here we synthesize the findings from two decades of genetic study of sleep in flies. We describe what is known for each sleep parameter, localize the genes to their morphological location of action, if known, examine the impact of different environments on the genes, and identify gaps in our existing knowledge that need to be filled. As we cover a lot of prior work in this short chapter, we do not discuss every aspect of sleep studied using the fly model. This work covers only the basics of genetic studies of sleep in flies. Accordingly, we consider this chapter a starting point for understanding the genetic basis of sleep in *Drosophila*. We encourage the interested reader to refer to the original literature for additional details.

Sleep Phenotypes Used in Genetic Studies

In order to map the genomic constituents of sleep, we need a comprehensive description of the phenomenon. Though the concept of sleep is intuitively familiar, it encompasses a broad range of measurable parameters [6]. The parameters measured are often a function of the system used to measure sleep. In flies, the predominant method of measuring sleep and activity is the *Drosophila* Activity Monitoring System (DAMS) (Trikinetics, Waltham, MA). This system relies on an infrared beam (or sometimes multiple beams) to detect motion. Each time the fly disrupts the infrared beam, the system records it as an activity count. The system tabulates the number of activity counts for a given interval, usually 1 minute. Within a given interval of time, if there is no activity count, the fly may be resting or asleep. Early work enabled researchers to distinguish between these two possibilities by noting the flies' response to gentle tapping after 1–12 minutes of inactivity [7]. With each successive minute of inactivity, fewer flies responded to the taps than the preceding minute. The number of flies responding reached a nadir after 5 minutes of inactivity and remained low for up to 12 minutes. This lack of response suggested that the flies were asleep after 5 or more minutes of inactivity. This "5-minute" rule of thumb defines sleep in the vast majority of fly studies. Based on the 5-minute definition of sleep, many sleep parameters can be calculated, as outlined below. Video-based systems are a second means of measuring sleep in flies [8, 9]. These systems promise to expand the ways we measure and interpret sleep in flies, to the degree that videography can pick up fine movements of the fly at high resolution, study the interactions among multiple flies, or assess their behavior in different environmental contexts [8, 9]. While 81 distinct sleep phenotypes have been studied over the past two decades, our emphasis will be on the traits most often measured in genetic studies.

The most basic types of sleep traits are duration, bout number, and average bout length. Sleep duration, or the time spent sleeping, is the most straightforward sleep trait to measure and the most thoroughly investigated. Sleep duration is usually calculated in minutes for the day or light period, the night or dark period, and over the 24-hour day. Sometimes sleep duration is expressed as a proportion, i.e., the percentage of time spent asleep for a period of interest. As mentioned previously, flies exhibit sleep homeostasis, which means that they will respond to sleep loss with an increase in sleep. Researchers disrupt sleep in flies using several methods to induce sleep loss. Most often mechanical means are applied, such as gentle tapping by hand [4, 5], giving the DAMS monitors a quick jolt to induce negative geotaxis with the sleep-nullifying apparatus (SNAP) [10], shaking DAMS monitors on a vortexer (Trikinetics, Waltham, MA), using small motors to deliver vibrations of varying intensity to trays of flies using the *Drosophila* ARousal Tracking (DART) system [9], or rotation of individual fly chambers based on sleep status (ethoscope) [8]. Other means to perturb flies include the use of drugs and transgenic strategies to activate specific subsets of neurons or to block their transmission [11–15]. Flies recover sleep over several days subsequent to the perturbation of sleep. The additional time spent sleeping after a fly has been deprived of sleep is often referred to as a sleep rebound. Sleep rebound in flies and mammals rarely recovers the full quantity of sleep lost, suggesting that rebound sleep may be more intense than sleep under non-deprived conditions (e.g., see [5]). Sleep duration, whether accompanied by a perturbation stimulus or not, is the most common sleep trait studied.

Sleep duration in flies rarely manifests as a single continuous sleep period during the night, as shown in Fig. 1a, b. Consequently, researchers employ two other parameters to describe the architecture of sleep: sleep bout number and average sleep bout length. Sleep bout number refers to the number of naps that a fly takes during a given period. As with sleep duration, bout number is usually calculated for day, night, and 24 hours. Bout number gives an experimenter a sense of the degree to which a fly's sleep is consolidated. Given two flies with the same overall sleep duration, if one fly has fewer sleep bouts than the other, its sleep would be more consolidated. Like bout number, average sleep bout length also measures sleep consolidation. Average sleep bout length is the quotient of sleep duration divided by sleep bout number. Longer, less frequent nap durations imply more consolidated sleep [16]. Average sleep bout length tends to increase after flies have been deprived of sleep [10] and can thus gauge the homeostatic response to sleep. Not surprisingly, a correlation among these three parameters exists: sleep duration tends to be positively correlated with average bout length and can be positively or negatively correlated with bout number [17, 18]. These three traits form the phenotypic basis of most genetic studies.

Some additional sleep traits calculated in flies originate from similar measures used in human studies, such as sleep latency, wake after sleep onset (WASO), and

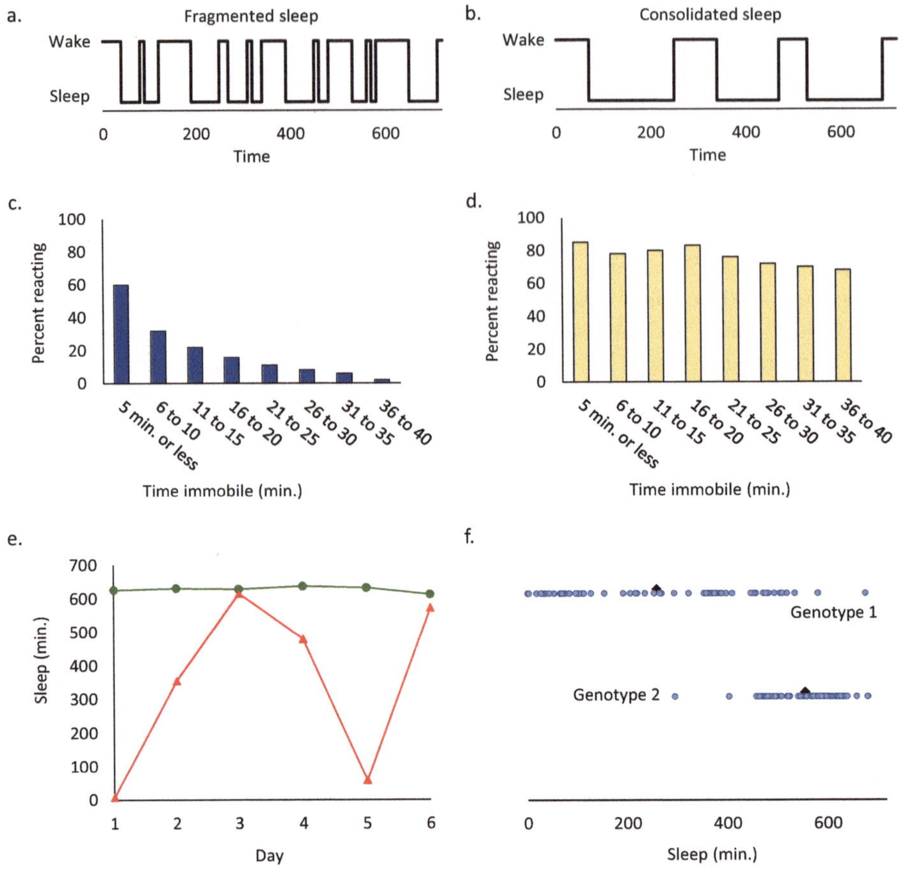

Fig. 1 Sleep architecture in *Drosophila*. (**a**) Hypnogram of sleep and activity during the night for a fly with fragmented sleep. (**b**) Hypnogram for a fly with more consolidated sleep. (**c**) Arousal threshold measured as the percentage of flies reacting to a stimulus at night plotted against the amount of time immobile. (**d**) Arousal threshold for flies reacting to a stimulus during the day. (**e**) Day-to-day variation in night sleep duration in individual flies. One fly has consistent sleep across time (green circles), while the other has large fluctuations in daily sleep (red triangles). (**f**) Fluctuations in night sleep duration among flies of two genotypes. Flies of genotype 1 vary considerably from one another, while flies of genotype 2 have similar sleep. The black diamond indicates the mean sleep duration for each genotype

sleep efficiency. Sleep latency is the number of minutes that it takes for the fly to fall asleep once the lights go off (night sleep latency) [19] or when the lights come on (day sleep latency) [20]. If a fly is already asleep during the transition period from day to night (or night to day), the sleep latency is zero. Sleep latency is one indicator of the need to sleep: a shorter sleep latency may indicate greater sleep need. Like sleep latency, WASO is a measure of sleep consolidation often used in human studies. WASO is the number of minutes a fly spends awake after having initially fallen asleep, that is, the total number of minutes spent awake during the night minus the sleep latency [21]. Sleep efficiency is a third measure that originates from human sleep studies. It is the amount of time spent sleeping divided by the fly's "opportunity" to sleep, for example, the 12-hour night period [21]. These measures function as additional indicators of sleep consolidation and need.

Thus far we have discussed sleep traits that describe the observed pattern of sleep and wake in flies. But measures involving the amount of time slept do not reflect the intensity of sleep. Like humans and mammals, sleeping flies experience periods of lighter and deeper sleep [22, 23]. One way to evaluate sleep intensity is to measure the percentage of flies that respond to a stimulus. To do this, sleeping flies are stimulated by tapping on the fly monitor tube over defined intervals of time with successively increasing force [5]. Flies that are awake will immediately respond to the stimulus; flies in light sleep will respond to a mild tapping stimulus; and flies in deep sleep will require a much stronger tapping stimulus to awaken. For example, Fig. 1c shows the percentage of flies responding to the tapping stimulus after a given interval of immobility during the night. As the time spent immobile increases, fewer flies respond, suggesting that they experience more intense sleep. Contrast this with Fig. 1d, where most flies respond to a tapping stimulus during the day no matter how long they were immobile. Arousal threshold is computed by counting the number of flies of a given genotype that awaken to each level of force [5, 9]. Another way to measure the intensity of sleep is respirometry as the fly's metabolic rate shows a linear decrease with the amount of time spent asleep [24]. These methods enable researchers to comprehensively describe sleep behavior.

Like humans and mice, sleep in flies varies from day to day [25–27]. These day-to-day differences can be quantified as the standard deviation (σ) of a sleep trait across days per fly [27]. Day-to-day differences can be extreme for some genotypes, as shown in Fig. 1e. However, variation exists not only within individual flies from day to day but among flies of identical genotype as well. For some genotypes, sleep is highly variable among flies, while, for others, the flies exhibit very similar behavior [18]. Figure 1f illustrates the differences among individual flies having the same inbred line genotype. Interestingly, genotypes with shorter mean sleep duration tend to have more variable sleep among individual flies [18]. This degree of variability can be quantified as the coefficient of environmental variation, or CV_E. The CV_E is the standard deviation of a phenotype divided by its mean, multiplied by 100 [28], and can be calculated for any trait. Day and night sleep duration is highly correlated with their respective CV_E; the other traits are correlated to a lesser degree [18]. Using these measures, variation in sleep can be quantified and studied.

Often sleep studies will include circadian rhythm parameters, as the circadian clock is thought to regulate sleep [29]. Typical traits include circadian period, which in flies usually refers to the free-running period of locomotor behavior during a 24-hour day [29]. Circadian period is calculated in free-running (constant darkness and temperature) conditions. A second parameter is rhythmicity index, which measures how similar a fly's activity patterns are from day to day. It is calculated by auto-regressing the fly's activity patterns and is defined as the third positive peak of the resulting auto-correlogram [30, 31]. A third trait is acrophase, the peak of circadian activity [32, 33]. A fourth type of trait measured is the start of activity prior to lights on in the morning and prior to lights off in the evening, which is referred to as morning and evening anticipation, respectively [34]. A fifth type of trait is the center of mass, which indicates the mean phase of circadian behavior and reflects patterns in the overall waveform [33]. Finally, the diurnality of locomotory behavior can be assessed as well [35]. These measures are used to relate circadian behavior to sleep.

The DAMS data can be used to make more sophisticated assessments of sleep architecture. For example, Wiggin and colleagues proposed that the probability that a moving fly will become quiescent estimates sleep pressure, while the probability that a quiescent fly will begin moving estimates sleep depth [36]. In addition, the DART system has been used to assess the loss of behavioral responsiveness as flies are falling asleep [37]. These studies suggest that the rich DAMS data sets can be further mined for additional parameters that describe sleep and activity.

Heritability of Sleep Phenotypes

With the above parameters in mind, let us turn to the subject of heritability. Heritability estimates the percentage of phenotypic variation in a trait that is due to genetic factors as opposed to environmental factors. Some heritability must be present in order to map genes to a trait. Despite its relevance to genetic mapping, few sleep studies in flies have estimated heritability. Table 1 lists heritability estimates for sleep and circadian traits. Heritabilities are moderate for most sleep traits. For example, the three most-often studied sleep traits, sleep duration, bout number, and average bout length, have heritabilities that range from 0.19 to 0.54. Interestingly, daily fluctuations in sleep parameters are largely heritable, as is the variability observed among individuals of the same genotype (σ and CV_E traits) (Table 1). Finally, circadian parameters are heritable as well (Table 1). The implication from this table is that the majority of sleep and circadian parameters studied have at least a partial genetic basis.

Genetic Mapping Approaches Used in Drosophila

Drosophila melanogaster is a versatile tool for genetic mapping. The fly's short generation time, low maintenance cost, and availability of mutations and other transgenic tools make it well suited to identify candidate genes for sleep. Mapping strategies in flies can be broken down into two general approaches: candidate gene or biased approaches and unbiased approaches. Candidate gene approaches are based on a survey of prior literature, which may implicate certain genes or pathways in sleep. Based on prior results, an experimenter may choose to survey genes involved in specific biological processes, localized to particular anatomical structures, belonging to certain classes of genes, or orthologs of genes putatively involved in human sleep or disease. This approach can be either large, as in a reverse genetic

Table 1 Heritability (H^2) estimates for sleep and circadian traits

Sleep trait	H^2 estimate	Ref.
24-hour sleep duration (min.)	0.45	[17]
Day sleep duration (min.)	0.43–0.49	[17, 18]
Night sleep duration (min.)	0.38–0.54	
24-hour bout number	0.29	[17]
Day bout number	0.33	[17, 18]
Night bout number	0.24–0.42	
24-hour avg. bout length (min.)	0.35	[17]
Day avg. bout length (min.)	0.19	[18]
Night avg. bout length (min.)	0.38	
Night sleep latency (min.)	0.44	[38]
Day sleep duration σ (min.)	0.49	[27]
Night sleep duration σ (min.)	0.26	
Day bout number σ	0.07	
Night bout number σ	0.13	
Day avg. bout length σ (min.)	0.30	
Night avg. bout length σ (min.)	0.30	
Day sleep duration CV_E	0.55	[18]
Night sleep duration CV_E	0.72	
Day bout number CV_E	0.47	
Night bout number CV_E	0.31	
Day avg. bout length CV_E	0.06	
Night avg. bout length CV_E	0.15	
Circadian period (hr.)	0.14–0.39	[39, 40]
Rhythmicity index	0.43	[40]

σ, standard deviation of day-to-day sleep
CV_E, coefficient of environmental variation

screen, or small in scale. For example, a screen of 1015 mutations of genes known to have sleep- or wake-specific expression or targets of the circadian clock uncovered a mutation in *inc* that resulted in short sleep and a mutation in *Tango10* that affected the strength of the circadian rhythm [41, 42]. Likewise, an RNA interference (RNAi) strategy was used to knock down 1000 genes having predicted expression within neurons to determine whether they function in sleep, and one gene, *pdm3*, was found to have a role in development that affected sleep at the adult stage [43]. Further, certain classes of genes can be surveyed, as was done with 143 "sponge" lines designed to inhibit expression of microRNA genes in the genome [44]. Mutations in fly orthologs confirmed the involvement of genes identified in human genome-wide association and exome sequencing studies of the Pittsburgh Sleep Quality Index, restless leg syndrome, insomnia, and sleep duration as well as autism spectrum disorder, attention deficit hyperactivity disorder, and schizophrenia [21, 45–53]. In some cases, studies used transgenic approaches to express wild-type or mutated forms of human genes implicated in disease. For example, transgenic models of Alzheimer's disease [54–57], Parkinson's disease [58, 59], autism spectrum disorders [60], Huntington's disease [61, 62], Rett syndrome/MeCP2 duplication syndrome [63], and epilepsy [64] demonstrated the effect of the human transgene on sleep in the fly. These approaches in all their permutations rely on prior information from the sleep literature to inform future work.

Unbiased approaches scan the genome for genes involved in sleep and include forward random mutagenesis screens, transcriptomic analysis, quantitative trait locus (QTL) mapping, genome-wide association studies (GWAS), and artificial selection. We will describe each of these strategies briefly. Random mutagenesis screens test the effect of mutated genes across the genome for changes in sleep parameters. Exposing flies to mutagens such as ethane methyl sulfonate (EMS) randomly induces mutations in the genome [65–67]. Another common strategy is to screen mutations created by the random insertion of transposable elements into the genome [65, 68–71]. A third strategy is to knock down the expression of genes using RNA interference [72–76], or to overexpress them [77], sometimes in specific anatomical regions [72, 73, 75, 77]. Sleep can be measured in the resulting collection of mutant flies. The experimenter then compares the sleep of these mutagenized flies to a wild-type control or to the mean of the resulting distribution in order to identify mutants that have the greatest effects on the sleep phenotype of interest. These strategies enabled researchers to find genes that produced extreme phenotypes; for example, extreme short sleep duration was observed in EMS mutations in *Sh* and *inc* [65, 66], a transposable element insertion in *qvr* (*sleepless*) [68], and RNAi knockdown of *CycA* [72].

GWAS and QTL mapping approaches represent another type of unbiased strategy. The two methods are similar in that they take advantage of naturally occurring polymorphic markers as landmarks in the genome [78]. Both strategies also rely on linkage disequilibrium between the polymorphic marker and the causal variant [78]. In GWAS, the data on individuals or lines are divided by allelic class for each marker. The mean sleep for one allelic class is compared to the other. A significant difference between allele classes implicates the marker in sleep. GWAS in flies uses nearly all available segregating polymorphisms for mapping, usually those that occur between 5% and 50% in the mapping population [18]. Consequently, the polymorphisms are very close to the causal variant or may in fact be the causal variant, as linkage disequilibrium decays rapidly in flies [79]. QTL mapping proceeds in a very similar fashion as GWAS but may use backcrossed, F2, or recombinant inbred populations [78, 80]. Markers for QTL mapping are more sparsely distributed, which results in larger regions of the genome being implicated in sleep [81]. Both GWAS and QTL mapping approaches may analyze polymorphic variants one at a time or consider many variants simultaneously [78]. GWAS and QTL mapping find naturally occurring genomic variation that contributes to sleep phenotypes.

Another unbiased approach used to identify candidate genes is to survey the transcriptome or proteome. Genetic differences may manifest in differential expression between wild-type controls and mutant or overexpression lines [82, 83], between flies altered by selective breeding and controls [84, 85], or among a panel of wild-type lines [86]. Other studies examined the changes in gene expression under different conditions to identify genes impacting sleep. Comparisons include those between sleep-deprived flies, flies experiencing rebound sleep after sleep deprivation, or handled controls [87], starved flies [88, 89], flies fed ethanol [90], flies isolated or exposed to other flies [91], or flies subjected to traumatic brain injury [92]. Additionally, RNA-Seq and TRAP-Seq targeted the differential expression over a 24-hour day in specific neuronal populations, such as mushroom body neurons [93], astrocytes [94, 95], and circadian neurons [83] to isolate micro-RNAs and genes in these tissues with roles in sleep duration, wakefulness, and daily sleep patterns. Finally, differences in protein expression identified genes as well [96, 97]. These methods identify the intermediaries between genome and phenotype that influence sleep, including sleep under different conditions.

Selective breeding is an additional means to identify candidate genes for sleep. With selective breeding, an experimenter breeds flies that have an extreme phenotype for one or more sleep parameters. Candidate genes can be identified by measuring gene expression [84, 85, 98] or by following allele frequency changes in the genome across selected generations [99]. This approach not only identifies candidate polymorphisms, it also uncovers variants that may be involved with fitness [100].

Trends Emerging from Genetic Studies

Work over the past two decades led to the identification of a multitude of genes affecting different aspects of sleep. One major conclusion to be drawn from this work is that the genes regulating sleep are not confined to a single pathway or process as was observed for circadian behavior [6]. Here we reflect on several trends that emerge from considering these studies as a whole: pleiotropy, epistasis, and sex specificity.

Pleiotropy in Sleep

Current methods enable the collection of data on multiple sleep phenotypes, as explained above. Interestingly, many of the genes tested for a role in sleep are pleiotropic, that is, they affect the phenotypes of more than one sleep trait. Figure 2 shows a schematic presenting the numbers of genes found in candidate gene and mutagenesis studies for eight of the most studied sleep traits. The figure reveals pervasive gene-level pleiotropy, as might be anticipated for genes with large effects [78]. Of the 185 unique genes in Fig. 2, 138 (74.6%) map to more than one sleep trait. The figure is surely an underestimate of the true level of pleiotropy, as some studies did not investigate all of the sleep traits listed. The presence of pleiotropy suggests some commonality among the genetic pathways underlying each trait.

In contrast, pleiotropy among natural variants implicated in genome-wide association studies is less frequent. The GWAS identifies polymorphisms associated with sleep, which can then be mapped to nearby gene regions; thus, there are two

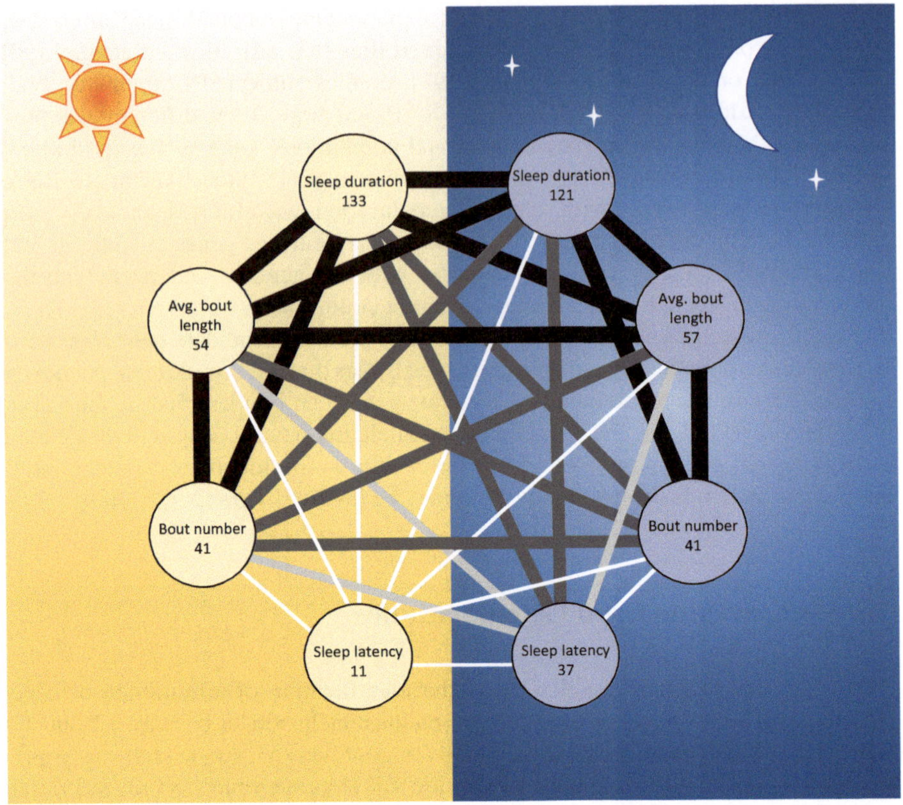

Fig. 2 Pleiotropy among genes identified in candidate gene and mutagenesis studies. The figure shows the number of genes identified for sleep traits commonly studied, as measured during the day or night. For each trait, the thickness and color of the bar linking it to other traits shows the amount of pleiotropy found thus far. Black bars: more than 30 genes common between traits. Dark gray bars: 21–30 genes common between traits. Light gray bars: 11–20 genes common between traits. White bars: less than ten genes common between traits

levels of the genome to consider. At the level of genomic polymorphisms, only eight variants out of 2763 (0.3%) associated with seven of the same sleep traits overlap (day sleep duration, night sleep duration, day avg. bout length, night avg. bout length, day bout number, night bout number, and night sleep latency). Yet these seven traits are genetically correlated with one another [18], which suggests that their underlying genetic architecture is similar [101]. A possible explanation is that the different polymorphisms associated with each trait are in close proximity to each other in the genome, implying linkage disequilibrium between the two variants [101]. This appears to be the case if we consider the gene level, as 66 genes are common to two or more sleep traits [18]. An additional explanation for reduced pleiotropy in the GWAS is the severity of the mutation under consideration. The GWAS considers the effects of naturally occurring polymorphisms, which are likely more subtle than the gene knockdowns/knockouts used in candidate gene and mutagenesis studies. Accordingly, when severe mutations of candidate genes were tested to verify GWAS predictions, extensive pleiotropy was observed [18, 27]. Thus, despite the complexity of sleep, there is an underlying commonality at the genetic level, and the list of genes common among traits has yet to be fully described.

Among these eight sleep traits, only 19 genes identified using the GWAS approach overlap with those identified via candidate gene and mutagenesis approaches. Candidate gene and mutagenesis approaches are thus complementary to GWAS. The GWAS identifies natural variants that contribute to the maintenance of genetic variation for sleep, while candidate gene/mutagenesis approaches identify genes for sleep that may be relatively invariant in natural populations.

Regulatory Variation and Epistasis in Sleep

Several studies address the sources of naturally occurring genetic variation in sleep using two population resources: the *Drosophila* Genetic Reference Panel (DGRP) [79, 102] and the *Drosophila* Synthetic Population Resource (DSPR) [80, 103]. Genome-wide association studies in the DGRP identified variants associated with 19 sleep and circadian traits [18, 27, 40]. Several trends emerged from the GWAS data. First, larger effects on sleep map to low-frequency (2.4–5%) alleles (Fig. 3a) [18]. Common variants with large effect sizes were not found. Short night sleep duration mapped exclusively to the rarer alleles in the population, while long sleep mapped to the more common alleles. This suggests that natural selection may act against short night sleep, a finding supported by recent evidence [100]. In fact, more low-frequency variants dominate the polymorphisms mapping to sleep phenotypes; few common variants map to sleep (Fig. 3b). In addition, significant variants tend to reside in regulatory rather than coding regions of the genome (Fig. 3c). These data indicate that large numbers of regulatory variants may impact sleep.

If the variants implicated via GWAS act strictly additively, one would expect that the results could be replicated across populations [101]. As *Drosophila* are easily intercrossed, the GWAS replication could be accomplished using a number of strategies: testing an outbred panel created by randomly shuffling the genomic variants [104], creating an advanced intercross line population by crossing the genotypes having the most extreme phenotypes [105], testing artificially selected populations derived from the original GWAS lines [99], or mapping candidate genes in a different population altogether [81]. Thus far, the latter two methods have been attempted in sleep studies. Interestingly, variants identified for night sleep duration using artificial selection did not overlap with GWAS variants [99]. Likewise, a QTL analysis using the DSPR found 12 genomic regions mapping to five sleep traits, but the number of GWAS variants overlapping these QTL was lower than expected by chance [81]. This suggests that epistasis among variants, which is a function of their allele frequencies in each population, has a potentially significant role in sleep [101, 106]. In support of this idea, thousands of epistatic interactions were predicted among variants associated with circadian behavior [40]; although yet to be

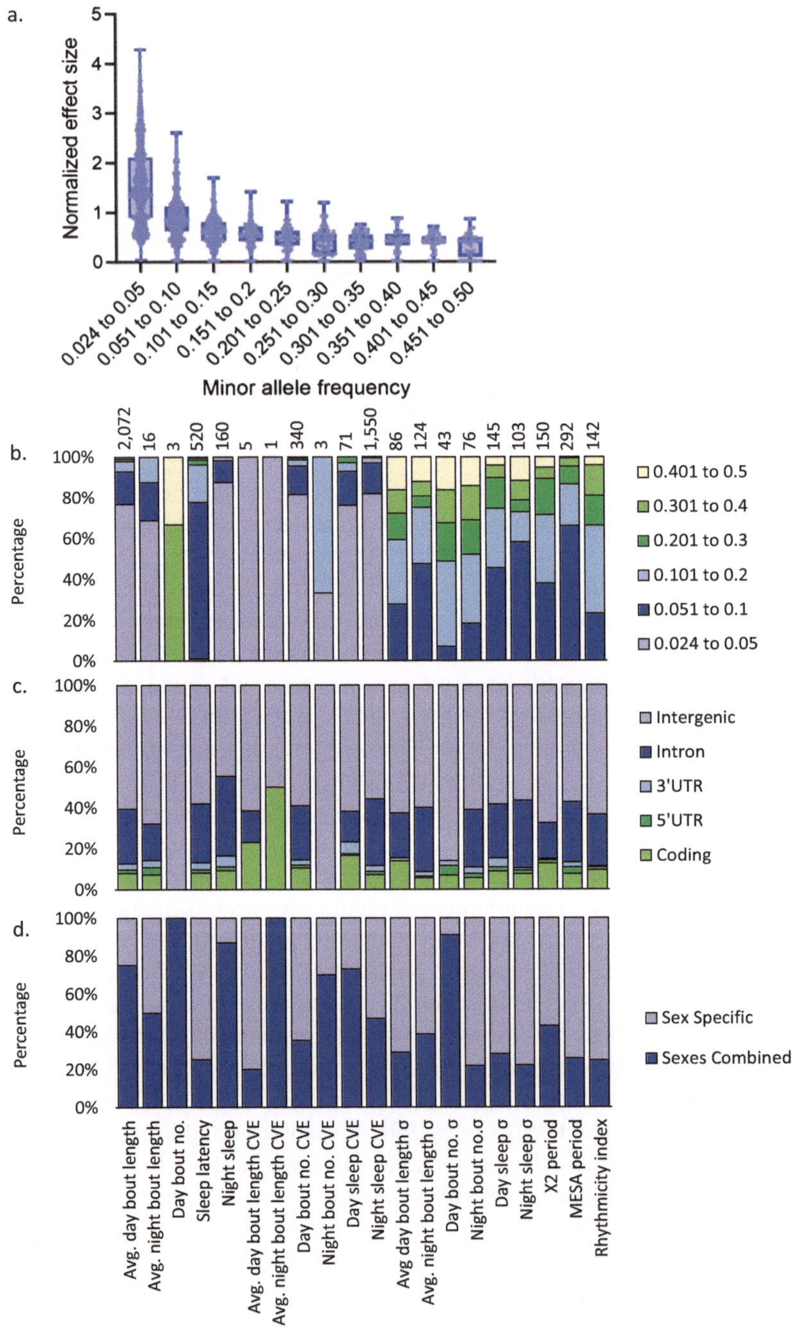

Fig. 3 Characteristics of natural variants associated with 19 sleep and circadian phenotypes. (**a**) Plot of effect size (normalized by the standard deviation of genetic variance) with minor allele frequency. (**b**) Percentage of variants at each allele frequency bin indicated. Above each bar is the total number of variants for each trait. (**c**) Percentage of variants in coding and noncoding genomic regions. (**d**) Percentage of sex-specific variants compared to those mapping to both sexes. Note that circadian period is calculated using two different methods

determined, sleep may have many epistatic interactions among variants as well. Other studies attempting functional validation of GWAS findings in flies reported similar observations, indicating that epistatic interactions among genomic variants are common in complex traits [104, 105, 107–109]. These data indicate that epistasis has a role in maintaining variation in sleep in natural populations.

Sex-Biased, Sex-Specific, and Sex-Antagonistic Effects

Early work discovered differences in male and female sleep patterns. Male flies characteristically sleep more during the day than females, and their 24-hour sleep duration is generally longer than that of females as well [7]. Genotype contributes to these differences. Wild-derived inbred lines show significant sex-by-line interaction effects, suggesting a clear involvement of genotype [18, 27, 40], and in many cases the number of sex-specific variants that map to a trait exceed those that map to both sexes (Fig. 3d) [18, 27, 40]. Mutational analyses also reflect these sex-specific differences. Histograms of ~9000 EMS mutagenesis and *P*-element insertion lines tested for sleep duration showed a dramatic difference between male and female 24-hour sleep duration, with males having characteristically longer sleep than females [65]. A ~ 12,000-line gain-of-function screen observed the same phenomenon [77]. Despite the presence of these clear sex-specific differences in sleep architecture, roughly two-thirds of the genetic studies conducted over the last two decades test the effects of genes in only one sex. The remaining studies, which examine the effects of genetic manipulation in both sexes, often find differential effects. Mutations having differential effects on sex can be divided up into three types: sex-biased, sex-specific, and sex-antagonistic [110]. Sex-biased effects occur when both sexes show the effect of the mutation, but it is stronger in one sex than another. Sex-specific effects occur when only one sex shows the effect. Sex-antagonistic effects occur when both sexes are affected, but the effect is in the opposite direction for each sex [110]. We will examine these three possibilities in more detail below.

Sex-biased effects of mutations on sleep reported thus far show stronger effects in females. Effects of mutant *Hk* on 24-hour bout number were more prominent in females [111]. A 24-hour sleep duration was more sensitive to the level of *Fmr1* in females than in males across the lifespan [112]. Knockdown of *Pka-C1* increased sleep consolidation in older flies by altering the distribution of night avg. bout length; the effect was more prominent in females than in males [113]. And sleep in females is more responsive to optogenetic activation of *SIFa* and *sNPF* neurons than sleep in males [11, 12]. That sex-biased effects are seen only in females begs the question of altered dosage compensation as a potential explanation for its existence; however, *Hk* is the only gene tested thus far that resides on the *X* chromosome.

Sex-specific effects of mutations frequently occur. Some mutations affect sleep only in males, as observed in pan-neuronal knockdown of *Lztr1* [114], ubiquitous knockdown of *bmm* [115], and mutations in *wake* [69]. Upregulation of *NPF* and

NPFR altered sleep only in males [116]. Sometimes the sex-specific effects are confined to the particular alleles tested. For example, three mutant alleles of *Nf1* consistently decreased 24-hour sleep duration in males relative to controls, while the effects on females varied [117]. Likewise, different mutant alleles of *CG17574* consistently affected males but not females [86]. Other mutations in contrast affect sleep only in females. *Acer* null mutations [118], *tara* mutants [70], *Sdc* mutants [119], *InR* and *Pi3K92E* mutants [120], and *Rel* mutants [87] impacted sleep in females alone. Effects of mutations may be consistent in females but not males across different alleles tested, as with *Tsp42Ef* [86]. Thus, for some genes, the effects of a given mutation manifest in one sex only.

Instances of sex-antagonistic effects are relatively rare. Flies with ubiquitous knockdown of *bmm* had significantly decreased 24-hour sleep bout number in females, but it was significantly increased in males [115]. Gene editing at three conserved sites of the gene *wake* produced increased sleep in female flies but decreased it in males [20]; however, a null allele of the same gene reduced sleep consistently in both sexes [69]. Flies overexpressing the enzyme *jhamt*, a key regulator in the juvenile hormone biosynthesis pathway, had decreases in day sleep duration and day avg. bout length in females, while there were increases in males [121]. Interestingly, the loss of *jhamt* produced the opposite effect on day avg. bout length, increasing it in females and decreasing it in males [121]. The differential effects on sleep were traced back to the genes of the sex determination pathway [121], which may ultimately underly sex-antagonistic differences in sleep.

Behavioral differences between males and females may reflect underlying differences in gene expression between sexes, which in flies are often profoundly different [122]. Differential expression between sexes was observed for insulin-like proteins *dilp3*, *dilp5*, and *Akt* in *Sdc* mutants [119] and in *bin3* and *Akt1* mutants [86]. *jhamt* exhibited sex-antagonistic changes in expression under sleep deprivation conditions as well [121]. In a recent experiment, Mokashi and colleagues found differences in sleep parameters between males and females fed ethanol during development. Single-cell RNA-Seq of wild-type male and female brains revealed differentially expressed genes that rarely overlapped between sexes, and for some of the genes that did overlap, expression was upregulated in one sex and downregulated in the other [90]. Though many individual genes were differentially expressed between males and females, they did map to similar Gene Ontology categories, suggesting some commonality at a higher level of integration such as a gene pathway [90]. Overall these findings suggest that an understanding of the differences in sleep among males and females is an important future goal.

Neural Circuit of Sleep in the *Drosophila* Brain

Classical experiments like brain stimulation and lesions and measurements of cFos expression and electrical activity identified many anatomical areas in mammalian brains that regulate sleep. Accordingly, transgenic methods that artificially activate and silence specific regions of the brain are used to map the sleep center(s) of the *Drosophila* brain. Similar to mammals, these studies found multiple regions in the brain that regulate sleep and wakefulness [123, 124], including the central complex, the mushroom bodies, and circadian clock neurons. Interestingly, recent evidence suggests that the activity of other cell types, including neuroendocrine cells, peripheral metabolic tissue, peripheral neurons, and glial cells have an essential role in sleep regulation [125–128]. Glial cells that physically and metabolically support neuronal cells can influence almost every function including sleep [127, 129]. In addition, the role of glial cells in transporting waste products out of the brain have a major impact in regulating sleep behavior [130]. In this section, we will discuss the specific role of different brain regions and genes required for their function in *Drosophila* sleep regulation.

The Role of the Central Complex in Sleep and Wake

One heavily investigated region of the fly brain in sleep behavior is the central complex (Fig. 4). On the basis of neuronal activity, the central complex can be further subdivided into three regions: the dorsal fan-shaped body, the ellipsoid body, and the helicon cells (Table 2). Overexpression of the bacterial sodium channel NaChBac to artificially activate specific neuronal circuits suggested that dorsal fan-shaped body (dFB) neurons promote sleep [141]. Developmental defects within or inactivation of dFB neurons cause loss of sleep [43, 45, 132, 133]. Sleep deprivation experiments revealed that these neurons get electrically activated (ON) and electrically silenced (OFF) to regulate sleep and wake, respectively [132]. Dopamine emerges as a neuromodulator that switches dFB neurons from the ON to the OFF state by upregulation of the potassium ion leak channel *sand* and downregulation of potassium voltage-gated A-type currents carried by *Sh* and *Shab* [133, 142, 143]. In this model, just altering expression of *Sh* and *sand* in the dFB neurons is enough to regulate the sleep and wake cycle. dFB-specific mutations in pathway members like *Dop1R1*, *Dop1R2*, and *DAT* alter sleep function [133, 134, 142]. We know little about the switch from the OFF to the ON state of dFB neurons. However, knockdown of the Rho-GTPase activating protein *cv-c* in these cells locks them in the OFF state, resulting in severe insomnia and a lack of rebound sleep following deprivation [132]. This suggests that Rho-GTPase activation is required for switching from OFF to ON state; however, the mechanism by which this occurs remains unknown. In another study, knockdown of a sleep-promoting serotonin receptor *5-HT2B* or ablation of *5-HT2B* neurons in the dFB decreased sleep and impaired sleep homeostasis [135]. There is a critical need to understand if these pathways work in parallel or in tandem to operate the ON/OFF switch in dFB neurons. In addition, dFB neurons may also have a role in regulating the intensity of sleep. Knockdown of a gap junction channel protein *Inx6* in the dFB increases arousal threshold in response to mechanical stimuli [37]. The dFB neurons also seem to respond to the external environment in other ways to regulate sleep (see section

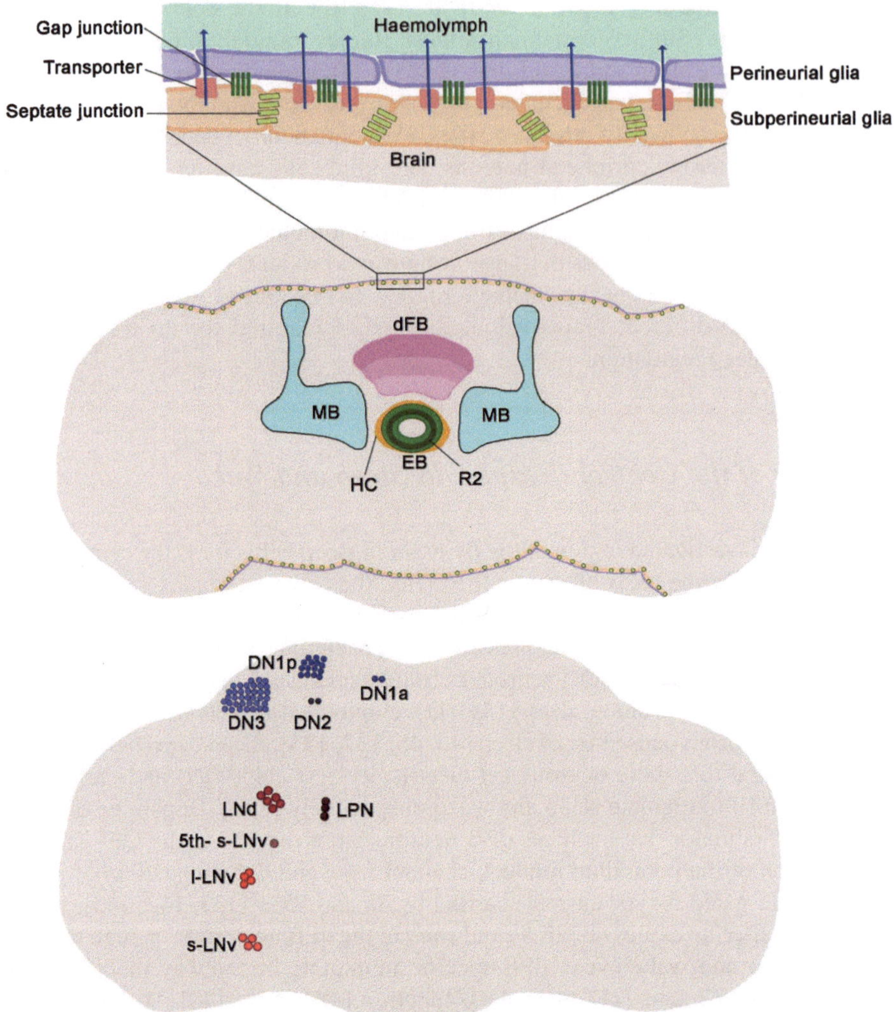

Fig. 4 Sleep control centers in the *Drosophila* brain. Middle panel shows the location of the central complex (dFB-pink and EB-green) and mushroom bodies (turquoise). The enlarged top panel shows the *Drosophila* blood-brain barrier (BBB) that includes two layers of perineurial and subperineurial cells. The junction and transporter proteins in the subperineurial cells help the movement of molecules from brain to hemolymph (arrow). The lower panel shows 75 clock neurons in one hemisphere of the fly brain. The top dorsal neurons (DNs) are in blue shades, while ventral neurons are in red shades

Table 2 Genes having expression or function in central complex cells

Gene	Tissue/cell type	Ref.
AstA	dFB	[131]
cv-c	dFB	[132]
Dop1R1	dFB	[133, 134]
DAT (fumin)	dFB	[134]
5-HT2B	dFB	[135]
Inx6	dFB	[37]
nur	dFB	[77]
per	dFB	[136]
ISWI	dFB	[45]
Pdm3	dFB + EB	[43]
dbt	Central complex	[137]
esc	EB	[138]
brp	EB-R2 neurons	[139]
Nmdar1 and 2	EB	[140]

"The Context-Dependent Nature of Sleep"). These observations suggest that dFB is a major center in the *Drosophila* brain that regulates the sleep-wake cycle. However, recent work has challenged this finding, suggesting that effects on sleep attributed to the dFB may instead map to the ventral nerve cord [144, 145]. Additional work is necessary to resolve the role of these structures in sleep.

The dFB neurons may also regulate sleep homeostasis through interaction with helicon cells and R2 ring neurons of the ellipsoid body (EB) (Fig. 4) [146]. Helicon cells transmit visual inputs to R2 ring neurons to suppress sleep by inducing locomotion. However, in response to high sleep pressure, dFB neurons inhibit activity of helicon cells through a range of inhibitory neurotransmitters that include the neuropeptide *allatostatin-A* (*AstA*), which leads to elevated visual thresholds (i.e., a reduced response to visual stimuli) and promotes immobility [131]. A recent study conflicts with this finding, as *AstA* expression was not detected in the dFB [144]. R2 neurons also exhibit sleep need-dependent plasticity. Expression of tetanus toxin to inhibit neurotransmitter release in R2 neurons significantly reduces rebound sleep after mechanical sleep deprivation. In response to sleep deprivation, the cytosolic Ca^{2+} levels increase in R2 cells that induce rebound sleep in flies. Reducing Ca^{2+} levels in R2 neurons either by knocking down the inositol-triphosphate receptor (*Itpr*) or NMDA receptors, *Nmdar1* and *Nmdar2* that mediate Ca^{2+}-permeability reduce the amount of rebound sleep, suggesting that Ca^{2+} signaling regulates sleep plasticity in R2 neurons [140]. Inhibition of R2 neurons only affects sleep rebound after deprivation and does not have an effect on baseline sleep, indicating the specific role of these neurons in sleep homeostasis. However, inhibition of R2 neurons does not completely eliminate sleep rebound indicating that additional centers of sleep homeostasis may exist in the brain. Taken together, these observations suggest that the dFB, EB neurons, and helicon cells of the central complex form an integrating circuit for regulation of sleep homeostasis.

The Role of the Mushroom Bodies in Sleep and Wake

Similar to the central complex, the mushroom bodies (MB) act as a hub that integrates multiple signals: sensory information, arousal, and sleep homeostasis (Fig. 4) [147].

The MBs are a paired structure classically associated with olfactory learning and memory [148, 149]. However, studies also identified both sleep- and wake-promoting neuronal populations in the MBs [14, 150, 151]. Artificial activation of MB neurons using the bacterial sodium channel NaChBac induces sleep, while their silencing—whether by chemical means, expressing K^+ channels (*Kir2.1* and *EKO*), or expressing the temperature-sensitive synaptic blocker *shi*—shows a loss of sleep [14, 150]. These early studies suggested that MBs may function in a similar way as the dFB, but more recent analyses revealed that separate subsets of MB cells can promote both sleep and wakefulness. MBs are comprised of 2500 Kenyon cells that organize to form three lobes, α/β, α'/β', and γ. Activation and silencing of different subsets of Kenyon cells revealed that γ_{dorsal} neurons promote sleep and α'/β' and γ_{main} neurons promote wakefulness, while the role of α/β neurons in sleep is more ambiguous [150, 151].

Mushroom body neurons are an attractive model to investigate the neural circuits controlling sleep because the input and output neurons to this region are well defined [152–154]. The Kenyon neurons get input signals from dopaminergic neurons (DANs) and dorsal paired medial neurons (DPM). Arborization of DAN and DPM neurons overlaps with wake- and sleep-promoting Kenyon cells of the MB. All DAN neurons studied so far have been found to be wake promoting, while DPM neurons are sleep promoting [155–157]. The sleep-promoting role of DPM neurons seems to be regulated through GABA and serotonin signaling (Table 3) [155, 164]. Postsynaptically, Kenyon cells connect to mushroom body output neurons (MBONs) to promote sleep and wakefulness. Wake-promoting Kenyon cells synapse with glutamatergic MBONs, while sleep-promoting Kenyon cells synapse with cholinergic MBONs [151, 153]. However, some sleep- and wake-promoting Kenyon cells are synapsed with both sleep- and wake-promoting MBONs, revealing more complex synaptic connections [151]. Genetic activation and repression of the MBONs revealed compartmentalized roles in sleep as some of them affect only sleep rebound after sleep deprivation and have no role in the regulation of baseline sleep [151, 168]. These findings suggest that MBONs integrate a diverse set of signals from Kenyon cells to drive either sleep or wakefulness.

The circadian regulation of gene expression in Kenyon cells is another mechanism that alters their activity [93]. Calcium levels in Kenyon cells decline when flies fall asleep and increase with wakefulness [169]. Higher calcium levels during the day correlate with greater activity of MB neurons, which is mediated by activating cAMP-PKA signaling through higher expression of *Nf1* and *Pka-C1* [93]. Mutants of *Nrx-1*, which regulates synaptic transmission by coupling calcium channels, show reduced nighttime sleep with lower bout duration and higher bout numbers, indicating poor sleep consolidation [158]. This reduction in night sleep duration

Table 3 List of genes having expression and/or function in the mushroom bodies (MB)

Gene	Tissue/cell type	Ref.
dbt	MB	[137]
Nrx-1	MB	[158]
Nca	MB	[46]
Nf1	MB	[93]
Alk	MB	[117]
unc79	MB	[76]
mir-276A	MB	[159]
Nkt	MB	[160]
let7	MB	[44]
inc	MB	[161]
faf	MB	[162]
hiw	MB	[162]
Rdl	MB	[162]
fru	MB	[163]
Dop1R1	MB	[157]
5-HT1A	MB	[164]
bun	MB	[165]
D	MB	[165]
bsk	MB	[166]
Go	MB	[167]
ISWI	MB	[45]

correlates with impaired output from the αβ neurons. Similarly, mutation of a calcium sensor *Nca* impacts only night sleep duration by suppressing synaptic release from wake-promoting neurons [46]. These observations suggest that circadian regulation of calcium levels in MB neurons could be a common mechanism to regulate sleep behavior. In addition, there are many other molecular pathways like protein degradation [162], sodium channels [76], *Go* [167], JNK [166], Notch signaling, etc. Genes in Table 3 are implicated in sleep regulation through their function in MB neurons. However, we still do not understand their direct role in modulating MB function.

The Role of Circadian Clock Cells in Sleep and Wake

The ~150 clock neurons are sparsely distributed across the two hemispheres of the *Drosophila* brain [170]. These neurons integrate circadian input to sleep behavior and regulate *Drosophila* bimodal locomotor activity in the morning and evening [19, 171, 172]. Based on their position in the fly brain, these neurons are subdivided in eight groups: three groups of dorsal neurons (DN1s, DN2s, and DN3s), three groups of ventral-lateral neurons (l-LNvs [large], four s-LNvs [small], and a fifth s-LNv), the dorsal-lateral neurons (LNds), and the lateral-posterior neurons (LPNs)

(Fig. 4) [170, 172–174]. Despite their distinct locations in the brain, these neurons are interconnected [175]. A molecular hallmark of all the clock cells is a feedback loop of transcriptional regulation that shows oscillations in their daily levels. A simplified version of the feedback loop operates between transcriptional activators *Clk* and *cyc* and transcriptional repressors *per* and *tim* [176]. While CLK and CYC proteins promote the expression of PER and TIM, PER and TIM proteins dimerize to repress CLK and CYC, which ultimately represses their own transcription [172, 177, 178]. Disruption of this feedback loop shows reduced activity and rhythmicity in all the clock neurons [179].

Past studies have clearly demonstrated that the circadian regulation of sleep-wake timing is mainly controlled by clock neurons [124, 180–184]. Until recently, the specific role of each of the subgroups in this process was not very well defined due to a limited ability to target only these subgroups. The creation of new molecular tools (i.e., GAL4 and split GAL4 driver lines, along with optogenetics) has helped to refine our understanding of the role of various clock neurons [185–187]. Each subgroup of the clock neurons has a specific role in regulating various aspects of sleep and circadian behavior.

The lateral neurons (l-LNvs, s-LNvs, fifth s-LNv, LNds, and LPNs) have a dominant role in the regulation of the clock network. Among these, the *pigment-dispersing factor* (*Pdf*) neuropeptide-positive s-LNvs are considered master regulators of the daily bimodal activity in *Drosophila* [172]. These cells drive the morning activity peak under light/dark conditions and help to sustain circadian rhythms in constant darkness [171, 188, 189]. The wake activity in these neurons is regulated through PDF activity and rhythmic GABAergic input (Table 4). Wakefulness is induced by *Fbxl4*-mediated ubiquitination and degradation of the GABA receptor, while sleep is promoted by the expression of the WAKE protein that upregulates GABA receptors in the s-LNvs [69, 212]. In addition to s-LNvs, the l-LNvs also express PDF. These l-LNvs fire with a higher rate in response to light to keep flies awake, which is dependent on photon capture by *cryptochrome* (*cry*) [223]. On the other hand, the timing of evening locomotor activity is mainly regulated by LNds along with the fifth s-LNv [188, 189]. These neurons also use *cry* for light perception. The CRY protein is UV/blue-light-sensitive protein that is activated by light to bind with TIM, initiating its degradation to reset the clock [224, 225]. However, a recent study found that under high-intensity light conditions, a CRY-independent network operates through s-LNvs to delay the afternoon siesta and to determine timing of the evening locomotor activity via PDF signaling [226]. These data suggest that morning and evening peaks of locomotor activity rely on two different types of clock neurons—the s-LNvs (morning oscillator) and LNds (evening oscillator), respectively. However, s-LNvs can reset the evening oscillator in response to environmental conditions. Additionally, the LPN neurons have been implicated in various roles in regulating sleep behaviors, including regulation of sleep and feeding, synchronization of activity to temperature cycles, and timing of evening activity and wakefulness [227–230]. The multiple roles of LPNs seem to be regulated by their synaptic connections to the other clock neurons as well as to neurons downstream of the clock that control these different behaviors [228]. Thus, the

Table 4 Genes having expression and/or function in clock neurons (CLN)

Gene	Tissue/cell type	Ref.
per	CLN	[190–194]
Clk	CLN	[176, 191]
cyc	CLN	[191, 195]
tim	CLN	[190, 191]
dbt	CLN	[137, 196]
NRXN1	CLN-PDF	[60]
Dop1R1 and *R2*	CLN-PDF	[197]
faf	CLN	[162]
hiw	CLN	[162]
Rdl	CLN	[162]
PDF	CLN	[198–200]
IPT	CLN	[198]
CycA	CLN	[72]
Rca1	CLN	[72]
cry	CLN_cry	[201]
NPF	CLN_cry	[202]
fru	CLN_DN1	[163]
Dh31	CLN_PDF + DN1 + DN2	[163, 203]
GABA-B-R2	CLN_l-LNvs	[204]
Nlg4	CLN_l-LNvs	[205]
ATF-2	CLN_l-LNvs +s-LNvs	[206]
Ilp1–3, 5–7	CLN_LNv + LNd	[207]
InR	CLN_LNv + LNd	[207]
Ih	CLN_LNvs	[208]
tau0N4R	CLN_LNvs	[209]
rut	CLN_LNvs	[193]
bs	CLN_LNvs	[193]
Tango10	CLN_PDF	[42]
14–3-3e	CLN_PDF	[210]
PERK	CLN_PDF	[211]
fbx14	CLN_PDF	[212]
ap	CLN_PDF	[213]
Chi	CLN_PDF	[213]
para	CLN_PDF	[64]
SPR	CLN_PDF	[214]
Mip	CLN_PDF	[214]
HisCl1	CLN_PDF	[215]
Hdc	CLN_PDF	[215]
Pdfr	CLN_PDF	[199]
Rdl	CLN_PDF	[199]
Tip60	CLN_s-LNv	[216]
APP	CLN_s-LNv	[216]
mir-375	CLN_l-LNv	[217]
Shal	CLN_PDF	[218]

(continued)

Table 4 (continued)

Gene	Tissue/cell type	Ref.
wake	CLN_PDF + l-LNvs	[69]
Tor	CLN_per	[219]
Atg7	CLN_per	[219]
foxo	CLN_s-LNv	[220]
mir-276A and *B*	CLN_tim	[159]
brun	CLN_tim	[50]
B52	CLN_tim	[221]
vari	CLN_PDF	[48]
cac	CLN_tim	[47]
mir-276B	CLN	[222]

lateral neurons have fundamental roles in the regulation of the circadian timing of sleep, with each group of neurons regulating different aspects of behavior.

The dorsal neurons (DN1s, DN2s, and DN3s) appear to be less important in controlling rhythmic sleep behavior under constant environmental conditions [231]. However, they integrate environmental information like light, temperature, and metabolic state with the circadian molecular program. There are two clusters of DN1s: the anterior DN1s (DN1as) are the downstream target of neurons that process cold temperatures [184, 232], while the posterior DN1s (DN1ps) are downstream of warm-sensing neurons [233]. DN1p activity is influenced by both increases and decreases in temperature [234]. In addition, the DN2s are also known to act as temperature-entrainable clock neurons that regulate temperature preference rhythm [235]. Temperature preference rhythm is a behavior where flies seek a colder temperature in the morning, a warm temperature in the evening, and a colder temperature in the night [235]. Accordingly, a recent study suggests that DN2s and DN1ps connect with each other to coordinate temperature preference rhythm throughout the day [187]. DN1s also seem to communicate with DN3s to promote sleep. These two neuronal groups show a similar Ca^{+} activity rhythm [236, 237]. In addition, a subset of DN3s express *AstC* and regulate the cold-induced dormancy of the flies, indicating a role for the DN3s in temperature-dependent behavior [238]. Taken together, these studies demonstrate that dorsal neurons integrate environmental inputs of light and temperature into the circadian clock and the DN1s seem to be at the center of this process.

The lateral and dorsal neurons communicate with each other through direct synaptic connections [239]. The clock neurons also communicate with each other using neuropeptides (PDF, NPF, sNPF, Dh31, CCHa1, and ITP) and neurotransmitters [240]. One of the best studied neuropeptides, PDF is expressed only in l-LNvs and s-LNvs, but its receptor is present on other clock neurons including the DN1s, DN2s, DN3s, and LNds. Disrupting PDF signaling affects anticipation of morning and evening peak locomotory behavior and makes flies arrhythmic in constant darkness [241]. This was further narrowed down to PDF signaling in DN1 and LNds that is required for anticipation of morning and evening activity in constant darkness and normal light/dark conditions, respectively [242–244]. Dh31, another neuropeptide, also acts through the PDF receptor to regulate morning anticipation [242]. Similarly, different neuropeptides are used by other clock neurons to communicate and regulate specific functions in circadian rhythms and sleep. For example, NPF is expressed in subsets of l-LNvs and LNds to govern evening wakefulness, sNPF is expressed in LNds and s-LNvs to promote sleep, ITP is expressed in LNds and the fifth s-LNv to promote evening activity and inhibit activity in the middle of the night, subsets of DNs express Dh31 to suppress late-night sleep, and subsets of DNs express CCHa1 to regulate day sleep bout length [29, 124, 245]. Similar to neuropeptides, neurotransmitters including glycine, glutamate, and acetylcholine communicate and regulate circadian activities via the clock neurons [240]. The daily clock cycle is set by a combination of neuronal function and interneuronal interactions through specific synaptic connectivity, neurotransmitters, and neuropeptides.

One of the major clock output neurons is the pars intercerebralis (PI). Early studies in crickets and cockroaches showed that surgical destruction of these neurons results in a loss of locomotor rhythms [246, 247]. Using more sophisticated approaches, the PI neurons were identified as a circadian output circuit for rest/activity rhythms [248]. The neuronal circuit extends from the master pacemaker s-LNvs to DN1s and onto PI cells representing a possible regulatory cascade [248]. Time-of-day input to PI cells are received from both CRY-negative LNd and DN1 clock neurons [249]. In addition to locomotion, PI cells also seem to act as a common relay for multiple circadian output circuits where each physiological function is subserved by discrete subpopulations of PI neurons. Locomotor activity is controlled by a subset of PI cells expressing DH44 neuropeptide, feeding behavior is controlled by a subset of PI cells expressing Dilp2 neuropeptide, and rhythmicity and sleep behavior is controlled by a subset of PI cells expressing SIFa neuropeptide [11, 248, 250, 251]. This suggests that the PI neurons regulate circadian timing of these distinct behaviors through secretion of neuropeptides. Accordingly, PI neurons are considered a major endocrine control center in the fly brain.

Other Sleep- and Wake-Regulating Neurons

Historically, sleep was thought to be controlled by neurons of the brain. Evidence now exists for a role of peripheral organs in the regulation of sleep. One indicator of the importance of sleep is that severe sleep deprivation leads to the premature death of animals [252–254]. Surprisingly, sleep deprivation causes buildup of reactive oxygen species (ROS) in both the mouse and the fly gut that may contribute to premature death [255]. Neutralizing ROS in the gut enables flies to live with little sleep, implying that excess ROS in the gut induces sleep. This study indicated that a strong connection between sleep and gut health may be required for clearance of gut ROS. Another connection between sleep and the gut is feeding behavior. As explained above, the rhythmic circadian feeding behavior in flies is regulated by insulin-producing cells (*Dilp2*-positive) of the PI through clock neurons. However, activation of posterior lateral protocerebrum (PLP) cells in brain and enteroendocrine cells of midgut suppresses feeding behaviors and promotes sleep [256, 257]. Similarly, knockdown of a secreted cytokine *upd2* in the fat body to mimic starvation-like conditions promotes feeding-related behaviors through insulin-producing cells and suppresses sleep [258]. This suggests that PLP and enteroendocrine cells promote satiety, while in starvation fat body cells promote feeding behavior. These findings identify a role for peripheral tissue-to-brain interactions in the simultaneous regulation of sleep quality and attention, to potentially promote adaptive behaviors necessary for survival in animals. Another set of peripheral neurons, *pickpocket* (*ppk*) neurons, which are ubiquitously distributed throughout the peripheral nervous system and specifically in the legs, are sufficient to drive sleep homeostasis [128]. These neurons project their axons to the central nervous system; however, we still do not know their targets in the brain. In addition, another study found that antennal cells help synchronize clock neurons in response to temperature changes [259]. These findings suggest that sleep is not only required and regulated by the central brain but also by other peripheral organs.

The Role of Glial Cells in Sleep and Wake

The essential role of glial cells in nervous system development, neural circuit function, and plasticity led to speculation that these cells impact sleep. Specialized glial cells contribute to activity of neurons at spatially distinct locations in the body including the central brain, blood-brain barrier (BBB), peripheral sensory organs, and neuromuscular junctions. Recent work demonstrates a role of glial cells in BBB and central regions of the brain in sleep.

Glial Cells of Blood-Brain Barrier (BBB)

One major anatomical role of glial cells in the brain is to create a blood-brain barrier (BBB) that separates the brain from circulating hemolymph, the equivalent of mammalian blood in flies (Fig. 4) [260, 261]. The inner layer of the BBB includes a layer of septate junctions connected to subperineurial glia (SPG) cells that cover the entire periphery of the central brain [262]. Perineurial glia (PG) and macrophage cells cover the SPG and secrete a dense carbohydrate-rich lamella to create a chemical and physical barrier [262, 263]. Gap junctions connect the SPG and PG cells and tightly regulate molecules moving into and out of the central nervous system (CNS) [264].

Recent studies describe the role of the BBB in sleep regulation. The expression of the circadian genes *per* and *Clk* cycle over 24 hours in glial cells of *Drosophila* BBB and regulate its permeability [173, 264]. This suggests that permeability of the

BBB cells is highly dynamic and regulated by daily circadian rhythms. In addition to circadian rhythms, sleep behavior also regulates endocytosis across the BBB [265]. Sleep loss enhances the permeability of the BBB; conversely, inhibition of endocytosis across the BBB increases the need for sleep. In flies, endocytosis is highest during early night sleep (ZT14) while lowest in the early morning (ZT2), indicating a strong positive correlation between sleep need and endocytosis [264]. The molecular components involved in the sleep-mediated regulation of endocytosis are not known; however, screening of GTPase proteins identified that the active form of *Rab11* in both SPG and PG cells facilitates endosomal vesicular transport [265]. Similar to flies, mammalian models also show a correlation between sleep and vesicular trafficking across the BBB, suggesting that vesicular trafficking could be one of the conserved functions of sleep in animals [266].

Another behavioral activity in *Drosophila* that correlates with endocytosis is repetitive proboscis extensions during sleep [267]. This so-called proboscis extension sleep is thought to be a deep sleep state in flies as it is characterized by increased arousal thresholds and reduced activity in the brain. Interestingly, preventing proboscis extension activity during sleep slows the clearance of injected dye from the hemolymph, suggesting that proboscis extensions during sleep have a functional role in clearing waste. A similar waste clearance system known as the glymphatic system is present in the mammalian brain as well. In a landmark study, the authors showed that clearance of waste byproducts from CNS is increased during sleep, suggesting a novel role of sleep in waste clearance [130]. This process has a potentially important role in various diseases because waste clearance is reduced in the aging brain probably because of lower, poor-quality sleep [268, 269].

Glial Cells of CNS

There are three major glial subtypes in the *Drosophila* CNS—cortex glia, ensheathing glia, and astrocytes—that interact with different parts of the neuron: the cell bodies, neurites, and synapse, respectively [270]. Astrocytes are the best studied glial subpopulation while the role of cortex and ensheathing glia in sleep behavior is at early stage. In an isolated study, authors show that inhibition of the *Drosophila* homolog of human amyloid precursor protein (*Appl*) in cortex glia resulted in increased night sleep duration and longer night avg. sleep bouts (Table 5) [271]. In addition, these sleep changes were correlated with alteration in the gap-junction component *Inx2* [271]. Interestingly, overexpressing the *Eaat2* glutamate transporter reverses the sleep phenotypes in *Appl* knockdown flies [74]. This suggests that App family proteins and glutamate recycling in glial cells are required for sleep regulation.

The ensheathing glia cells isolate neuronal projections and also mediate phagocytosis to engulf degenerating neurons [280, 281]. These cells express PER protein; however, whether they have a role in circadian rhythm is unknown [282]. Interestingly, manipulation of amino acid transport in these cells impacts sleep

Table 5 Genes having expression and/or function in glial cells

Gene	Tissue/cell type	Ref.
Appl	Cortex glia	[271]
Gs2/Inx2	Cortex and astrocyte-like glia	[271]
Eaat1	Cortex and astrocyte-like glia	[271]
rumpel	Ensheathing glia	[272]
kis	Subperineural glia	[21]
Eaat2	Ensheathing glia	[74]
DmMANF	Ensheathing and astrocyte-like glia	[273]
Smg6	Astrocytes	[94]
nec	Astrocytes	[94]
Toll-7	Astrocytes	[94]
TyRII	Astrocytes	[75]
Spatzle	Astrocytes	[75]
speck	Astrocytes	[274]
egr	Astrocytes	[275]
FABP7	Astrocytes	[276]
dFabp	Glia	[277]
Smg6	Glia	[94]
BomBC2	Glia	[94]
Nkt (CG14141)	Glia	[160]
AP-2α	Glia	[95]
Ncc69	Glia	[95]
CG9657	Glia	[95]
Mtk	Glia	[278]
Gabat	Glia	[97]
DAT	Glia	[279]

behavior. For example, feeding flies with a GABA agonist and an abundant amino acid, taurine, promotes sleep in flies, while knockdown of the transporter *Eaat2* increases daytime sleep [74]. These studies suggest that removal of synaptic taurine amino acid promotes sleep.

Astrocytes are important for regulating various aspect of sleep behavior [165]. Altering *Notch* signaling in astrocytes and mushroom body neurons eliminated homeostatic sleep recovery, while knockdown of *egr*, a homolog of mammalian *TNFα*, in astrocytes reduces 24-hour sleep duration [165, 275]. Overexpression of the mutant form of human fabp7 protein, a fatty acid binder and transporter, or its selective disruption in flies greatly fragments day and night sleep by decreasing avg. bout length and increasing bout number [276, 277, 283]. These observations suggest that astrocytes act in conjunction with neurons to regulate various aspects of sleep behavior including neural development, metabolism, and immunity [284, 285].

The Context-Dependent Nature of Sleep

Effect of Aging on Sleep

Like humans, sleep in flies becomes more fragmented with age; older flies have increased numbers of sleep bouts and decreased sleep bout length [286, 287]. Not only do older flies experience more fragmented sleep, but weakening of the intensity of their sleep and impairment of their response to sleep deprivation also occur [287]. Several studies explored whether the sleep fragmentation seen in older flies has a genetic basis. One-week-old flies carrying a null mutation of *mGluR*, the sole metabotropic glutamate receptor in flies, have redistributed daily sleep, with higher day sleep and lower night sleep duration, mimicking the sleep profile of aged flies [288]. Aged *mGluR* mutant flies lose sleep and die earlier than aged controls as well [288]. While mutations in *Hk* reduced sleep over time and shortened lifespan relative to controls, they lacked a trend for increased sleep fragmentation with age [289]. A reduction of insulin/insulin-like growth factor signaling using a null of three insulin-like peptide (*dilp2–3,5*) genes or a dominant negative insulin receptor (*InR*) mutant, which have increased lifespan, actually reduced sleep fragmentation in older flies [290]. Loss of function in *amn* leads to fragmented sleep and short sleep latency, but the flies did not show any significant changes in sleep with age [156]. These studies suggest that changes in sleep architecture with age may be ameliorated in part by altering gene signaling.

Effect of Developmental Processes on Sleep

Gene activity during fly developmental stages has important consequences for adult sleep as well. Like young mammals, flies sleep more during their first days after eclosion [5, 291, 292]. Restricting sleep during this critical stage has consequences on courtship behavior and on learning [291, 292]. Key genes expressed at particular stages of development contribute to this increase in sleep. Disruption of *Pdm3* during the mid-pupal stage prevents increased sleep in young flies [43]. Further, knockdown of the chromatin remodeler ISWI through the mid-third-instar stage decreases sleep duration and fragments sleep in the adult [45]. Interestingly, some molecules exert effects on sleep in both larvae and adults. For example, the neurotransmitter octopamine (*OA*) plays a wake-promoting role in adult flies [293] and also modulates larval sleep [294]. Excitation of the *OA* neurons in larvae reduced total sleep, and *OA* synthesis mutants (*Tβh* and *Tdc2*) and the *OA* receptor mutant (*Oamb*) increased larval sleep. Thus, the timing of the activity of key genes may have long-lasting effects on sleep across the flies' lifespan, and some genes affect sleep across developmental stages.

Effect of Immune Response and Injury on Sleep

Sleep changes in response to immune challenge in humans and other animals; conversely, sleep loss impairs the immune response [295]. In 2007, a survey of the head transcriptome revealed that immune response-related genes, including antimicrobial peptides, components of the *Toll* and *Imd* signaling pathways, peptidoglycan recognition pathways, and all of the nuclear factor kappa B (NFκB) genes (*dl*, *Dif*, and *Rel*), had increased expression during sleep deprivation [87], confirming an interplay between sleep and immune response in *Drosophila*. Similarly, analysis of the transcriptome of glia in flies subjected to traumatic brain injury revealed upregulation of immune response-related genes, including antimicrobial peptides [92]. Infection with *E. coli, P. aeruginosa*, or injury promoted sleep [296], yet traumatic brain injury resulted in reduced, fragmented sleep [92]. These alterations in sleep involve the NFκB *Rel* [92, 296], and its effect on sleep manifests in part through the fat body [87, 296]. Sleep loss prior to infection improved survivability, depending on the species of infection, and improved infection clearance; sleep deprivation immediately after infection produced a rebound that also improved survival, but *Dif*/*Rel* double mutants lacked these improvements [297]. *Wolbachia*, a bacteria known to infect flies, increases total sleep by increasing night sleep bout number, perhaps through the modulation of the dopaminergic pathway genes *ple* and *Ddc* [298]. Finally, *nur* is the only somnogen (a secreted molecule transmitting homeostatic sleep need) identified in flies thus far; interestingly, its role in sleep promotion may be important for host defense [77]. Flies overexpressing *nur* in the dFBs have increased sleep, improved survival, and greater bacterial clearance against infection by *S. marcescens* and *S. pneumoniae*. These studies show how genes modulate the relationship between sleep and immune response and injury.

Effect of Starvation/Feeding Regulation on Sleep

Like infection and injury, cellular and physiological stresses impact sleep. One major cause for cellular and physiological stress is exposure to starvation conditions. Starvation induces hyperactivity and reduces sleep in *Drosophila* [299, 300]. Despite their increased sleep propensity, *OA* receptor (*Oamb*) mutant flies show this starvation-induced hyperactivity [301]; *NPF*, another wake-promoting molecule, was shown to affect the sleep-wake behavior in a metabolic state-dependent manner, such that starvation-induced sleep loss is absent with the loss of *NPF* [202]. *NPF*, like *OA*, drives wakefulness and metabolic effects separately and acts as a hunger signal to keep the flies awake. *Trsn* acting in leucokinin neurons is a key regulator of the change in sleep behavior [73]; further, leucokinin signaling through the *Lkr* receptor processes both metabolic and sleep signals [302]. Serine metabolism in cholinergic neurons as mediated by *aay* and *stdh* also influences sleep suppression during starvation [89]. Interestingly, starvation-induced sleep may not be as detrimental as other kinds of sleep loss. For example, downregulation of the cytokine *upd2* and its receptor *dome* function as a signal of starvation, reducing sleep, yet these mutants do not suffer from decreased attention [258]. A subset of clock neurons, the small ventral lateral neurons (s-LNvs), can be remodeled by the *adipokinetic hormone* (*AKH*)-*Fork head Box-O* (*FOXO*) pathway in response to energy change and affect sleep [220]. Knockdown of *AKH* or the AKH receptor (*AKHR*) and a downstream factor *FOXO* does not show starvation-induced sleep loss, thus acting as a starvation response factor. *unc79*, which codes for an auxiliary subunit of the sodium leak channel *na*, is implicated in both sleep and starvation resistance. *unc79* mutations or genetic knockdowns abolish the starvation-induced sleep loss and provide starvation resistance by increasing lifespan of the flies under starved conditions [76]. Interestingly, an environmental context-specific trade-off occurs between naturally occurring alleles of the *for* gene [303]. Flies with the for^R allele lack a sleep rebound when deprived of sleep yet do not show decreased short-term memory. If starved, however, for^R flies have impaired short-term memory. Flies with the for^S or for^{S2} allele show the opposite response: when sleep deprived, their memory suffers; when starved, their memory is not impaired [303]. These studies show the relationship between sleep and stress and suggest potential trade-offs that might maintain genetic variation for sleep in natural populations.

Effect of Sleep Deprivation

Sleep consists of two components, according to prevailing theory: a circadian component that regulates the timing of sleep and a homeostatic component that indicates sleep need [304, 305]. A major goal is to understand the genetic networks regulating the homeostatic component of sleep [6, 29, 306–308]. One approach is to perturb the genome and observe the effect on compensatory sleep after sleep deprivation, referred to as the sleep rebound. Parameters manipulated to deprive flies of sleep include the timing and duration of the sleep deprivation period and the frequency, intensity, and duration of the depriving stimulus. Mechanical, environmental, genetic, and pharmacological stimuli have been used to deprive flies of sleep. In this section, we focus on mechanical and environmental methods of sleep deprivation. Stimulating the flies mechanically to deprive them of sleep is the most common method used (for a description of particular mechanical methods used, see section "Sleep Phenotypes Used in Genetic Studies"). While the knockdown or overexpression of many genes typically reduces or increases sleep rebound relative to controls (Table 6), we will emphasize the cases where a change in the gene's activity abolishes sleep rebound completely, which are relatively rare by comparison. Such cases may establish the critical building blocks of the sleep homeostat, or they may actually be common as the effects of sleep deprivation remain relatively understudied in flies.

One postulated function of sleep is that it has a role in learning and memory. Evidence gathered thus far indicates that sleep facilitates long-term memory

Table 6 Sleep deprivation in *Drosophila* mutants

Method	Time of day	Duration (hrs)	Frequency/ intensity	Gene	Effect on rebound sleep	Ref.
SNAP	Night	12	–	*bun*	Abolished	[165]
	Night	12	–	*Dl*	Abolished	[165]
	Night	12	10/min	*WT*	a	[193]
	Night	12	6/min	*bgm*	Reduced	[88]
	Night	12	6/min	*cue*	Reduced	[88]
	Night	12	6/min	*hll*	Reduced	[88]
	Night	12	–	*SPR*	Abolished	[214]
	Night	12	–	*Mip*	Abolished	[214]
	Night	12	6/min	*Lsd-2*	Abolished	[309]
	Night	12	10/min	*cyc*	Increased	[252]
	Night	12	10/min	*Hsp83*	Increased	[252]
	Day and night	12 and 24	–	*wgn*	Reduced	[275]
	Night	12	–	*pdgy*	Reduced	[310]
	Night	12	–	*AstA*	Reduced	[131]
	Night	12	–	*hiw*	Abolished	[162]
	Night	12	–	*Rdl*	Abolished	[162]
	Night	12		*Trh*	Reduced	[135]
	Night	12	20 s/3 min	*5-HT2B*	Reduced	[135]
	Night	9 and 12	–	*Nrx-1*	Reduced	[158]
	Night	12	–	*Htt*	Reduced	[62]
Timer-controlled rotating shaker	Night	12	2 s/min	*sNPF*	No effect	[311]
	Night	12	2 s/min	*NPF*	Reduced	[116]
	Night	12	2 s/min	*NPFR*	Reduced	[116]
	Night	12	1 s/min[b]	*Atf-2*	Increased	[206]
TriKinetics vortexer mounting plate	Night	12	2 s/20 s	*kis*	Abolished	[21]
	Night	6	2 s/20 s	*nur*	c	[77]
	Night	12	20 s,every 5 s	*hAPP*	Reduced	[312]
	Night	12	2–10 s/min	*tyrRII*	Reduced	[75]
	Night	12	2 s/20 s	*AANAT1*	d	[274]
	Night	9	1.2 s/20 s[e]	*Nmdar1/2*	Reduced	[313]
	Night	6	2 s/min	*alc*	Abolished	[314]
	Night	12	2 s/20 s	*ISWI*	Abolished	[45]
Framed box[f]	Day and night	24	6 cm	*para*	Abolished	[64]
	Day and night	24	1 cm	*dFmra1*	Abolished	[112]
	Day and night	24	1 cm	*amn*	Abolished	[156]

(continued)

Table 6 (continued)

Method	Time of day	Duration (hrs)	Frequency/ intensity	Gene	Effect on rebound sleep	Ref.
Gentle handling	Night	–		*Elp3*	No effect	[315]
Attaching activity monitors to a multi-tube vortexer	Night	12	1 s/2–20 s	*Rel + Dif*	g	[297]
Silica gel holder of tubes in a box	Night	12	–	*nAChRα3*	Reduced	[316]
DART	Night	12	5 s/1 hr	*cv-c*	Reduced	[132]
Custom	Night	12	2 s/20 s	*CycA*	Reduced	[72]
Shaken	Night	12	2 s or 10 s	*Mir-let7*	Reduced	[44]
Programmable rotor	Night	12	Directional shifts	*Sap/Syn*	Reduced	[317]
Stepper motor	Night	12	–	*dCREB2*	Increased, prolonged	[318]
Shaker	Night	12	3 s/min	*Hsp70A/ Hsp70B*	Increased	[138]
Rotating box	Night	12	–	*inc*	Abolished	[41]
Custom	Night	12	2 s/20 s	*qvr*	Reduced	[68]
Manual dropping	Night	12	7 mm or 21 mm drops[h]	*fumin*	Abolished	[319]
Manual tapping	Night	4 and 6	–	*homer*	i	[321]
	Night	3 and 6	–	*BiP*	j	[96]
	Night	12	–	*AANAT1*	Increased	[5]

[a]No. of synaptic terminals reduced during sleep and prevented by sleep deprivation
[b]Level 1 (1000 rpm), 2 (1500 rpm), and 3 (2000 rpm)
[c]Kinetics of sleep recovery the following morning changed, such that sleep increases were delayed in *nur* mutants, although overall sleep amounts were not affected
[d]Increased daytime recovery sleep following overnight sleep deprivation
[e]Light-induced waking—flies under mechanical disruption for 9 hours and then 10-minute light stimulus
[f]Rotates 180° at a speed of 2–3 rpm every ~5 minute monitors dropped N cm each rotation, mechanical shock
[g]Effect of early sleep deprivation on postinfection sleep and survival was abolished in double mutants
[h]For mild and moderate stimulus—once time, for strong stimulus—consecutively six times
[i]Longer recovery period ~24 hours compared to controls; no increase in average sleep bout length as seen in control
[j]Overexpression increased rebound sleep/dominant negative reduced rebound sleep

consolidation, while sleep deprivation disrupts learning and memory (reviewed in [321]). Mutations in *Lsd-2*, a gene involved in lipid metabolism, do not show any rebound sleep in the 48 hours following 12 hours of sleep deprivation during the night. Further investigation suggested that disruption of *Lsd-2* actually protects flies from sleep loss as they maintain normal levels of learning after sleep deprivation [309]. Along the same lines, knocking down both *hiw*, a gene involved in the ubiquitin-proteasome system, and *Rdl*, a GABA receptor, in the mushroom bodies leads to complete loss of rebound sleep following 12 hours of nighttime sleep deprivation yet ameliorates deficits in short-term memory [162]. A dopamine transporter (*DAT*) mutant, *fumin*, lacks a sleep rebound after 2,4, or 6 hours of nighttime sleep deprivation [319] and has impaired memory retention due to an abundance of dopamine at synaptic terminals [322]. Disruption of *bun*, a negative regulator of *Notch* signaling, eliminates sleep rebound; however, combining the *bun* mutation with the amorphic *Dl* mutant restores normal sleep homeostasis. Overexpression of *Dl* in the mushroom bodies eliminates sleep rebound and provides resistance to learning impairment following sleep deprivation [165]. *amn* is a gene involved in middle-term memory, and mutants do not have a sleep rebound after 12 and 24 hours of sleep deprivation [156, 323]. Reduced AMPK signaling in the brain by pan-neuronal knockdown of *alc* severely affects learning and disrupts sleep architecture; reduction of *alc* also abolishes any sleep rebound following 6 hours of late-night mechanical sleep deprivation [314]. This disparate group of genes regulating the effect of sleep on learning and memory maps to broad metabolic, neurotransmitter, and developmental pathways.

Fly models of human diseases provide a unique opportunity to investigate effect of these mutations on sleep along with the disease symptoms. A fly model of human generalized epilepsy with febrile seizures plus (GEFS+) is recreated by mutating the sole *Drosophila* voltage-gated sodium channel gene *para*, which results in enhanced GABAergic signaling. These mutants have no sleep rebound after 24 hours of sleep deprivation and are less susceptible to heat-induced seizures [64]. Both amorphs and hypermorphs of the *Drosophila* fragile X mental retardation gene *Fmr1* (*dFmra1*)—which show defects in neurite structures and neurotransmissions—also lack a sleep rebound after 24 hours of sleep deprivation and have more intense baseline sleep compared to wild-type flies [112]. Thus, these models show that abnormal sleep homeostasis may accompany disease.

Apart from these, null/knockdown mutants of genes important for sleep, synaptic transmission, putative circadian function, and mating responses also affect sleep homeostasis. Twelve hours of mechanical sleep deprivation produced no response in flies lacking *inc* [41], flies with pan-neuronal knockdown of *Mip* and *ISWI* using RNAi [45, 214], flies lacking *SPR*, the receptor of *Mip* or *SPR* knockdown in PDF neurons [214], and flies with reduced expression of *kis* in subperineurial glia (SPG) using RNAi [21]. These results suggest that genes with very diverse functions are essential for adequate rebound sleep following mechanical sleep deprivation.

Abiotic environmental factors like light or temperature affect sleep quality and amount; however, these factors can also cause sleep deprivation or ameliorate its effects. A dim light of 10 lux throughout the night induces sleep deprivation and rebound next day in wild-type flies [324]; accordingly, reduced night sleep duration and increased day sleep duration can result from continuous light exposure [325]. When exposed to 10 lux light throughout the night, a *Drosophila* tauopathy model overexpressing the human Tau protein (*hTau*) shows severe defects in rebound sleep. Abiotic stimuli can also be protective. For example, *cyc* mutant flies pre-treated with 36 °C for 3 hours are protected against the increased mortality due to sleep deprivation [252]. Similarly, an hour-long heat shock of 37 °C at the end of the daytime shows an increase in daytime sleep the next day in flies where *Egfr* signaling pathway is activated, even though no sleep loss is observed [326]. Environmental modalities such as light and temperature enable the study of sleep homeostasis in a noninvasive manner; however, they may impact the molecular circadian clock directly or indirectly; thus, inferences drawn from these experiments may not always reflect the true contribution of the sleep homeostat.

Conclusions

Drosophila melanogaster are a vital model system for the genetics of sleep. The relatively short generation time and rapid phenotyping using the DAMS system or video recording have enabled the survey of tens of thousands of mutations and candidate genes and millions of genomic variants for their effects on sleep. Some important themes emerge from sleep studies conducted over the past two decades.

The first theme concerns the number and complexity of sleep phenotypes measured in flies. While we noted that as many as 81 distinct sleep and circadian traits have been studied, most work centers around sleep duration, average sleep bout length, sleep bout number, and sleep latency. Nonzero heritability is present for all sleep traits where it has been calculated, indicating that sleep is at least partially genetic. To determine the genes impacting these traits, researchers applied both biased (candidate gene and reverse genetic screens) and unbiased approaches (forward mutagenesis screens, GWAS, QTL mapping, and artificial selection). Pleiotropy among sleep traits is a common feature of major-effect genes. Though fewer polymorphisms map to more than one sleep trait in GWAS/QTL studies, variants associated with different traits often map to the same gene, indicating that "gene-level" pleiotropy is more common. Further, many sleep traits are strongly genetically correlated, suggesting that they have a common genetic basis [27]. Accordingly, increasingly sophisticated methods for phenotypic measurement and calculation as well as approaches to combine or cluster phenotypes may reduce the number of traits to be considered, streamlining the search for the underlying genetic architecture of sleep [23, 36, 327, 328].

The second theme is that the results of different approaches used to discover genes are largely complementary, with relatively low levels of overlap among identified genes. This is indicative of a large genetic network affecting sleep rather than a single regulatory pathway [6]. The lack of overlap among genomic variants identified in GWAS, QTL mapping, and artificial selection studies in particular is also indicative of epistatic interactions among variants [101], a topic that deserves further investigation. Even within a large genetic network, however, highly interconnected (i.e., "hub") genes may exert a disproportionate degree of control over sleep and should be a focus.

The third theme is that sex has a significant role in the genetic architecture of sleep. Where both sexes have been studied, sex-specific and even sex-antagonistic effects of gene manipulation have been observed. Evidence for these differences exists at the level of genomic variants, gene expression, and neural connectivity. To fully understand the genetic architecture of sleep, a more comprehensive understanding of the underlying genetic networks influencing sex-specific differences in sleep is needed.

The fourth theme is the apparent lack of a single brain structure that controls all aspects of sleep. Instead, sleep behavior is non-uniformly distributed across the brain, and certain parts of the brain may be more important in sleep regulation than others [123, 124, 329, 330]. Accordingly, brain structures influencing sleep tend to affect multiple sleep traits rather than being exclusive to a single sleep trait (Fig. 5). Interestingly, sleep regulatory neurons have been mapped to regions that subserve different functions. For example, sleep-regulating regions in *Drosophila* include the mushroom bodies (required for learning and memory), ellipsoid body (required for orientation, locomotion, and navigation), and fan-shaped body (required for visual learning) [14, 140, 141, 150]. This suggests the intriguing possibility that different sleep-promoting loci could serve different functions and be activated under different circumstances, controlling specific aspects of sleep [331]. Adding to this complexity are neurons outside the brain that influence sleep [128, 144, 145]. How these multiple regions integrate the requirement for sleep and generate sleep behavior remains a challenge for the field.

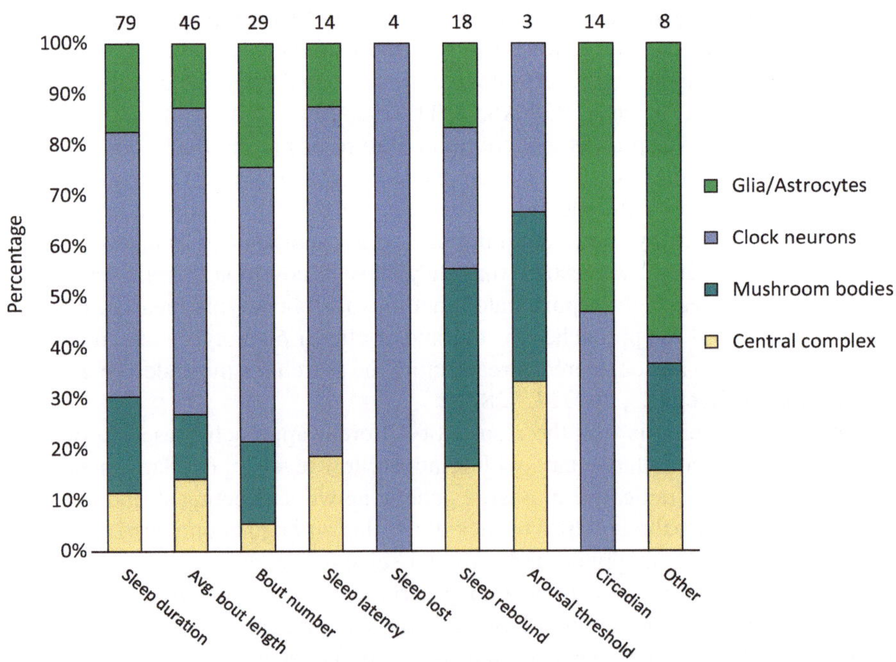

Fig. 5 Sleep traits map to multiple brain regions. The plot shows the percentage of genes mapping to each trait in each of four major brain regions/cell types. The number of genes per trait is shown at the top of the bar

A fifth theme is the physical and genetic relationship between sleep and circadian rhythms. Sleep is considered to be regulated by two processes, a circadian process and a homeostatic process [304]. The circadian process of sleep is regulated by clock neurons, while sleep homeostasis is mainly regulated by the dFB and EB [131, 313, 332, 333]. However, we do not know if these two components of the sleep interact with each other to set up the final sleep requirement and output. A general observation is that the circadian process tightly regulates sleep even after sleep homeostasis is disrupted through sleep deprivation, which suggests that it may be higher in the hierarchy to regulate daily sleep. Several studies identified connections between clock neurons LPN and DN1s with central complex cells like dFB and EB, raising the possibility that homeostasis centers may take synaptic inputs from upstream circadian clock neurons [229, 236, 334]. These connections and the underlying regulatory gene network that links these two processes are an important focus area for future research.

Lastly, the influence of genes on sleep is context dependent. Age, the presence of infection and injury, food availability, and environmental stressors have profound effects on sleep. Sleep regulation flexibly integrates these environmental signals into the core program. Accordingly, there should be specific gene networks and circuitry that integrate these requirements into the core sleep circuit. For example, the role of glial cells in supporting the activities of the neurons adds an indirect manipulator of sleep. A comprehensive delineation of genotype-by-environment interactions is needed, including a description of genes affecting sleep rebound in flies, a relatively understudied aspect of sleep behavior.

Acknowledgments This work was supported by the Intramural Research Program of the NIH, the National Heart, Lung, and Blood Institute.

References

1. Campbell SS, Tobler I. Animal sleep: a review of sleep duration across phylogeny. Neurosci Biobehav Rev. 1984;8:269–300.
2. Tobler I. Effect of forced locomotion on the rest-activity cycle of the cockroach. Behav Brain Res. 1983;8(3):351–60.
3. Kaiser W, Steiner-Kaiser J. Neuronal correlates of sleep, wakefulness and arousal in a diurnal insect. Nature. 1983;301:707–9.
4. Hendricks JC, Finn SM, Panckeri KA, et al. Rest in Drosophila is a sleep-like state. Neuron. 2000;25:129–38.
5. Shaw PJ, Cirelli C, Greenspan RJ, Tononi G. Correlates of sleep and waking in Drosophila melanogaster. Science. 2000;287:1834–7.
6. Jan M, O'Hara BF, Franken P. Recent advances in understanding the genetics of sleep. F1000Res. 2020;9:F1000 Faculty Rev-214.
7. Huber R, Hill SL, Holladay C, Biesiadecki M, Tononi G, Cirelli C. Sleep homeostasis in Drosophila melanogaster. Sleep. 2004;27:628–39.
8. Geissmann Q, Garcia Rodriguez L, Beckwith EJ, French AS, Jamasb AR, Gilestro GF. Ethoscopes: an open platform for high-throughput ethomics. PLoS Biol. 2017;15(10):e2003026.

9. Faville R, Kottler B, Goodhill GJ, Shaw PJ, van Swinderen B. How deeply does your mutant sleep? Probing arousal to better understand sleep defects in Drosophila. Sci Rep. 2015;5:8454.
10. Melnattur K, Morgan E, Duong V, Kalra A, Shaw PJ. The sleep nullifying apparatus: a highly efficient method of sleep depriving Drosophila. J Vis Exp. 2020;(166).
11. Huang H, Possidente DR, Vecsey CG. Optogenetic activation of SIFamide (SIFa) neurons induces a complex sleep-promoting effect in the fruit fly Drosophila melanogaster. Physiol Behav. 2021;239:113507.
12. Juneau BA, Stonemetz JM, Toma RF, Possidente DR, Heins RC, Vecsey CG. Optogenetic activation of short neuropeptide F (sNPF) neurons induces sleep in Drosophila melanogaster. Physiol Behav. 2019;206:143–56.
13. Nitabach MN, Wu Y, Sheeba V, et al. Electrical hyperexcitation of lateral ventral pacemaker neurons desynchronizes downstream circadian oscillators in the fly circadian circuit and induces multiple behavioral periods. J Neurosci. 2006;26(2):479–89.
14. Pitman JL, McGill JJ, Keegan KP, Allada R. A dynamic role for the mushroom bodies in promoting sleep in Drosophila. Nature. 2006;441(7094):753–6.
15. Zemelman BV, Lee GA, Ng M, Miesenbock G. Selective photostimulation of genetically chARGed neurons. Neuron. 2002;33(1):15–22.
16. Kirszenblat L, Yaun R, van Swinderen B. Visual experience drives sleep need in Drosophila. Sleep. 2019;42(7):zsz102.
17. Harbison ST, Sehgal A. Quantitative genetic analysis of sleep in Drosophila melanogaster. Genetics. 2008;178:2341–60.
18. Harbison ST, McCoy LJ, Mackay TF. Genome-wide association study of sleep in Drosophila melanogaster. BMC Genomics. 2013;14:281.
19. Agosto J, Choi JC, Parisky KM, Stilwell G, Rosbash M, Griffith LC. Modulation of GABAA receptor desensitization uncouples sleep onset and maintenance in Drosophila. Nat Neurosci. 2008;11(3):354–9.
20. Zhang S, Ross KD, Seidner GA, et al. Nmf9 Encodes a highly conserved protein important to neurological function in mice and flies. PLoS Genet. 2015;11(7):e1005344.
21. Coll-Tane M, Gong NN, Belfer SJ, et al. The CHD8/CHD7/Kismet family links blood-brain barrier glia and serotonin to ASD-associated sleep defects. Sci Adv. 2021;7(23):eabe2626.
22. van Alphen B, Yap MH, Kirszenblat L, Kottler B, van Swinderen B. A dynamic deep sleep stage in Drosophila. J Neurosci. 2013;33(16):6917–27.
23. Jagannathan SR, Jeans T, Van De Poll, MN, van Swinderen B. Multivariate classification of multichannel long-term electrophysiology data identifies different sleep stages in fruit flies. Science Advances. 2024;10(8) 10.1126/sciadv.adj4399
24. Stahl BA, Slocumb ME, Chaitin H, DiAngelo JR, Keene AC. Sleep-dependent modulation of metabolic rate in Drosophila. Sleep. 2017;40(8):zsx084.
25. Bei B, Wiley JF, Trinder J, Manber R. Beyond the mean: a systematic review on the correlates of daily intraindividual variability of sleep/wake patterns. Sleep Med Rev. 2016;28:108–24.
26. Keenan BT, Galante RJ, Lian J, et al. High-throughput sleep phenotyping produces robust and heritable traits in Diversity Outbred mice and their founder strains. Sleep. 2020;43(5):zsz278.
27. Wu KJ, Kumar S, Serrano Negron YL, Harbison ST. Genotype influences day-to-day variability in sleep in Drosophila melanogaster. Sleep. 2018;41(2):zsx205. https://doi.org/10.1093/sleep/zs1205.
28. Mackay TF, Lyman RF. Drosophila bristles and the nature of quantitative genetic variation. Philos Trans R Soc B. 2005;360:1513–27.
29. Dubowy C, Sehgal A. Circadian rhythms and sleep in Drosophila melanogaster. Genetics. 2017;205(4):1373–97.
30. Chatfield C. The analysis of time series. 6th ed. Boca Raton: CRC Press, Taylor and Francis Group; 2003.
31. Levine JD, Funes P, Dowse HB, Hall JC. Signal analysis of behavioral and molecular cycles. BMC Neurosci. 2002;3:1.
32. Pegoraro M, Sayegh Rezek E, Fishman B, Tauber E. Nucleotide variation in Drosophila cryptochrome is linked to circadian clock function: an association analysis. Front Physiol. 2022;13:781380.

33. Ghosh A, Sharma P, Dansana S, Sheeba V. Evidence for co-evolution of masking with circadian phase in Drosophila melanogaster. J Biol Rhythm. 2021;36(3):254–70.
34. Valadas JS, Esposito G, Vandekerkhove D, et al. ER lipid defects in neuropeptidergic neurons impair sleep patterns in Parkinson's disease. Neuron. 2018;98(6):1155–1169 e1156.
35. Kumar S, Chen D, Sehgal A. Dopamine acts through cryptochrome to promote acute arousal in Drosophila. Genes Dev. 2012;26(11):1224–34.
36. Wiggin TD, Goodwin PR, Donelson NC, et al. Covert sleep-related biological processes are revealed by probabilistic analysis in Drosophila. Proc Natl Acad Sci USA. 2020;117(18):10024–34.
37. Troup M, Yap MH, Rohrscheib C, et al. Acute control of the sleep switch in Drosophila reveals a role for gap junctions in regulating behavioral responsiveness. elife. 2018;7:e37105.
38. Eiman MN, Kumar S, Serrano Negron YL, Tansey TR, Harbison ST. Genome-wide association in Drosophila identifies a role for Piezo and Proc-R in sleep latency. Sci Rep. 2024;14(1) https://doi.org/10.1038/s41598-023-50552-z
39. Emery PT, Birley AJ, Morgan E. Investigation of natural genetic variation in the circadian system of Drosophila melanogaster. II. Biometrical analyses of locomotor activity rhythms recorded in constant darkness. Chronobiol Int. 1995;12(2):77–86.
40. Harbison ST, Kumar S, Huang W, McCoy LJ, Smith KR, Mackay TFC. Genome-wide association study of circadian behavior in Drosophila melanogaster. Behav Genet. 2019;49(1):60–82.
41. Pfeiffenberger C, Allada R. Cul3 and the BTB adaptor Insomniac are key regulators of sleep homeostasis and a dopamine arousal pathway in Drosophila. PLoS Genet. 2012;8:e1003003.
42. Lee J, Lim C, Han TH, et al. The E3 ubiquitin ligase adaptor Tango10 links the core circadian clock to neuropeptide and behavioral rhythms. Proc Natl Acad Sci USA. 2021;118(47):e2110767118.
43. Chakravarti Dilley L, Szuperak M, Gong NN, et al. Identification of a molecular basis for the juvenile sleep state. elife. 2020;9:e52676.
44. Goodwin PR, Meng A, Moore J, et al. MicroRNAs regulate sleep and sleep homeostasis in Drosophila. Cell Rep. 2018;23(13):3776–86.
45. Gong NN, Dilley LC, Williams CE, et al. The chromatin remodeler ISWI acts during Drosophila development to regulate adult sleep. Sci Adv. 2021;7(8):eabe2597.
46. Chen KF, Lowe S, Lamaze A, Kratschmer P, Jepson J. Neurocalcin regulates nighttime sleep and arousal in Drosophila. elife. 2019;8:e38114.
47. Hidalgo S, Campusano JM, Hodge JJL. The Drosophila ortholog of the schizophrenia-associated CACNA1A and CACNA1B voltage-gated calcium channels regulate memory, sleep and circadian rhythms. Neurobiol Dis. 2021;155:105394.
48. Khoury S, Wang QP, Parisien M, et al. Multi-ethnic GWAS and meta-analysis of sleep quality identify MPP6 as a novel gene that functions in sleep center neurons. Sleep. 2021;44(3):zsaa211.
49. Singgih EL, van der Voet M, Schimmel-Naber M, Brinkmann EL, Schenck A, Franke B. Investigating cytosolic 5′-nucleotidase II family genes as candidates for neuropsychiatric disorders in Drosophila (114/150 chr). Transl Psychiatry. 2021;11(1):55.
50. Klein M, Singgih EL, van Rens A, et al. Contribution of intellectual disability-related genes to ADHD risk and to locomotor activity in Drosophila. Am J Psychiatry. 2020;177(6):526–36.
51. Harich B, Klein M, Ockeloen CW, et al. From man to fly – convergent evidence links FBXO25 to ADHD and comorbid psychiatric phenotypes. J Child Psychol Psychiatry. 2020;61(5):545–55.
52. Allebrandt KV, Amin N, Muller-Myhsok B, et al. A K(ATP) channel gene effect on sleep duration: from genome-wide association studies to function in Drosophila. Mol Psychiatry. 2013;18(1):122–32.
53. Freeman A, Pranski E, Miller RD, et al. Sleep fragmentation and motor restlessness in a Drosophila model of Restless Legs Syndrome. Curr Biol. 2012;22(12):1142–8.
54. Wang YY, Ma WW, Peng IF. Screening of sleep assisting drug candidates with a Drosophila model. PLoS One. 2020;15(7):e0236318.

55. Song Q, Feng G, Huang Z, Chen X, Chen Z, Ping Y. Aberrant axonal arborization of PDF neurons induced by Abeta42-mediated JNK activation underlies sleep disturbance in an Alzheimer's model. Mol Neurobiol. 2017;54(8):6317–28.
56. Berlandi J, Lin FJ, Ambree O, Rieger D, Paulus W, Jeibmann A. Swing boat: inducing and recording locomotor activity in a Drosophila melanogaster model of Alzheimer's disease. Front Behav Neurosci. 2017;11:159.
57. Gerstner JR, Lenz O, Vanderheyden WM, Chan MT, Pfeiffenberger C, Pack AI. Amyloid-beta induces sleep fragmentation that is rescued by fatty acid binding proteins in Drosophila. J Neurosci Res. 2017;95(8):1548–64.
58. Asjad HMM, Kasture A, El-Kasaby A, et al. Pharmacochaperoning in a Drosophila model system rescues human dopamine transporter variants associated with infantile/juvenile parkinsonism. J Biol Chem. 2017;292(47):19250–65.
59. Sun X, Ran D, Zhao X, et al. Melatonin attenuates hLRRK2-induced sleep disturbances and synaptic dysfunction in a Drosophila model of Parkinson's disease. Mol Med Rep. 2016;13(5):3936–44.
60. Liu Y, Shen L, Zhang Y, et al. Rare NRXN1 missense variants identified in autism interfered protein degradation and Drosophila sleeping. J Psychiatr Res. 2021;143:113–22.
61. Farago A, Zsindely N, Bodai L. Mutant huntingtin disturbs circadian clock gene expression and sleep patterns in Drosophila. Sci Rep. 2019;9(1):7174.
62. Gonzales ED, Tanenhaus AK, Zhang J, Chaffee RP, Yin JC. Early-onset sleep defects in Drosophila models of Huntington's disease reflect alterations of PKA/CREB signaling. Hum Mol Genet. 2016;25(5):837–52.
63. Gupta T, Morgan HR, Bailey JA, Certel SJ. Functional conservation of MBD proteins: MeCP2 and Drosophila MBD proteins alter sleep. Genes Brain Behav. 2016;15(8):757–74.
64. Petruccelli E, Lansdon P, Kitamoto T. Exaggerated nighttime sleep and defective sleep homeostasis in a Drosophila knock-in model of human epilepsy. PLoS One. 2015;10(9):e0137758.
65. Cirelli C, Bushey D, Hill S, et al. Reduced sleep in Drosophila Shaker mutants. Nature. 2005;434:1087–92.
66. Stavropoulos N, Young MW. Insomniac and Cullin-3 regulate sleep and wakefulness in Drosophila. Neuron. 2011;72:964–76.
67. Bedont JL, Toda H, Shi M, et al. Short and long sleeping mutants reveal links between sleep and macroautophagy. elife. 2021;10:e64140.
68. Koh K, Joiner WJ, Wu MN, Yue Z, Smith CJ, Sehgal A. Identification of SLEEPLESS, a sleep-promoting factor. Science. 2008;321:372–6.
69. Liu S, Lamaze A, Liu Q, et al. WIDE AWAKE mediates the circadian timing of sleep onset. Neuron. 2014;82:151–66.
70. Afonso DJ, Liu D, Machado DR, et al. TARANIS functions with cyclin A and Cdk1 in a novel arousal center to control sleep in Drosophila. Curr Biol. 2015;25(13):1717–26.
71. Dai X, Zhou E, Yang W, Zhang X, Zhang W, Rao Y. D-Serine made by serine racemase in Drosophila intestine plays a physiological role in sleep. Nat Commun. 2019;10(1):1986.
72. Rogulja D, Young MW. Control of sleep by Cyclin A and its regulator. Science. 2012;335:1617–21.
73. Murakami K, Yurgel ME, Stahl BA, et al. Translin is required for metabolic regulation of Sleep. Curr Biol. 2016;26(7):972–80.
74. Stahl BA, Peco E, Davla S, et al. The Taurine transporter Eaat2 functions in ensheathing glia to modulate sleep and metabolic rate. Curr Biol. 2018;28(22):3700–3708.e4.
75. Blum ID, Keles MF, Baz ES, et al. Astroglial calcium signaling encodes sleep need in Drosophila. Curr Biol. 2021;31(1):150–162.e7.
76. Murakami K, Palermo J, Stanhope BA, Gibbs AG, Keene AC. A screen for sleep and starvation resistance identifies a wake-promoting role for the auxiliary channel unc79. G3 (Bethesda). 2021;11(8):jkab199.
77. Toda H, Williams JA, Gulledge M, Sehgal A. A sleep-inducing gene, nemuri, links sleep and immune function in Drosophila. Science. 2019;363(6426):509–15.

78. Falconer DS, Mackay TF. Introduction to quantitative genetics. 4th ed. Edinburgh Gate, Harlow: Addison Wesley Longman Limited; 1996.
79. Mackay TF, Richards S, Stone EA, et al. The Drosophila melanogaster genetic reference panel. Nature. 2012;482(7384):173–8.
80. King E, Merkes CM, McNeil CL, et al. Genetic dissection of a model complex trait using the Drosophila synthetic population resource. Genome Res. 2012;22(8):1558–66.
81. Smith BR, Macdonald SJ. Dissecting the genetic basis of variation in Drosophila sleep using a multiparental QTL mapping resource. Genes (Basel). 2020;11(3):294.
82. Tomita J, Ueno T, Mitsuyoshi M, Kume S, Kume K. The NMDA receptor promotes sleep in the fruit fly, Drosophila melanogaster. PLoS One. 2015;10(5):e0128101.
83. Chen X, Rosbash M. MicroRNA-92a is a circadian modulator of neuronal excitability in Drosophila. Nat Commun. 2017;8:14707.
84. Seugnet L, Suzuki Y, Thimgan M, et al. Identifying sleep regulatory genes using a Drosophila model of insomnia. J Neurosci. 2009;29:7148–57.
85. Souto-Maior C, Serrano Negron YL, Harbison ST. Nonlinear expression patterns and multiple shifts in gene network interactions underlie robust phenotypic change in Drosophila melanogaster selected for night sleep duration. PLOS Comput Biol. 2023;19(8):e1011389. https://doi.org/10.1371/journal.pcbi.1011389
86. Harbison ST, Carbone MA, Ayroles JF, Stone EA, Lyman RF, Mackay TF. Co-regulated transcriptional networks contribute to natural genetic variation in Drosophila sleep. Nat Genet. 2009;41(3):371–5.
87. Williams JA, Sathyanarayanan S, Hendricks JC, Sehgal A. Interaction between sleep and the immune response in Drosophila: a role for the NFkappaB relish. Sleep. 2007;30(4):389–400.
88. Thimgan MS, Seugnet L, Turk J, Shaw PJ. Identification of genes associated with resilience/vulnerability to sleep deprivation and starvation in Drosophila. Sleep. 2016;38:801–14.
89. Sonn JY, Lee J, Sung MK, et al. Serine metabolism in the brain regulates starvation-induced sleep suppression in Drosophila melanogaster. Proc Natl Acad Sci USA. 2018;115(27):7129–34.
90. Mokashi SS, Shankar V, MacPherson RA, Hannah RC, Mackay TFC, Anholt RRH. Developmental alcohol exposure in Drosophila: effects on adult phenotypes and gene expression in the brain. Front Psych. 2021;12:699033.
91. Li W, Wang Z, Syed S, et al. Chronic social isolation signals starvation and reduces sleep in Drosophila. Nature. 2021;597(7875):239–44.
92. van Alphen B, Stewart S, Iwanaszko M, et al. Glial immune-related pathways mediate effects of closed head traumatic brain injury on behavior and lethality in Drosophila. PLoS Biol. 2022;20(1):e3001456.
93. Machado Almeida P, Lago Solis B, Stickley L, Feidler A, Nagoshi E. Neurofibromin 1 in mushroom body neurons mediates circadian wake drive through activating cAMP-PKA signaling. Nat Commun. 2021;12(1):5758.
94. You S, Yu AM, Roberts MA, Joseph IJ, Jackson FR. Circadian regulation of the Drosophila astrocyte transcriptome. PLoS Genet. 2021;17(9):e1009790.
95. Ng FS, Sengupta S, Huang Y, et al. TRAP-seq profiling and RNAi-based genetic screens identify conserved Glial genes required for adult Drosophila behavior. Front Mol Neurosci. 2016;9:146.
96. Naidoo N, Casiano V, Cater J, Zimmerman J, Pack AI. A role for the molecular chaperone protein BiP/GRP78 in Drosophila sleep homeostasis. Sleep. 2007;30(5):557–65.
97. Chen WF, Maguire S, Sowcik M, Luo W, Koh K, Sehgal A. A neuron-glia interaction involving GABA transaminase contributes to sleep loss in sleepless mutants. Mol Psychiatry. 2015;20(2):240–51.
98. Pegoraro M, Flavell LMM, Menegazzi P, et al. The genetic basis of diurnal preference in Drosophila melanogaster. BMC Genomics. 2020;21(1):596.
99. Harbison ST, Serrano Negron YL, Hansen NF, Lobell AS. Selection for long and short sleep duration in Drosophila melanogaster reveals the complex genetic network underlying natural variation in sleep. PLoS Genet. 2017;13(12):e10007098.

100. Souto-Maior C, Serrano Negron YL, Harbison ST. Natural selection on sleep duration in Drosophila melanogaster. Sci Rep. 2020;10:20652.
101. Mackay TFC, Huang W. Charting the genotype-phenotype map: lessons from the Drosophila melanogaster Genetic Reference Panel Wiley Interdiscip Rev Dev Biol. 2017;7(1):10.1002/wdev.289.
102. Huang W, Massouras A, Inoue Y, et al. Natural variation in genome architecture among 205 Drosophila melanogaster Genetic Reference Panel lines. Genome Res. 2014;24(7):1193–208.
103. King EG, Macdonald SJ, Long AD. Properties and power of the Drosophila Synthetic Population Resource for the routine dissection of complex traits. Genetics. 2012;191(3):935–49.
104. Huang W, Richards S, Carbone MA, et al. Epistasis dominates the genetic architecture of Drosophila quantitative traits. Proc Natl Acad Sci USA. 2012;109:15553–9.
105. Swarup S, Huang W, Mackay TF, Anholt RR. Analysis of natural variation reveals neurogenetic networks for Drosophila olfactory behavior. Proc Natl Acad Sci USA. 2013;110:1017–22.
106. Mackay TF. Epistasis and quantitative traits: using model organisms to study gene-gene interactions. Nat Rev Genet. 2014;15(1):22–33.
107. Shorter J, Couch C, Huang W, et al. Genetic architecture of natural variation in Drosophila melanogaster aggressive behavior. Proc Natl Acad Sci USA. 2015;112:E3555–63.
108. Morozova TV, Huang W, Pray VA, Whitham T, Anholt RR, Mackay TF. Polymorphisms in early neurodevelopmental genes affect natural variation in alcohol sensitivity in adult drosophila. BMC Genomics. 2015;16:865.
109. Carbone MA, Yamamoto A, Huang W, et al. Genetic architecture of natural variation in visual senescence in Drosophila. Proc Natl Acad Sci USA. 2016;113(43):E6620–9.
110. Harbison ST, Chang S, Kamdar KP, Mackay TF. Quantitative genomics of starvation stress resistance in Drosophila. Genome Biol. 2005;6(4):R36.
111. Bushey D, Huber R, Tononi G, Cirelli C. Drosophila hyperkinetic mutants have reduced sleep and impaired memory. J Neurosci. 2007;27(20):5384–93.
112. Bushey D, Tononi G, Cirelli C. The Drosophila Fragile X mental retardation gene regulates sleep need. J Neurosci. 2009;29:1948–61.
113. Luo W, Chen WF, Yue Z, et al. Old flies have a robust central oscillator but weaker behavioral rhythms that can be improved by genetic and environmental manipulations. Aging Cell. 2012;11(3):428–38.
114. Maurer GW, Malita A, Nagy S, et al. Analysis of genes within the schizophrenia-linked 22q11.2 deletion identifies interaction of night owl/LZTR1 and NF1 in GABAergic sleep control. PLoS Genet. 2020;16(4):e1008727.
115. Nazario-Yepiz NO, Fernandez Sobaberas J, Lyman R, et al. Physiological and metabolomic consequences of reduced expression of the Drosophila brummer triglyceride Lipase. PLoS One. 2021;16(9):e0255198.
116. He C, Yang Y, Zhang M, Price JL, Zhao Z. Regulation of sleep by neuropeptide Y-like system in Drosophila melanogaster. PLoS One. 2013;8(9):e74237.
117. Bai L, Sehgal A. Anaplastic lymphoma kinase acts in the Drosophila mushroom body to negatively regulate sleep. PLoS Genet. 2015;11(11):e1005611.
118. Glover Z, Hodges MD, Dravecz N, et al. Loss of angiotensin-converting enzyme-related (ACER) peptidase disrupts behavioural and metabolic responses to diet in Drosophila melanogaster. J Exp Biol. 2019;222(Pt 8):jeb194332.
119. De Luca M, Klimentidis YC, Casazza K, et al. A conserved role for syndecan family members in the regulation of whole-body energy metabolism. PLoS One. 2010;5(6):e11286.
120. Alvarez-Rendon JP, Riesgo-Escovar JR. Circadian and rhythmic-related behavioral comorbidities of the diabetic state in Drosophila melanogaster. Gen Comp Endocrinol. 2020;295:113477.
121. Wu B, Ma L, Zhang E, et al. Sexual dimorphism of sleep regulated by juvenile hormone signaling in Drosophila. PLoS Genet. 2018;14(4):e1007318.
122. Brown JB, Boley N, Eisman R, et al. Diversity and dynamics of the Drosophila transcriptome. Nature. 2014;512:393–9.

123. Artiushin G, Sehgal A. The Drosophila circuitry of sleep-wake regulation. Curr Opin Neurobiol. 2017;44:243–50.
124. Shafer OT, Keene AC. The regulation of Drosophila sleep. Curr Biol. 2021;31(1):R38–49.
125. Morgan D, Tsai SC. Sleep and the endocrine system. Sleep Med Clin. 2016;11(1):115–26.
126. Spiegel K, Leproult R, Van Cauter E. Impact of sleep debt on metabolic and endocrine function. Lancet. 1999;354:1435–9.
127. Artiushin G, Sehgal A. The glial perspective on sleep and circadian rhythms. Annu Rev Neurosci. 2020;43:119–40.
128. Satterfield LK, De J, Wu M, Qiu T, Joiner WJ. Inputs to the sleep homeostat originate outside the brain. J Neurosci. 2022;42(29):5695–704.
129. Cuddapah VA, Zhang SL, Sehgal A. Regulation of the blood-brain barrier by circadian rhythms and sleep. Trends Neurosci. 2019;42(7):500–10.
130. Xie L, Kang H, Xu Q, et al. Sleep drives metabolite clearance from the adult brain. Science. 2013;342:373–7.
131. Donlea JM, Thimgan MS, Suzuki Y, Gottschalk L, Shaw PJ. Inducing sleep by remote control facilitates memory consolidation in Drosophila. Science. 2011;332(6037):1571–6.
132. Liu Q, Liu S, Kodama L, Driscoll MR, Wu MN. Two dopaminergic neurons signal to the dorsal fan-shaped body to promote wakefulness in Drosophila. Curr Biol. 2012;22(22):2114–23.
133. Donlea JM, Pimentel D, Miesenbock G. Neuronal machinery of sleep homeostasis in Drosophila. Neuron. 2014;81(4):860–72.
134. Pimentel D, Donlea JM, Talbot CB, Song SM, Thurston AJ, Miesenbock G. Operation of a homeostatic sleep switch. Nature. 2016;536(7616):333–7.
135. Andretic R, van Swinderen B, Greenspan RJ. Dopaminergic modulation of arousal in Drosophila. Curr Biol. 2005;15(13):1165–75.
136. Ueno T, Tomita J, Tanimoto H, et al. Identification of a dopamine pathway that regulates sleep and arousal in Drosophila. Nat Neurosci. 2012;15(11):1516–23.
137. Qian Y, Cao Y, Deng B, et al. Sleep homeostasis regulated by 5HT2b receptor in a small subset of neurons in the dorsal fan-shaped body of drosophila. elife. 2017;6:e26519.
138. Omoto JJ, Nguyen BM, Kandimalla P, Lovick JK, Donlea JM, Hartenstein V. Neuronal constituents and putative interactions within the Drosophila ellipsoid body neuropil. Front Neural Circuits. 2018;12:103.
139. Donlea JM, Pimentel D, Talbot CB, et al. Recurrent circuitry for balancing sleep need and sleep. Neuron. 2018;97(2):378–389.e4.
140. Liu S, Liu Q, Tabuchi M, Wu MN. Sleep drive is encoded by neural plastic changes in a dedicated circuit. Cell. 2016;165(6):1347–60.
141. Yagi R, Mabuchi Y, Mizunami M, Tanaka NK. Convergence of multimodal sensory pathways to the mushroom body calyx in Drosophila melanogaster. Sci Rep. 2016;6:29481.
142. Fropf R, Zhou H, Yin JCP. The clock gene period differentially regulates sleep and memory in Drosophila. Neurobiol Learn Mem. 2018;153(Pt A):2–12.
143. Wang J, Fan JY, Zhao Z, Dissel S, Price J. DBT affects sleep in both circadian and non-circadian neurons. PLoS Genet. 2022;18(2):e1010035.
144. De J, Wu M, Lambatan V, Hua Y, Joiner WJ. Re-examining the role of the dorsal fan-shaped body in promoting sleep in Drosophila. Curr Biol. 2023;33(17):3660–3668.e4. https://doi.org/10.1016/j.cub.2023.07.043
145. Zhao X, Li Y, Zhao Z, Du J. Extra sex combs buffers sleep-related stresses through regulating Heat shock proteins. FASEB J. 2021;35(1):e21190.
146. Jones JD, Holder BL, Eiken KR, Vogt A, Velarde AI, Elder AJ, McEllin JA, Dissel S. Regulation of sleep by cholinergic neurons located outside the central brain in Drosophila. e3002012. PLOS Biol. 2023;21(3) https://doi.org/10.1371/journal.pbio.3002012
147. Huang S, Piao C, Beuschel CB, Gotz T, Sigrist SJ. Presynaptic active zone plasticity encodes sleep need in Drosophila. Curr Biol. 2020;30(6):1077–1091.e5.
148. Heisenberg M, Borst A, Wagner S, Byers D. Drosophila mushroom body mutants are deficient in olfactory learning. J Neurogenet. 1985;2(1):1–30.

149. McGuire SE, Le PT, Davis RL. The role of Drosophila mushroom body signaling in olfactory memory. Science. 2001;293(5533):1330–3.
150. Joiner WJ, Crocker A, White BH, Sehgal A. Sleep in Drosophila is regulated by adult mushroom bodies. Nature. 2006;441(7094):757–60.
151. Sitaraman D, Aso Y, Jin X, et al. Propagation of homeostatic sleep signals by segregated synaptic microcircuits of the Drosophila mushroom body. Curr Biol. 2015;25(22):2915–27.
152. Aso Y, Hattori D, Yu Y, et al. The neuronal architecture of the mushroom body provides a logic for associative learning. elife. 2014;3:e04577.
153. Aso Y, Sitaraman D, Ichinose T, et al. Mushroom body output neurons encode valence and guide memory-based action selection in Drosophila. elife. 2014;3:e04580.
154. Caron SJ, Ruta V, Abbott LF, Axel R. Random convergence of olfactory inputs in the Drosophila mushroom body. Nature. 2013;497(7447):113–7.
155. Haynes PR, Christmann BL, Griffith LC. A single pair of neurons links sleep to memory consolidation in Drosophila melanogaster. elife. 2015;4:e03868.
156. Liu W, Guo F, Lu B, Guo A. amnesiac regulates sleep onset and maintenance in Drosophila melanogaster. Biochem Biophys Res Commun. 2008;372(4):798–803.
157. Jiang Y, Pitmon E, Berry J, Wolf FW, McKenzie Z, Lebestky TJ. A genetic screen to assess Dopamine Receptor (DopR1) dependent sleep regulation in Drosophila. G3 (Bethesda). 2016;6(12):4217–26.
158. Yuan Q, Joiner WJ, Sehgal A. A sleep-promoting role for the Drosophila serotonin receptor 1A. Curr Biol. 2006;16:1051–62.
159. Hobin M, Dorfman K, Adel M, et al. The Drosophila microRNA bantam regulates excitability in adult mushroom body output neurons to promote early night sleep. iScience. 2022;25(9):104874.
160. Bushey D, Tononi G, Cirelli C. Sleep- and wake-dependent changes in neuronal activity and reactivity demonstrated in fly neurons using in vivo calcium imaging. Proc Natl Acad Sci USA. 2015;112(15):4785–90.
161. Tong H, Li Q, Zhang ZC, Li Y, Han J. Neurexin regulates nighttime sleep by modulating synaptic transmission. Sci Rep. 2016;6:38246.
162. Seugnet L, Dissel S, Thimgan M, Cao L, Shaw PJ. Identification of genes that maintain behavioral and structural plasticity during sleep loss. Front Neural Circuits. 2017;11:79.
163. Guo F, Yi W, Zhou M, Guo A. Go signaling in mushroom bodies regulates sleep in Drosophila. Sleep. 2011;34(3):273–81.
164. Takahama K, Tomita J, Ueno T, Yamazaki M, Kume S, Kume K. Pan-neuronal knockdown of the c-Jun N-terminal Kinase (JNK) results in a reduction in sleep and longevity in Drosophila. Biochem Biophys Res Commun. 2012;417(2):807–11.
165. Zhang R, Zhao X, Du J, Wei L, Zhao Z. Regulatory mechanism of daily sleep by miR-276a. FASEB J. 2021;35(1):e21222.
166. Sengupta S, Crowe LB, You S, Roberts MA, Jackson FR. A secreted Ig-domain protein required in both astrocytes and neurons for regulation of Drosophila night sleep. Curr Biol. 2019;29(15):2547–2554.e2.
167. Li Q, Jang H, Lim KY, Lessing A, Stavropoulos N. insomniac links the development and function of a sleep-regulatory circuit. elife. 2021;10:e65437.
168. Chen D, Sitaraman D, Chen N, et al. Genetic and neuronal mechanisms governing the sex-specific interaction between sleep and sexual behaviors in Drosophila. Nat Commun. 2017;8(1):154.
169. Shimada N, Inami S, Sato S, Kitamoto T, Sakai T. Modulation of light-driven arousal by LIM-homeodomain transcription factor Apterous in large PDF-positive lateral neurons of the Drosophila brain. Sci Rep. 2016;6:37255.
170. Seugnet L, Suzuki Y, Merlin G, Gottschalk L, Duntley SP, Shaw PJ. Notch signaling modulates sleep homeostasis and learning after sleep deprivation in Drosophila. Curr Biol. 2011;21(10):835–40.
171. Shafer OT, Helfrich-Forster C, Renn SC, Taghert PH. Reevaluation of Drosophila melanogaster's neuronal circadian pacemakers reveals new neuronal classes. J Comp Neurol. 2006;498(2):180–93.

172. Renn SCP, Park JH, Rosbash M, Hall JC, Taghert PH. A pdf neuropeptide gene mutation and ablation of PDF neurons each cause severe abnormalities of behavioral circadian rhythms in Drosophila. Cell. 1999;99:791–802.
173. Allada R, Chung BY. Circadian organization of behavior and physiology in Drosophila. Annu Rev Physiol. 2010;72:605–24.
174. Ewer J, Frisch B, Hamblen-Coyle MJ, Rosbash M, Hall JC. Expression of the period clock gene within different cell types in the brain of Drosophila adults and mosaic analysis of these cells' influence on circadian behavioral rhythms. J Neurosci. 1992;12(9):3321–49.
175. Nitabach MN, Taghert PH. Organization of the Drosophila circadian control circuit. Curr Biol. 2008;18(2):R84–93.
176. Helfrich-Forster C, Yoshii T, Wulbeck C, et al. The lateral and dorsal neurons of Drosophila melanogaster: new insights about their morphology and function. Cold Spring Harb Symp Quant Biol. 2007;72:517–25.
177. Allada R, White NE, So WV, Hall JC, Rosbash M. A mutant Drosophila homolog of mammalian clock disrupts circadian rhythms and transcription of period and timeless. Cell. 1998;93(5):791–804.
178. Sehgal A, Rothenfluh-Hilfiker A, Hunter-Ensor M, Chen Y, Myers MP, Young MW. Rhythmic expression of timeless: a basis for promoting circadian cycles in period gene autoregulation. Science. 1995;270(5237):808–10.
179. Hardin PE, Hall JC, Rosbash M. Feedback of the Drosophila period gene product on circadian cycling of its messenger RNA levels. Nature. 1990;343(6258):536–40.
180. Liang X, Holy TE, Taghert PH. Synchronous Drosophila circadian pacemakers display nonsynchronous Ca(2)(+) rhythms in vivo. Science. 2016;351(6276):976–81.
181. Schlichting M, Diaz MM, Xin J, Rosbash M. Neuron-specific knockouts indicate the importance of network communication to Drosophila rhythmicity. elife. 2019;8:e48301.
182. Parisky KM, Agosto J, Pulver SR, et al. PDF cells are a GABA-responsive wake-promoting component of the Drosophila sleep circuit. Neuron. 2008;60(4):672–82.
183. Shang Y, Griffith LC, Rosbash M. Light-arousal and circadian photoreception circuits intersect at the large PDF cells of the Drosophila brain. Proc Natl Acad Sci USA. 2008;105(50):19587–94.
184. Fujiwara Y, Hermann-Luibl C, Katsura M, et al. The CCHamide1 neuropeptide expressed in the anterior dorsal neuron 1 conveys a circadian signal to the ventral lateral neurons in Drosophila melanogaster. Front Physiol. 2018;9:1276.
185. Alpert MH, Frank DD, Kaspi E, et al. A circuit encoding absolute cold temperature in Drosophila. Curr Biol. 2020;30(12):2275–2288.e5.
186. Sun L, Jiang RH, Ye WJ, Rosbash M, Guo F. Recurrent circadian circuitry regulates central brain activity to maintain sleep. Neuron. 2022;110(13):2139–2154.e5.
187. Jones JR, Tackenberg MC, McMahon DG. Optogenetic methods for the study of circadian rhythms. Methods Mol Biol. 2021;2130:325–36.
188. Chen SC, Tang X, Goda T, et al. Dorsal clock networks drive temperature preference rhythms in Drosophila. Cell Rep. 2022;39(2):110668.
189. Grima B, Chelot E, Xia R, Rouyer F. Morning and evening peaks of activity rely on different clock neurons of the Drosophila brain. Nature. 2004;431(7010):869–73.
190. Stoleru D, Peng Y, Agosto J, Rosbash M. Coupled oscillators control morning and evening locomotor behaviour of Drosophila. Nature. 2004;431(7010):862–8.
191. Li Q, Li Y, Wang X, et al. Fbxl4 serves as a clock output molecule that regulates sleep through promotion of rhythmic degradation of the GABAA receptor. Curr Biol. 2017;27(23):3616–3625.e5.
192. Sheeba V, Gu H, Sharma VK, O'Dowd DK, Holmes TC. Circadian- and light-dependent regulation of resting membrane potential and spontaneous action potential firing of Drosophila circadian pacemaker neurons. J Neurophysiol. 2008;99(2):976–88.
193. Kistenpfennig C, Nakayama M, Nihara R, Tomioka K, Helfrich-Forster C, Yoshii T. A Tug-of-War between cryptochrome and the visual system allows the adaptation of evening activity to long photoperiods in Drosophila melanogaster. J Biol Rhythm. 2018;33(1):24–34.

194. Vinayak P, Coupar J, Hughes SE, et al. Exquisite light sensitivity of Drosophila melanogaster cryptochrome. PLoS Genet. 2013;9(7):e1003615.
195. Schlichting M, Menegazzi P, Rosbash M, Helfrich-Forster C. A distinct visual pathway mediates high-intensity light adaptation of the circadian clock in Drosophila. J Neurosci. 2019;39(9):1621–30.
196. Diaz MM, Schlichting M, Abruzzi KC, Long X, Rosbash M. Allatostatin-C/AstC-R2 is a novel pathway to modulate the circadian activity pattern in Drosophila. Curr Biol. 2019;29(1):13–22.e13.
197. Reinhard N, Bertolini E, Saito A, et al. The lateral posterior clock neurons of Drosophila melanogaster express three neuropeptides and have multiple connections within the circadian clock network and beyond. J Comp Neurol. 2022;530(9):1507–29.
198. Ni JD, Gurav AS, Liu W, et al. Differential regulation of the Drosophila sleep homeostat by circadian and arousal inputs. elife. 2019;8:e40487.
199. Miyasako Y, Umezaki Y, Tomioka K. Separate sets of cerebral clock neurons are responsible for light and temperature entrainment of Drosophila circadian locomotor rhythms. J Biol Rhythm. 2007;22(2):115–26.
200. Nettnin EA, Sallese TR, Nasseri A, Saurabh S, Cavanaugh DJ. Dorsal clock neurons in Drosophila sculpt locomotor outputs but are dispensable for circadian activity rhythms. iScience. 2021;24(9):103001.
201. Marin EC, Buld L, Theiss M, et al. Connectomics analysis reveals first-, second-, and third-order thermosensory and hygrosensory neurons in the adult Drosophila brain. Curr Biol. 2020;30(16):3167–3182.e4.
202. Zhang Y, Liu Y, Bilodeau-Wentworth D, Hardin PE, Emery P. Light and temperature control the contribution of specific DN1 neurons to Drosophila circadian behavior. Curr Biol. 2010;20(7):600–5.
203. Yadlapalli S, Jiang C, Bahle A, Reddy P, Meyhofer E, Shafer OT. Circadian clock neurons constantly monitor environmental temperature to set sleep timing. Nature. 2018;555(7694):98–102.
204. Kaneko H, Head LM, Ling J, et al. Circadian rhythm of temperature preference and its neural control in Drosophila. Curr Biol. 2012;22(19):1851–7.
205. Guo F, Holla M, Diaz MM, Rosbash M. A circadian output circuit controls sleep-wake arousal in Drosophila. Neuron. 2018;100(3):624–635.e4.
206. Liang X, Ho MCW, Zhang Y, et al. Morning and evening circadian pacemakers independently drive premotor centers via a specific dopamine relay. Neuron. 2019;102(4):843–857.e4.
207. Meiselman MR, Alpert MH, Cui X, et al. Recovery from cold-induced reproductive dormancy is regulated by temperature-dependent AstC signaling. Curr Biol. 2022;32(6):1362–1375.e8.
208. Scheffer LK, Xu CS, Januszewski M, et al. A connectome and analysis of the adult Drosophila central brain. elife. 2020;9:e57443.
209. Ahmad M, Li W, Top D. Integration of circadian clock information in the Drosophila circadian neuronal network. J Biol Rhythm. 2021;36(3):203–20.
210. Shafer OT, Taghert PH. RNA-interference knockdown of Drosophila pigment dispersing factor in neuronal subsets: the anatomical basis of a neuropeptide's circadian functions. PLoS One. 2009;4(12):e8298.
211. Goda T, Umezaki Y, Alwattari F, Seo HW, Hamada FN. Neuropeptides PDF and DH31 hierarchically regulate free-running rhythmicity in Drosophila circadian locomotor activity. Sci Rep. 2019;9(1):838.
212. Duhart JM, Herrero A, de la Cruz G, Ispizua JI, Pirez N, Ceriani MF. Circadian structural plasticity drives remodeling of E cell output. Curr Biol. 2020;30(24):5040–5048.e5.
213. Herrero A, Yoshii T, Ispizua JI, et al. Coupling neuropeptide levels to structural plasticity in Drosophila clock neurons. Curr Biol. 2020;30(16):3154–3166.e4.
214. Dubowy CM, Cavanaugh DJ. Sleep: a neuropeptidergic wake-up call for flies. Curr Biol. 2014;24(22):R1092–4.
215. Sokolove PG, Loher W. Role of eyes, optic lobes, and pars intercerebralis in locomotory and stridulatory circadian rhythms of Teleogryllus commodus. J Insect Physiol. 1975;21(4):785–99.

216. Matsui T, Matsumoto T, Ichihara N, et al. The pars intercerebralis as a modulator of locomotor rhythms and feeding in the American cockroach, Periplaneta americana. Physiol Behav. 2009;96(4–5):548–56.
217. Cavanaugh DJ, Geratowski JD, Wooltorton JR, et al. Identification of a circadian output circuit for rest:activity rhythms in Drosophila. Cell. 2014;157(3):689–701.
218. Barber AF, Fong SY, Kolesnik A, Fetchko M, Sehgal A. Drosophila clock cells use multiple mechanisms to transmit time-of-day signals in the brain. Proc Natl Acad Sci USA. 2021;118(10):e2019826118.
219. King AN, Barber AF, Smith AE, et al. A peptidergic circuit links the circadian clock to locomotor activity. Curr Biol. 2017;27(13):1915–1927.e5.
220. Barber AF, Erion R, Holmes TC, Sehgal A. Circadian and feeding cues integrate to drive rhythms of physiology in Drosophila insulin-producing cells. Genes Dev. 2016;30(23):2596–606.
221. Meyer P, Saez L, Young MW. PER-TIM interactions in living Drosophila cells: an interval timer for the circadian clock. Science. 2006;311(5758):226–9.
222. Kadener S, Menet JS, Schoer R, Rosbash M. Circadian transcription contributes to core period determination in Drosophila. PLoS Biol. 2008;6(5):e119.
223. Yu W, Zheng H, Houl JH, Dauwalder B, Hardin PE. PER-dependent rhythms in CLK phosphorylation and E-box binding regulate circadian transcription. Genes Dev. 2006;20(6):723–33.
224. Donlea JM, Ramanan N, Shaw PJ. Use-dependent plasticity in clock neurons regulates sleep need in Drosophila. Science. 2009;324(5923):105–8.
225. Konopka RJ, Benzer S. Clock mutants of Drosophila melanogaster. Proc Natl Acad Sci USA. 1971;68(9):2112–6.
226. Rutila JE, Suri V, Le M, So WV, Rosbash M, Hall JC. CYCLE is a second bHLH-PAS clock protein essential for circadian rhythmicity and transcription of Drosophila period and timeless. Cell. 1998;93(5):805–14.
227. Yu W, Zheng H, Price JL, Hardin PE. DOUBLETIME plays a noncatalytic role to mediate CLOCK phosphorylation and repress CLOCK-dependent transcription within the Drosophila circadian clock. Mol Cell Biol. 2009;29(6):1452–8.
228. Fernandez-Chiappe F, Hermann-Luibl C, Peteranderl A, et al. Dopamine signaling in wake-promoting clock neurons is not required for the normal regulation of sleep in Drosophila. J Neurosci. 2020;40(50):9617–33.
229. Hermann-Luibl C, Yoshii T, Senthilan PR, Dircksen H, Helfrich-Forster C. The ion transport peptide is a new functional clock neuropeptide in the fruit fly Drosophila melanogaster. J Neurosci. 2014;34(29):9522–36.
230. Chung BY, Kilman VL, Keath JR, Pitman JL, Allada R. The GABA(A) receptor RDL acts in peptidergic PDF neurons to promote sleep in Drosophila. Curr Biol. 2009;19(5):386–90.
231. Sheeba V, Sharma VK, Gu H, Chou YT, O'Dowd DK, Holmes TC. Pigment dispersing factor-dependent and -independent circadian locomotor behavioral rhythms. J Neurosci. 2008;28(1):217–27.
232. Kawasaki H, Okano H, Nedachi T, Nakagawa-Yagi Y, Hara A, Ishida N. Effects of an electric field on sleep quality and life span mediated by ultraviolet (UV)-A/blue light photoreceptor CRYPTOCHROME in Drosophila. Sci Rep. 2021;11(1):20543.
233. Chung BY, Ro J, Hutter SA, et al. Drosophila neuropeptide F signaling independently regulates feeding and sleep-wake behavior. Cell Rep. 2017;19(12):2441–50.
234. Kunst M, Hughes ME, Raccuglia D, et al. Calcitonin gene-related peptide neurons mediate sleep-specific circadian output in Drosophila. Curr Biol. 2014;24(22):2652–64.
235. Gmeiner F, Kolodziejczyk A, Yoshii T, Rieger D, Nassel DR, Helfrich-Forster C. GABA(B) receptors play an essential role in maintaining sleep during the second half of the night in Drosophila melanogaster. J Exp Biol. 2013;216(Pt 20):3837–43.
236. Li Y, Zhou Z, Zhang X, et al. Drosophila neuroligin 4 regulates sleep through modulating GABA transmission. J Neurosci. 2013;33(39):15545–54.

237. Shimizu H, Shimoda M, Yamaguchi T, Seong KH, Okamura T, Ishii S. Drosophila ATF-2 regulates sleep and locomotor activity in pacemaker neurons. Mol Cell Biol. 2008;28(20):6278–89.
238. Cong X, Wang H, Liu Z, He C, An C, Zhao Z. Regulation of sleep by insulin-like peptide system in Drosophila melanogaster. Sleep. 2015;38(7):1075–83.
239. Fernandez-Chiappe F, Frenkel L, Colque CC, et al. High-frequency neuronal bursting is essential for circadian and sleep behaviors in Drosophila. J Neurosci. 2021;41(4):689–710.
240. Buhl E, Higham JP, Hodge JJL. Alzheimer's disease-associated tau alters Drosophila circadian activity, sleep and clock neuron electrophysiology. Neurobiol Dis. 2019;130:104507.
241. Vanderheyden WM, Gerstner JR, Tanenhaus A, Yin JC, Shaw PJ. ERK phosphorylation regulates sleep and plasticity in Drosophila. PLoS One. 2013;8(11):e81554.
242. Wei Y, Du J, Zhao Z. Integrative role of 14-3-3epsilon in sleep regulation. Int J Mol Sci. 2021;22(18):9748.
243. Ly S, Lee DA, Strus E, Prober DA, Naidoo N. Evolutionarily conserved regulation of sleep by the protein translational regulator PERK. Curr Biol. 2020;30(9):1639–1648.e3.
244. Oh Y, Yoon SE, Zhang Q, et al. A homeostatic sleep-stabilizing pathway in Drosophila composed of the sex peptide receptor and its ligand, the myoinhibitory peptide. PLoS Biol. 2014;12(10):e1001974.
245. Oh Y, Jang D, Sonn JY, Choe J. Histamine-HisCl1 receptor axis regulates wake-promoting signals in Drosophila melanogaster. PLoS One. 2013;8(7):e68269.
246. Pirooznia SK, Chiu K, Chan MT, Zimmerman JE, Elefant F. Epigenetic regulation of axonal growth of Drosophila pacemaker cells by histone acetyltransferase tip60 controls sleep. Genetics. 2012;192(4):1327–45.
247. Xia X, Fu X, Du J, et al. Regulation of circadian rhythm and sleep by miR-375-timeless interaction in Drosophila. FASEB J. 2020;34(12):16536–51.
248. Feng G, Zhang J, Li M, et al. Control of sleep onset by Shal/Kv4 channels in Drosophila circadian neurons. J Neurosci. 2018;38(42):9059–71.
249. Kijak E, Pyza E. TOR signaling pathway and autophagy are involved in the regulation of circadian rhythms in behavior and plasticity of L2 interneurons in the brain of Drosophila melanogaster. PLoS One. 2017;12(2):e0171848.
250. He Q, Du J, Wei L, Zhao Z. AKH-FOXO pathway regulates starvation-induced sleep loss through remodeling of the small ventral lateral neuron dorsal projections. PLoS Genet. 2020;16(10):e1009181.
251. Zhang Z, Cao W, Edery I. The SR protein B52/SRp55 regulates splicing of the period thermosensitive intron and mid-day siesta in Drosophila. Sci Rep. 2018;8(1):1872.
252. Zhang R, Du J, Zhao X, Wei L, Zhao Z. Regulation of circadian behavioural output via clock-responsive miR-276b. Insect Mol Biol. 2021;30(1):81–9.
253. Shaw PJ, Tononi G, Greenspan RJ, Robinson DF. Stress response genes protect against lethal effects of sleep deprivation in Drosophila. Nature. 2002;417(6886):287–91.
254. Bentivoglio M, Grassi-Zucconi G. The pioneering experimental studies on sleep deprivation. Sleep. 1997;20(7):570–6.
255. Rechtschaffen A, Bergmann BM. Sleep deprivation in the rat by the disk-over-water method. Behav Brain Res. 1995;69(1–2):55–63.
256. Vaccaro A, Kaplan Dor Y, Nambara K, et al. Sleep loss can cause death through accumulation of reactive oxygen species in the gut. Cell. 2020;181(6):1307–1328.e5.
257. Chen J, Reiher W, Hermann-Luibl C, et al. Allatostatin A signalling in Drosophila regulates feeding and sleep and is modulated by PDF. PLoS Genet. 2016;12(9):e1006346.
258. Hergarden AC, Tayler TD, Anderson DJ. Allatostatin-A neurons inhibit feeding behavior in adult Drosophila. Proc Natl Acad Sci USA. 2012;109(10):3967–72.
259. Ertekin D, Kirszenblat L, Faville R, van Swinderen B. Down-regulation of a cytokine secreted from peripheral fat bodies improves visual attention while reducing sleep in Drosophila. PLoS Biol. 2020;18(8):e3000548.
260. Roessingh S, Rosing M, Marunova M, et al. Temperature synchronization of the Drosophila circadian clock protein PERIOD is controlled by the TRPA channel PYREXIA. Commun Biol. 2019;2:246.

261. DeSalvo MK, Mayer N, Mayer F, Bainton RJ. Physiologic and anatomic characterization of the brain surface glia barrier of Drosophila. Glia. 2011;59(9):1322–40.
262. Schirmeier S, Klambt C. The Drosophila blood-brain barrier as interface between neurons and hemolymph. Mech Dev. 2015;138(Pt 1):50–5.
263. Stork T, Engelen D, Krudewig A, Silies M, Bainton RJ, Klambt C. Organization and function of the blood-brain barrier in Drosophila. J Neurosci. 2008;28(3):587–97.
264. Stork T, Bernardos R, Freeman MR. Analysis of glial cell development and function in Drosophila. Cold Spring Harb Protoc. 2012;2012(1):1–17.
265. Zhang SL, Yue Z, Arnold DM, Artiushin G, Sehgal A. A circadian clock in the blood-brain barrier regulates xenobiotic efflux. Cell. 2018;173(1):130–139.e10.
266. Artiushin G, Zhang SL, Tricoire H, Sehgal A. Endocytosis at the Drosophila blood-brain barrier as a function for sleep. elife. 2018;7:e43326.
267. Gomez-Gonzalez B, Hurtado-Alvarado G, Esqueda-Leon E, Santana-Miranda R, Rojas-Zamorano JA, Velazquez-Moctezuma J. REM sleep loss and recovery regulates blood-brain barrier function. Curr Neurovasc Res. 2013;10(3):197–207.
268. van Alphen B, Semenza ER, Yap M, van Swinderen B, Allada R. A deep sleep stage in Drosophila with a functional role in waste clearance. Sci Adv. 2021;7(4):eabc2999.
269. Kress BT, Iliff JJ, Xia M, et al. Impairment of paravascular clearance pathways in the aging brain. Ann Neurol. 2014;76(6):845–61.
270. Ohayon MM, Carskadon MA, Guilleminault C, Vitiello MV. Meta-analysis of quantitative sleep parameters from childhood to old age in healthy individuals: developing normative sleep values across the human lifespan. Sleep. 2004;27(7):1255–73.
271. Kremer MC, Jung C, Batelli S, Rubin GM, Gaul U. The glia of the adult Drosophila nervous system. Glia. 2017;65(4):606–38.
272. Farca Luna AJ, Perier M, Seugnet L. Amyloid precursor protein in Drosophila glia regulates sleep and genes involved in glutamate recycling. J Neurosci. 2017;37(16):4289–300.
273. MacDonald JM, Beach MG, Porpiglia E, Sheehan AE, Watts RJ, Freeman MR. The Drosophila cell corpse engulfment receptor Draper mediates glial clearance of severed axons. Neuron. 2006;50(6):869–81.
274. Doherty J, Logan MA, Tasdemir OE, Freeman MR. Ensheathing glia function as phagocytes in the adult Drosophila brain. J Neurosci. 2009;29(15):4768–81.
275. Long DM, Giebultowicz JM. Age-related changes in the expression of the circadian clock protein PERIOD in Drosophila glial cells. Front Physiol. 2017;8:1131.
276. Vanderheyden WM, Goodman AG, Taylor RH, Frank MG, Van Dongen HPA, Gerstner JR. Astrocyte expression of the Drosophila TNF-alpha homologue, Eiger, regulates sleep in flies. PLoS Genet. 2018;14(10):e1007724.
277. Gerstner JR, Perron IJ, Riedy SM, et al. Normal sleep requires the astrocyte brain-type fatty acid binding protein FABP7. Sci Adv. 2017;3(4):e1602663.
278. Gerstner JR, Vanderheyden WM, Shaw PJ, Landry CF, Yin JC. Fatty-acid binding proteins modulate sleep and enhance long-term memory consolidation in Drosophila. PLoS One. 2011;6(1):e15890.
279. Jang S, Choi B, Lim C, Lee B, Cho KS. Roles of Drosophila fatty acid-binding protein in development and behavior. Biochem Biophys Res Commun. 2022;599:87–92.
280. Jackson FR, You S, Crowe LB. Regulation of rhythmic behaviors by astrocytes. Wiley Interdiscip Rev Dev Biol. 2020;9(4):e372.
281. Ingiosi AM, Frank MG. Goodnight, astrocyte: waking up to astroglial mechanisms in sleep. FEBS J. 2023;290(10):2553–64.
282. Yildirim K, Winkler B, Pogodalla N, et al. Redundant functions of the SLC5A transporters Rumpel, Bumpel, and Kumpel in ensheathing glial cells. Biol Open. 2022;11(1):bio059128.
283. Walkowicz L, Krzeptowski W, Krzeptowska E, et al. Glial expression of DmMANF is required for the regulation of activity, sleep and circadian rhythms in the visual system of Drosophila melanogaster. Eur J Neurosci. 2021;54(5):5785–97.
284. Davla S, Artiushin G, Li Y, et al. AANAT1 functions in astrocytes to regulate sleep homeostasis. elife. 2020;9:e53994.

285. Dissel S, Seugnet L, Thimgan MS, et al. Differential activation of immune factors in neurons and glia contribute to individual differences in resilience/vulnerability to sleep disruption. Brain Behav Immun. 2015;47:75–85.
286. Ueno T, Kume K. Functional characterization of dopamine transporter in vivo using Drosophila melanogaster behavioral assays. Front Behav Neurosci. 2014;8:303.
287. Koh K, Evans JM, Hendricks JC, Sehgal A. A Drosophila model for age-associated changes in sleep:wake cycles. Proc Natl Acad Sci USA. 2006;103:13843–7.
288. Vienne J, Spann R, Guo F, Rosbash M. Age-related reduction of recovery sleep and arousal threshold in Drosophila. Sleep. 2016;39(8):1613–24.
289. Ly S, Naidoo N. Loss of DmGluRA exacerbates age-related sleep disruption and reduces lifespan. Neurobiol Aging. 2019;80:83–90.
290. Bushey D, Hughes KA, Tononi G, Cirelli C. Sleep, aging, and lifespan in Drosophila. BMC Neurosci. 2010;11:56.
291. Metaxakis A, Tain LS, Gronke S, et al. Lowered insulin signalling ameliorates age-related sleep fragmentation in Drosophila. PLoS Biol. 2014;12:e1001824.
292. Kayser MS, Yue Z, Sehgal A. A critical period of sleep for development of courtship circuitry and behavior in Drosophila. Science. 2014;344:269–74.
293. Seugnet L, Suzuki Y, Donlea J, Gottschalk L, Shaw PJ. Sleep deprivation during early-adult development results in long-lasting learning deficits in adult Drosophila. Sleep. 2011;34:137–46.
294. Crocker A, Shahidullah M, Levitan IB, Sehgal A. Identification of a neural circuit that underlies the effects of octopamine on sleep:wake behavior. Neuron. 2010;65(5):670–81.
295. Szuperak M, Churgin MA, Borja AJ, Raizen DM, Fang-Yen C, Kayser MS. A sleep state in Drosophila larvae required for neural stem cell proliferation. elife. 2018;7:e33220.
296. Irwin MR, Opp MR. Sleep health: reciprocal regulation of sleep and innate immunity. Neuropsychopharmacology. 2017;42(1):129–55.
297. Kuo TH, Pike DH, Beizaeipour Z, Williams JA. Sleep triggered by an immune response in Drosophila is regulated by the circadian clock and requires the NFkappaB Relish. BMC Neurosci. 2010;11:17.
298. Kuo TH, Williams JA. Acute sleep deprivation enhances post-infection sleep and promotes survival during bacterial infection in Drosophila. Sleep. 2014;37(5):859–69.
299. Bi J, Sehgal A, Williams JA, Wang YF. Wolbachia affects sleep behavior in Drosophila melanogaster. J Insect Physiol. 2018;107:81–8.
300. Keene AC, Duboue ER, McDonald DM, et al. Clock and cycle limit starvation-induced sleep loss in Drosophila. Curr Biol. 2010;20(13):1209–15.
301. Lee G, Park JH. Hemolymph sugar homeostasis and starvation-induced hyperactivity affected by genetic manipulations of the adipokinetic hormone-encoding gene in Drosophila melanogaster. Genetics. 2004;167(1):311–23.
302. Erion R, DiAngelo JR, Crocker A, Sehgal A. Interaction between sleep and metabolism in Drosophila with altered octopamine signaling. J Biol Chem. 2012;287(39):32406–14.
303. Yurgel ME, Kakad P, Zandawala M, Nassel DR, Godenschwege TA, Keene AC. A single pair of leucokinin neurons are modulated by feeding state and regulate sleep-metabolism interactions. PLoS Biol. 2019;17(2):e2006409.
304. Donlea J, Leahy A, Thimgan MS, et al. Foraging alters resilience/vulnerability to sleep disruption and starvation in Drosophila. Proc Natl Acad Sci USA. 2012;109(7):2613–8.
305. Borbely AA. A two process model of sleep regulation. Hum Neurobiol. 1982;1(3):195–204.
306. Borbely AA, Daan S, Wirz-Justice A, Deboer T. The two-process model of sleep regulation: a reappraisal. J Sleep Res. 2016;25(2):131–43.
307. Hill VM, O'Connor RM, Shirasu-Hiza M. Tired and stressed: examining the need for sleep. Eur J Neurosci. 2020;51(1):494–508.
308. Crocker A, Sehgal A. Genetic analysis of sleep. Genes Dev. 2010;24(12):1220–35.
309. Vassalli A, Dijk DJ. Sleep function: current questions and new approaches. Eur J Neurosci. 2009;29(9):1830–41.

310. Dissel S. Drosophila as a model to study the relationship between sleep, plasticity, and memory. Front Physiol. 2020;11:533.
311. Thimgan MS, Suzuki Y, Seugnet L, Gottschalk L, Shaw PJ. The perilipin homologue, lipid storage droplet 2, regulates sleep homeostasis and prevents learning impairments following sleep loss. PLoS Biol. 2010;8(8):e1000466.
312. Kume K, Kume S, Park SK, Hirsh J, Jackson FR. Dopamine is a regulator of arousal in the fruit fly. J Neurosci. 2005;25:7377–84.
313. Zhang S, Yin Y, Lu H, Guo A. Increased dopaminergic signaling impairs aversive olfactory memory retention in Drosophila. Biochem Biophys Res Commun. 2008;370(1):82–6.
314. Quinn WG, Sziber PP, Booker R. The Drosophila memory mutant amnesiac. Nature. 1979;277(5693):212–4.
315. Thimgan MS, Kress N, Lisse J, Fiebelman C, Hilderbrand T. The acyl-CoA Synthetase, pudgy, Promotes Sleep and Is Required for the Homeostatic Response to Sleep Deprivation. Front Endocrinol (Lausanne). 2018;9:464.
316. Chen W, Shi W, Li L, et al. Regulation of sleep by the short neuropeptide F (sNPF) in Drosophila melanogaster. Insect Biochem Mol Biol. 2013;43(9):809–19.
317. Shao L, Zhang Y, Hao Y, Ping Y. Upregulation of IP3 receptor mediates APP-induced defects in synaptic downscaling and sleep homeostasis. Cell Rep. 2022;38(13):110594.
318. Raccuglia D, Huang S, Ender A, et al. Network-specific synchronization of electrical slow-wave oscillations regulates sleep drive in Drosophila. Curr Biol. 2019;29(21):3611–3621 e3613.
319. Nagy S, Maurer GW, Hentze JL, Rose M, Werge TM, Rewitz K. AMPK signaling linked to the schizophrenia-associated 1q21.1 deletion is required for neuronal and sleep maintenance. PLoS Genet. 2018;14(12):e1007623.
320. Singh N, Lorbeck MT, Zervos A, Zimmerman J, Elefant F. The histone acetyltransferase Elp3 plays in active role in the control of synaptic bouton expansion and sleep in Drosophila. J Neurochem. 2010;115(2):493–504.
321. Dai X, Zhou E, Yang W, Mao R, Zhang W, Rao Y. Molecular resolution of a behavioral paradox: sleep and arousal are regulated by distinct acetylcholine receptors in different neuronal types in Drosophila. Sleep. 2021;44(7):zsab017.
322. Blanco-Redondo B, Nuwal N, Kneitz S, et al. Implications of the Sap47 null mutation for synapsin phosphorylation, longevity, climbing proficiency and behavioural plasticity in adult Drosophila. J Exp Biol. 2019;222(Pt 19):jeb203505.
323. Hendricks JC, Williams JA, Panckeri K, et al. A non-circadian role for cAMP signaling and CREB activity in Drosophila rest homeostasis. Nat Neurosci. 2001;4(11):1108–15.
324. Naidoo N, Ferber M, Galante RJ, et al. Role of Homer proteins in the maintenance of sleep-wake states. PLoS One. 2012;7(4):e35174.
325. Rodrigues NR, Macedo GE, Martins IK, et al. Short-term sleep deprivation with exposure to nocturnal light alters mitochondrial bioenergetics in Drosophila. Free Radic Biol Med. 2018;120:395–406.
326. Kim M, Subramanian M, Cho YH, Kim GH, Lee E, Park JJ. Short-term exposure to dim light at night disrupts rhythmic behaviors and causes neurodegeneration in fly models of tauopathy and Alzheimer's disease. Biochem Biophys Res Commun. 2018;495(2):1722–9.
327. Foltenyi K, Greenspan RJ, Newport JW. Activation of EGFR and ERK by rhomboid signaling regulates the consolidation and maintenance of sleep in Drosophila. Nat Neurosci. 2007;10:1160–7.
328. Diessler S, Jan M, Emmenegger Y, et al. A systems genetics resource and analysis of sleep regulation in the mouse. PLoS Biol. 2018;16(8):e2005750.
329. Geissmann Q, Beckwith EJ, Gilestro GF. Most sleep does not serve a vital function: evidence from Drosophila melanogaster. Sci Adv. 2019;5(2):eaau9253.
330. Saper CB, Fuller PM, Pedersen NP, Lu J, Scammell TE. Sleep state switching. Neuron. 2010;68(6):1023–42.
331. Scammell TE, Arrigoni E, Lipton JO. Neural circuitry of wakefulness and sleep. Neuron. 2017;93(4):747–65.

332. Liu D, Dan Y. A motor theory of sleep-wake control: arousal-action circuit. Annu Rev Neurosci. 2019;42:27–46.
333. King AN, Sehgal A. Molecular and circuit mechanisms mediating circadian clock output in the Drosophila brain. Eur J Neurosci. 2020;51(1):268–81.
334. Ki Y, Lim C. Sleep-promoting effects of threonine link amino acid metabolism in Drosophila neuron to GABAergic control of sleep drive. elife. 2019;8:e40593.
335. Lamaze A, Kratschmer P, Chen KF, Lowe S, Jepson JEC. A wake-promoting circadian output circuit in Drosophila. Curr Biol. 2018;28(19):3098–3105.e3.

The Genetics of Sleep in *Caenorhabditis elegans*

Madison Honer and Matthew D. Nelson

Abstract Essential functions have preserved sleep throughout evolution, and the molecular pathways controlling sleep are conserved, making model organisms powerful tools for exploring the mechanisms of sleep. *Caenorhabditis elegans* is a microscopic nematode, which has been leveraged to study many biological processes, including sleep. *C. elegans* is ideal for mechanistic sleep studies because of their simple, well-defined nervous system, conserved neurochemistry, and genetic tractability. Two distinct sleep states have been the primary focus of *C. elegans* researchers over the past decade, developmentally timed sleep and stress-induced sleep, because of their similarities to mammalian circadian sleep and sleep during sickness, respectively. Here we review the genetics and molecular signaling pathways that govern these and other sleep states of *C. elegans*.

Keywords *C. elegans* · Lethargus · Sleep · Quiescence · Neuropeptides

The History of *C. elegans* as a Sleep Model

Caenorhabditis elegans is a powerful model organism for biomedical research because of its genetic tractability, rapid life cycle, simple and transparent body, and fully mapped nervous system [1]. Sydney Brenner adopted *C. elegans* for research in the 1960s to explore mechanisms of development and the physiology of nervous systems. Mutations were quickly identified which altered worm behavior, opening the door for mechanistic behavioral studies [2]. *C. elegans* behavioral research took huge leaps forward with the groundbreaking reconstruction of their simple

M. Honer
Saint Joseph's University, Department of Biology, Philadelphia, PA, USA

Temple University, Lewis Katz School of Medicine, Philadelphia, PA, USA

M. D. Nelson (✉)
Saint Joseph's University, Department of Biology, Philadelphia, PA, USA
e-mail: mnelson@sju.edu

P. Gehrman et al. (eds.), *Genetics of Sleep and Sleep Disorders*,
https://doi.org/10.1007/978-3-031-62723-1_3

302-celled nervous system [3] and the sequencing of their genome [4]. The first behavioral studies focused on the genetics of movement [2], egg laying [5], mechanosensation [6], and chemosensation [7]. Other work explored learning and memory [8], feeding [9], and defecation [10]. These foundational studies established *C. elegans* as a system for behavioral research, and because of conserved neurogenetics (Fig. 1) [11], work in these nematodes is relevant to understanding the behavior of all animals.

Sleep was defined behaviorally in the 1980s [12]; however, the use of nonmammalian sleep models for sleep research did not become commonplace until a number of years later. *Drosophila melanogaster* had been established for decades as a powerful model for exploring circadian and daily locomotory rhythms [13]; however, sleep was first defined in the year 2000, through two independent studies [14, 15]. Soon after, the zebrafish *Danio rerio* was established as a genetic model for sleep research [16]. These foundational discoveries set the groundwork for genetic sleep research in *C. elegans* that occurred later that decade. However, sleep research in *C. elegans* was made possible by earlier work in which Ramandeep Singh and John Sulston (MRC Laboratory of Molecular Biology) described their observations of the *C. elegans* molting process [17]. They described a life stage called lethargus, named for the worms' lethargic behavior displayed during this time. Lethargus occurs during larval transitions, of which there are four, and is coincident with molting or ecdysis. Genetic dissection of these behaviors did not begin until many years later though. In the late 2000s, two studies published within a year from one another defined sleep-like behaviors in *C. elegans* [18, 19]. Cheryl Van Buskirk and Paul Sternberg (Howard Hughes Medical Institute, Division of Biology 156–29, California Institute of Technology) described a quiescent state in which animals ceased movement and feeding following overexpression of the epidermal growth factor (EGF), encoded by the gene *lin-3* [19].

This work demonstrated that the conserved EGF signaling pathway can function as a somnogen in *C. elegans*. This study was one of the first lines of evidence that

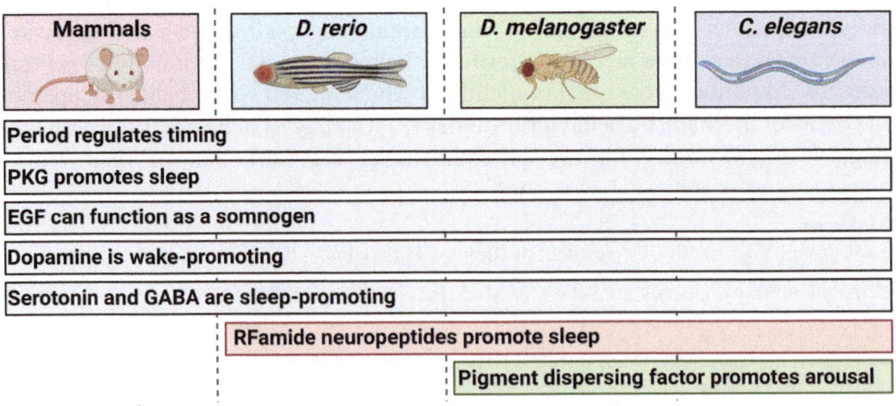

Fig. 1 Conserved molecular pathways regulate sleep. (Created with BioRender.com)

C. elegans sleep is conserved genetically, considering EGF enhances sleep in *Drosophila* [20], *D. rerio* [21], and mammals [22, 23]. Remarkably, this EGF-induced sleep is controlled in large part by two neuropeptidergic interneurons called ALA and RIS [19, 24], revealing an avenue for dissecting the mechanisms of sleep with cellular resolution [3, 19]. Around the same time, David Raizen and colleagues (Center for Sleep and Respiratory Neurobiology, University of Pennsylvania School of Medicine) looked more closely at the behaviors of lethargus, using a sleep research perspective. Using sleep criteria for animals in which EEG analyses are not possible [12, 18], they described lethargus as a reversible quiescent state, in which animals display an increased threshold for arousal, a stereotypic posture, and an increased drive to sleep following deprivation (i.e., homeostasis) [18]. Thus, lethargus is a sleep-like state in *C. elegans.* Compelling evidence that sleep during lethargus is related evolutionarily to sleep of more complex animals is the underlying molecular mechanism. Sleep in mammals, *Drosophila*, and *C. elegans* is timed by molecular components of the biological clock [25]; however, sleep during lethargus is developmentally timed and linked with the molt [26]. Despite these key differences, the timing of these sleep states is related at the molecular level. Expression of the circadian protein Period (PER), called LIN-42 in *C. elegans* [27, 28], cycles with the 24 h light-dark cycle in mammals and insects and during larval progressions in *C. elegans*. Mutations affecting *lin-42*/PER cause arrhythmic, short-period, or long-period sleep phenotypes across phyla, including *C. elegans* [13, 26]. Other conserved sleep-regulating genes include those encoding components of cyclic nucleotide pathways, neurotransmitter signaling, and many others [18, 25]. This conservation supports sleep during lethargus (developmentally timed sleep (DTS)) as being evolutionarily related to the circadian sleep of other animals and worthy of investigation.

Around this same time period, Young-jai You and colleagues (Department of Molecular Biology, University of Texas Southwestern Medical Center) described a distinct sleep state in *C. elegans* that resembles satiety in mammals [29]. This satiety quiescence is dependent on feeding history and is induced following the consumption of high-quality food. After a period of eating, animals slow their feeding behaviors and enter a reversible quiescent state which resembles a nap after a big meal. This sleep requires insulin and TGF-β signals and fulfills the behavioral criteria of satiety in mammals [29]. In the same year, Rajarshi Ghosh and Scott Emmons (Department of Molecular Genetics, Albert Einstein College of Medicine) described another sleep-like state named episodic swimming quiescence. Animals immersed in liquid displayed periodic bouts of reversible inactivity, in which they displayed stereotypical body postures [30]. In the wild, nematodes likely encounter conditions that require them to swim; this sleep state may allow them to recover following increased metabolic demand. However, the function of this sleep is still unclear.

Several years later, Andrew Hill and colleagues (Department of Biology, California State University, Northridge) explored a sleep state displayed by animals who had been damaged by noxious heat, cold, and ethanol, or injured by pore-forming toxins. Following cellular damage, *C. elegans* enters a state of reversible quiescence which is homeostatically regulated, and characterized by a heightened arousal threshold [31, 32]. This stress-induced sleep (SIS) functions to promote recovery, akin to the heightened sleep experienced during illness seen in other animals [33, 34]. EGF signaling is indispensable for this form of sleep [31], which largely signals through the ALA and RIS neurons [24, 32]. With the reconstruction of the nervous system, it was shown that the ALA neuron possesses an abundance of dense core vesicles [3], indicating that it functions using neuropeptidergic mechanisms. Matthew Nelson and colleagues (Department of Neurology, Perelman School of Medicine, University of Pennsylvania) identified a neuropeptide-encoding gene called *flp-13*, which is expressed in the ALA and is required for stress-induced sleep. Intriguingly, overexpression of *flp-13* induces ectopic quiescence, like EGF [32], evidence that it functions as a somnogen. Considering the FLP-13 neuropeptides are a part of a broader class of FMRFamides (Phe-Met-Arg-Phe-NH2) neuropeptides, found broadly throughout the animal kingdom [35], this motivated others to explore the function of related peptides. It was discovered that FMRFamides also regulate sleep in *D. melanogaster* [36] and *D. rerio* [37], supporting a conserved sleep role across phyla. Building off of this work, Ravi Nath and colleagues (Division of Biology and Biological Engineering, California Institute of Technology) transcriptionally profiled the ALA and identified a broad array of neuropeptide genes in addition to *flp-13*, including other FMRFamides required for stress-induced sleep [38]. Because of its simple circuitry, stress-induced sleep has become a popular model for identifying sleep mechanisms.

The discovery of multiple sleep-like states and key sleep-controlling neurons [19, 39] set the groundwork for *C. elegans* to serve as a genetic sleep model. This early work demonstrated that sleep is evolutionarily conserved across even distantly related phyla and challenged sleep researchers to expand their definitions of sleep [33]. The rapid advancement of genetic techniques has allowed *C. elegans* research to move forward quickly with mechanistic studies. The worm model can be leveraged to monitor and manipulate cellular activity in vivo and rapidly identify novel gene functions. Optogenetics [40], calcium imaging [41], and CRISPR gene editing [42, 43] have solidified the *C. elegans* system as a powerful model for sleep research. This chapter focuses on the genetics of developmentally timed sleep and stress-induced sleep; however, other sleep-like states including satiety quiescence [29], episodic swimming quiescence [30], microfluidic-induced sleep [44], spontaneous adult sleep [45], and post-response quiescence [46] are also reviewed (Fig. 2).

The Genetics of Developmentally Timed Sleep

Animals of the Ecdysozoa group molt their exoskeleton, a physiological process essential for growth [47]. In *C. elegans*, this occurs at the end of lethargus following developmentally timed sleep [17, 48], which is characterized by periods of behavioral quiescence which fulfill the widely viewed definitions of sleep [12, 18]. These include reversible quiescence, increased sensory arousal [18], stereotypical postures [49, 50], homeostatic sleep drive following acute deprivation [18, 51], and lethality

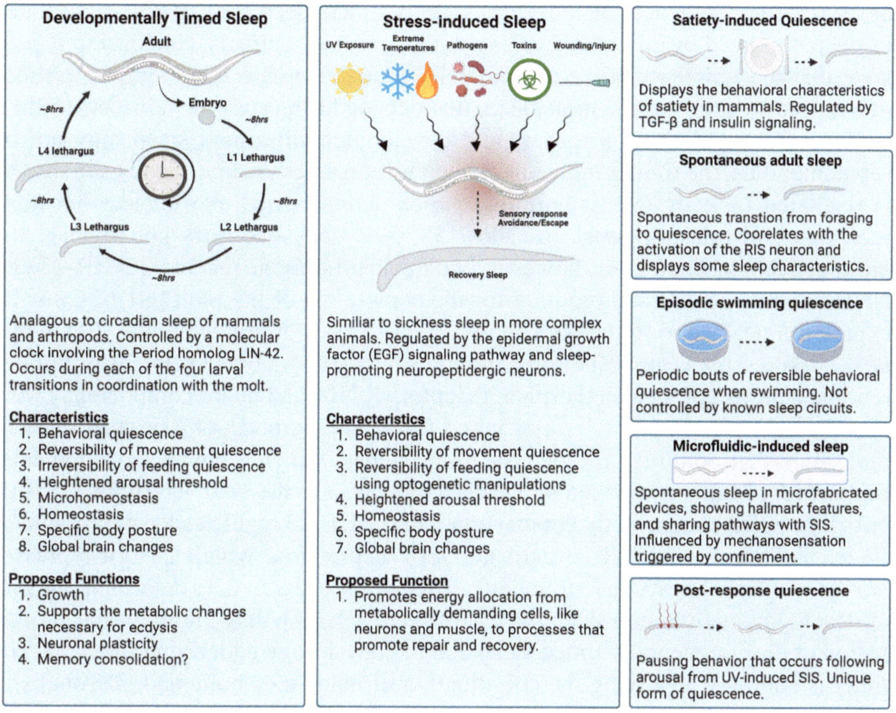

Fig. 2 Sleep behaviors of *C. elegans*. Developmentally timed sleep (left panel), stress-induced sleep (middle panel), and satiety-induced quiescence, spontaneous adult sleep, episodic swimming quiescence, microfluidic-induced sleep, and post-response quiescence. (Created with BioRender.com)

in response to chronic deprivation [52]. The name developmentally timed sleep [48] reflects the precise temporal control underlying these behaviors. Unlike sleep of insects and mammals that is linked to the 24-hour circadian cycle, developmentally timed sleep cycles every 7–9 hours when animals are cultivated at 20 °C (Fig. 2—*left panel*) [33, 53]. The molecules that time developmentally timed sleep are related to those of the circadian clock, most prominently LIN-42, the *C. elegans* homolog for PERIOD [26, 27]. Expression of *lin-42* cycles with developmental time, peaking prior to the initiation of and decreasing throughout developmentally timed sleep [27]. Loss-of-function of *lin-42* disrupts the timing of developmentally timed sleep and impairs the molt, which can be lethal [26]. *lin-42* loss-of-function animals show a highly variable period ranging from as little as 1 hour to greater than 20 [26]. A wild-type, synchronous period is restored when the *lin-42a* isoform is expressed specifically in seam cells, a group of cells embedded in the syncytial hypoderm (i.e., skin), which aid in the secretion of the cuticle [17]. Other central clock molecules, like *timeless* [54] and *doubletime* [55], encoded by *tim-1* and *kin-20* in *C. elegans*, also regulate developmental timing [56], although their role during developmentally timed sleep has not been extensively explored. Regulation by a conserved molecular clock is strong evidence that developmentally timed sleep is evolutionarily related to sleep regulated by circadian cycles of more complex animals. However, it is possible that the functions of developmentally timed sleep are more similar to those of a quiescent state termed molt sleep, that occurs during ecdysis in insects [57].

Temporal regulation of ecdysis and developmentally timed sleep may not be separable; thus, the timing molecules might be one and the same. Ecdysis in insects is regulated by ecdysone, a hormone which activates multiple nuclear hormone receptors and transcriptional cascades [58, 59]. Ecdysone does not regulate the molting cycle in *C. elegans*; however, two nuclear hormone receptors NHR-23 and NHR-25 are essential and required for the completion of the molt [60–62]. *nhr-23* expression peaks during the inter-molts and decreases during lethargus [61], while *nhr-25* expression is highest during lethargus [63]. Mammalian orthologs of *nhr-23*, retinoic acid receptor-related orphan receptors RORα and β, are components of the circadian clock [64]. *nhr-23*, *nhr-25*, and *lin-42* are members of a broader class of molecules that comprise the heterochronic genes, which regulate developmental timing [65]. Major components of these pathways are the *let-7* family of microRNAs [66, 67]. A recent study demonstrated that NHR-23 regulates the expression of *lin-42*. Additionally, NHR-23 activates *let-7* expression, which in turn represses *nhr-23* and *lin-42* posttranscriptionally. This inhibitory step may determine the end of developmentally timed sleep [68]. This feedback loop may allow for the coordination of developmentally timed sleep and the physiology underlying the molt. The timer is complex, involving the coordination of many metabolic and neuronal signaling pathways; moreover, the mechanisms connecting the timer to sleep regulation are largely unknown.

Exploring these mechanisms first involves the identification of sleep genes. One study identified gene expression changes during L4 lethargus, in which ~1800 were upregulated and ~1000 downregulated, by microarray analyses [69]. Another study profiled gene expression during the L3 and L4 lethargus stages, identifying 434 upregulated and 191 downregulated genes that were common to both stages. Of these shared genes, a subset oscillate with the molt [70], like *lin-42* [27]. Based on GO ontology, most of these genes likely regulate molting physiology, pharyngeal cuticular formation, and grinder reconstruction [69]. However, neuronally expressed genes encoding neuropeptides, G-protein-coupled receptors, and components of neurotransmitter metabolism, like *dat-1*, may represent sleep genes. Loss-of-function mutations in *dat-1*, a dopamine transporter, reduced quiescence during developmentally timed sleep [70, 71]. Cyclical gene expression was also documented for the tolloid/bone morphogenic-like gene *nas-38* [72], which influences the expression of many other genes [73].

Another study took a candidate-gene approach, in which mutants for orthologs of known sleep-regulating genes of *D. melanogaster* were examined for sleep deficits [71]. Twenty genes with cross-species sleep functions were identified. These include the potassium channel *shk-1*/Shaker [74], components of the cAMP/PKA pathway [14], calcineurin [75], and several neurotransmitter signaling pathways. Disruption of any of these 20 genes positively or negatively influenced sleep. This study highlighted the genetic conservation of sleep across phyla (Fig. 1) and further supports using *C. elegans* as a sleep model. Additionally, it expanded an already complex gene network underlying *C. elegans* sleep.

Mechanistic dissection of developmentally timed sleep is underway, and evidence suggests that the sleep behaviors can be separated genetically. Some genes have broad functions, regulating many aspects of developmentally timed sleep, while others have more specific roles [33]. These behaviors include quiescence of movement and feeding, increased sensory arousal, stereotypic body posture, and homeostatic regulation. Table 1 lists genes known to regulate quiescence of movement and/or sensory arousal, which encompasses most developmentally timed sleep genes. In fact, no mutations have been identified which strongly influence the quiescence of feeding or posture.

The Genetics of Quiescence

RIS/FLP-11 Developmentally timed sleep begins with slowed forward movement, followed by sleep bouts of complete immobility (i.e., quiescence of movement) [17, 18]. How LIN-42 mechanistically times sleep is unknown; however, important work by Michal Turek and colleagues (Max Planck Institute for Biophysical Chemistry) identified a central "sleep-on" neuron [39]. Whole nervous system recordings of neural activity using calcium imaging demonstrated that developmentally timed sleep is a global brain state. Most neurons decrease their activity when transitioning to bouts of quiescence; however, a subset maintain or increase their activity as sleep is initiated [76], like is seen in the hypothalamic sleep-active neurons of mammals [77, 78]. This brain state is rapidly reversible in response to arousing cues, like hyperoxia [76]. During quiescent bouts, neuropeptidergic and GABAergic neurons, including RIS, RMEL, RMER, and ALA, maintain tonic activity or display spontaneous depolarization [76], with RIS showing the most consistent sleep-on activity. Several lines of evidence support the RIS functioning as a central sleep control neuron [39, 79]. RIS activity immediately precedes cessation of movement, which correlates with sleep pressure [39, 76]. Optogenetic depolarization of RIS induces quiescence of movement in awake animals [39]. Loss-of-function of *aptf-1*, which encodes an AP-2 transcription factor required for gene expression in the RIS, strongly reduces quiescence of movement, with anterior body movements being most affected [80]. RIS secretes somnogenic FLP-11 neuropeptides and GABA, although GABA is dispensable for quiescence [81]. Optogenetic-induced quiescence of the RIS is attenuated in *flp-11* loss-of-function animals but not those with diminished GABA signaling [79]. Moreover, overexpression of *flp-11* is somnogenic and *flp-11* is required for quiescence [81]. *flp-11*-induced sleep is partially dependent upon three redundantly acting GPCRs, *npr-22*, *npr-4*, and *frpr-3*, each of which bind FLP-11 peptides in cell culture systems [82]. However, triple *npr-22;npr-4;frpr-3* mutants display only modest reductions in quiescence, suggesting FLP-11 acts through other receptors as well [81]. Global brain imaging describes developmentally timed sleep as having features of a fixed point attractor, reflecting

Table 1 Genes regulating developmentally timed sleep

C. elegans	*Drosophila*	Human	Description	Sleep effects	Sensory gating	Site of action*
lin-42	*Per*	Per2	Clock protein	Timer [26]	?	Hypodermal seam cells [26]
nhr-23	*DHR3*	NR1F1	Nuclear hormone receptor	Timer [61]	?	?
nhr-25	*FTZ-F1*	SF-1	Nuclear hormone receptor	Timer [62]	?	?
egl-4	*for-PA*	PRKG1	cGMP-dependent protein kinase	+ [18]	+ [18]	Sensory neurons expressing *tax-4* and/or *odr-3* [18]
shk-1	*Shaker*	KCNA3	Voltage-gated potassium channel	+ [71]	+ [71]	?
daf-16	*Foxo*	FOXO1 FOXO3 FOXO4	FOXO transcription factor—sleep homeostasis	+ [52]#	+ [52]#	Neurons [155], muscle [52, 155]
aptf-1	*TfAP-2*	TFAP2A	AP-2 transcription factor	+ [39, 80]	?	RIS neuron [39]
flp-11	*FMRFa*	?	RFamide neuropeptides	+ [81]	?	RIS neuron [81]
npr-1	*sNPF-R*	NPY	Neuropeptide Y receptor	+ [137]	+ [51, 138, 146]	RMG circuit [137]
pdf-1	*Pdf*	?	Pigment-dispersing factor	– [137]	– [146]	RMG circuit, ASK, ASI [137]
pdfr-1	*Pdfr*	?	Pigment-dispersing factor receptor	– [137]	– [146]	Touch neurons, muscle [137]
flp-2	*FMRFa*	?	RFamide neuropeptides	– [145]	– [145]	ASI, head neurons [145]
frpr-18	*FMRFaR*	HCRTR2	Orexin-like receptor	– [145]	– [145]	ASI, other neurons [145]
nlp-2	?	GNRH?	RPamide neuropeptide	– [85]	No [85]	AWA, other neurons [85]
nlp-22	?	NMS?	RPamide neuropeptide	+ [83]	+ [83]	RIA neuron [83]
gnrr-3 *gnrr-6*	*AkhR*	GNRHR	GnRH/AKH-like receptor	– [85]	?	Neurons [85]
nlp-14	*orcokinin*	?	Orcokinin-like neuropeptides	+ [156]	?	ALA neuron [156]
ceh-17	*Phdp*	PHOX2A	Homeodomain protein	+ [19]	?	ALA neuron [19]
lin-3	*spitz*	EGF	Epidermal growth factor (EGF)	+ [19]	?	?
let-23	*Egfr*	EGFR	EGF receptor	+ [19]	?	ALA neuron [19]
plc-3	*small wing*	PLCG1	Phospholipase C	+ [19]	?	ALA neuron [19]
kin-29	*SIK2*	SIK3	Salt-inducible kinase	+ [97, 98]	?	Neurons [97], ASK, ASJ [98]

(continued)

Table 1 (continued)

C. elegans	*Drosophila*	Human	Description	Sleep effects	Sensory gating	Site of action*
acy-1	*Ac13E*	ADCY9	Adenylyl cyclase	– [71]§ [105]	– [18, 71]	?
kin-2	*Pka-R1*	PRKAR1B	Protein kinase A	– [71, 99]	– [71]	Neurons [80], motor neurons [41]
crh-1	*CrebB*	CREB	cAMP-responsive transcription factor	– [71, 107]	– [71]	?
pde-4	*dunce*	PDE4B	Phosphodiesterase	+ [71]	+ [71]	?
dat-1	*DAT*	SLC6A2	Dopamine transporter	+ [71]	+ [71]	Dopaminergic neurons[&]
dop-1	*DoPR*	DRD1	Dopamine receptor	– [71]	– [71]	Dopaminergic neurons[&]
ser-4	*5HT1A*	HTR1A	Serotonin receptor	+ [71]	+ [71]	?
rcn-1	*sarah*	RCAN1	Calcineurin regulator	+ [71]	+ [71]	?
cnb-1	*CanB2*	PPP3R1	Calcineurin B	+ [71]	+ [71]	?
tax-6	*Pp2B-14D*	PPP3CB	Calcium-dependent phosphatase	+ [71]	+ [71]	?
unc-25	*GAD1*	GAD1	GABA synthesis	+ [71]	+ [71]	GABAergic neurons[&]
tdc-1	*Tdc1*	DDC	Tyrosine decarboxylase	– [71]	– [71]	Octopaminergic neurons[&]
tbh-1	*TBH*	DBH	Tyramine beta hydroxylase	– [71]	– [71]	Octopaminergic neurons[&]
goa-1	*Gαo-PI*	GNAO1	G-protein subunit alpha O1	+ [71]	+ [71]	?
gpb-2	*Gβ5*	GNB5	G protein, beta subunit	+ [122]	?	?
egl-30	*Gαq*	GNA11	G protein, alpha subunit	+ [122]	?	?
egl-10	*Rsg7*	RGS7	Regulator of G-protein signaling	– [122]	?	?
eat-16	*CG42450*	RGS9 RGS11	Regulator of G-protein signaling	+ [122]	?	?
rgs-1 *rgs-2*	*Dhit*	RGS17 RGS19	Regulator of G-protein signaling	– [122]	?	?
jnk-1	*Basket*	MAPK10	Jun N-terminal kinase	+ [71]	+ [71]	?
inso-1	*Insomniac*	KCTD5	BTB domain-containing protein	+ [71]	+ [71]	?
cul-3	*Cul3*	CUL3	Cullin protein	+ [71]	+ [71]	?
cya-1	*CycA*	CCNA1	Cyclin	– [71]	– [71]	?

(continued)

Table 1 (continued)

C. elegans	*Drosophila*	Human	Description	Sleep effects	Sensory gating	Site of action*
osm-7 *osm-11*	*Delta*	JAG1	DOS-motif (Notch)	+/– [115]£	+ [115]	Seam cells&
lag-2	*Delta*	JAG1	DOS-motif (Notch)	– [115, 155]	+ [115]	Seam cells&
lin-12	*Notch*	NOTCH1	Notch receptor	+/– [115]£	+ [115]	Neurons [157]
glp-1	*Notch*	NOTCH1	Notch receptor	+/– [115]£	+ [115]	Ciliated neurons [115]
unc-7, unc-9	*Inx3*	?	Gap junctions	+ [123]	+ [123]	Neurons [123]
unc-79 *unc80*	*Unc79* *Unc80*	UNC79 UNC80	NALCN channel complex subunit	– [123]	?	?
nca-1 *nca-2*	*Na*	NALCN	Sodium leak channel	– [123]	?	?
nas-38	*CG6974*	BMP1	Tolloid/bone morphogenic protein	+ [72]	?	Cyclical expression in epidermal cells [72]
nlp and *cnc* genes	?	?	Antimicrobial peptides	+ [72]	?	Cyclical expression in epidermal cells [72]
slo-1	*Slo*	KCNMA1	BK channel	+ [132]	?	Neurons [132]
homt-1	*Ntmt*	NTMT2	Hydroxyindole-O-methyltransferase	+ [132]	?	Intestine [132]
pcdr-1	*DopEcR*	MTNR1A	Melatonin-like receptor	+ [132]	?	Neurons [132]

+ Indicates that this gene promotes quiescence of movement
– Indicates that this gene inhibits quiescence of movement and promotes arousal
*Site of action based on tissue-specific rescue or RNAi
#Increased quiescence of movement was observed following sleep deprivation but not during normal developmentally timed sleep
§Gain-of-function mutations in *acy-1* disrupts microhomeostasis
&This has not been specifically tested experimentally
£*osm-7* and *osm-11* single mutants display increased quiescence of movement, but sleep was reduced in double mutants. Loss-of-function and gain-of-function alleles of *lin-12* or *glp-1* increased quiescence

the downregulation of most neurons, a global change stabilized by a subset of neurons that retain or increase their activity, with RIS being a central component [76]. Outside of developmentally timed sleep, RIS activation correlates with stopping behavior [79], microfluidic-induced sleep [44], and stress-induced sleep [24]. Taken together, the RIS functions as a central command neuron for the quiescence of movement.

NLP-2/NLP-22 Other pathways must function redundantly with RIS signaling, highlighted by the fact that *aptf-1* loss-of-function mutants still display movement quiescence, however, this is restricted primarily to the posterior half of their body [80]. Overexpression of the neuropeptide gene *nlp-22* is somnogenic, inducing quiescence of movement and feeding. *nlp-22* loss-of-function animals or wild-type animals treated with *nlp-22* RNAi display movement quiescence defects [83]. NLP-22 function was mapped to a pair of interneurons called the RIAs; genetic ablation of the RIAs reduces movement quiescence [83], but paradoxically, optogenetic activation of the RIAs during developmentally timed sleep arouses animals [83]. These latter observations suggest that the RIAs play a more complicated role during developmentally timed sleep than the RIS. In fact, the RIA is a highly connected HUB interneuron receiving substantial sensory input and coordinating motor output and displays compartmentalized calcium dynamics [84]. On which circuits the NLP-22 peptides function is unclear; however, the overexpression and loss-of-function phenotypes indicate they play a key role during developmentally timed sleep.

NLP-22 peptides are a part of a larger class of RPamides [35], which also include those encoded by *nlp-2* and *nlp-23*. These peptides display sequence similarities to gonadotropin-releasing hormone (GnRH)/adipokinetic hormone (AKH)-like peptides [85]. Of these, NLP-2 and NLP-22 are most important during developmentally timed sleep and function by antagonizing one another through the same GnRH/AKH-like receptors, coded by the genes *gnrr-3* and *gnrr-6*. *nlp-2* loss-of-function mutants display heightened quiescence, while overexpression of *nlp-2* or *gnrr-3* reduces it [85]. Both *nlp-2* and *nlp-22* cycle their expression and ectopic expression of *lin-42* enhances *nlp-22* expression, suggesting they are both regulated by the developmental clock [83, 85]. While NLP-22 peptides may be released from the RIAs to promote quiescence, it is unclear which cells release the NLP-2 peptides. It is also unclear how these peptides function downstream of their receptors; however, mutations in genes of the cAMP/PKA pathway, including *kin-2*, *pde-4*, *acy-1*, *gsa-1*, and *goa-1*, suppress *nlp-22*-induced quiescence [48].

LIN-3/EGF and the ALA Overexpression of *lin-3*, which encodes epidermal growth factors (EGF), is somnogenic [19]. EGF signaling regulates sleep in *Drosophila* [20], zebrafish [21], and mammals [22, 23], thus may represent an ancient sleep mechanism. During *C. elegans* sleep, LIN-3 ligands are released from unknown cells and signal through the EGF receptor (EGFR) LET-23 [86]. Reduction-of-function mutations in *let-23* impair EGF-induced sleep, which can be restored by expression of *let-23* in the ALA neuron [19]. *let-23* mutant animals also display quiescence defects during developmentally timed sleep [19]. Several genes are required for EGF-induced sleep, including *ceh-17*, which encodes a homeodomain protein required for ALA development [87], *egl-4*, encoding a cGMP-dependent protein kinase [88], *plc-3*, encoding phospholipase C [89], genes required for clear vesicle release, *unc-13* [90], *unc-18* [91], and *unc-64* [92], and *unc-31* which is required for dense core vesicle release [93]. Gain-of-function mutations in *deg-3*, encoding an acetylcholine receptor [94], also impair EGF-induced sleep.

ceh-17, *plc-3*, *let-23*, and *egl-4* are also required for the quiescence of movement of developmentally timed sleep [18, 19]. Taken together, these data suggest a model in which LIN-3 ligands activate LET-23 on the ALA neuron, triggering the PLC second messenger system, ultimately causing a rise in intracellular calcium levels which promotes the release of neuropeptides. Neuropeptide release may also be mediated by membrane depolarization [32]. While EGF and ALA are involved in developmentally timed sleep, they play a much more prominent role during the regulation of stress-induced sleep.

Although it is thought that the ALA functions primarily as a neuroendocrine cell [32, 38], it forms five direct synaptic connections, four of them with the AVE interneuron [3]. The ALA-AVE synapse is encapsulated by processes from the CEPsh glial cells [3]. Genetic ablation of these cells causes precocious periods of behavioral quiescence leading up to lethargus and enhanced developmentally timed sleep [95]. These phenotypes require the ALA neuron and can be partially recapitulated by expressing tetanus toxin in the AVE. Thus, the CEPsh glia negatively affect neurotransmission between ALA and AVE to regulate developmentally timed sleep initiation and sleep amounts [95].

Further evidence that the ALA neuron regulates quiescence of movement has been provided by calcium imaging. During developmentally timed sleep, the ALA displays spontaneous activity while maintaining levels similar to those during wakefulness [76]. However, the role of the ALA is more complicated than functioning solely as a locomotion stop neuron, like RIS [79]. ALA activity correlates with sleep drive and may be more involved in controlling sleep homeostasis during developmentally timed sleep (see below: The genetics of homeostasis) [96].

KIN-29/SIK A forward genetic screen for sleep-regulating genes in mice identified a gain-of-function allele of the gene *SIK3*, which encodes a salt-inducible kinase. This mutation caused a "sleepy" phenotype [97]. A hypomorphic allele of the *Drosophila* ortholog *Sik3* and a loss-of-function allele of the *C. elegans* ortholog *kin-29* reduce sleep in each species [97]. Restoration of *kin-29* in a subset of *odr-4*-expressing sensory neurons rescues the sleep defects [98]. *kin-29* is not required for either EGF-induced sleep or the quiescence induced by optogenetic activation of the RIS [97]. This places the salt-inducible kinase KIN-29 upstream of both the ALA and RIS. KIN-29 action in sensory neurons is activated in response to a reduction in cellular energy levels. Phosphorylated KIN-29 localizes to the nucleus and phosphorylates the class II histone deacetylase HDA-4, leading to the derepression of genes required for sleep upstream of the ALA and RIS neurons. Additionally, KIN-29's action in sensory neurons promotes fat mobilization from nonneuronal tissues which enhances sleep through the ALA and RIS [98]. Thus, cellular energetics and sensory modulation play an important role in coordinating sleep neuron activation during developmentally timed sleep.

cAMP/PKA/CREB Heightened PKA signaling in *C. elegans* [99], *Drosophila* [100], and mammals [101] is wake-promoting. The components of this conserved signaling pathway in *C. elegans* are encoded by *gsa-1* (G_s alpha subunit) [102], *acy-1* (adenylyl cyclase [AC]) [103], *kin-2* (regulatory subunit of PKA), and *kin-1* (catalytic subunit of PKA) [104]. cAMP levels are opposed by the action of phosphodiesterases, coded by the *pde* genes in *C. elegans*. Mutations in any gene that increase the cAMP/PKA pathway's activity disrupt developmentally timed sleep. *acy-1* gain-of-function animals are hyperactive when awake [105] and more responsive to stimuli during developmentally timed sleep [18]. Gain-of-function mutations in *gsa-1* [71], reduction-of-function mutations in *kin-2*, and loss-of-function of *pde-4* substantially reduce quiescence of movement during developmentally timed sleep [71, 99]. Optogenetic stimulation of cAMP in the nervous system using a red light-activated AC called IlaC22 [106] reduces quiescence [107]. Thus, the wake-promoting effects of cAMP/PKA are conserved across phyla.

How and where PKA elicits wake-promoting effects are not clear; however, CRH-1 (cyclic AMP-response element-binding protein [CREB]) [108] is one downstream component of this pathway. Loss-of-function of *crh-1* enhances movement quiescence [71] and attenuates the wake-promoting effects of optogenetic induction of cAMP in the nervous system [107]. Thus, CRH-1 functions downstream of cAMP/PKA; however, the site of action is still unclear. Rescue experiments for *kin-2* have shown that it functions in the nervous system [99]. Additionally, reduction-of-function of *kin-2* suppresses *nlp-22*-induced quiescence, which can be restored when *kin-2* is expressed in cholinergic motor neurons [48]. Taken together, these experiments support a model in which the cAMP/PKA pathway is inhibited in motor neurons during developmentally timed sleep.

EGL-4/PKG The cGMP/PKG pathway plays an opposing role to cAMP/PKA; PKG promotes sleep in *C. elegans* [18, 109], *Drosophila* [18], and mammals [110]. Gain-of-function mutations in *egl-4* cause wake animals to slow and sometimes stop their movement in the presence of food [109], while *egl-4* loss-of-function animals move more frequently [111]. Consistent with these phenotypes, loss of *egl-4* reduces quiescence of movement during developmentally timed sleep, and *egl-4* gain-of-function mutations enhance it. *egl-4* loss-of-function mutants also display sensory arousal defects, which are rescued by the expression of *egl-4* in a subset of *odr-3*- and *tax-4*-expressing sensory neurons [18]. Where EGL-4 functions to regulate quiescence is unclear, but it may be connected with its role in regulating sensory arousal.

Notch Notch signaling is highly conserved, controlling developmental events across phyla [112]. Notch is a cell surface receptor, encoded by *lin-12* and *glp-1*, that is activated by cell surface and diffusible notch ligands. Multiple ligands exist in *C. elegans*, encoded by *osm-7*, *lag-2*, and *osm-11*, the latter of which is likely diffusible [113]. Receptor activation leads to intracellular signaling culminating in transcriptional activation of target genes [112]. While LIN-12 activation leads to expressional changes of many genes, one predicted target is *egl-4* [114], implicating notch with sleep regulation [115]. Overexpression of *osm-11* is somnogenic, inducing quiescence of movement and feeding, and increasing arousal threshold [115]. Removal of both *glp-1* and *lin-12* suppresses these phenotypes, suggesting that they act in parallel. Multiple sleep genes are required for *osm-11*-induced quiescence, including *egl-4*, *ceh-17*, *let-23*, and *deg-3*, placing PKG and ALA signaling downstream of notch. However, *osm-7;osm-11* double mutants suppress the sleep-promoting effects of *lin-3* overexpression, suggesting that the role of notch during sleep regulation is complex [115]. Loss-of-function of *osm-7*, *osm-9*, *lin-12*, or *glp-1* increases quiescence of movement, which is the opposite of what would be expected based on the *osm-11* overexpression experiments, further highlighting the complexity. Paradoxically, *osm-7;osm-11* double mutants display decreased quiescence and a shortened lethargus period [115]. The mechanisms of notch signaling are better explained in the context of sensory arousal, which is discussed below.

Neurotransmitters Multiple genes controlling neurotransmission have been implicated in developmentally timed sleep; how they fit into a larger network of genes is far from understood. Moreover, even though the *C. elegans* connectome is relatively simple [3], identifying the complete sleep circuitry has proven to be a quite challenging, yet achievable goal. Nonetheless, genes regulating neurotransmission are conserved; similar neurochemistry underlies sleep in *C. elegans*, *Drosophila*, fish, and mammals. Dopamine and octopamine signaling have wake-promoting effects in *Drosophila* [116, 117]; similarly, loss of *dat-1*, *tdc-1*, tyrosine decarboxylase, or *tbh-1*, tyramine beta hydroxylase, increases quiescence of movement during developmentally timed sleep [71]. Serotonin and GABA are sleep-promoting in *Drosophila* [118, 119], and loss of *ser-4*, a serotonin receptor, the GABA receptor genes *unc-49* and *lgc-38*, or *unc-25*, a glutamate decarboxylase required for GABA synthesis [120], reduces quiescence.

Gap Junctions/Innexins A forward genetic sleep screen identified the gene *unc-7*, which encodes an innexin, a gap junctional component [121, 122]. Loss of either *unc-7* or *unc-9*, another innexin, disrupts quiescence of movement and arousal threshold, and overexpression of *unc-7* or *unc-9* is somnogenic. Rescue experiments map innexin function to somewhere in the nervous system, with UNC-7 likely functioning in motor interneurons and UNC-9 in cholinergic motor neurons. *unc-7* and *unc-9* are required for the enhanced sleep seen in *egl-4* gain-of-function animals, supporting a model in which EGL-4 action in sensory neurons is functioning upstream of the innexins in motor neurons and interneurons [123].

NAS-38 and Immune Response Genes A reverse genetic sleep screen found that *nas-38* gain-of-function animals display heightened quiescence and significantly longer sleep stages, and loss-of-function of *nas-38* causes a reduction in developmentally timed sleep [72]. A transcriptomic analysis showed that genes involved in cuticular formation and innate immunity are differentially expressed in *nas-38* gain-of-function animals. In particular, antimicrobial peptide (AMP) genes like *nlp-29*, known to be upregulated in the epidermis upon wounding [124], are highly expressed during developmentally timed sleep [72]. This suggests that sleep is regulated in part by innate immune genes. The innate immune genes of the p38 MAP kinase and the TGF-β-SMAD pathways are required for the enhanced sleep of *nas-38* gain-of-function mutants. In addition, simultaneous knockdown of multiple AMP genes of the *nlp* and *cnc* classes reduces sleep in *nas-38* gain-of-function animals. Moreover, overexpression of the *nlp* and *cnc* genes induces quiescence in an RIS-dependent manner. Taken together, these data suggest that innate immunity pathways regulate developmentally timed sleep duration via the action of NAS-38. NAS-38 function was mapped to the epidermis, indicating that a collection of AMPs from the epidermis signal through the RIS neuron to promote sleep [72].

Other Signaling Pathways The calcineurin signaling pathway alters gene expression in response to intracellular rises in calcium [125]; disrupting this pathway reduces sleep in *Drosophila* [75, 126]. Genes of this pathway are required for developmentally timed sleep in *C. elegans;* this includes *rcn-1*, *cnb-1*, and *tax-6* [71]. The potassium channel gene *shk-1*, an ortholog of the known sleep regulator *Shaker* in *Drosophila* [71, 74], is also required for developmentally timed sleep. Genes encoding components of the ubiquitin-proteasome system, *inso-1* and *cul-3*, a cyclin A gene *cya-1*, the c-Jun N-terminal kinase gene *jnk-1*, and the gene encoding a Goα subunit protein, *goa-1*, also positively regulate developmentally timed sleep [71], like in *Drosophila* [127–130]. Melatonin promotes sleep in *C. elegans*, like it does across phyla [131], which is dependent upon genes encoding the melatonin-like receptor, *pcdr-1*, a hydroxyindole-O-methyltransferase, *homt-1*, and a BK channel *slo-1* [132]. HOMT-1 regulates melatonin release from the intestine, which signals through PCDR-1 in the nervous system, which in turn modulates the function of SLO-1 [132]. This study further highlights the molecular conservation of sleep across the animal kingdom.

The Genetics of Sensory Arousal

Sensory stimuli can arouse animals from sleep states; however, a landmark of sleep is a heightened arousal threshold, referred to as sensory gating [12]. This characteristic of sleep allows animals to balance their need for sustained sleep with an ability to rapidly respond to potential threats. In mammals, sensory gating is a result of subcortical processing of sensory information in thalamic neurons. This modulates an animal's perception and responsiveness to stimuli [133]. In *C. elegans*, sensory gating is controlled at the level of the sensory and interneurons by multiple signaling pathways. A touch neuron called ALM displays reduced activity and responsiveness to mild stimulation during developmentally timed sleep, evident by calcium imaging [134]. The nociceptive and chemosensory ASH neurons are less responsive to 1 mM Cu^{2+}, an aversive stimulus [135]. These studies show that sensory gating is at least in part controlled by the sensory neurons; however, a more complex level of regulation likely exists. The ASH has two synaptic connections with the interneurons AVD and AVA, which stimulate aversive reversal movements [136]. AVD and AVA activity fluctuates during wake periods and is synchronously activated in response to stimuli. During developmentally timed sleep, both neurons are relatively inactive, and stimulus-driven experiments show a loss of synchrony [135]. Coordinated optogenetic activation of AVD and AVA causes rapid waking, while activation of sensory neurons only induces wakefulness in combination with prior, waking mechanical stimuli [135]. Sustained depolarization of sensory neurons can sensitize command interneurons allowing them to respond to a diverse array of stimuli and ensuring rapid reversibility from sleep [135]. This work supports a model in which reduced neurotransmission from sensory neurons and a lack of coordination of interneurons are responsible for sensory gating. These experiments focused on aversive chemical cues; however, mechanical and light energies also wake animals but utilize different circuits. How sleeping animals coordinate these multiple stimuli and the molecular mechanisms that govern them are still being explored.

An Arousal Circuit: NPR-1/Neuropeptide Y, PDF, FLP-2, and Glutamate The neuropeptide Y receptor NPR-1 regulates responsiveness to environmental stimuli, including oxygen levels and mechanical energy [137], two cues that robustly arouse sleeping animals [18, 76]. *npr-1* loss-of-function animals have mild defects in quiescence during developmentally timed sleep under conditions where sensory stimuli are low and quiescence is monitored continually [51, 138]. However, when sensory input is higher and quiescence is measured less frequently, *npr-1* mutants display little to no movement quiescence [137]. Solitary *npr-1* mutant animals are rarely quiescent, while animals found in aggregates or along the bacterial food border display robust bouts of movement quiescence [76]. Under these latter conditions, O_2 tension is lower than atmospheric levels. Experiments utilizing O_2 flow arenas discovered that wild-type and *npr-1* loss-of-function animals display only minor differences in quiescence at 10% O_2 [76]. However, in atmospheric conditions, 21% O_2, wild-type animals sleep normally, while *npr-1* mutants do not sleep at all. Moreover, raising levels during developmentally timed sleep from 10% to 21% O_2 arouses wild-type and *npr-1* animals. However, aroused wild-type animals display rapid rebound sleep, re-entering a quiescent state, while *npr-1* mutants remain active for the duration of the exposure [76]. These studies highlight a central role for NPR-1 signaling as a mechanism for heightening the arousal threshold.

Arousal in mammals is controlled by wake-promoting orexin neuropeptides and in *Drosophila* by the pigment-dispersing factor (PDF) [139, 140]. And, glutamate neurotransmission is a conserved arousal signal [141, 142]. In *C. elegans,* neuropeptides encoded by *flp-2* signal through an orexin-like receptor, FRPR-18. Peptides encoded by *pdf-1*, an ortholog of *Drosophila Pdf*, signal via the PDFR-1 receptor [143, 144]. FLP-2/FRPR-18 and PDF-1/PDFR-1 signaling is required for the heightened arousal of *npr-1* mutants [137, 145]. The vesicular glutamate transporter *eat-4* and the glutamate receptor *glr-2* are also required [146]. Thus, orexin-like, PDF, and glutamate signaling promote sensory arousal in *C. elegans.*

Rescue experiments support a model in which PDF-1 and FLP-2 neuropeptides are released from the ASI interneuron, which signals to mechanosensory neurons and muscle to promote arousal [145]. Additionally, autocrine FLP-2 and PDF-1 signaling of the ASI promotes positive feedback during active periods. *eat-4* function was mapped to the nociceptive ASH neuron and mechanosensory touch neurons [146]. GLR-2 functions partially through the interneurons DVA and AIA. The touch neuron PLM synapses with DVA, and ASH connects with AIA, highlighting a potential glutamatergic arousal circuit [3, 146]. Epistasis experiments place glutamate singling downstream of *flp-2* and *pdf-1*. In summary, a positive feedback loop involving PDF-1 and FLP-2 neuropeptide release from the ASI neuron strengthens an arousal circuit which acts through PDFR-1 and FRPR-18 on mechanosensory neurons and muscle. In turn, mechanosensory neurons and the ASH neuron release glutamate which promotes arousal through the AIA and DVA interneurons [146].

Where and how does NPR-1 act to support changes in sensory arousal during developmentally timed sleep? During sleep in wild-type animals, secretion of PDF-1 and FLP-2 peptides is low. In *npr-1* mutants, the levels of secretion remain high [137, 145]. Restoration of *npr-1* function in the RMG sensory circuit rescues the heightened arousal of *npr-1* mutants and significantly decreases the secretion rate of PDF-1 and FLP-2 neuropeptides during sleep [137, 145]. Additionally, rescue of *npr-1* in RMG or O_2-sensing neurons restores the rebound sleep that occurs following O_2 arousal [76]. Thus, NPR-1 likely functions to heighten the arousal threshold in the RMG sensory circuit by inhibiting the release of wake-promoting PDF-1 and FLP-2 neuropeptides.

Notch Mutations that decrease notch signaling reduce arousal threshold, while those that enhance notch signaling have the opposite effect [115]. Unlike quiescence of movement, where notch signaling is more complex, sensory arousal correlates with notch signaling [115]. Loss-of-function of *osm-7*, *osm-11*, *lin-12*, or *glp-1* reduces the arousal threshold, while gain-of-function alleles of *lin-12* and *glp-1* increase it. Rescue experiments and tissue-specific RNAi support a model in which LIN-12 functions in interneurons (AVA, RIM, AIY, and RIG), while GLP-1 functions in ciliated sensory neurons. Like with quiescence of movement, notch functions downstream of EGL-4 [115]. It is unclear how notch and NPR-1 signaling are connected mechanistically.

Other Signaling Pathways Of those pathways known to regulate quiescence of movement, a subset have specifically been tested for defects in sensory arousal. Overexpression of *nlp-22* induces quiescence with a heightened arousal threshold to blue light and mechanical stimulation, while loss-of-function of *nlp-22* decreases arousal threshold during sleep [83]. Mutations that disrupt calcineurin, dopamine, GABA, cAMP/PKA, *goa-1*, and *jnk-1* signaling reduce arousal thresholds to mechanical stimulation. Mutations in genes affecting octopamine signaling have the opposite effect and make animals even less responsive [71]. In summary, many genes have been implicated in regulating sensory arousal; however, further work needs to be done to understand where and how they function.

The Genetics of Homeostasis

In mammals and *Drosophila*, sleep drive is increased following acute deprivation, and chronic sleep loss can be lethal [147, 148]. In *C. elegans,* forced movement by mechanical stimulation during developmentally timed sleep can also be lethal; however, most animals survive chronic sleep deprivation [52]. Those that die display evidence of molting defects, getting trapped in a cuticle that they failed to shed [52], suggesting that sleep is an essential component of the molting process. However, mutations in genes that reduce sleep, even substantially, do not necessarily cause lethality, except for those that also strongly influence ecdysis [26, 56, 60, 62]. So, sleep levels are highly plastic during developmentally timed sleep, and it is unclear how much sleep is needed for survival. However, developmentally timed sleep is homeostatically controlled, and molecular components regulating sleep homeostasis have been identified.

Forced deprivation by mechanical stimulation or genetic manipulations induces rebound sleep. This can be assessed by quantifying movement quiescence or sensory arousal following deprivation [18, 51]. Using these approaches, *daf-16*, which encodes a highly conserved FOXO transcription factor [149], was identified as a key mediator of sleep homeostasis. Loss-of-function of *daf-16* eliminates sleep rebound following deprivation. Rescue experiments support a model in which DAF-16 functions in body-wall muscles [52], which may modulate sensory arousal by signaling to sensory neurons or by directly regulating the neuromuscular junction.

Sleep homeostasis can be separated into two distinct phases, microhomeostasis, which occurs in response to routine sleep disturbances (weak stimulation), and homeostatic response to strong stimuli [51]. Developmentally timed sleep is characterized by bouts of movement and quiescence of variable lengths; microhomeostasis refers to the correlation between the length of movement bouts and the sleep rebound that occurs thereafter [105]. Microhomeostasis is measured by exposing sleeping animals to weak mechanical stimulation. It is abolished in *acy-1* gain-of-function animals; thus, it is cAMP-dependent [105], and it requires *npr-1* [51]. However, *daf-16* is not required for the maintenance of microhomeostasis [51]. Strong stimulation disrupts movement and quiescent bouts more profoundly, inducing prolonged periods of forward movement, followed by enhanced sleep (i.e., homeostasis) [51, 52]. In this case, *daf-16* is required, while *npr-1* is dispensable [51]. Taken together, sleep bout architecture is stabilized by NPR-1 under conditions of weak stimulation, while DAF-16, functioning in muscle [52] and potentially neurons [51], promotes rebound sleep following strong sleep disturbances.

Calcium imaging revealed that microhomeostasis is regulated by the ALA [96]. During sleep bouts, the RIS neuron is activated, while the ALA displays a gradual decay in activity. During motion bouts, RIS activity is reduced and ALA activity shows a measured increase, which correlates with the length of the bout [96]. *ceh-17* mutants, who have altered ALA activity, have longer motion bouts coupled with a decrease in sleep bouts. While total levels of developmentally timed sleep are reduced, these observations suggest that the ALA neuron regulates sleep homeostasis and that the levels of activation of ALA reflect sleep drive [96]. At this point, it is unclear how the ALA, cAMP, NPR-1, and DAF-16 pathways are connected mechanistically to regulate homeostasis.

The Genetics of Feeding Quiescence

Feeding in *C. elegans* is accomplished by rhythmic contractions of the pharynx, an organ composed of muscle, neurons, glands, an epithelium, and marginal cells [150, 151]. Muscular contractions lead to the movement of an extracellular toothlike grinder, which is easily visualized with a stereomicroscope [152]. The grinder movement ceases and is subsequently shed and rebuilt during developmentally timed sleep [69, 153]. No mutations have been identified which inhibit quiescence of feeding during developmentally timed sleep, which could indicate that multiple genes function redundantly or that the pharynx is not capable of contracting at this time. This latter possibility is supported by several lines of evidence. First, direct optogenetic stimulation of pharyngeal muscle does not induce feeding during developmentally timed sleep [48]. Second, electron microscopic reconstruction of the pharynx and grinder during developmentally timed sleep showed pharyngeal muscle trans-differentiation, in which the contractile elements lose their organization, and the cytoplasm resembles that of secretory cells. These two studies strongly suggest that the pharynx is not a functional muscle during developmentally timed sleep [153].

However, quiescence of feeding begins prior to these developmental changes at the onset of lethargus, evident by slow, lethargic pumping. Thus, signaling pathways may be involved initially to ensure quiescence of feeding is established. There are several molecules and/or cells capable of inducing quiescence of feeding, which have also been implicated in other aspects of developmentally timed sleep regulation. First, optogenetic stimulation of the RIS neuron induces both quiescence of movement and feeding [39]; however, *aptf-1* mutants still cease pumping during lethargus. Overexpression of the neuropeptide gene *nlp-22* induces quiescence of feeding, but while *nlp-22* is required for quiescence of movement, it is not required for quiescence of feeding [83]. Overexpression of *osm-11* [115] or *lin-3* [19] induces quiescence of feeding, the latter of which is dependent upon *ceh-17*, *egl-4*, *plc-3*, *unc-13*, *unc-18*, *unc-64*, and *unc-31* [19]. *let-23* or *plc-3* loss-of-function animals initiate feeding quiescence later than wild-type animals, suggesting that EGF signaling may play a role during these initial stages [19]. At this point, it is unclear how quiescence of feeding is initiated at the onset of developmentally timed sleep. Furthermore, based on the ultrastructural changes of the pharynx [153] and the essential nature of the grinder [69, 154], it seems unlikely that a mutant with significant defects in feeding quiescence would be viable; however, this is yet to be determined.

The Genetics of Stress-Induced Sleep

When animals encounter noxious stimuli, they activate avoidance behaviors which increase their chance of survival. These stimuli include extreme temperature, ultraviolet (UV) irradiation, toxins, pathogenic bacteria, injurious substrates, and osmotic stress [6, 158–160]. Avoiding these damaging stimuli is not always possible; thus, cellular damage may be incurred, requiring a restorative healing response. During these times, energy is allocated from energetically demanding cells, like neurons and muscle, to promote recovery [34]. It is believed that these needs reflect the function of stress-induced sleep. Animals with reduced quiescence during stress-induced sleep have shorter lifespans [24, 31], demonstrating that stress-induced sleep is an essential recovery behavior. Interestingly, the protective functions of stress-induced sleep are outweighed by the drive to forage, in that food deprivation reduces stress-induced sleep but also mitigates the detrimental effects of sleep loss [161]. Also, avoidance behaviors, particularly those initiated during conditions of hyperoxia, can override stress-induced sleep [162]. Thus, stress-induced sleep levels are plastic based on the need to balance sleep with other essential behaviors. Stress-induced sleep pressure dominates when ATP-dependent repair mechanisms are needed for recovery [163]. Stress-induced sleep is an intriguing genetic model of the heightened sleep during illness observed in more complex animals.

Stress-induced sleep fulfills the criteria of sleep described by Campbell and Tobler [12]: behavioral quiescence with a stereotypic posture, reversibility, reduced responsiveness to stimuli, and homeostatic rebound following deprivation [12, 31, 162]. Stress-induced sleep is comprised of distinct behavioral subprograms which include quiescence of movement, feeding, and defecation quiescence [38]. Mutants have been identified which suppress all of these behaviors simultaneously, like those of the genes *ceh-14* and *ceh-17* [38], while others have been found to only suppress specific sub-behaviors [33]. A common EGF/ALA signaling pathway underlies sleep induced by all stressors [31, 164]; however, some of the genetics may be distinct based on the type of environmental stress.

Of note, stress-induced sleep can occur at any point in the life of the animal, distinguishing it from developmentally timed sleep which is linked with a biological clock. Genes controlling the timing of these two sleep behaviors are distinct; however, some general mechanisms regulating quiescent behaviors are shared. But the roles these molecules play during each sleep state are not identical, suggesting that the behaviors of sleep are controlled by a common set of genes, but new pathways have evolved, reflecting the unique functions of both developmentally timed and stress-induced sleep [48].

LIN-3/EGF, ALA, and RIS EGF signaling via the neuropeptidergic ALA neuron plays a prominent role in the regulation of stress-induced sleep. Reduction-of-function of *lin-3* or *let-23* and loss-of-function of *plc-3* cause significant impairments in quiescence of both movement and feeding during heat-induced stress-induced sleep and UV-induced stress-induced sleep [31, 164]. Loss-of-function of *ceh-14* or *ceh-17* also severely impairs quiescence of movement and feeding, which was assessed following exposure to heat, cold, ethanol, hyperosmotic conditions, UV, or pore-forming Cry5B proteins [31, 38, 164]. Overexpression of *lin-3* inhibits movement, feeding [19], sensory arousal [162], and defecation [156], showing that EGF signaling controls all sub-behaviors of stress-induced sleep.

Multiple lines of evidence support the ALA neuron as being a primary target of EGF signaling during stress-induced sleep. First, *ceh-17* loss-of-function or ALA-ablated animals do not display prominent quiescence of movement or feeding following *lin-3* overexpression. In other words, the ALA is required for the sleep-promoting effects of EGF. Second, ALA-specific rescue of the EGF receptor *let-23* restores quiescence of movement and feeding following *lin-3* overexpression. *lin-3* overexpression also induces quiescence of defecation, which requires *nlp-14* neuropeptides released from the ALA [156]. Taken together, EGF signaling via the ALA neuron regulates all sub-behaviors of stress-induced sleep.

Quiescence of movement during stress-induced sleep is further regulated by the RIS neuron, which expresses *let-23* as well. Removal of *let-23* from the RIS using the FLP-FRT system or loss-of-function of *aptf-1* substantially reduces quiescence, most prominently head movement [24, 80]. *ceh-17;aptf-1* double mutants display even less quiescence than either single mutant, suggesting that ALA and RIS work together to promote quiescence. Calcium imaging demonstrated that both the ALA and RIS neurons increase their activity following heat exposure and during stress-induced sleep, in which ALA activity remains high for an extended period of time and RIS activity correlates with each quiescent bout. The activity of the RIS is dependent upon the ALA neuron, evident by reduced RIS activity in *ceh-17* or *ceh-14* mutants [24]. Second messenger pathways involving phospholipase C regulate the action of the ALA during stress-induced sleep, and many of the same genes are expressed in RIS [19, 24]. In addition to this, both neurons act by membrane depolarization [32, 39]. Of note, unlike developmentally timed sleep, the CEPsh glial cells that encompass the ALA-AVE synaptic connection play little to no role during stress-induced sleep [165]. Taken together, ALA and RIS work in concert, via neuropeptidergic signaling, to control the quiescent behaviors of stress-induced sleep, with RIS and ALA regulating quiescence of movement and ALA controlling all sub-behaviors of stress-induced sleep.

What molecules are released by the ALA and RIS? Single-cell transcriptomics of the ALA revealed the expression of a wide array of neuropeptide genes with *flp-24*, *flp-7*, *nlp-8*, and *flp-13* being the most highly enriched [38]. *flp-13* neuropeptides were previously shown to be sleep-promoting and released by the ALA [32]. RNA-sequencing of transcripts isolated from the RIS confirmed the presence of *flp-11* [81], one of the most enriched genes, with a number of other neuropeptide genes expressed at lower levels [24]. The importance of neuropeptide signaling during stress-induced sleep is highlighted by the fact that genes required for neuropeptide processing, like *egl-21*, and dense core vesicle processing and release, like *unc-31* and *unc-108*, are also indispensable for stress-induced sleep [19, 164, 166].

Moreover, some specific functions regulating the sub-behaviors of neuropeptides and their GPCRS have been described.

Neuropeptides and Other Pathways A collection of neuropeptides and GPCR genes regulating quiescence of movement during stress-induced sleep were identified by cross-species comparison of gene expression, transcriptomics, loss-of-function, and overexpression studies. Quiescence during stress-induced sleep has been analyzed by manual observation and with microfabricated imaging devices and tracking systems [138]. Neuropeptides encoded by *flp-13* are some of the most important molecules controlling quiescence during stress-induced sleep. *flp-13* loss-of-function animals are partially defective in heat-induced [32, 38] and UV-induced movement quiescence [164]. Additionally, *flp-13* is required for EGF-induced sleep, which can be rescued when *flp-13* is expressed in the ALA [32]. However, *flp-13* mutants display a wild-type arousal threshold [38]. Thus, *flp-13* is required in the ALA to promote quiescence of movement but not for the reduced responsiveness to stimuli. Overexpression of *flp-13* is somnogenic; it induces quiescence of all sub-behaviors of stress-induced sleep [32, 38]. Quiescence of movement induced by *flp-13* is dependent on genes of the Gαq and Gαs signaling pathways [48], and optogenetic activation of motor neurons stimulates arousal. This supports a model in which FLP-13 neuropeptides from the ALA signal to the nervous system, modulating second messenger systems and altering motor neuron excitability [48].

FLP-13 signaling from the ALA occurs largely via the GPCR DMSR-1. Animals harboring loss-of-function alleles of *dmsr-1* were isolated from a suppressor screen of *flp-13*-induced sleep. *dmsr-1* is required for quiescence during stress-induced sleep and is likely to function in the nervous system [167]. Chemogenetic hyperpolarization of *dmsr-1*-expressing neurons enhances sleep, suggesting it functions by inhibiting wake neurons, potentially the AIY [167]. *dmsr-1;flp-13* double mutants display less quiescence than either *flp-13* or *dmsr-1* single mutants, suggesting that FLP-13 neuropeptides act through multiple receptors. This is supported by the fact that FLP-13 neuropeptides can activate multiple receptors in cell culture, including DMSR-1, FRPR-4, and NPR-22 [167–169]. However, *frpr-4* is not required for stress-induced sleep or *flp-13*-induced sleep alone; but, some evidence indicate that FLP-13 and FRPR-4 are ligand-receptor pairs in vivo. Overexpression of *frpr-4* induces quiescence with an exaggerated bend posture, which partially requires *flp-13* [169]. The posture phenotype is recapitulated by overexpression of *frpr-4* specifically in the wake-promoting DVA interneuron [107], a phenotype dependent upon *flp-13* [169]. Thus, FRPR-4 may be a FLP-13 receptor, but be more involved in regulating body posture during stress-induced sleep. Taken together, these data suggest that FLP-13 from the ALA signals through multiple receptors, including DMSR-1 and FRPR-4. Whether NPR-22 is an additional FLP-13 receptor during stress-induced sleep is unknown.

Other ALA-released neuropeptides regulating quiescence are encoded by *nlp-14*, *flp-24*, and *nlp-8* [38, 156]. Like *flp-13*, overexpression of any one of these three genes induces quiescence of movement. *nlp-14* is required for the full induction of movement quiescence during stress-induced sleep [156], but *flp-24* and *nlp-8* are not required alone. However, *flp-13;flp-24* and *nlp-8;flp-13* double mutants display less quiescence than *flp-13* single mutants, and they display sensory arousal defects [38]. Interestingly, *flp-13* and *flp-24* have been found to be expressed in the ALA neuron of the parasitic nematode *Ascaris suum* [170], suggesting that their redundant function may be evolutionarily conserved. Presently, receptors for FLP-24, NLP-8, and NLP-14 peptides have not been identified.

The RIS neuron also regulates quiescence of movement during stress-induced sleep via the action of FLP-11 neuropeptides [24]. Removal of the EGF receptor gene *let-23*, specifically from the RIS using the FLP-FRT system, reduces quiescence of movement, suggesting that the RIS is directly activated by EGF. Additionally, the RIS is modulated by signaling from the ALA. Overexpression of *flp-24* increases RIS activity, evident by calcium imaging studies [24]. Taken together, EGF may directly activate the ALA which releases FLP-13, FLP-24, NLP-8, and NLP-24 neuropeptides and the RIS which releases FLP-11 neuropeptides. The ALA further activates the RIS enhancing its sleep-promoting effects. To more fully understand how quiescence of movement is regulated during stress-induced sleep, it will be important to determine the identity of the likely many receptors of these neuropeptides and their precise location of action.

Despite not knowing all of the receptors at this moment in time, most evidence suggests that quiescence of movement is regulated in the nervous system via the modulation of membrane potential and second messenger signaling [48, 107, 167]. As stated above, hyperpolarization of *dmsr-1*-expressing neurons enhances quiescence of movement, while optogenetic depolarization of motor neurons disrupts it [48, 167]. Additionally, multiple genes in the Gαq and Gαs signaling pathways, including those regulating cAMP/PKA, are required for quiescence of movement [48]. Pan-neuronal optogenetic induction of cAMP using IlaC22 [106] substantially reduces quiescence during stress-induced sleep [107]. These wake-promoting effects of cAMP do not require *crh-1*; in fact, *crh-1* mutants are defective in quiescence. Thus, CREB is not a likely downstream target of cAMP signaling during stress-induced sleep [107]. More selective expression of IlaC22 using promoters from the genes *twk-16*, *hlh-17*, and *pdfr-1*, expressed in the DVA, CEPsh glia, and RIF interneurons, respectively, also reduces quiescence. During stress-induced sleep, cAMP levels are reduced in the DVA and RIF, in an ALA-dependent manner [107]. And, overexpression of *flp-13* is capable of reducing cAMP in the DVA neuron. Moreover, the knockdown of *kin-2* in *twk-16*-expressing cells, which includes the DVA, reduces quiescence [107]. Last, heightened Gαs/cAMP/PKA signaling inhibits quiescence following *flp-13* or *nlp-14* overexpression [48, 156]. Taken together, neuropeptide signaling from the ALA and RIS may function by reducing the excitability of motor and interneurons, in part by reducing cAMP levels (Fig. 3).

Like with developmentally timed sleep, distinguishing between quiescence of movement and arousal threshold is challenging, considering mild stimulation can arouse wild-type animals during stress-induced sleep [162]. *npr-1* loss-of-function animals are more active and have a reduced arousal threshold during stress-induced sleep induced by exposure to UV or the pore-forming protein Cry5B [162]. *npr-1* loss-of-function animals display wild-type levels of sleep during low 10% O_2 conditions but are rapidly aroused and remain awake upon exposure to 21% O_2. These arousal effects are partially dependent upon *pdf-1* [162]. Wild-type animals, on the other hand, are aroused but then display rebound sleep. Thus, *npr-1* is required for the coordination of hyperoxia avoidance and sleep. This work highlights the reversibility and plasticity of stress-induced sleep. An animal's survival hinges on its ability to successfully escape from damaging stimuli and recover following cellular damage [162].

Lastly, some genetic pathways are only involved when responding to specific noxious insults. This has become evident through studies examining stress-induced sleep induced by UV irradiation and wounding. Like with all other stressors, UV exposure and wounding activate EGF signaling, the ALA, and RIS. For UV-induced stress-induced sleep, the gene *cep-1*, encoding a homolog of the mammalian p53 tumor suppressor protein [171], is required in the nervous system, downstream or in parallel with EGF signaling [164]. Wounding causes cuticular and epidermal damage leading to the upregulation of *nlp* and *cnc* AMP genes. These AMPs promote stress-induced sleep in a manner independent of *nas-38* but dependent upon EGF signaling [72]. AMPs encoded by *nlp-29* signal to their receptor NPR-12 [172] in the nervous system which in turn promotes RIS activity (Fig. 3) [72]. Thus, different noxious stimuli activate distinct pathways which promote the canonical EGF/ALA/RIS mechanisms.

Neuropeptides Regulating Feeding Quiescence Quiescence of feeding during stress-induced sleep is a regulated behavioral state, not a developmental one [48]. Quiescence of feeding is almost entirely controlled by neuropeptides released from the ALA. Overexpression of *lin-3* inhibits pumping [19], and optogenetic depolarization of the ALA induces quiescence of feeding [32]. Also, *ceh-14* and *ceh-17* mutants are fully defective in feeding quiescence [38]. *flp-13* loss-of-function animals are partially defective in feeding quiescence and resistant to the feeding quiescence induced by optogenetic activation of the ALA. Additionally, *flp-13* overexpression induces prolonged quiescence of feeding [32]. In fact, the direct application of several *flp*-encoded neuropeptides to the pharynx, including *flp-13*, inhibits pumping [173]. Overexpression of *nlp-8*, *nlp-14*, or *flp-24* does not induce quiescence of feeding; however, *flp-13;flp-24* and *nlp-8;flp-13* double mutants display less feeding quiescence than *flp-13* single mutants. Thus, several peptides act redundantly with *flp-13* (Fig. 3).

FLP-13 neuropeptides act by inhibiting pharyngeal, cholinergic motor neurons, which consist of the paired MC and M2 neurons, and the single M4 [174]. Outside of stress-induced sleep, optogenetic activation of MC, M2, or M4 increases pumping [175], with MC being the only neuron required for rapid pumping [174]. Optogenetic activation of MC, M2, or M4 stimulates pumping following *flp-13* overexpression, and activation of all cholinergic neurons stimulates feeding during stress-induced sleep [48]. Additionally, serotonin administration increases pumping during stress-induced sleep and in animals overexpressing *flp-13* [48]. Serotonin stimulates pharyngeal neurons by acting on the SER-7 receptor and activating the

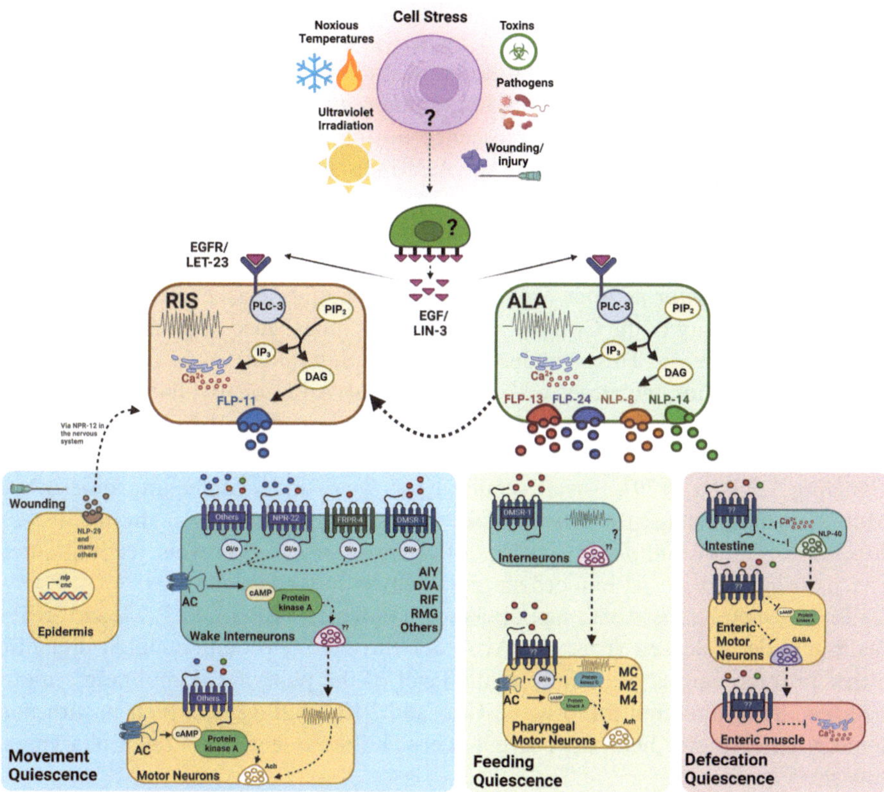

Fig. 3 A molecular model of stress-induced sleep. Cell damage as a result of noxious insults is probably sensed in different locations based on the stimulus. This causes LIN-3 and other EGF peptides to be released from unknown cells, which activate the LET-23 receptors on the RIS and ALA neurons. This leads to the activation of the IP_3/DAG pathways, a change in membrane potential, and the release of numerous neuropeptides. The ALA enhances the activity of the RIS through an unknown mechanism. *Movement quiescence*: FLP-11, FLP-13, FLP-24, NLP-14, and NLP-8 peptides signal through multiple GPCRs on interneurons (FLP-13→DMSR-1 and FRPR-4; FLP-11→NPR-22). Some evidence suggests that they signal directly to motor neurons. Inhibitory G-proteins inactivate adenylyl cyclase (AC) which reduces intracellular cAMP, possibly hyperpolarizing wake-promoting neurons. Following wounding, the expression of antimicrobial peptide genes, like *nlp-29*, are upregulated in the epidermis. NLP-29 peptides activate the NPR-12 GPCR in the nervous system, indirectly promoting RIS activation. *Feeding quiescence:* FLP-13 neuropeptides signal through DMSR-1 on wake-promoting interneurons, which indirectly reduce pharyngeal motor neuron activity. FLP-24 and NLP-8 peptides act redundantly with FLP-13. Ultimately, the cAMP/PKA pathways are inhibited in motor neurons. *Defecation quiescence:* NLP-8, NLP-14, and FLP-13 regulate quiescence of defecation; however, the specific sites of action are unknown. Possibilities include inhibiting NLP-40 release from the intestine, inactivation of enteric motor neurons, or inactivating enteric muscles. (Created with BioRender.com)

Gαs signaling pathway [176]. Taken together, neuropeptides from the ALA likely inhibit feeding by modulating the activity of the pharyngeal motor neurons.

How do the ALA-released peptides control the motor neurons? Feeding quiescence induced by *flp-13* overexpression requires *dmsr-1* but not *frpr-4* [167, 169]. *dmsr-1* is required for feeding quiescence during stress-induced sleep; however, while DMSR-1 acts in part through the AIY interneuron to regulate quiescence of movement, the site of action has not been identified for feeding quiescence. So, it is still unclear how DMSR-1 signaling is mechanistically connected to the modulation of the cholinergic motor neurons. However, genes known to either be expressed or function in the motor neurons are required for feeding quiescence. Mutations that increase the activity of the GαS pathway suppress quiescence of feeding induced by *flp-13* overexpression. This includes gain-of-function mutations of *gsa-1* and *acy-1* and loss-of-function mutations of *kin-2* and *pde-4*. Additionally, the gap junctional component UNC-7 [121], the voltage-gated calcium channel EGL-19 [177], the protein kinase C TPA-1, the G-protein GPA-12 [178], and the regulator of G-protein signaling EAT-16 [179] are all required for *flp-13*-induced feeding quiescence. Mutations that increase the activity of the Gαq signaling pathway, including loss-of-function of *goa-1* and *dkg-1* and loss-of-function of *egl-4*, prevent feeding quiescence during both *flp-13*-induced sleep and stress-induced sleep [48].

Taken together, these data support a model in which FLP-13, FLP-24, and NLP-8 neuropeptides released from the ALA act through G-protein-coupled signaling either in interneurons or directly at the level of the pharyngeal cholinergic motor neurons. This signaling inhibits the Gαq and PKG pathways, which in turn may hyperpolarize motor neurons, prevent acetylcholine release, and relax pharyngeal muscle (Fig. 3).

Neuropeptides Regulating Quiescence of Defecation The defecation cycle is executed every 50–60 seconds and consists of the following behavioral programs: a posterior body contraction (pBoc), followed by an anterior body contraction (aBoc) and an expulsion (Exp) [10]. These events are timed by the rhythmic release of NLP-40 neuropeptides from the intestine which activates excitatory GABAergic neurons, via GαS signaling. GABA stimulates the enteric muscles which execute the expulsion step [180, 181]. Exp events are completely and partially inhibited during heat- and UV-induced stress-induced sleep, respectively [38, 156]. Overexpression of *lin-3* fully inhibits defecation [156]. Loss-of-function of *ceh-17* or *ceh-14* suppresses quiescence of defecation during heat-induced stress-induced sleep but not to the levels seen in wild-type awake animals. These data suggest that the ALA neuron is required for quiescence of defecation, but other ALA-independent pathways probably exist [38]. *flp-13;flp-24;nlp-8* triple mutants display mild defects in defecation quiescence [38], while *nlp-14* mutants display stronger defects [156]. Thus, the ALA neuron and multiple neuropeptide genes are required for defecation quiescence.

Overexpression of *flp-13*, *nlp-8*, or *nlp-14* induces quiescence of defecation outside of stress-induced sleep [38, 156]. NLP-40 peptides stimulate cAMP production and intracellular calcium release in GABAergic neurons via the GPCR AEX-2 [180]. Optogenetic stimulation of cAMP in all neurons does not suppress feeding quiescence during UV-induced stress-induced sleep, and *kin-2* mutants show mild defects in defecation quiescence [107]. However, *kin-2* mutations do not suppress quiescence of defecation induced by *nlp-14* overexpression [156]. These data suggest that the ALA neuropeptides act downstream of cAMP signaling; the specific site of action and mechanisms are unknown.

Satiety-Induced Quiescence

Satiety-induced quiescence shares qualities with satiety in mammals [29]. Satiety follows a typical behavioral sequence consisting of a gradual reduction in feeding and movement, followed by complete cessation, and postprandial sleep [182]. *C. elegans* displays quiescence of movement and feeding when grown on high-quality food like the *E. coli* strain HB101 and in response to refeeding following starvation [183]. Satiety-induced quiescence is dependent upon the nutritional state of the animal, considering quiescence is reduced in *eat-2* loss-of-function animals that feed less effectively [9] and *act-5* loss-of-function animals that have impaired intestinal absorption [184]. When fasted and then refed the *E. coli* strains DA837 or HB101, worms undergo the satiety sequence, with the higher-quality HB101 causing a more pronounced response. This begins with vigorous feeding, evident by an increased pumping rate, followed by a gradual reduction throughout the duration of the refeeding event. Ultimately, animals cease movement and feeding, entering sleep bouts of varying length and quantity [29]. Thus, satiety-induced quiescence in *C. elegans* resembles satiety of mammals.

Satiety-induced quiescence shares some mechanisms with developmentally timed sleep and stress-induced sleep, but is initiated by nutritional cues. Quiescence begins following the activation of the chemosensory ASI neuron, which releases TGF-β/DAF-7 ligands [185]. DAF-7 promotes satiety-induced quiescence by acting through the DAF-1 receptor, together with insulin via the DAF-2 receptor [29]. Satiety-induced quiescence is reversible; the nociceptive ASH neurons directly antagonize ASI, allowing animals to wake and escape [186]. During favorable conditions, DAF-7/DAF-1 signal through two neurons, RIM and RIC [187]. *daf-1* is required for quiescence; expression of *daf-1* in RIM and RIC rescues the *daf-1* loss-of-function quiescence phenotype [187]. Ablation of RIM and RIC in *daf-1* loss-of-function animals restores quiescence, providing evidence that DAF-7 inactivates RIM and RIC. Downstream of DAF-1, components of the canonical TGF-β signaling pathway, including the DAF-8 and DAF-14 SMAD transcription factors, regulate RIM and RIC activity [187]. How insulin and TGF-β signaling is coordinated is unclear.

ASI/RIM/RIC connect with known sleep circuitry through the RIS neuron. *flp-11* is required for satiety-induced quiescence, and epistasis analyses place *flp-11* downstream of *daf-7*. Additionally, *egl-4* is required downstream of the RIS [188]. This supports a model in which nutritional cues activate the ASI causing the release of DAF-7, which inactivates RIM and RIC through DAF-1, indirectly leading to FLP-11 release from RIS, which promotes sleep in an EGL-4-dependent manner.

How nutritional status is mechanistically connected with satiety-induced quiescence is not fully understood. However, insulin signaling and fat storage play important roles. The insulin receptor DAF-2 is required for quiescence, showing that insulin is a nutritional signal promoting sleep [29]. However, knockdown of the insulin gene *ins-1* enhances quiescence [188, 189]; INS-1 peptides antagonize DAF-2 [189]. In fact, multiple *ins* genes antagonize satiety-induced quiescence, while others enhance it, suggesting a complex mechanism underlies quiescence and foraging. Also, multiple components of fat biosynthetic pathways are required for satiety-induced quiescence, such as *sbp-1*, which encodes a homolog of the sterol regulatory element-binding protein-1C, *pod-2*, which encodes a homolog of acetyl-CoA carboxylase, *fat-6*, and *fat-7*, which encode stearoyl-CoA desaturases. Providing animals with an exogenous fat source, oleic acid, rescues the quiescence defects [190]. It has been proposed that a "satiety quotient" is set in intestinal cells based on fat levels and food availability. Loss-of-function of *ets-5*, which encodes a homolog of the FEV/Pet transcription factors, causes an increase in intestinal fat which enhances quiescence. ETS-5 function was mapped to the BAG neurons, suggesting it functions cell non-autonomously. The BAG neurons release neuropeptides which regulate behavioral state switching based on food availability and quality and intestinal fat levels [191]. Thus, insulin signaling and fat storage likely regulate the transition from foraging to quiescence.

Other Sleep-Like States of *C. elegansCaenorhabditis elegans*

Episodic Swimming Quiescence Behavioral state switching is a common occurrence in swimming animals [30]. Suspension in liquid causes animals to swim with sinusoidal thrashing movements. Following an initial bout of prolonged swimming, animals straighten their body posture and enter regular quiescent bouts of variable length, a behavior called episodic swimming quiescence. The length of the initial swim is dependent upon food availability; when food is abundant, the initial swim spans ~70 minutes, followed by regular intervals of quiescence, which last ~9 minutes. When food is unavailable, the initial swim is longer (~90 minutes) and followed by shorter ~6-minute quiescent bouts [30]. In a study that monitored animals for 4 hours, regular bouts of episodic swimming quiescence were separated by ~15 minutes of swimming [30]. A study that followed animals for 16 hours found a consistent pattern up to 10 hours. Following the 10-hour period, the length of active bouts remained relatively unchanged; however, quiescent bouts lengthened ~fourfold [192]. Episodic swimming quiescence is reversible upon mechanical stimulation, which causes animals to swim for a length of time similar to what would have occurred spontaneously. These observations suggest that episodic swimming quiescence is a controlled neural state that is not explained as simple fatigue [30].

However, loss-of-function of *daf-16* decreases the initial swim, suggesting that stress adaptation and/or fatigue may contribute to episodic swimming quiescence [193]. Despite not knowing the exact cause or function of this quiescent state, it is clear that it differs from other sleep states. For example, quiescence of feeding is not a prominent feature of episodic swimming quiescence. During quiescent bouts, animals continue to feed while remaining immobile. If pharyngeal pumping does cease, it only stops at the beginning of a quiescent bout and does not span the entire length [193]. Thus, episodic swimming quiescence is a unique quiescent state of *C. elegans*.

The mechanisms regulating episodic swimming quiescence are distinct from other sleep states as well but do share some common genes. Loss-of-function of *ceh-17* or *aptf-1* does not alter the pattern of episodic swimming quiescence seen in wild-type animals [193]. Thus, sleep-promoting neuropeptides are not required for episodic swimming quiescence. Loss-of-function of *unc-25*, which is required for GABA synthesis [120], increases the length of quiescent bouts, and treatment with the GABA agonist muscimol causes premature termination of quiescence. Thus, GABA promotes swimming and antagonizes quiescence. Cholinergic signaling has the opposite effect; treatment with the acetylcholinesterase inhibitor aldicarb [194] increases quiescence and decreases swimming [30]. *ace-1;ace-2* double mutants, which have severely impaired acetylcholinesterase function [195], and *dgk-1* loss-of-function animals, which have enhanced acetylcholine release [196], display decreased swimming and enhanced quiescence. Acetylcholine but not GABA is required for the stereotypic posture of episodic swimming quiescence [30]. Taken together, these data suggest that GABAergic signaling promotes swimming [30] and cholinergic signaling promotes quiescence.

Like the other sleep states previously discussed, *egl-4* is required for episodic swimming quiescence. Loss-of-function of *egl-4* reduces quiescence [30, 197] and suppresses the enhanced quiescence induced by aldicarb [197]. *egl-4* is not required for the increased quiescence observed in *unc-25* mutants. This supports a model in which cholinergic motor neurons function upstream of *egl-4*, which in turn function upstream of GABAergic signaling [197]. Rescue experiments indicate that *egl-4* functions in *tax-4*-expressing sensory neurons, similar to developmentally timed sleep [18, 197]. Additionally, *osm-3* and *osm-6* mutants, who have defective sensory function [198], show enhanced quiescence dependent on *egl-4* [197]. This further supports EGL-4 functioning in sensory neurons.

Calcineurin signaling is required for developmentally timed sleep; however, *tax-6* loss-of-function and *cnb-1* loss-of-function animals display pronounced enhancement of quiescence. Quiescence is restored to wild-type levels when *tax-6* is rescued in sensory neurons. Epistasis experiments place *tax-6* upstream of *egl-4* [197]. Taken together, episodic swimming quiescence is regulated by cholinergic motor neurons that function upstream of calcineurin and EGL-4 signaling in sensory neurons. This quiescence-promoting circuitry functions to antagonize GABAergic signaling which normally promotes swimming.

Spontaneous Adult Sleep Adult animals display circadian rhythms of locomotory activity following light-dark entrainment [199]. However, prominent circadian sleep, like that of mammals and insects [25], does not take place in *C. elegans*. Foraging is the dominant behavior of adult animals. Animals transition between roaming and dwelling foraging states [200]. A careful analysis of freely moving adults in microfabricated behavioral arenas observed regular transitions from active foraging states to quiescent bouts that spanned seconds to minutes, a behavior called spontaneous adult sleep [45]. A transition to this sleep state involves a slowing of locomotion followed by a complete stop accompanied with a relaxed body posture. Arousal occurs spontaneously when animals resume forward locomotion. Spontaneous adult sleep fulfills some criteria of sleep including a stereotypic posture, reversibility, a heightened arousal threshold, and homeostasis. Moreover, spontaneous adult sleep levels are plastic; sleep levels are reduced in the presence of environmental stimuli, including high oxygen, fluid flow, food, and odorants. Because of this, it is difficult to determine how spontaneous adult sleep occurs in the natural environment. However, it can be assumed that in the absence of sensory input, spontaneous adult sleep likely takes place allowing for periods of rest between episodes of foraging.

Spontaneous adult sleep correlates with RIS activation but does not require the ALA neuron; in fact, *ceh-17* loss-of-function animals have elevated sleep. Thus, some mechanisms of spontaneous adult sleep are shared with other sleep states. However, some unique aspects of sensory arousal characterize spontaneous adult sleep. Calcium imaging showed that sensory neurons are as responsive as those of awake worms. Interneurons, however, are slow to respond to stimuli. This localizes the sensory modulation of spontaneous adult sleep to synaptic transmission of sensory neurons or interneuron responsiveness [45]. The specific mechanisms governing sleep and sensory arousal of spontaneous adult sleep are yet to be determined.

Microfluidic-Induced Sleep Microfabricated devices of differing sizes have become powerful tools for measuring sleep [99, 115, 201, 202] and correlating it with neural activity [39, 76, 134]. A unique sleep behavior occurs when animals are confined in microfluidic devices [44]. This microfluidic-induced sleep is characterized by frequent, short bouts of quiescence and fulfills some characteristics of a sleep state [12], including rapid reversibility, heightened sensory arousal, and homeostasis. When animals transition into sleep under these conditions, they display a global brain state akin to what occurs during developmentally timed sleep [76], which includes the activation of the RIS neuron. *ceh-17* and *aptf-1* are required for this sleep state, implicating the ALA and RIS neurons. Thus, microfluidic-induced sleep is more like stress-induced sleep than episodic swimming quiescence.

Microfluidic-induced sleep is highly plastic; starved animals display prominent quiescence, while well-fed worms do not. Higher temperatures (25 °C) increase quiescence, and lower temperatures (18 °C) completely abolish it. Increased quiescence under warmer conditions requires a functioning thermosensory AFD neuron [44, 203]. Additionally, microfluidic-induced sleep occurs more robustly in small microfluidic devices where animals are partially compressed. This form of quiescence requires the gene *mec-10* which is essential for mechanosensation [6]. Thus, microfluidic-induced sleep is a sleep state that shares mechanisms and qualities with stress-induced sleep; however, the function of this sleep state is still poorly understood.

Post-response Quiescence Stress modulates an animal's ability to respond appropriately to sensory information. During UV-induced stress-induced sleep, sensory arousal to mechanical stimuli is dampened [46, 164]. Regular and repeated vibrations from an audio loudspeaker can arouse animals during UV-induced stress-induced sleep. Upon waking, instead of displaying a stereotypic sensory response of prolonged movement, animals enter a brief period of quiescence lasting seconds. This form of quiescence was termed post-response quiescence and was only observed during UV-induced stress-induced sleep and not sleep in response to other stressors. Non-sleeping worms did not display post-response quiescence. This form of quiescence is enhanced following *lin-3* overexpression and requires functioning mechanosensory neurons, as *mec-4* mutants do not display it. Post-response quiescence is dependent on the ALA and RIS neurons and controlled by neuropeptide signaling. While post-response quiescence shares some common pathways with other forms of sleep, there are some key differences. Depriving animals of stress-induced sleep does not alter post-response quiescence levels, suggesting that it is disconnected from the homeostasis of stress-induced sleep. In fact, post-response quiescence itself is not regulated homeostatically, and there is no evidence of micro-homeostasis. Also, post-response quiescence is not characterized by a heightened arousal threshold, another hallmark of sleep. Taken together, post-response quiescence is a unique sleep-like state of *C. elegans*; however, its function is not understood.

Closing Remarks

With the description of lethargus as a sleep-like state [18] and the discovery of multiple sleep-like behaviors [30, 31, 44, 46, 187], *C. elegans* has emerged as a powerful genetic model system for sleep research. Transcriptomics, genetic screens, and candidate-gene approaches have amassed greater than 100 genes that have been experimentally validated as bona fide sleep genes, with likely many more yet to be discovered. The molecules they encode regulate the timing, amount, and duration of sleep, sensory arousal, and homeostasis, and most of them have homologs in mammals. The simple, completely mapped, nervous system of the worm combined with in vivo calcium imaging and the ease of genetic manipulation has allowed for huge steps forward in a short period of time. A challenge, but an important goal, will be to determine the specific temporal-spatial interactions of these molecules within this simple nervous system. Because of the conserved nature of these molecules, this will provide valuable insight into sleep in animals with more complex nervous systems.

References

1. Corsi AK, Wightman B, Chalfie M. A transparent window into biology: a primer on Caenorhabditis elegans. Genetics. 2015;200:387–407. https://doi.org/10.1534/genetics.115.176099.
2. Brenner S. The genetics of Caenorhabditis elegans. Genetics. 1974;77:71–94. https://doi.org/10.1093/genetics/77.1.71.
3. White JG, Southgate E, Thomson JN, Brenner S. The structure of the nervous system of the nematode Caenorhabditis elegans. Philos Trans R Soc Lond Ser B Biol Sci. 1986;314:1–340. https://doi.org/10.1098/rstb.1986.0056.
4. C. elegans Sequencing Consortium. Genome sequence of the nematode C. elegans: a platform for investigating biology. Science. 1998;282:2012–8. https://doi.org/10.1126/science.282.5396.2012.
5. Trent C, Tsuing N, Horvitz HR. Egg-laying defective mutants of the nematode Caenorhabditis elegans. Genetics. 1983;104:619–47. https://doi.org/10.1093/genetics/104.4.619.
6. Chalfie M, Sulston J. Developmental genetics of the mechanosensory neurons of Caenorhabditis elegans. Dev Biol. 1981;82:358–70. https://doi.org/10.1016/0012-1606(81)90459-0.
7. Lewis JA, Hodgkin JA. Specific neuroanatomical changes in chemosensory mutants of the nematode Caenorhabditis elegans. J Comp Neurol. 1977;172:489–510. https://doi.org/10.1002/cne.901720306.
8. Rankin CH, Beck CD, Chiba CM. Caenorhabditis elegans: a new model system for the study of learning and memory. Behav Brain Res. 1990;37:89–92. https://doi.org/10.1016/0166-4328(90)90074-o.
9. Avery L. The genetics of feeding in Caenorhabditis elegans. Genetics. 1993;133:897–917. https://doi.org/10.1093/genetics/133.4.897.
10. Thomas JH. Genetic analysis of defecation in Caenorhabditis elegans. Genetics. 1990;124:855–72. https://doi.org/10.1093/genetics/124.4.855.
11. Bargmann CI. Neurobiology of the Caenorhabditis elegans genome. Science. 1998;282:2028–33. https://doi.org/10.1126/science.282.5396.2028.
12. Campbell SS, Tobler I. Animal sleep: a review of sleep duration across phylogeny. Neurosci Biobehav Rev. 1984;8:269–300. https://doi.org/10.1016/0149-7634(84)90054-x.
13. Konopka RJ, Benzer S. Clock mutants of Drosophila melanogaster. Proc Natl Acad Sci USA. 1971;68:2112–6. https://doi.org/10.1073/pnas.68.9.2112.
14. Hendricks JC, et al. Rest in Drosophila is a sleep-like state. Neuron. 2000;25:129–38. https://doi.org/10.1016/s0896-6273(00)80877-6.
15. Shaw PJ, Cirelli C, Greenspan RJ, Tononi G. Correlates of sleep and waking in Drosophila melanogaster. Science. 2000;287:1834–7. https://doi.org/10.1126/science.287.5459.1834.
16. Zhdanova IV, Wang SY, Leclair OU, Danilova NP. Melatonin promotes sleep-like state in zebrafish. Brain Res. 2001;903:263–8. https://doi.org/10.1016/s0006-8993(01)02444-1.
17. Singh RN, Sulston JE. Some Observations On Moulting in Caenorhabditis Elegans. Nematologica. 1978;24:63–71. https://doi.org/10.1163/187529278X00074.
18. Raizen DM, et al. Lethargus is a Caenorhabditis elegans sleep-like state. Nature. 2008;451:569–72. https://doi.org/10.1038/nature06535.
19. Van Buskirk C, Sternberg PW. Epidermal growth factor signaling induces behavioral quiescence in Caenorhabditis elegans. Nat Neurosci. 2007;10:1300–7. https://doi.org/10.1038/nn1981.
20. Foltenyi K, Greenspan RJ, Newport JW. Activation of EGFR and ERK by rhomboid signaling regulates the consolidation and maintenance of sleep in Drosophila. Nat Neurosci. 2007;10:1160–7. https://doi.org/10.1038/nn1957.
21. Lee DA, et al. Evolutionarily conserved regulation of sleep by epidermal growth factor receptor signaling. Sci Adv. 2019;5:eaax4249. https://doi.org/10.1126/sciadv.aax4249.

22. Kushikata T, Fang J, Chen Z, Wang Y, Krueger JM. Epidermal growth factor enhances spontaneous sleep in rabbits. Am J Phys. 1998;275:R509–14. https://doi.org/10.1152/ajpregu.1998.275.2.R509.
23. Kramer A, et al. Regulation of daily locomotor activity and sleep by hypothalamic EGF receptor signaling. Science. 2001;294:2511–5. https://doi.org/10.1126/science.1067716.
24. Konietzka J, et al. Epidermal growth factor signaling promotes sleep through a combined series and parallel neural circuit. Curr Biol. 2020;30:1–16.e13. https://doi.org/10.1016/j.cub.2019.10.048.
25. Crocker A, Sehgal A. Genetic analysis of sleep. Genes Dev. 2010;24:1220–35. https://doi.org/10.1101/gad.1913110.
26. Monsalve GC, Van Buskirk C, Frand AR. LIN-42/PERIOD controls cyclical and developmental progression of C. elegans molts. Curr Biol. 2011;21:2033–45. https://doi.org/10.1016/j.cub.2011.10.054.
27. Jeon M, Gardner HF, Miller EA, Deshler J, Rougvie AE. Similarity of the C. elegans developmental timing protein LIN-42 to circadian rhythm proteins. Science. 1999;286:1141–6. https://doi.org/10.1126/science.286.5442.1141.
28. Siwicki KK, Eastman C, Petersen G, Rosbash M, Hall JC. Antibodies to the period gene product of Drosophila reveal diverse tissue distribution and rhythmic changes in the visual system. Neuron. 1988;1:141–50. https://doi.org/10.1016/0896-6273(88)90198-5.
29. You YJ, Kim J, Raizen DM, Avery L. Insulin, cGMP, and TGF-beta signals regulate food intake and quiescence in C. elegans: a model for satiety. Cell Metab. 2008;7:249–57. https://doi.org/10.1016/j.cmet.2008.01.005.
30. Ghosh R, Emmons SW. Episodic swimming behavior in the nematode C. elegans. J Exp Biol. 2008;211:3703–11. https://doi.org/10.1242/jeb.023606.
31. Hill AJ, Mansfield R, Lopez JM, Raizen DM, Van Buskirk C. Cellular stress induces a protective sleep-like state in C. elegans. Curr Biol. 2014;24:2399–405. https://doi.org/10.1016/j.cub.2014.08.040.
32. Nelson MD, et al. FMRFamide-like FLP-13 neuropeptides promote quiescence following heat stress in Caenorhabditis elegans. Curr Biol. 2014;24:2406–10. https://doi.org/10.1016/j.cub.2014.08.037.
33. Trojanowski NF, Raizen DM. Call it Worm Sleep. Trends Neurosci. 2016;39:54–62. https://doi.org/10.1016/j.tins.2015.12.005.
34. Davis KC, Raizen DM. A mechanism for sickness sleep: lessons from invertebrates. J Physiol. 2017;595:5415–24. https://doi.org/10.1113/JP273009.
35. Li C, Nelson LS, Kim K, Nathoo A, Hart AC. Neuropeptide gene families in the nematode Caenorhabditis elegans. Ann N Y Acad Sci. 1999;897:239–52. https://doi.org/10.1111/j.1749-6632.1999.tb07895.x.
36. Lenz O, Xiong J, Nelson MD, Raizen DM, Williams JA. FMRFamide signaling promotes stress-induced sleep in Drosophila. Brain Behav Immun. 2015;47:141–8. https://doi.org/10.1016/j.bbi.2014.12.028.
37. Lee DA, et al. Genetic and neuronal regulation of sleep by neuropeptide VF. elife. 2017;6:e25727. https://doi.org/10.7554/eLife.25727.
38. Nath RD, Chow ES, Wang H, Schwarz EM, Sternberg PW. C. elegans stress-induced sleep emerges from the collective action of multiple neuropeptides. Curr Biol. 2016;26:2446–55. https://doi.org/10.1016/j.cub.2016.07.048.
39. Turek M, Lewandrowski I, Bringmann H. An AP2 transcription factor is required for a sleep-active neuron to induce sleep-like quiescence in C. elegans. Curr Biol. 2013;23:2215–23. https://doi.org/10.1016/j.cub.2013.09.028.
40. Nagel G, et al. Channelrhodopsin-2, a directly light-gated cation-selective membrane channel. Proc Natl Acad Sci USA. 2003;100:13940–5. https://doi.org/10.1073/pnas.1936192100.
41. Kerr RA. Imaging the activity of neurons and muscles. WormBook. 2006:1–13. https://doi.org/10.1895/wormbook.1.113.1.

42. Hsu PD, Lander ES, Zhang F. Development and applications of CRISPR-Cas9 for genome engineering. Cell. 2014;157:1262–78. https://doi.org/10.1016/j.cell.2014.05.010.
43. Jinek M, et al. A programmable dual-RNA-guided DNA endonuclease in adaptive bacterial immunity. Science. 2012;337:816–21. https://doi.org/10.1126/science.1225829.
44. Gonzales DL, Zhou J, Fan B, Robinson JT. A microfluidic-induced C. elegans sleep state. Nat Commun. 2019;10:5035. https://doi.org/10.1038/s41467-019-13008-5.
45. Lawler DE, et al. Sleep analysis in adult C. elegans reveals state-dependent alteration of neural and behavioral responses. J Neurosci. 2021;41:1892–907. https://doi.org/10.1523/JNEUROSCI.1701-20.2020.
46. McClanahan PD, et al. A quiescent state following mild sensory arousal in Caenorhabditis elegans is potentiated by stress. Sci Rep. 2020;10:4140. https://doi.org/10.1038/s41598-020-60994-4.
47. Telford MJ, Bourlat SJ, Economou A, Papillon D, Rota-Stabelli O. The evolution of the Ecdysozoa. Philos Trans R Soc Lond Ser B Biol Sci. 2008;363:1529–37. https://doi.org/10.1098/rstb.2007.2243.
48. Trojanowski NF, Nelson MD, Flavell SW, Fang-Yen C, Raizen DM. Distinct mechanisms underlie quiescence during two Caenorhabditis elegans sleep-like states. J Neurosci. 2015;35:14571–84. https://doi.org/10.1523/JNEUROSCI.1369-15.2015.
49. Schwarz J, Spies JP, Bringmann H. Reduced muscle contraction and a relaxed posture during sleep-like Lethargus. Worm. 2012;1:12–4. https://doi.org/10.4161/worm.19499.
50. Tramm N, Oppenheimer N, Nagy S, Efrati E, Biron D. Why do sleeping nematodes adopt a hockey-stick-like posture? PLoS One. 2014;9:e101162. https://doi.org/10.1371/journal.pone.0101162.
51. Nagy S, et al. Homeostasis in C. elegans sleep is characterized by two behaviorally and genetically distinct mechanisms. elife. 2014;3:e04380. https://doi.org/10.7554/eLife.04380.
52. Driver RJ, Lamb AL, Wyner AJ, Raizen DM. DAF-16/FOXO regulates homeostasis of essential sleep-like behavior during larval transitions in C. elegans. Curr Biol. 2013;23:501–6. https://doi.org/10.1016/j.cub.2013.02.009.
53. Byerly L, Cassada RC, Russell RL. The life cycle of the nematode Caenorhabditis elegans. I. Wild-type growth and reproduction. Dev Biol. 1976;51:23–33. https://doi.org/10.1016/0012-1606(76)90119-6.
54. Myers MP, Wager-Smith K, Wesley CS, Young MW, Sehgal A. Positional cloning and sequence analysis of the Drosophila clock gene, timeless. Science. 1995;270:805–8. https://doi.org/10.1126/science.270.5237.805.
55. Kloss B, et al. The Drosophila clock gene double-time encodes a protein closely related to human casein kinase Iepsilon. Cell. 1998;94:97–107. https://doi.org/10.1016/s0092-8674(00)81225-8.
56. Banerjee D, Kwok A, Lin SY, Slack FJ. Developmental timing in C. elegans is regulated by kin-20 and tim-1, homologs of core circadian clock genes. Dev Cell. 2005;8:287–95. https://doi.org/10.1016/j.devcel.2004.12.006.
57. Reinecke J, Buckner JS, Grugel SR. Life cycle of laboratory-reared tobacco hornworms, *Manduca sexta*, a study of development and behavior, using time-lapse cinematography. Biol Bull. 1980;158:129. https://doi.org/10.2307/1540764.
58. Riddiford LM, Cherbas P, Truman JW. Ecdysone receptors and their biological actions. Vitam Horm. 2000;60:1–73. https://doi.org/10.1016/s0083-6729(00)60016-x.
59. Yao TP, et al. Functional ecdysone receptor is the product of EcR and Ultraspiracle genes. Nature. 1993;366:476–9. https://doi.org/10.1038/366476a0.
60. Gissendanner CR, Sluder AE. nhr-25, the Caenorhabditis elegans ortholog of ftz-f1, is required for epidermal and somatic gonad development. Dev Biol. 2000;221:259–72. https://doi.org/10.1006/dbio.2000.9679.
61. Kostrouchova M, Krause M, Kostrouch Z, Rall JE. Nuclear hormone receptor CHR3 is a critical regulator of all four larval molts of the nematode Caenorhabditis elegans. Proc Natl Acad Sci USA. 2001;98:7360–5. https://doi.org/10.1073/pnas.131171898.

62. Asahina M, et al. The conserved nuclear receptor Ftz-F1 is required for embryogenesis, moulting and reproduction in Caenorhabditis elegans. Genes Cells. 2000;5:711–23. https://doi.org/10.1046/j.1365-2443.2000.00361.x.
63. Gissendanner CR, Crossgrove K, Kraus KA, Maina CV, Sluder AE. Expression and function of conserved nuclear receptor genes in Caenorhabditis elegans. Dev Biol. 2004;266:399–416. https://doi.org/10.1016/j.ydbio.2003.10.014.
64. Akashi M, Takumi T. The orphan nuclear receptor RORalpha regulates circadian transcription of the mammalian core-clock Bmal1. Nat Struct Mol Biol. 2005;12:441–8. https://doi.org/10.1038/nsmb925.
65. Ambros V, Horvitz HR. Heterochronic mutants of the nematode Caenorhabditis elegans. Science. 1984;226:409–16. https://doi.org/10.1126/science.6494891.
66. Reinhart BJ, et al. The 21-nucleotide let-7 RNA regulates developmental timing in Caenorhabditis elegans. Nature. 2000;403:901–6. https://doi.org/10.1038/35002607.
67. Ambros V, Ruvkun G. Recent Molecular Genetic Explorations of Caenorhabditis elegans MicroRNAs. Genetics. 2018;209:651–73. https://doi.org/10.1534/genetics.118.300291.
68. Patel R, Galagali H, Kim JK, Frand AR. Feedback between a retinoid-related nuclear receptor and the let-7 microRNAs controls the pace and number of molting cycles in C. elegans. elife. 2022;11:e80010. https://doi.org/10.7554/eLife.80010.
69. George-Raizen JB, Shockley KR, Trojanowski NF, Lamb AL, Raizen DM. Dynamically-expressed prion-like proteins form a cuticle in the pharynx of Caenorhabditis elegans. Biol Open. 2014;3:1139–49. https://doi.org/10.1242/bio.20147500.
70. Turek M, Bringmann H. Gene expression changes of Caenorhabditis elegans larvae during molting and sleep-like lethargus. PLoS One. 2014;9:e113269. https://doi.org/10.1371/journal.pone.0113269.
71. Singh K, Ju JY, Walsh MB, DiIorio MA, Hart AC. Deep conservation of genes required for both Drosophila melanogaster and Caenorhabditis elegans sleep includes a role for dopaminergic signaling. Sleep. 2014;37:1439–51. https://doi.org/10.5665/sleep.3990.
72. Sinner MP, Masurat F, Ewbank JJ, Pujol N, Bringmann H. Innate Immunity Promotes Sleep through Epidermal Antimicrobial Peptides. Curr Biol. 2021;31:564–577.e12. https://doi.org/10.1016/j.cub.2020.10.076.
73. Hendriks GJ, Gaidatzis D, Aeschimann F, Grosshans H. Extensive oscillatory gene expression during C. elegans larval development. Mol Cell. 2014;53:380–92. https://doi.org/10.1016/j.molcel.2013.12.013.
74. Cirelli C, et al. Reduced sleep in Drosophila Shaker mutants. Nature. 2005;434:1087–92. https://doi.org/10.1038/nature03486.
75. Tomita J, et al. Pan-neuronal knockdown of calcineurin reduces sleep in the fruit fly, Drosophila melanogaster. J Neurosci. 2011;31:13137–46. https://doi.org/10.1523/JNEUROSCI.5860-10.2011.
76. Nichols ALA, Eichler T, Latham R, Zimmer M. A global brain state underlies C. elegans sleep behavior. Science. 2017;356:eaam6851. https://doi.org/10.1126/science.aam6851.
77. Bringmann H. Sleep-active neurons: conserved motors of sleep. Genetics. 2018;208:1279–89. https://doi.org/10.1534/genetics.117.300521.
78. Saper CB, Scammell TE, Lu J. Hypothalamic regulation of sleep and circadian rhythms. Nature. 2005;437:1257–63. https://doi.org/10.1038/nature04284.
79. Steuer Costa W, et al. A GABAergic and peptidergic sleep neuron as a locomotion stop neuron with compartmentalized Ca2+ dynamics. Nat Commun. 2019;10:4095. https://doi.org/10.1038/s41467-019-12098-5.
80. Robinson B, Goetting D, Cisneros Desir J, Van Buskirk C. aptf-1 mutants are primarily defective in head movement quiescence during C. elegans sleep. MicroPubl Biol. 2019;2019. https://doi.org/10.17912/micropub.biology.000148.
81. Turek M, Besseling J, Spies JP, Konig S, Bringmann H. Sleep-active neuron specification and sleep induction require FLP-11 neuropeptides to systemically induce sleep. elife. 2016;5:e12499. https://doi.org/10.7554/eLife.12499.

82. Frooninckx L, et al. Neuropeptide GPCRs in C. elegans. Front Endocrinol (Lausanne). 2012;3:167. https://doi.org/10.3389/fendo.2012.00167.
83. Nelson MD, et al. The neuropeptide NLP-22 regulates a sleep-like state in Caenorhabditis elegans. Nat Commun. 2013;4:2846. https://doi.org/10.1038/ncomms3846.
84. Hendricks M, Ha H, Maffey N, Zhang Y. Compartmentalized calcium dynamics in a C. elegans interneuron encode head movement. Nature. 2012;487:99–103. https://doi.org/10.1038/nature11081.
85. Van der Auwera P, et al. RPamide neuropeptides NLP-22 and NLP-2 act through GnRH-like receptors to promote sleep and wakefulness in C. elegans. Sci Rep. 2020;10:9929. https://doi.org/10.1038/s41598-020-66536-2.
86. Aroian RV, Koga M, Mendel JE, Ohshima Y, Sternberg PW. The let-23 gene necessary for Caenorhabditis elegans vulval induction encodes a tyrosine kinase of the EGF receptor subfamily. Nature. 1990;348:693–9. https://doi.org/10.1038/348693a0.
87. Pujol N, Torregrossa P, Ewbank JJ, Brunet JF. The homeodomain protein CePHOX2/CEH-17 controls antero-posterior axonal growth in C. elegans. Development. 2000;127:3361–71. https://doi.org/10.1242/dev.127.15.3361.
88. Hirose T, et al. Cyclic GMP-dependent protein kinase EGL-4 controls body size and lifespan in C elegans. Development. 2003;130:1089–99. https://doi.org/10.1242/dev.00330.
89. Yin X, Gower NJ, Baylis HA, Strange K. Inositol 1,4,5-trisphosphate signaling regulates rhythmic contractile activity of myoepithelial sheath cells in Caenorhabditis elegans. Mol Biol Cell. 2004;15:3938–49. https://doi.org/10.1091/mbc.e04-03-0198.
90. Maruyama IN, Brenner S. A phorbol ester/diacylglycerol-binding protein encoded by the unc-13 gene of Caenorhabditis elegans. Proc Natl Acad Sci USA. 1991;88:5729–33. https://doi.org/10.1073/pnas.88.13.5729.
91. Hosono R, et al. The unc-18 gene encodes a novel protein affecting the kinetics of acetylcholine metabolism in the nematode Caenorhabditis elegans. J Neurochem. 1992;58:1517–25. https://doi.org/10.1111/j.1471-4159.1992.tb11373.x.
92. Ogawa H, Harada S, Sassa T, Yamamoto H, Hosono R. Functional properties of the unc-64 gene encoding a Caenorhabditis elegans syntaxin. J Biol Chem. 1998;273:2192–8. https://doi.org/10.1074/jbc.273.4.2192.
93. Speese S, et al. UNC-31 (CAPS) is required for dense-core vesicle but not synaptic vesicle exocytosis in Caenorhabditis elegans. J Neurosci. 2007;27:6150–62. https://doi.org/10.1523/JNEUROSCI.1466-07.2007.
94. Treinin M, Chalfie M. A mutated acetylcholine receptor subunit causes neuronal degeneration in C. elegans. Neuron. 1995;14:871–7. https://doi.org/10.1016/0896-6273(95)90231-7.
95. Katz M, Corson F, Iwanir S, Biron D, Shaham S. Glia Modulate a Neuronal Circuit for Locomotion Suppression during Sleep in C. elegans. Cell Rep. 2018;22:2575–83. https://doi.org/10.1016/j.celrep.2018.02.036.
96. Miyazaki S, Kawano T, Yanagisawa M, Hayashi Y. Intracellular Ca(2+) dynamics in the ALA neuron reflect sleep pressure and regulate sleep in Caenorhabditis elegans. iScience. 2022;25:104452. https://doi.org/10.1016/j.isci.2022.104452.
97. Funato H, et al. Forward-genetics analysis of sleep in randomly mutagenized mice. Nature. 2016;539:378–83. https://doi.org/10.1038/nature20142.
98. Grubbs JJ, Lopes LE, van der Linden AM, Raizen DM. A salt-induced kinase is required for the metabolic regulation of sleep. PLoS Biol. 2020;18:e3000220. https://doi.org/10.1371/journal.pbio.3000220.
99. Belfer SJ, et al. Caenorhabditis-in-drop array for monitoring C. elegans quiescent behavior. Sleep. 2013;36:689–698G. https://doi.org/10.5665/sleep.2628.
100. Hendricks JC, et al. A non-circadian role for cAMP signaling and CREB activity in Drosophila rest homeostasis. Nat Neurosci. 2001;4:1108–15. https://doi.org/10.1038/nn743.
101. Graves LA, et al. Genetic evidence for a role of CREB in sustained cortical arousal. J Neurophysiol. 2003;90:1152–9. https://doi.org/10.1152/jn.00882.2002.

102. Korswagen HC, Park JH, Ohshima Y, Plasterk RH. An activating mutation in a Caenorhabditis elegans Gs protein induces neural degeneration. Genes Dev. 1997;11:1493–503. https://doi.org/10.1101/gad.11.12.1493.
103. Berger AJ, Hart AC, Kaplan JM. G alphas-induced neurodegeneration in Caenorhabditis elegans. J Neurosci. 1998;18:2871–80.
104. Schade MA, Reynolds NK, Dollins CM, Miller KG. Mutations that rescue the paralysis of Caenorhabditis elegans ric-8 (synembryn) mutants activate the G alpha(s) pathway and define a third major branch of the synaptic signaling network. Genetics. 2005;169:631–49. https://doi.org/10.1534/genetics.104.032334.
105. Iwanir S, et al. The microarchitecture of C. elegans behavior during lethargus: homeostatic bout dynamics, a typical body posture, and regulation by a central neuron. Sleep. 2013;36:385–95. https://doi.org/10.5665/sleep.2456.
106. Ryu MH, et al. Engineering adenylate cyclases regulated by near-infrared window light. Proc Natl Acad Sci USA. 2014;111:10167–72. https://doi.org/10.1073/pnas.1324301111.
107. Cianciulli A, et al. Interneurons regulate locomotion quiescence via cyclic adenosine monophosphate signaling during stress-induced sleep in Caenorhabditis elegans. Genetics. 2019;213:267–79. https://doi.org/10.1534/genetics.119.302293.
108. Kimura Y, et al. A CaMK cascade activates CRE-mediated transcription in neurons of Caenorhabditis elegans. EMBO Rep. 2002;3:962–6. https://doi.org/10.1093/embo-reports/kvf191.
109. Raizen DM, Cullison KM, Pack AI, Sundaram MV. A novel gain-of-function mutant of the cyclic GMP-dependent protein kinase egl-4 affects multiple physiological processes in Caenorhabditis elegans. Genetics. 2006;173:177–87. https://doi.org/10.1534/genetics.106.057380.
110. Langmesser S, et al. cGMP-dependent protein kinase type I is implicated in the regulation of the timing and quality of sleep and wakefulness. PLoS One. 2009;4:e4238. https://doi.org/10.1371/journal.pone.0004238.
111. Fujiwara M, Sengupta P, McIntire SL. Regulation of body size and behavioral state of C. elegans by sensory perception and the EGL-4 cGMP-dependent protein kinase. Neuron. 2002;36:1091–102. https://doi.org/10.1016/s0896-6273(02)01093-0.
112. Kopan R. Notch signaling. Cold Spring Harb Perspect Biol. 2012;4(10):a011213. https://doi.org/10.1101/cshperspect.a011213.
113. Komatsu H, et al. OSM-11 facilitates LIN-12 Notch signaling during Caenorhabditis elegans vulval development. PLoS Biol. 2008;6:e196. https://doi.org/10.1371/journal.pbio.0060196.
114. Yoo AS, Bais C, Greenwald I. Crosstalk between the EGFR and LIN-12/Notch pathways in C. elegans vulval development. Science. 2004;303:663–6. https://doi.org/10.1126/science.1091639.
115. Singh K, et al. C. elegans Notch signaling regulates adult chemosensory response and larval molting quiescence. Curr Biol. 2011;21:825–34. https://doi.org/10.1016/j.cub.2011.04.010.
116. Kume K, Kume S, Park SK, Hirsh J, Jackson FR. Dopamine is a regulator of arousal in the fruit fly. J Neurosci. 2005;25:7377–84. https://doi.org/10.1523/JNEUROSCI.2048-05.2005.
117. Crocker A, Sehgal A. Octopamine regulates sleep in drosophila through protein kinase A-dependent mechanisms. J Neurosci. 2008;28:9377–85. https://doi.org/10.1523/JNEUROSCI.3072-08a.2008.
118. Yuan Q, Joiner WJ, Sehgal A. A sleep-promoting role for the Drosophila serotonin receptor 1A. Curr Biol. 2006;16:1051–62. https://doi.org/10.1016/j.cub.2006.04.032.
119. Parisky KM, et al. PDF cells are a GABA-responsive wake-promoting component of the Drosophila sleep circuit. Neuron. 2008;60:672–82. https://doi.org/10.1016/j.neuron.2008.10.042.
120. McIntire SL, Jorgensen E, Kaplan J, Horvitz HR. The GABAergic nervous system of Caenorhabditis elegans. Nature. 1993;364:337–41. https://doi.org/10.1038/364337a0.

121. Starich TA, Lee RY, Panzarella C, Avery L, Shaw JE. eat-5 and unc-7 represent a multigene family in Caenorhabditis elegans involved in cell-cell coupling. J Cell Biol. 1996;134:537–48. https://doi.org/10.1083/jcb.134.2.537.
122. Huang H, Zhu CT, Skuja LL, Hayden DJ, Hart AC. Genome-wide screen for genes involved in Caenorhabditis elegans developmentally timed sleep. G3 (Bethesda). 2017;7:2907–17. https://doi.org/10.1534/g3.117.300071.
123. Huang H, et al. Gap Junctions and NCA cation channels are critical for developmentally timed sleep and arousal in Caenorhabditis elegans. Genetics. 2018;210:1369–81. https://doi.org/10.1534/genetics.118.301551.
124. Pujol N, et al. Distinct innate immune responses to infection and wounding in the C. elegans epidermis. Curr Biol. 2008;18:481–9. https://doi.org/10.1016/j.cub.2008.02.079.
125. Klee CB, Ren H, Wang X. Regulation of the calmodulin-stimulated protein phosphatase, calcineurin. J Biol Chem. 1998;273:13367–70. https://doi.org/10.1074/jbc.273.22.13367.
126. Nakai Y, et al. Calcineurin and its regulator sra/DSCR1 are essential for sleep in Drosophila. J Neurosci. 2011;31:12759–66. https://doi.org/10.1523/JNEUROSCI.1337-11.2011.
127. Guo F, Yi W, Zhou M, Guo A. Go signaling in mushroom bodies regulates sleep in Drosophila. Sleep. 2011;34:273–81. https://doi.org/10.1093/sleep/34.3.273.
128. Takahama K, et al. Pan-neuronal knockdown of the c-Jun N-terminal Kinase (JNK) results in a reduction in sleep and longevity in Drosophila. Biochem Biophys Res Commun. 2012;417:807–11. https://doi.org/10.1016/j.bbrc.2011.12.040.
129. Stavropoulos N, Young MW. insomniac and Cullin-3 regulate sleep and wakefulness in Drosophila. Neuron. 2011;72:964–76. https://doi.org/10.1016/j.neuron.2011.12.003.
130. Rogulja D, Young MW. Control of sleep by cyclin A and its regulator. Science. 2012;335:1617–21. https://doi.org/10.1126/science.1212476.
131. Nath RD, et al. The jellyfish Cassiopea exhibits a sleep-like state. Curr Biol. 2017;27:2984–2990.e3. https://doi.org/10.1016/j.cub.2017.08.014.
132. Niu L, et al. Melatonin promotes sleep by activating the BK channel in C. elegans. Proc Natl Acad Sci USA. 2020;117:25128–37. https://doi.org/10.1073/pnas.2010928117.
133. McCormick DA, Bal T. Sensory gating mechanisms of the thalamus. Curr Opin Neurobiol. 1994;4:550–6. https://doi.org/10.1016/0959-4388(94)90056-6.
134. Schwarz J, Lewandrowski I, Bringmann H. Reduced activity of a sensory neuron during a sleep-like state in Caenorhabditis elegans. Curr Biol. 2011;21:R983–4. https://doi.org/10.1016/j.cub.2011.10.046.
135. Cho JY, Sternberg PW. Multilevel modulation of a sensory motor circuit during C. elegans sleep and arousal. Cell. 2014;156:249–60. https://doi.org/10.1016/j.cell.2013.11.036.
136. Hilliard MA, Bargmann CI, Bazzicalupo PC. elegans responds to chemical repellents by integrating sensory inputs from the head and the tail. Curr Biol. 2002;12:730–4. https://doi.org/10.1016/s0960-9822(02)00813-8.
137. Choi S, Chatzigeorgiou M, Taylor KP, Schafer WR, Kaplan JM. Analysis of NPR-1 reveals a circuit mechanism for behavioral quiescence in C. elegans. Neuron. 2013;78:869–80. https://doi.org/10.1016/j.neuron.2013.04.002.
138. Nagy S, Raizen DM, Biron D. Measurements of behavioral quiescence in Caenorhabditis elegans. Methods. 2014;68:500–7. https://doi.org/10.1016/j.ymeth.2014.03.009.
139. Renn SC, Park JH, Rosbash M, Hall JC, Taghert PH. A pdf neuropeptide gene mutation and ablation of PDF neurons each cause severe abnormalities of behavioral circadian rhythms in Drosophila. Cell. 1999;99:791–802. https://doi.org/10.1016/s0092-8674(00)81676-1.
140. Lin L, et al. The sleep disorder canine narcolepsy is caused by a mutation in the hypocretin (orexin) receptor 2 gene. Cell. 1999;98:365–76. https://doi.org/10.1016/s0092-8674(00)81965-0.
141. Zimmerman JE, et al. Glutamate is a wake-active neurotransmitter in Drosophila melanogaster. Sleep. 2017;40:zsw046. https://doi.org/10.1093/sleep/zsw046.

142. John J, Ramanathan L, Siegel JM. Rapid changes in glutamate levels in the posterior hypothalamus across sleep-wake states in freely behaving rats. Am J Physiol Regul Integr Comp Physiol. 2008;295:R2041–9. https://doi.org/10.1152/ajpregu.90541.2008.
143. Janssen T, et al. Functional characterization of three G protein-coupled receptors for pigment dispersing factors in Caenorhabditis elegans. J Biol Chem. 2008;283:15241–9. https://doi.org/10.1074/jbc.M709060200.
144. Mertens I, Meeusen T, Janssen T, Nachman R, Schoofs L. Molecular characterization of two G protein-coupled receptor splice variants as FLP2 receptors in Caenorhabditis elegans. Biochem Biophys Res Commun. 2005;330:967–74. https://doi.org/10.1016/j.bbrc.2005.03.071.
145. Chen D, Taylor KP, Hall Q, Kaplan JM. The neuropeptides FLP-2 and PDF-1 act in concert to arouse Caenorhabditis elegans locomotion. Genetics. 2016;204:1151–9. https://doi.org/10.1534/genetics.116.192898.
146. Choi S, et al. Sensory neurons arouse C. elegans locomotion via both glutamate and neuropeptide release. PLoS Genet. 2015;11:e1005359. https://doi.org/10.1371/journal.pgen.1005359.
147. Rechtschaffen A, Bergmann BM, Everson CA, Kushida CA, Gilliland MA. Sleep deprivation in the rat: X. Integration and discussion of the findings. Sleep. 1989;12:68–87.
148. Shaw PJ, Tononi G, Greenspan RJ, Robinson DF. Stress response genes protect against lethal effects of sleep deprivation in Drosophila. Nature. 2002;417:287–91. https://doi.org/10.1038/417287a.
149. Kenyon C, Chang J, Gensch E, Rudner A, Tabtiang R. A C. elegans mutant that lives twice as long as wild type. Nature. 1993;366:461–4. https://doi.org/10.1038/366461a0.
150. Albertson DG, Thomson JN. The pharynx of Caenorhabditis elegans. Philos Trans R Soc Lond Ser B Biol Sci. 1976;275:299–325. https://doi.org/10.1098/rstb.1976.0085.
151. Avery L, You YJ. C. elegans feeding. WormBook. 2012:1–23. https://doi.org/10.1895/wormbook.1.150.1.
152. Raizen D, Song BM, Trojanowski N, You YJ. Methods for measuring pharyngeal behaviors. WormBook. 2012:1–13. https://doi.org/10.1895/wormbook.1.154.1.
153. Sparacio AP, Trojanowski NF, Snetselaar K, Nelson MD, Raizen DM. Teething during sleep: ultrastructural analysis of pharyngeal muscle and cuticular grinder during the molt in Caenorhabditis elegans. PLoS One. 2020;15:e0233059. https://doi.org/10.1371/journal.pone.0233059.
154. Zhang Y, Foster JM, Nelson LS, Ma D, Carlow CK. The chitin synthase genes chs-1 and chs-2 are essential for C. elegans development and responsible for chitin deposition in the eggshell and pharynx, respectively. Dev Biol. 2005;285:330–9. https://doi.org/10.1016/j.ydbio.2005.06.037.
155. Bennett HL, et al. Normal sleep bouts are not essential for C. elegans survival and FoxO is important for compensatory changes in sleep. BMC Neurosci. 2018;19:10. https://doi.org/10.1186/s12868-018-0408-1.
156. Honer M, et al. Orcokinin neuropeptides regulate sleep in Caenorhabditis elegans. J Neurogenet. 2020;34:440–52. https://doi.org/10.1080/01677063.2020.1830084.
157. Chao MY, Larkins-Ford J, Tucey TM, Hart AC. lin-12 Notch functions in the adult nervous system of C. elegans. BMC Neurosci. 2005;6:45. https://doi.org/10.1186/1471-2202-6-45.
158. Dusenbery DB, Sheridan RE, Russell RL. Chemotaxis-defective mutants of the nematode Caenorhabditis elegans. Genetics. 1975;80:297–309. https://doi.org/10.1093/genetics/80.2.297.
159. Edwards SL, et al. A novel molecular solution for ultraviolet light detection in Caenorhabditis elegans. PLoS Biol. 2008;6:e198. https://doi.org/10.1371/journal.pbio.0060198.
160. Wittenburg N, Baumeister R. Thermal avoidance in Caenorhabditis elegans: an approach to the study of nociception. Proc Natl Acad Sci USA. 1999;96:10477–82. https://doi.org/10.1073/pnas.96.18.10477.

161. Goetting DL, Soto R, Van Buskirk C. Food-dependent plasticity in Caenorhabditis elegans stress-induced sleep is mediated by TOR-FOXA and TGF-beta signaling. Genetics. 2018;209:1183–95. https://doi.org/10.1534/genetics.118.301204.
162. Soto R, Goetting DL, Van Buskirk C. NPR-1 modulates plasticity in C. elegans stress-induced sleep. iScience. 2019;19:1037–47. https://doi.org/10.1016/j.isci.2019.08.050.
163. Goetting DL, Mansfield R, Soto R, Buskirk CV. Cellular damage, including wounding, drives C. elegans stress-induced sleep. J Neurogenet. 2020;34:430–9. https://doi.org/10.1080/01677063.2020.1752203.
164. DeBardeleben HK, Lopes LE, Nessel MP, Raizen DM. Stress-induced sleep after exposure to ultraviolet light is promoted by p53 in Caenorhabditis elegans. Genetics. 2017;207:571–82. https://doi.org/10.1534/genetics.117.300070.
165. Soto R, Van Buskirk C. The C. elegans CEPsh glia are largely dispensable for stress-induced sleep. MicroPubl Biol. 2020;2020. https://doi.org/10.17912/micropub.biology.000261.
166. Robinson B, Van Buskirk C. UNC-108/RAB-2 is required for C. elegans stress-induced sleep. MicroPubl Biol. 2019;2019. https://doi.org/10.17912/micropub.biology.000112.
167. Iannacone MJ, et al. The RFamide receptor DMSR-1 regulates stress-induced sleep in C. elegans. elife. 2017;6:e19837. https://doi.org/10.7554/eLife.19837.
168. Mertens I, Clinckspoor I, Janssen T, Nachman R, Schoofs L. FMRFamide related peptide ligands activate the Caenorhabditis elegans orphan GPCR Y59H11AL.1. Peptides. 2006;27:1291–6. https://doi.org/10.1016/j.peptides.2005.11.017.
169. Nelson MD, et al. FRPR-4 Is a G-Protein Coupled Neuropeptide Receptor That Regulates Behavioral Quiescence and Posture in Caenorhabditis elegans. PLoS One. 2015;10:e0142938. https://doi.org/10.1371/journal.pone.0142938.
170. Jarecki JL, et al. Mapping neuropeptide expression by mass spectrometry in single dissected identified neurons from the dorsal ganglion of the nematode Ascaris suum. ACS Chem Neurosci. 2010;1:505–19. https://doi.org/10.1021/cn1000217.
171. Derry WB, Putzke AP, Rothman JH. Caenorhabditis elegans p53: role in apoptosis, meiosis, and stress resistance. Science. 2001;294:591–5. https://doi.org/10.1126/science.1065486.
172. E, L., et al. An antimicrobial peptide and its neuronal receptor regulate dendrite degeneration in aging and infection. Neuron. 2018;97:125–138.e5. https://doi.org/10.1016/j.neuron.2017.12.001.
173. Rogers CM, Franks CJ, Walker RJ, Burke JF, Holden-Dye L. Regulation of the pharynx of Caenorhabditis elegans by 5-HT, octopamine, and FMRFamide-like neuropeptides. J Neurobiol. 2001;49:235–44. https://doi.org/10.1002/neu.1078.
174. Avery L, Horvitz HR. Pharyngeal pumping continues after laser killing of the pharyngeal nervous system of C. elegans. Neuron. 1989;3:473–85. https://doi.org/10.1016/0896-6273(89)90206-7.
175. Trojanowski NF, Padovan-Merhar O, Raizen DM, Fang-Yen C. Neural and genetic degeneracy underlies Caenorhabditis elegans feeding behavior. J Neurophysiol. 2014;112:951–61. https://doi.org/10.1152/jn.00150.2014.
176. Hobson RJ, et al. SER-7, a Caenorhabditis elegans 5-HT7-like receptor, is essential for the 5-HT stimulation of pharyngeal pumping and egg laying. Genetics. 2006;172:159–69. https://doi.org/10.1534/genetics.105.044495.
177. Lee RY, Lobel L, Hengartner M, Horvitz HR, Avery L. Mutations in the alpha1 subunit of an L-type voltage-activated Ca2+ channel cause myotonia in Caenorhabditis elegans. EMBO J. 1997;16:6066–76. https://doi.org/10.1093/emboj/16.20.6066.
178. van der Linden AM, Moorman C, Cuppen E, Korswagen HC, Plasterk RH. Hyperactivation of the G12-mediated signaling pathway in Caenorhabditis elegans induces a developmental growth arrest via protein kinase C. Curr Biol. 2003;13:516–21. https://doi.org/10.1016/s0960-9822(03)00164-7.
179. Hajdu-Cronin YM, Chen WJ, Patikoglou G, Koelle MR, Sternberg PW. Antagonism between G(o)alpha and G(q)alpha in Caenorhabditis elegans: the RGS protein EAT-16 is necessary for

G(o)alpha signaling and regulates G(q)alpha activity. Genes Dev. 1999;13:1780–93. https://doi.org/10.1101/gad.13.14.1780.
180. Wang H, et al. Neuropeptide secreted from a pacemaker activates neurons to control a rhythmic behavior. Curr Biol. 2013;23:746–54. https://doi.org/10.1016/j.cub.2013.03.049.
181. Wang H, Sieburth D. PKA controls calcium influx into motor neurons during a rhythmic behavior. PLoS Genet. 2013;9:e1003831. https://doi.org/10.1371/journal.pgen.1003831.
182. Gibbs J, Young RC, Smith GP. Cholecystokinin elicits satiety in rats with open gastric fistulas. Nature. 1973;245:323–5. https://doi.org/10.1038/245323a0.
183. Shtonda BB, Avery L. Dietary choice behavior in Caenorhabditis elegans. J Exp Biol. 2006;209:89–102. https://doi.org/10.1242/jeb.01955.
184. MacQueen AJ, et al. ACT-5 is an essential Caenorhabditis elegans actin required for intestinal microvilli formation. Mol Biol Cell. 2005;16:3247–59. https://doi.org/10.1091/mbc.e04-12-1061.
185. Ren P, et al. Control of C. elegans larval development by neuronal expression of a TGF-beta homolog. Science. 1996;274:1389–91. https://doi.org/10.1126/science.274.5291.1389.
186. Davis KC, Choi YI, Kim J, You YJ. Satiety behavior is regulated by ASI/ASH reciprocal antagonism. Sci Rep. 2018;8:6918. https://doi.org/10.1038/s41598-018-24943-6.
187. Gallagher T, Kim J, Oldenbroek M, Kerr R, You YJ. ASI regulates satiety quiescence in C. elegans. J Neurosci. 2013;33:9716–24. https://doi.org/10.1523/JNEUROSCI.4493-12.2013.
188. Makino M, Ulzii E, Shirasaki R, Kim J, You YJ. Regulation of Satiety Quiescence by Neuropeptide Signaling in Caenorhabditis elegans. Front Neurosci. 2021;15:678590. https://doi.org/10.3389/fnins.2021.678590.
189. Pierce SB, et al. Regulation of DAF-2 receptor signaling by human insulin and ins-1, a member of the unusually large and diverse C. elegans insulin gene family. Genes Dev. 2001;15:672–86. https://doi.org/10.1101/gad.867301.
190. Hyun M, et al. Fat Metabolism Regulates Satiety Behavior in C. elegans. Sci Rep. 2016;6:24841. https://doi.org/10.1038/srep24841.
191. Juozaityte V, et al. The ETS-5 transcription factor regulates activity states in Caenorhabditis elegans by controlling satiety. Proc Natl Acad Sci USA. 2017;114:E1651–8. https://doi.org/10.1073/pnas.1610673114.
192. McCloskey RJ, Fouad AD, Churgin MA, Fang-Yen C. Food responsiveness regulates episodic behavioral states in Caenorhabditis elegans. J Neurophysiol. 2017;117:1911–34. https://doi.org/10.1152/jn.00555.2016.
193. Schuch KN, et al. Discriminating between sleep and exercise-induced fatigue using computer vision and behavioral genetics. J Neurogenet. 2020;34:453–65. https://doi.org/10.1080/01677063.2020.1804565.
194. Nguyen M, Alfonso A, Johnson CD, Rand JB. Caenorhabditis elegans mutants resistant to inhibitors of acetylcholinesterase. Genetics. 1995;140:527–35. https://doi.org/10.1093/genetics/140.2.527.
195. Johnson CD, Rand JB, Herman RK, Stern BD, Russell RL. The acetylcholinesterase genes of C. elegans: identification of a third gene (ace-3) and mosaic mapping of a synthetic lethal phenotype. Neuron. 1988;1:165–73. https://doi.org/10.1016/0896-6273(88)90201-2.
196. Nurrish S, Segalat L, Kaplan JM. Serotonin inhibition of synaptic transmission: Galpha(0) decreases the abundance of UNC-13 at release sites. Neuron. 1999;24:231–42. https://doi.org/10.1016/s0896-6273(00)80835-1.
197. Ghosh R, Emmons SW. Calcineurin and protein kinase G regulate C. elegans behavioral quiescence during locomotion in liquid. BMC Genet. 2010;11:7. https://doi.org/10.1186/1471-2156-11-7.
198. Perkins LA, Hedgecock EM, Thomson JN, Culotti JG. Mutant sensory cilia in the nematode Caenorhabditis elegans. Dev Biol. 1986;117:456–87. https://doi.org/10.1016/0012-1606(86)90314-3.
199. Saigusa T, et al. Circadian behavioural rhythm in Caenorhabditis elegans. Curr Biol. 2002;12:R46–7. https://doi.org/10.1016/s0960-9822(01)00669-8.

200. Flavell SW, et al. Serotonin and the neuropeptide PDF initiate and extend opposing behavioral states in C. elegans. Cell. 2013;154:1023–35. https://doi.org/10.1016/j.cell.2013.08.001.
201. Churgin MA, et al. Longitudinal imaging of Caenorhabditis elegans in a microfabricated device reveals variation in behavioral decline during aging. elife. 2017;6:e26652. https://doi.org/10.7554/eLife.26652.
202. Spies J, Bringmann H. Automated detection and manipulation of sleep in C. elegans reveals depolarization of a sleep-active neuron during mechanical stimulation-induced sleep deprivation. Sci Rep. 2018;8:9732. https://doi.org/10.1038/s41598-018-28095-5.
203. Clark DA, Biron D, Sengupta P, Samuel AD. The AFD sensory neurons encode multiple functions underlying thermotactic behavior in Caenorhabditis elegans. J Neurosci. 2006;26:7444–51. https://doi.org/10.1523/JNEUROSCI.1137-06.2006.

The Genetics of Sleep in Zebrafish

Amber Zimmerman

Abstract *Danio rerio* (zebrafish) has become a popular model for assaying a variety of conserved behaviors and traits. As a simple vertebrate, zebrafish provide insight into development of skeletal and nervous system processes in a small, transparent system that is easily observable. They have become a useful model for studying neural development, effects of drugs, and even behaviors once thought to be uniquely mammalian, including sleep. Over the last decade, advanced techniques for molecular imaging and automated behavioral acquisition have allowed researchers to identify rest states in the zebrafish that highly resemble mammalian sleep, which has provided an excellent and simple model system for examining sleep-wake behaviors in a tractable vertebrate. This chapter will describe how the use of zebrafish as a model organism is contributing to the rapidly progressing field of sleep research.

Keywords Zebrafish · Sleep · Genetics · Model organism

Zebrafish as a Model Organism

The zebrafish is a freshwater vertebrate, established as a tractable model for biomedical research [1, 2]. The zebrafish genome has been fully sequenced allowing for easy genetic manipulation [3], and approximately 70 percent of zebrafish genes have a corresponding human ortholog encoding a protein with similar function [4]. Zebrafish have become a popular system for assaying multiple conserved traits

A. Zimmerman (✉)
Department of Medicine, Division of Sleep, University of Pennsylvania Perelman School of Medicine, Philadelphia, PA, USA

Center for Spatial and Functional Genomics, Children's Hospital of Philadelphia, Philadelphia, PA, USA
e-mail: Amber.Zimmerman1@pennmedicine.upenn.edu

P. Gehrman et al. (eds.), *Genetics of Sleep and Sleep Disorders*,
https://doi.org/10.1007/978-3-031-62723-1_4

including metabolic function [5], vertebrate development of circadian rhythms [2], and sleep [6–8].

Advantages for Biomedical Research

Several factors make zebrafish particularly useful to the biomedical research community including their small size, body transparency through early development, large numbers of offspring, and short developmental time [9–12]. Zebrafish development occurs rapidly following ex vivo fertilization, progressing from a single-cell embryo to larva in only 3 days [11] (Fig. 1). Zebrafish are photoperiodic breeders and mate just after sunrise [13], which makes collection of embryos reliable and embryo staging consistent [14]. Zebrafish breeding pairs are set up in horizontal tanks with a divider separating males and females, and the divider is removed at lights-on to allow for spawning [14]. Fertilized eggs can typically be collected within 1 h of lights-on, and development can readily be assessed with standard microscopy techniques [15]. The timing is critical for genetic manipulation since it needs to be performed at the single-cell stage to maximize efficiency and minimize mosaicism.

Zebrafish remain translucent for the first several weeks of development, which facilitates imaging throughout development of structures that would be inaccessible in most other model organisms. Researchers have developed mutant lines that remain transparent into adulthood, allowing for imaging of full organ systems

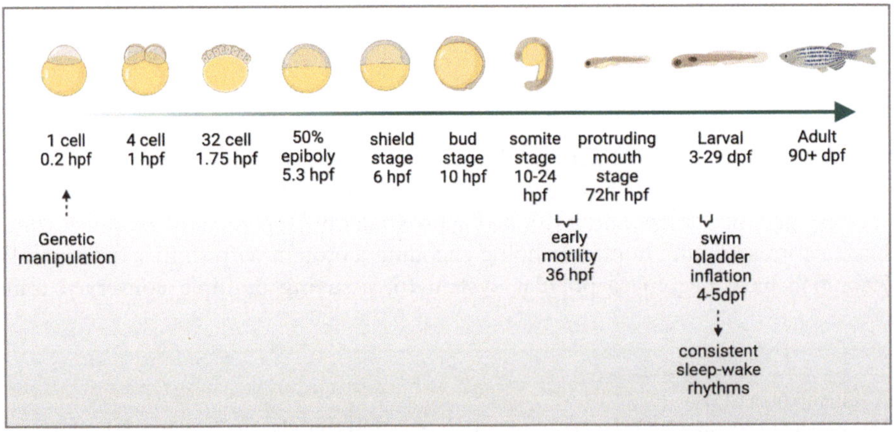

Fig. 1 Zebrafish developmental timeline. Zebrafish development occurs rapidly, progressing from a single cell to larva in 72 h. Genetic manipulation is typically performed at the single cell stage to allow incorporation of the mutation into all cells. Early motility begins approximately 36 h post fertilization (hpf). Consistent sleep-wake rhythms are present after 4 days post-fertilization (dpf). This is when larval sleep assays are performed. Adult zebrafish can also be tested, but their developmental timeline is protracted, similar to mouse

across all stages of life [16, 17]. While the worm and fruit fly have historically been the best high-throughput model organisms for assaying genetic regulation and conserved behaviors, zebrafish have now become commonplace for large-scale screening of behavioral traits following genetic manipulation providing a simple vertebrate system for biomedical research.

Immense effort has been put forth to develop a detailed cellular map with single-cell resolution of the zebrafish nervous system to reveal cellular circuitry within a vertebrate brain [18, 19]. The relatively small brain size and transparency of the larval zebrafish paired with the genetic and behavioral tractability make it an ideal model system for examining the complex dynamics of the nervous system that drive behavioral characteristics [20–23]. Whole-brain imaging of the transparent larval zebrafish has provided detailed insight into the patterns of neural activity and circuit-level mechanisms associated with different behaviors [23–28].

Application to Genetics Research

The zebrafish genome has been fully sequenced and annotated allowing for direct comparison to the human genome [4]. This sequencing effort revealed high conservation of protein-coding genes with approximately 70% of human genes having an orthologous zebrafish gene [4, 29]. Moreover, approximately 85% of disease-causing genes in humans have a conserved ortholog in zebrafish, which makes them a promising model for drug screening and therapeutic development [30, 31]. The zebrafish genome consists of approximately 26, 000 genes distributed across 25 pairs of chromosomes [4, 32, 33]. As a teleost fish, the zebrafish underwent a separate whole-genome duplication event producing multiple copies of each gene [34, 35]. This is of some benefit to genetic manipulation studies because embryonic lethality is less common following gene mutation in zebrafish relative to other model organisms such as mouse [29, 36].

High conservation between protein-coding genes of humans and zebrafish enables the interrogation of genetic traits using common tools including clustered regularly interspaced short palindromic repeats (CRISPR), transcription activator-like effector nucleases (TALENS), antisense morpholino oligonucleotides, and other forms of genetic manipulation [37–41]. Since embryonic development occurs externally, genetic manipulation can be applied at the single-cell stage without invasive approaches.

Application to Sleep Research

Until relatively recently, it was not well accepted that sleep existed outside of mammalian and complex vertebrate species [42, 43]; however, over the last two decades, researchers have identified sleep-like states in nearly every living organism [43–46].

Robust circadian rhythms exist all the way down to single-cell organisms as everything has adapted to the approximate 24-h cycle of the day [47, 48]. Rest-activity periods are common across all species [43, 47], with many being tightly linked to light-dark phases and others fluctuating back and forth throughout the day, but advancements in technology including molecular characterization, advanced imaging, and electroencephalography (EEG) have allowed researchers to reveal consistent biology across simple organisms suggesting they undergo a sleep-like state similar to mammalian species. The criteria accepted for sleep in a simple organism include (1) the absence of voluntary movement, (2) spontaneous and circadian regulated, (3) reversibility, (4) postural changes, (5) increased arousal threshold relative to active periods, (6) homeostatically regulated, such that restriction results in a rebound effect, and (7) state-related changes in neural function [8, 42, 43]. These traits should be stable and common within each organism for them to be representative of sleep.

These criteria can be applied to the zebrafish indicating they do, in fact, exhibit sleep behaviors consistent with mammalian species [6, 8, 43, 49, 50]. While *C. elegans* and *D. melanogaster* have become tractable, high-throughput model organisms to assay sleep-like behaviors, zebrafish provide an attractive vertebrate model with higher genetic sequence homology to human relative to invertebrates. Moreover, within the first few days of development, zebrafish have an intact nervous system and demonstrate core properties of sleep [8]. Both adult and larval zebrafish exhibit these stable behavioral criteria [6, 7, 49–54]; however, additional properties of larval zebrafish, including body transparency and small size, facilitate the assessment of neural development and molecular regulation through microscopy.

Unlike the commonly used laboratory mouse model, zebrafish exhibit diurnal sleep rhythms similar to humans, with the majority of their sleep consolidated to the dark [49, 50]. These rhythms are robust and tightly regulated by light-dark transitions. Zebrafish do, however, sleep in bouts throughout the light phase as well. Zebrafish also express endogenous melatonin in a circadian fashion unlike popular laboratory strains of mice, making them an excellent system to model sleep response to pathways involved in melatonin signaling [55] (Fig. 2).

In mammals, temperature can play a significant role in regulating sleep, and environmental fluctuations in temperature may mask or amplify sleep-wake phenotypes [56, 57]. Zebrafish are poikilothermic, meaning their body temperature is regulated by their external environment; however, their external environment is tightly regulated through temperature control of the surrounding water. This can be advantageous when examining the effects of genes on sleep to circumvent the role of temperature regulation in sleep and arousal [58].

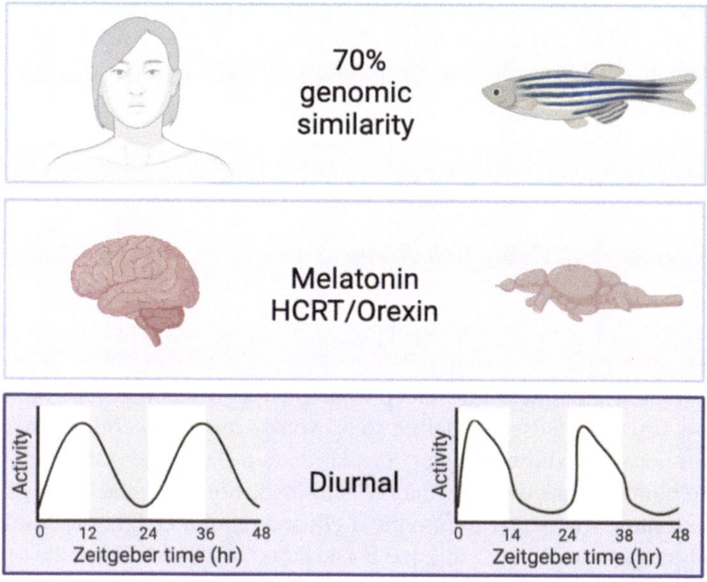

Fig. 2 Zebrafish as a model organism to examine vertebrate sleep. Zebrafish offer many advantages as a model organism to assay sleep-wake behaviors including their genetic similarity to humans. Completion of the sequencing of the zebrafish genome demonstrated approximately 70 percent of genes in zebrafish have a known human ortholog. Further, more than 80 percent of disease-causing genes in human have an orthologous gene in zebrafish facilitating disease research and therapeutic discovery. Zebrafish also share many of the same neuromodulatory mechanisms that promote sleep in humans including melatonin and hypocretin/orexin systems. In addition to molecular modulators of sleep, zebrafish also share analogous brain structures with humans that promote sleep and arousal. Unlike the commonly studied C57BL/6 J laboratory mouse strain, zebrafish are diurnal vertebrates with sleep that is circadian-regulated and primarily consolidated to the night. These features make zebrafish an excellent model to understand the genetic underpinnings of sleep and arousal in a simple vertebrate

Disadvantages of Using Zebrafish to Measure Sleep Behaviors

Like other simple model organisms, the methods for measuring sleep in zebrafish heavily depend on activity. Polysomnography or EEG and electromyography (EMG) are considered the gold standard for measuring mammalian sleep because they provide insight into brain-specific states that are considered to represent a true sleep state including slow-wave activity or non-rapid eye movement (NREM) sleep and rapid eye movement (REM) sleep, characterized by higher frequency brain activity paired with muscle atonia. Advancements in molecular imaging, such as calcium imaging [24, 26, 28] or imaging of endogenous sensors [59, 60], provide a promising avenue for measuring neuronal activity [21]; however, a stable electrical probe cannot be reliably placed in the cortex of a zebrafish brain to examine similar neuronal activity patterns. Present measures also require fish to be restrained during neuronal imaging [53], which can distort true sleep-like behaviors. Despite these drawbacks, many aspects of rest-activity regulation can still be reliably assessed in zebrafish, and future advancements in technology will likely reveal common brain mechanisms underlying zebrafish sleep states.

Sleep in Zebrafish

Characteristics of Zebrafish Sleep

Present studies rely on activity tracking to measure sleep using several criteria. Larval zebrafish sleep is generally characterized by 1 minute (min) of inactivity [7, 54, 61], whereas adult zebrafish sleep is defined as 6 seconds (s) of inactivity [6, 51]. These criteria are used to define sleep versus quiet wakefulness because they have been shown to exhibit a higher correlation to a sleep-like state when compared to shorter bouts of inactivity. In larval zebrafish, bouts of inactivity greater than 1 min occur more frequently during the night and exhibit a higher arousal threshold compared to bouts less than 1 min [61]. The same is true for 6-s bouts of inactivity in adult zebrafish [51]. Adult sleep is more fragmented; however, these 6-s bouts have been shown to represent an appropriate threshold for defining a sleep-like state [51]. Zebrafish show similar sleep ontogeny to humans across developmental trajectories as measured by sleep bout and duration across age, making them a good model for examining neural circuitry associated with age-related sleep changes [8].

Circadian Regulation of Activity in Zebrafish

Circadian regulation of transcription and behavior is common across virtually all species examined to date, and zebrafish are no exception [2, 62–66]. They possess the core circadian clock machinery including *clock*, *period*, *bmal*, and *cryptochrome* [62, 66–74] and exhibit robust circadian rhythms in their daily activity rhythms [2, 64]. In addition, circadian regulation in zebrafish is influenced by pharmacological agents known to phase-shift or blunt rhythms in mammals [75]. As a diurnal species, their activity is highest during the light phase and is gated by light-dark transitions. Zebrafish exhibit circadian regulation as early as 2 days postfertilization; however, this is dependent on light entrainment during early development [62, 63]. Cells and tissues of zebrafish undergo rhythmic expression influenced by the circadian clock but also exposure to light [66, 68, 76]. Zebrafish possess endogenous oscillators in peripheral tissues that can be rhythmically entrained by light [68, 76–78].

For zebrafish raised on a standard 14-h/10-h light-dark cycle, melatonin regulation becomes circadian-linked with increased nocturnal expression [55, 64]. Nocturnal regulation of melatonin is an important aspect of the conserved neuromodulatory regulation in zebrafish and will be discussed in further detail in the next section.

Since larval zebrafish are transparent, they have been shown to be especially sensitive to changing light even in the absence of vision or pineal signaling because they have populations of cells that are capable of sensing light [79–81]. This is an important consideration when examining sleep in zebrafish as they do not exhibit a strong homeostatic rebound to sleep deprivation in the presence of light [43, 51]. Activity of zebrafish is heavily influenced by their environmental light, and even the presence of low intensity or infrared light may mask behavioral abnormalities that would occur in true dark conditions [82, 83]. When examining the molecular and genetic factors influencing sleep in larval zebrafish, it is important to tightly regulate environmental conditions to mitigate its influence on rest-activity rhythms.

Neuromodulatory Systems that Promote Arousal and Sleep in Zebrafish

Sleep and arousal states are highly conserved across phylogeny [43, 44, 47, 84]; however, particular aspects of sleep-wake regulation differ between vertebrates and invertebrates including sleep timing and neuromodulatory mechanisms promoting arousal. While *Drosophila* are well suited for high-throughput sleep assays, they lack important regulators of human sleep including hypocretin/orexin signaling.

Mammalian vertebrates rely on input through the eyes to signal through the suprachiasmatic nucleus (SCN) and the pineal gland to receive photic stimuli and time-of-day information [85, 86]. Although zebrafish lack an analogous structure to the mammalian SCN, they contain cells in the lateral hypothalamus that express several markers of the mammalian SCN [87]. Moreover, zebrafish possess light-sensitive neuropeptides that interact with sleep- and wake-promoting systems to influence behavior [87]. Because they lack an SCN, zebrafish rely on the pineal as a master coordinator of circadian rhythms [88–90] by integrating light cues to influence gene expression. Apart from the lack of SCN, zebrafish exhibit remarkable conservation with mammalian species of neuroanatomy and neurotransmitters [6, 91] (Fig. 3) making them a well-suited model organism for examining the molecular underpinnings of sleep and arousal. Sleep and arousal states in zebrafish are controlled by similar brain regions to those in mammals and humans including the hypothalamus and locus coeruleus, and the neurotransmitters acting at these sites to promote sleep and arousal are also conserved [6].

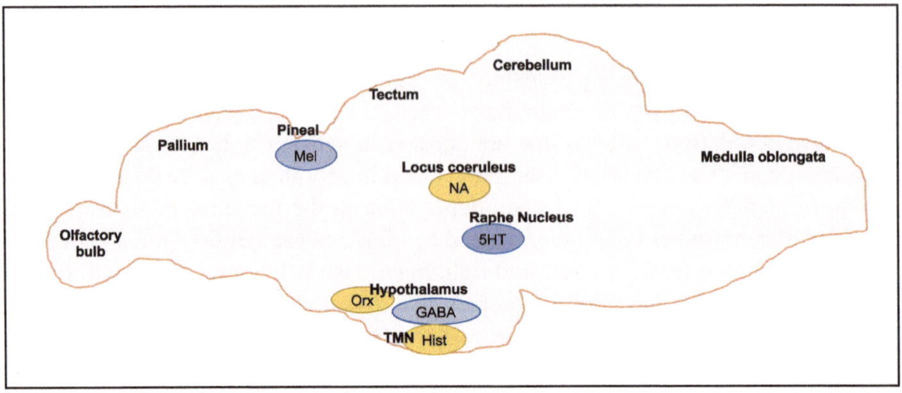

Fig. 3 Conservation of primary sleep and arousal neuromodulatory systems in the zebrafish brain. Zebrafish brain outline depicted as rostral (left) to caudal (right). Primary neuromodulatory circuits that promote sleep and arousal in humans are conserved in zebrafish including the pineal gland, hypothalamus, tuberomammilary nucleus (TMN). locus coeruleus, and raphe nucleus. Wake-promoting (yellow) neuromodulators include noradrenaline (NA), orexin/hypocretin (Orx), and histamine (Hist). Sleep-promoting (blue) neuromodulators include melatonin (Mel), serotonin (5HT), and gamma-aminobutyric acid (GABA)

Hypocretin/Orexin Signaling

The hypocretin/orexin (HCRT) system is a conserved neuromodulatory mechanism influencing wake and arousal [92, 93]. HCRT is a neuropeptide produced by a subset of hypothalamic neurons that project to wake-promoting regions including the locus coeruleus, tuberomammillary nucleus, pontine reticular formation, raphe nuclei, preoptic area, and the laterodorsal tegmental nucleus [93–97]. In mammals, HCRT is wake-promoting, and loss of HCRT signaling produces narcolepsy, which manifests as excessive and often uncontrolled daytime sleepiness [93, 98–104]. Hcrt has been extensively studied in zebrafish [52, 61, 105, 106] demonstrating Hcrt-expressing neurons within the lateral hypothalamus influence wakefulness in zebrafish possibly through their interaction with aminergic and cholinergic systems [107]. There is, however, controversy as to whether Hcrt signaling in zebrafish is wake-promoting or sleep-inducing [6, 7, 51, 52, 61]. Zebrafish express only one *hcrt* ortholog that is selectively expressed in a subset of hypothalamic neurons that project to wake-promoting regions similar to humans [52, 108]. Overexpression of Hcrt in larval zebrafish was shown to increase wakefulness and decrease rest [61]. Hcrt receptor zebrafish mutants exhibit fragmented sleep similar to mammalian models, indicating HCRT is important for maintenance of wake states across phylogeny. Compelling evidence supporting the role of Hcrt in waking activity comes from bioluminescence imaging, which targets a Ca2+-sensitive photoprotein to Hcrt-positive neurons within the hypothalamus [109]. During periods of heightened locomotor activity, fluorescence of these cells increased, implying heightened signaling within Hcrt neurons is associated with increased wakefulness [109]. While Hcrt receptor mutant adult zebrafish were found to have fragmented sleep (more sleep-wake transitions), consistent with a narcoleptic phenotype, mutant adult fish also displayed shorter sleep and increased activity [51], opposing the majority of literature that has implicated Hcrt signaling in the promotion of wakefulness. These differences were masked by constant light conditions demonstrating light is a particularly strong activity-inducing cue even in adult zebrafish [51]. Despite these inconsistencies in adult zebrafish, it is clear that Hcrt signaling is critical to sleep-wake maintenance in zebrafish as has been demonstrated in mammals. RNA-sequencing performed on Hcrt-expressing neurons has revealed substantial heterogeneity in zebrafish and identified important subtypes of Hcrt neurons that influence sleep and wake behaviors. Specifically, Hcrt neurons expressing the conserved voltage-gated potassium channel, *kcnh4a*, are involved in nighttime sleep maintenance, and CRISPR-mediated knockout of *kcnh4a* significantly reduced sleep duration and resulted in poorly consolidated sleep during the night [110]. While Hcrt neurons serve diverse biological functions, these results demonstrate that specific subtypes underlie the control of sleep-wake behaviors.

Melatonin

Melatonin is a hormone synthesized from tryptophan within pineal cells [111]. A series of enzymatic reactions catalyzes the conversion of tryptophan to serotonin. Two enzymes, arylalkylamine N-acetyltransferase (*aanat*) and acetylserotonin methyltransferase (*asmt*), catalyze the conversion from serotonin into melatonin [112]. Serotonin and melatonin are rhythmically expressed with serotonin peaking during the day and melatonin peaking during the night in zebrafish [112, 113]. While other vertebrates only express one *aanat* gene, zebrafish encode two copies with tissue-specific expression [112]. *Aanat1* is expressed in the retina, while *aanat2* is expressed predominantly in the pineal gland, with lower expression in the retina [112–114]. Spatiotemporal segregation of duplicated genomic transcripts is a common feature of zebrafish transcripts following the whole-genome duplication [36]. Loss of *aanat2* in zebrafish either through genetic manipulation or drug-induced cell ablation reduced sleep at night, while loss of *aanat1* did not, suggesting the ortholog expressed within the pineal gland is most critical for sleep regulation in zebrafish [55].

The pineal gland is conserved across vertebrates and is formed by an invagination of the roof of the diencephalon situating it directly beneath the frontal skull surface in zebrafish and making it accessible to light [115–117]. In zebrafish, the pineal gland contains photoreceptive cells similar to those of the retina, which sense light and convey photoperiod information to the brain via melatonin signaling [118, 119]. The fish pineal gland contains an intrinsic circadian oscillator that drives rhythmic melatonin production even in the absence of external input [88]. While circadian regulation of endogenous melatonin has been shown to be critical for nighttime sleep in zebrafish, loss of endogenous melatonin does not alter circadian rhythms in zebrafish, suggesting melatonin acts downstream of the circadian system [55]. In addition to the conservation of the molecular regulation of melatonin production, polymerase chain reaction amplification has detected melatonin receptor fragments that are conserved across vertebrates including zebrafish, implying a conserved genetic mechanism for the regulation of sleep timing in diurnal vertebrates [120].

In humans, melatonin is rhythmically expressed, and an increase in melatonin expression correlates with the onset of sleep [121–123]. Pharmacological administration of melatonin reduces latency to sleep onset and increases total sleep time when given under physiological relevant conditions indicating it serves an important role in promoting sleep [124–126]. Given that zebrafish are a diurnal organism, rhythmic expression of melatonin also coincides with their sleep [54, 64], unlike the commonly studied C57BL/6 J lab mouse strain, which lack the enzymes required for the production of melatonin [127]. Because zebrafish are a diurnal vertebrate expressing melatonin, they have become a popular model for understanding the hormonal regulation of sleep [52, 54, 64].

Melatonin is a popular target for behavioral intervention of sleep, and zebrafish can be used for drug discovery by applying high-throughput drug screening to a model endogenously expressing melatonin.

Other Sleep- and Wake-Promoting Neuromodulators

Many other neurotransmitters and neuropeptides are responsible for the control of sleep and arousal in both mammalian and nonmammalian vertebrates including noradrenaline, serotonin, dopamine, histamine, and melanin-concentrating hormone, all of which are conserved regulators of arousal and sleep in zebrafish [6, 53, 91, 128–132]. The catecholaminergic system, another system with a known role in the regulation of sleep-wake maintenance, is also present in zebrafish with conservation of the genes encoding tyrosine hydroxylase, monoamine oxidase, and catechol-o-methyltransferase [129, 133].

Zebrafish larvae yield themselves to high-throughput pharmacological screens since many larvae can be tested simultaneously, they possess a permeable blood-brain barrier, and pharmacological agents can be added directly to the water in which they are swimming [30, 31, 134]. Sedative drug application to zebrafish has revealed they possess many of the same receptors associated with sedative hypnotic pathways including gamma-amino-n-butyric acid (GABA) and histaminergic receptors indicating these are conserved pathways associated with rest-like states [135]; however, genetic manipulation of histamine signaling has provided mixed results [136, 137] and does not appear to significantly impact sleep in zebrafish [137, 138]. Large-scale drug screens have demonstrated that zebrafish are similarly impacted by sleep- and wake-promoting pharmacological agents [31]. In particular, alpha2-adrenergic receptor agonists were sedating, while beta-adrenergic agonists were arousing similar to mammals, and inflammatory signaling pathways are critical to rest-wake rhythms as in humans [31]. Zebrafish express multiple alpha-adrenoreceptors, which mediate the effects of norepinephrine and epinephrine on waking states [139–141] and are sensitive to the effects of prazosin, which inhibits alpha-adrenorecptors [141, 142]. Zebrafish are also sensitive to the effects of other clinically applicable wake-promoting agents including modafinil, which acts on dopaminergic and orexinergic systems [143].

These pharmacological and genetic screens demonstrate many of the molecular mechanisms that promote sleep-wake behaviors in humans are conserved in zebrafish, which makes them a valuable resource to examine the present therapeutics developed to target human sleep and arousal disorders.

Genetic Control of Sleep Homeostasis

Research into the molecular correlates of sleep homeostasis in zebrafish have identified critical regulators of sleep need in vertebrates. Sleep is described as a two-process model: [144] one process is circadian driven, while the other is homeostatically driven to fulfill a still-unknown need [145]. One hypothesis underlying the homeostatic regulation of sleep suggests that high neuronal activity necessary to promote wakefulness must be compensated through a general downscaling of synapses, whereby receptor composition is altered to reduce synaptic firing potential [146, 147]. Appelbaum and colleagues leveraged the transparency of larval zebrafish to examine synaptic regulation during sleep and wake, demonstrating that both circadian and homeostatic mechanisms influence structural synaptic plasticity, at least within hypocretin neurons [148], suggesting that neuronal circuitry is dynamically regulated across sleep and wake states.

It has repeatedly been described that prolonged wakefulness is associated with an increase in adenosine, a metabolic marker of neuronal activity [146, 149–151], which dissipates during sleep suggesting sleep serves a purpose to reduce metabolic load that accumulated during wake [152]. Caffeine is an adenosine receptor antagonist that promotes wakefulness by blocking adenosine receptors in the brain, preventing the sleep-initiating effects [153]. Adenosine receptors have been identified in zebrafish, which are sensitive to caffeine [154, 155]. Moreover, caffeine promotes prolonged wakefulness in zebrafish and can be used to pharmacologically induce rebound sleep, demonstrating adenosine receptors are conserved regulators of homeostatic sleep drive across vertebrates [156].

Although the regulation of sleep is still a topic of debate, studies have indicated sleep is not evenly distributed across the brain but rather selectively influences brain oscillatory activity in a state-dependent manner. A primary characteristic of rebound sleep is enhanced cortical slow-wave activity, which dissipates as sleep progresses [157] and is associated with region-specific waking activity [158–160]. One hypothesis for this regional bias proposes that certain networks of neurons selectively promote localized sleep through a homeostatic mechanism to differentially regulated neuronal assemblies [161–164]. It has been suggested that active neuronal assemblies release sleep-promoting factors, which in turn signals for their refinement during sleep [165].

Alternatively, homeostatic sleep drive may be sensed at the level of specific neurons but promoted by sleep-specific neurons through network communication to produce an organism-wide sleep state. Sleep-promoting neurons have been identified across model organisms [166, 167], with many converging upon the hypothalamus, specifically subregions including the preoptic area (POA) [168–171]. Work in zebrafish has determined galanin-expressing neurons in the POA are required for homeostatic rebound sleep in vertebrates [156]. Galanin is a conserved inhibitory neuropeptide expressed in GABAergic neurons of the POA [172–174], whose transcription is upregulated during rebound sleep in zebrafish in a dose-dependent manner associated with prior neuronal activity [156]. Importantly, this study demonstrated that homeostatic sleep rebound is responsive to the intensity of neuronal activity independent of physical waking activity [156]. Together, these experiments have identified conserved neural circuitry in zebrafish responsible for the homeostatic control of sleep and demonstrated that homeostatic sleep in response to neural activity represents a separate process from circadian-regulated sleep.

Methods for Assaying Sleep in Zebrafish

One major difficulty of sleep research in model organisms is determining whether the animal is truly sleeping or in a state of rest with reduced activity. The gold standard for characterizing sleep is through polysomnography, or the electrical recording of brain and muscle activity, which captures transitions between wake and sleep by monitoring electrical signal changes. Frequency characterization of these signals can be used to differentiate stages of sleep including rapid eye movement (REM) and non-rapid eye movement (NREM) sleep. Electroencephalography (EEG) captured at the skull and electromyography (EMG) (muscle recording) can be used in freely moving mammalian model organisms due to their brain size and environment. However, EEG/EMG cannot be used to determine sleep-wake states in zebrafish due to size constraints and environmental limitations. Before the advent of accessible electrophysiological measurements in freely behaving animals, the characterization of sleep has depended solely on behavioral criteria [175]. Today, many sleep studies across model organisms still rely on behavioral criteria because it has been shown to be reliably correlated with electrophysiological data characteristic of sleep, and it is more applicable to large-scale studies due to cost and labor limitations [175–178].

Locomotor-Based Quantification of Sleep-Activity States

To capture sleep-wake states in zebrafish, researchers rely on observable changes to activity and posture that are consistent with a sleep-like state. Standardized protocols exist for characterizing sleep traits in both larval and adult zebrafish [179]. These include high-speed automated video tracking using infrared cameras paired

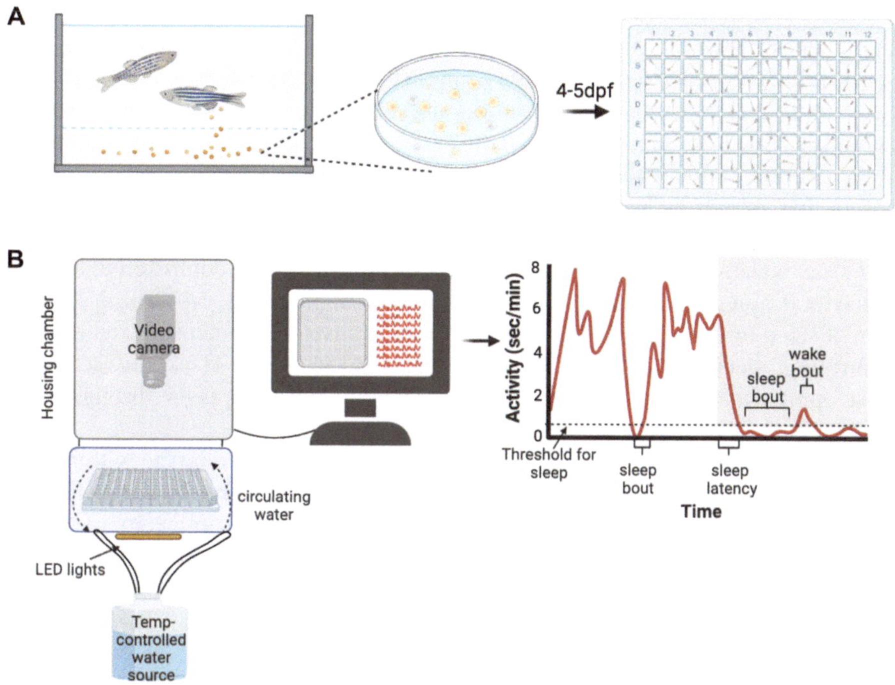

Fig. 4 Monitoring sleep and activity in zebrafish. The majority of zebrafish sleep research has been performed in larval zebrafish because of their small size that allows for dozens-to-hundreds of fish to be examined simultaneously. Sleep and activity can also be monitored in adult zebrafish using a modified approach that can capture activity of six to eight fish simultaneously using larger tanks. (**a**) Embryos are collected from zebrafish breeding pairs within 1 h from light-onset (typically 9 am). Embryos are collected in standard embryonic growth media in a petri dish. If genetic manipulation is being applied, it is typically performed at the single cell stage, which is approximately 15 min post fertilization (see Fig. 1). Injected embryos (F0) can either be tested or raised and bred to create stable mutant lines (F1, etc.). Following embryo collection, zebrafish are kept in embryo media for 4 days to allow hatching and development of motor behaviors and consistent sleep-wake rhythms, which arise at approximately 4 days post fertilization. Lights are kept on a 14 h:10 h light:dark cycle at a constant temperature of 28.5C to ensure entrainment to circadian cycles and optimal growth. On day 4 or 5 post fertilization (dpf), individual larval zebrafish are then transferred to separate wells of a 96-well plate and placed within a sound-attenuating chamber (**b**) with circulating water and automated lights. Constant video is captured for 24–72 h to examine sleep and activity rhythms. Data are analyzed using defined thresholds to differentiate sleep behaviors from behavioral quiescence

with computational image and pixel analysis to quantify locomotor behavior across prolonged periods. Multiple video-capture systems including Noldus DanioVision (https://www.noldus.com) and ViewPoint ZebraBox (https://www.vplsi.com) exist for the automated tracking of activity in model organisms with parameters specifically designed to differentiate sleep from quiet wakefulness in zebrafish [179] (Fig. 4).

Sleep is differentiated from lack of movement by multiple behavioral criteria. First, zebrafish sleep primarily occurs in a diurnal manner, with the majority consolidated to the night [6, 7, 50]. This is marked by a profound increase in a state of behavioral quiescence during the diurnal night [51, 54, 61]. It has been shown that zebrafish raised on an alternating light-dark cycle will continue to exhibit sleep-wake patterns in a circadian manner even when maintained in constant dark, indicating that zebrafish sleep is circadian regulated, much like other species [2, 64]. Second, arousal threshold, measured as the threshold at which an external stimulus provokes wakefulness, is significantly increased during sleep compared to bouts of behavioral quiescence that are not representative of sleep [51, 54]. Third, deprivation of sleep in zebrafish leads to a rebound suggestive of a homeostatic mechanism controlling sleep need similar to other species [156, 180]. These characteristics arise approximately 4 days into zebrafish development and persist through adulthood [7].

The benefit of using automated video tracking of individual larvae within transparent wells is that the data represent cumulative activity patterns that do not rely on beam breaks, which can overestimate sleep [181]. Multiple parameters of activity can be captured to create a behavioral "fingerprint" consisting of patterns across the night and day that represent changes to sleep and wake duration, fragmentation of sleep bouts, and latency to sleep, along with general activity [7, 31]. Together, these data can provide detailed insight into behavioral characteristics associated with human conditions including hyperactivity, hypersomnia, and insomnia.

Fluorescent Imaging of Brain Activity to Measure Sleep-like States

Due to the difficulty of capturing electrical brain activity in zebrafish, it has been unclear whether zebrafish brain activity is consistent with that of mammals when they are sleeping. Analysis of sleep in model organisms such as mouse relies on electrical recordings of cortical activity. Although zebrafish brains are simpler than their mammalian counterparts, they exhibit much of the same distribution of neuronal circuitry [6]. Zebrafish possess a region known as the dorsal pallium, which is considered to be analogous to the mammalian neocortex [182]. Activity across this region can be monitored to provide a readout similar to cortical slow-wave activity captured in mammals [53].

Leveraging fluorescent transgenic reporter lines, real-time neuronal activity can be captured in larval zebrafish using whole-brain light-sheet imaging [183]. This transgenic line pan-neuronally expresses the nuclear-localized, genetically encoded Ca2+ indicator GCaMP6s, which fluoresces upon activity induction [156, 183]. The transparent nature of larval zebrafish facilitates this imaging to capture neuronal activity during behavior and provides insight into brain activity during different behavioral states including sleep. Using the neuronally expressed GCaMP6F calcium indicator, Leung and colleagues demonstrated neural signatures in zebrafish that are reminiscent of sleep signatures in other organisms [53]. Importantly, they demonstrate that slow-bursting activity is sensitive to sleep deprivation and that deprivation leads to an increase in slow-bursting activity across the dorsal pallium consistent with the increase in cortical slow-wave oscillations recorded in sleep-deprived mammals [53]. These data suggest that zebrafish do, indeed, undergo a homeostatic sleep-like state. Moreover, authors applied pharmacological agents and recorded neural activity, demonstrating that antagonism of histamine receptors increases slow synchronous activity in the dorsal pallium, while application of a hypnotic GABA-receptor agonist silences activity in the dorsal pallium [53], albeit to a degree that may not represent a true sleep-like state but rather an anesthetic state. In order to develop a polysomnographic-like metric to describe zebrafish sleep, Leung et al. combined the Gal4/UAS system with a fluorescent calcium indicator GCaMP7a to measure fluorescence-based polysomnographic features including brain activity, muscle activity, heart rate, and eye movement. Application of hypnotic agents acting on GABAergic and cholinergic receptors produced propagating waves across the telencephalon, referred to as ponto-midbrain-telencephalic waves, that coincided with muscle atonia, spontaneous eye movements, and increased heart rate variance [53]. These characteristics appear to mimic certain characteristics of a REM-like brain state; however, future work is needed to confirm this brain activity is a naturally occurring phenomenon in zebrafish.

Another imaging technique that capitalizes on neuronal activity uses the endogenous sensor, phosphorylated extracellular signal-regulated kinase (pERK). Upon neuronal depolarization, ERK is phosphorylated and can be examined as a proxy to identify active neurons with high temporal resolution [59, 60, 184, 185]. By registering pERK signals against a zebrafish reference brain atlas, activated-neural circuitry can be defined in the context of sleep-wake behaviors [186]. This approach can be paired with optogenetic stimulation to identify critical cells contributing to sleep-wake behaviors [60].

These techniques have been used to demonstrate the brains of zebrafish do appear to undergo neuronal activity-dependent regulation involved in the homeostatic regulation of sleep [156], providing strong evidence for a conserved brain state once considered to be only present in mammalian systems.

Protocols for Sleep Deprivation

Zebrafish possess light-sensitive cells throughout their bodies and are, therefore, particularly sensitive to light exposure [65, 76, 77, 79–81, 85]. Because of this, constant light exposure does not appear to have a profound impact on homeostatic sleep drive [51, 187]. While sleep duration is reduced by the presence of light, protocols for sleep deprivation and circadian disruption using constant light conditions cannot be reliably used to promote homeostatic rebound sleep in zebrafish. To promote deprivation-induced rebound sleep in zebrafish, a common protocol is to use a continuous swim paradigm, which gently rotates a water-filled conical containing fish to continuously wake them through turbulence [53, 180]. Other deprivation techniques include vibration [54], pharmacological application [156], electrical stimulation [51], and optomotor response, which presents larvae with a moving patterned platform that induces persistent swimming [156].

Genetic Manipulation of Zebrafish Sleep

As a simple vertebrate that undergoes ex vivo fertilization, zebrafish yield themselves as a model for genetic interrogation [10, 36, 188]. Following completion of the whole zebrafish genome sequence [4], there was a great movement to integrate zebrafish into the realm of genetic screening and manipulation. Over the last two decades, tools for genetic manipulation have substantially improved, especially in their efficiency and specificity [36]. Their large clutch sizes and rapid development facilitate forward genetic screens to a level only previously possible using invertebrate species [36].

Techniques for Genetic Manipulation in Zebrafish

ENU Mutagenesis

The toolbox for genetic modification of zebrafish is vast and includes techniques for inducible gene expression and suppression as well as targeted base changes [36, 38, 189]. Zebrafish genetics established a foothold for implementing forward genetic screens through the application of N-ethylnitrosourea (ENU), which can be used to induce random point mutations throughout the genome [36, 190–193]. Large-scale ENU mutagenesis identified hundreds of genes associated with cellular migration, embryo and nervous system development, and other key genes regulating conserved processes [190, 191, 193, 194]. This approach demonstrated the capacity for large-scale forward genetic screens in zebrafish to unravel complex genetic regulation of vertebrate development and behavior [190, 191, 193, 194]. This approach was largely used before knowledge of the complete zebrafish genome and provided the basis for identification of many genes involved in developmental processes. A subsequent reverse genetic approach, called targeted induced local lesions in genomes (TILLING), leverages ENU mutagenesis data to identify rare mutations in specific genes of interest [37, 195]. This approach can be used to create libraries of genes that produce specific phenotypes of interest. While these approaches are complementary in nature and have been used to identify missense mutations contributing to a variety of important phenotypes, they are nonspecific and it is difficult to pinpoint which mutation is driving the primary phenotype of interest [195].

Morpholino Oligonucleotides

A popular approach for introducing more targeted mutations in zebrafish is through the use of antisense oligonucleotides, also known as morpholinos [36, 196]. These synthetic nucleotides are injected directly into the embryo and bind the target sequence interfering with splicing or translation of the messenger RNA, producing reduced or loss of expression of the target gene [36, 196, 197]. Morpholino studies have been applied to examine the role of genes involved in zebrafish sleep. Recently, Holm et al. transiently knocked down the highly conserved microRNA, miR-137, using morpholino oligonucleotides in zebrafish, demonstrating that loss of miR-137 increased wakefulness and reduced sleep, similar to the effects observed in mice [198]. These data were also consistent with GWAS data collected from humans, which implicate a role for miR-137 in sleep duration [198]. Srdanovic and colleagues performed targeted knockdown of two genes known to impact sleep, *kcna2* and *dec2* [199]. *Kcna2* encodes a voltage-gated potassium channel in zebrafish. These voltage-gated channels have been identified in *Drosophila* forward genetic screens to produce short sleepers [200, 201] and similarly disrupt sleep and increase wake in mice [202]. Indeed, morpholino-directed knockdown of *kcna2* in zebrafish also produced short sleepers demonstrating membrane excitability as a conserved mechanism for the control of sleep. A point mutation in *DEC2* implicates this gene as one of the few identified in humans to confer short sleep and has been shown to also reduce sleep in mice [203]. Knockout of *Dec2* in mice disrupts sleep-wake amplitude [204]. Morpholino knockdown of *dec2* in zebrafish resulted in fragmented and shorter sleep in zebrafish [199]. Importantly, authors of this study demonstrated morpholino-based knockdown is transient and persists only through early development. This has been a point of scrutiny for the use of morpholino oligonucleotides to examine the effects of genes on specific behaviors as it is unclear whether the phenotypes are driven by developmental abnormalities [205]. Concerns have also been raised regarding the specificity of morpholinos, and caution should be used when designing and analyzing these studies [36, 205–207]. Despite these concerns, the described morpholino knockdown studies have produced results consistent with human and mammalian studies.

RNA Interference

RNA interference (RNAi) is a widely used technique in multiple model organisms including *Drosophila* and mice [208]. This approach has been adopted for zebrafish genetics as well [209]. RNAi is a method of posttranscriptional gene modification whereby small RNA molecules can be targeted to specific sequences to generate stable knockdowns. RNAi allows for cell-specific genetic manipulation and can be used to target noncoding RNA species such as microRNAs (miRNAs) to examine conservation of function. Although the use of RNAi technology has not been as widely applied to zebrafish as other model organisms, it serves as a promising tool for tissue- or cell-specific genetic manipulation [209, 210].

Gal4/UAS

To precisely regulate gene expression in a spatiotemporal manner, multiple genetic tools can be applied in zebrafish. Several model organisms utilized the yeast Gal4/UAS system, which can drive transcription in a site-directed manner using the upstream activator sequence (UAS). The Gal4 system has been leveraged in combination with cell-specific regulatory factors to drive gene expression locally without impacting global gene expression [211]. Given the realization that multiple different cell types contribute to the nuanced regulation of sleep across development and in the context of disease, the Gal4/UAS system provides a promising approach to manipulate local gene expression to reveal the cell-specific roles in the control of sleep [211]. Moreover, the Gal4 system can be used to drive neuronal expression of calcium indicators to examine neuronal activation in the context of sleep [53].

Heat Shock

Heat shock-inducible gene expression is a simple tool that can be readily used in zebrafish because their environmental conditions are tightly controlled. By transiently raising the temperature, gene expression can be induced via activation of a heat shock-inducible promoter [212]. This method is beneficial to circumvent embryonic lethality or other undesirable effects due to gene disruption during early development [212]. Lee et al. used a transgenic zebrafish model paired with a heat shock-inducible promoter to drive overexpression of *npvf*, which increase sleep in larval zebrafish [213]. Further, they demonstrated transgenic overexpression of multiple *npvf*-encoded peptides acts synergistically to promote sleep. To complement this technique, Lee and colleagues used CRISPR-Cas9 mutagenesis to generate a *npvf* mutant line lacking the capacity to generate two of the three *npvf*-encoded peptides, which significantly reduced sleep demonstrating *npvf* signaling is both necessary and sufficient to regulate sleep duration [213].

Zinc Finger Nuclease

Zinc finger nuclease-induced (ZFN) mutation introduces targeted mutations by creating double-stranded DNA breaks that are then repaired by nonhomologous end joining (NHEJ) often resulting in an insertion or deletion [214, 215]. Singh and colleagues used ZFN-mediated targeted mutation in the dopamine beta-hydroxylase gene (*dbh*) to genetically ablate norepinephrine [142]. This mutation created a premature stop codon truncating the protein and completely eliminating norepinephrine production in homozygous mutants resulting in reduced overall activity and increased sleep [142], consistent with the role of norepinephrine in the promotion of waking activity [216].

TALEN

Transcription activator-like effector nucleases (TALENs) have been shown to be more efficient at inducing somatic mutations than ZFNs [217]. TALENs also create a double-stranded break relying on NHEJ but can be more reliably targeted to a specific nucleotide sequence than ZFNs [217]. Lee and colleagues created a genetic knockout of the EGFR ligand, transforming growth factor alpha (*tgfalpha*) using TALEN-guided genetic manipulation to interfere with EGFR signaling [218]. TALEN-mediated mutation produced a seven-base-pair deletion that shifted the reading frame resulting in a premature stop codon predicted to eliminate the regulatory regions of Tgfalpha and significantly reduced sleep in homozygous mutants [218]. These results were consistent with heat shock-induced overexpression of *tgfalpha*, which increased sleep duration and genetic perturbation of *egf* and *egfralpha,* demonstrating not only the specificity of this method but also the importance of Egfr-mediated signaling in sleep [218].

CRISPR-Cas

Clustered regularly interspaced short palindromic repeats (CRISPR), utilizing the Crispr-associated (Cas)9 endonuclease has become one of the most popular tools for inducing stable genetic changes with high precision [40, 41, 189]. Like ZFNs and TALENs, CRISPR-Cas9 can be used to introduce double-stranded breaks at a target DNA sequence that relies on DNA repair mechanisms to rejoin the sequences, ultimately producing a shift in the reading frame and a premature stop codon; however, CRISPR uses a short guide RNA sequence to specifically target and cut a genomic region of interest and can be employed with simple RNA injection protocols [36]. One advantage of CRISPR is the high-efficiency production of mutant alleles. Biallelic mutations can be induced in founder (F0) generations producing a phenotype without the need for extensive breeding [189]. Although CRISPR mutations can be introduced into F0 animals, crossbreeding is often required to create stable transgenics expressing a selected homozygous mutation. Because zebrafish development occurs completely ex vivo and developmental staging is predictable, single-cell embryos can be mutated by injecting preformed complexes containing the necessary CRISPR guide RNA and the Cas9 protein directly into the cell [189, 219–222]. This approach has been shown to create high-efficiency, stable mutations in zebrafish founder lines that phenocopy stable transgenic lines [221, 223]. Injection of multiple guide RNAs into a single-cell embryo targeting a single gene [223] and/or targeting highly conserved regulatory exons [224, 225] produces high (>90%) mutation rates in zebrafish founder generations. Pairing CRISPR-based gene editing with high-throughput behavioral screening in zebrafish is a rapid and cost-efficient practice for employing large-scale genetic screens [226, 227]. Multiple groups have used this method to examine genes influencing sleep and activity. Kroll and colleagues demonstrated CRISPR mutation of *scn1lab* in F0 lines phenocopied stable transgenic finish with increased activity and highly similar characteristics across ten behavioral measurements [223]. Palermo et al. mutated a conserved regulatory region of a gene implicated in human insomnia, *pigq*, in zebrafish producing the same sleep-activity phenotype observed in *Drosophila* RNAi knockdown lines [224], demonstrating CRISPR mutations of genes outside of core sleep-regulating pathways exhibit reproducible behavioral phenotypes.

Optogenetics and Chemogenetics

Advanced genetic manipulation techniques including optogenetics and chemogenetics allow for high-precision spatiotemporal regulation of molecular processes. Optogenetic manipulation utilizes light-sensing ion channels that can be selectively activated or inhibited under different wavelengths of light. The transparent nature of zebrafish during early development facilitates noninvasive optogenetic manipulation and imaging of distinct neural circuits associated with behaviors [142, 228]. Gal4/UAS transgenic zebrafish lines have been developed to stably express light-sensitive opsins in specific cell types [229]. External light illumination of different wavelengths can be applied to multiple zebrafish larvae at once inducing or inhibiting multiple behaviors [229]. Singh et al. applied blue light to Hcrt neurons expressing the light-sensitive channelrhodopshin-2, which induced locomotor activity consistent with the observation that Hcrt neurons are wake-promoting in zebrafish larvae [142]. Pairing optogenetic stimulation of Hcrt neurons with in vivo calcium imaging revealed Hcrt neurons likely influence arousal through the activation of *dbh*-expressing locus coeruleus neurons, demonstrating functional connectivity between two neuronal populations.

Zebrafish larvae are particularly sensitive to light [79, 83, 230], which may confound optogenetic studies if not tightly controlled. To avoid confounding effects of light application to zebrafish, chemical manipulation of gene expression has also been applied through the addition of receptor agonists capable of driving gene expression [231]. Chen and colleagues used the capsaicin-sensitive TRPV1 channel in sensory neurons to induce neuronal and locomotor activity in larval zebrafish [231]. By expressing *trpv1* in *hcrt*-expressing hypothalamic neurons, they were able to either stimulate or ablate *hcrt*-expressing neurons within the hypothalamus through application of low or high concentrations of capsaicin, respectively. Stimulation of *hcrt*-expressing neurons with capsaicin enhanced waking activity and reduced sleep, while ablation of *hcrt-expressing* neurons reduced waking activity and increased sleep [231], consistent with previous work showing Hcrt is wake-promoting in zebrafish [61]. This approach is easily applicable in free-swimming zebrafish and can be used to alter activity in a dose-dependent manner.

Genetic Screening of Sleep-Wake Behaviors

Large-scale genetic screens for sleep- and wake-promoting genes have successfully been deployed in invertebrate model organisms such as *Drosophila* [200] for decades; however, many of the genes identified through these screens lack a human ortholog limiting their translational relevance to human sleep and disease states. Recently, these genetic screens have been carried out using zebrafish due to their large clutch sizes, rapid development, and genetic similarity to humans. Forward genetic screens offer an unbiased approach to identify novel regulators of sleep and arousal as they identify a phenotype (i.e., short sleep duration) and the genes significantly associated with the phenotype. Additionally, reverse genetic approaches can be completed by first identifying genes that are likely to produce a phenotype and then examining their influence. These types of screens can be leveraged to examine long lists of genetic candidates identified through human GWAS efforts to facilitate the translation of GWAS findings into a functional outcome [224, 225, 227].

A high-throughput overexpression screen in larval zebrafish revealed the neuropeptide, neuromedin U, as a regulator of arousal that influences sleep-wake regulation through modulation of corticotrophin-releasing hormone receptor-expressing neurons in the brainstem [232] implicating a novel arousal pathway. Moreover, neuropeptide Y (NPY), a neuropeptide known to regulate feeding and waking activity in mammals, was also identified as a conserved regulator of sleep [58]. NPY is highly expressed in the mammalian brain with robust expression across multiple regions of the hypothalamus [233]. It has been shown to regulate circadian rhythms and stress and is a potent orexigenic neuropeptide promoting feeding behavior [233, 234]. In mammals, it has been difficult to segregate the effects of NPY on feeding and waking behaviors since feeding status can mask sleep drive. While NPY has been shown to promote feeding in zebrafish [235], sleep behavior can be assessed independent of feeding status in larval zebrafish since they survive on their yolk sac during early development and do not require additional feeding [236]. Singh and colleagues leveraged this benefit to disentangle the role of NPY in sleep regulation in the absence of feeding cues demonstrating that, in zebrafish, Npy promotes sleep likely through the inhibition of noradrenergic signaling [58].

Insights from Zebrafish Sleep

Mechanisms of Vertebrate Sleep

Genetic manipulation of sleep in zebrafish has unveiled novel mechanisms influencing sleep, and forward genetic approaches have determined many of the genes influencing sleep in humans have conserved roles in zebrafish. Importantly, zebrafish allow for rapid screening within a diurnal vertebrate organism that has advantages over common laboratory mouse strains. Targeted manipulation of sleep disorder risk genes has demonstrated that many of the genes we know to influence human sleep dysfunction also disrupt sleep in zebrafish in a similar manner.

Sleep deprivation of zebrafish has uncovered important mechanistic insights into why an organism may require sleep. Recent work demonstrated that loss of sleep in zebrafish alters chromosome dynamics and DNA repair mechanisms leading to accumulation of double-stranded DNA breaks [237] and that this is due to initiation of the DNA damage response by PARP1 [238]. Prolonged wakefulness leads to accumulation of these double-stranded DNA breaks triggering a homeostatic response to initiate *PARP1* expression, which promoted sleep, demonstrating DNA damage is, at least in part, involved in the homeostatic sleep response [238]. As the genes influencing DNA repair are highly conserved and circadian-regulated and their expression is altered by sleep, these zebrafish studies confirm that sleep serves a critical role in nuclear maintenance and stability [237, 238].

Application to Human Sleep Research

Ultimately, the goal is to be able to utilize zebrafish as a reliable model for human disease. With the accumulation of human genome-wide association data, researchers are now poised with an opportunity to identify novel disease-relevant mechanisms by integrating human genomics data and model organism genetics at scale. Zebrafish have rapidly become a promising tool for probing GWAS-implicated genes for conserved behavioral traits including sleep [227]. Human GWAS for sleep have implicated variants associated with epidermal growth factor receptor (EGFR) signaling [218, 239]. Moreover, EGF signaling has independently been identified in invertebrates *C. elegans* and *Drosophila* to be associated with sleep [240–245], indicating deep conservation of this molecular pathway in regulating rest-wake behavior. CRISPR deletion of *egfr* in zebrafish decreased sleep, while heat shock-induced overexpression of *egfr* increased sleep [218]. Pharmacological inhibition of Egfr similarly reduced sleep [218] indicating this signaling pathway plays a key role in regulating sleep across phylogeny and represents a potential therapeutic target for human sleep dysfunction.

Human GWAS have repeatedly identified variants within the *MEIS1* gene that confer risk to insomnia and restless leg syndrome (RLS). Through CRISPR-Cas9 manipulation of the zebrafish *MEIS1* ortholog, researchers have confirmed this gene serves a conserved role in the regulation of sleep consolidation, whereby mutation of the most conserved ortholog, *meis1b*, promoted night-specific sleep fragmentation and prolonged latency to sleep onset [225].

A simple vertebrate model with molecular and structural conservation to humans is valuable to study human disease. Since zebrafish are diurnal with conservation of many of the wake- and sleep-promoting neuromodulatory systems, they can be used to study the contribution of sleep to human disease. Zebrafish also exhibit other conserved behaviors such as social preference and sensory gating, which make them a valuable model for examining sleep in the context of behavioral disorders such as autism spectrum disorder and schizophrenia [246–248]. Ozcan and colleagues have shown a bidirectional relationship between amyloid beta oligomers and sleep in zebrafish, revealing a complex relationship between amyloid beta cleavage and sleep regulation [249] and demonstrating the utility of zebrafish to model complex disorders such as Alzheimer's disease with impaired sleep maintenance [250]. Given that zebrafish also possess non-neuronal cell types that often contribute to these disease states such as microglia, astrocyte-like radial glia, and oligodendrocytes, they will be an important resource to examine the role of these cells in vertebrate sleep regulation in the context of human disease [251].

Future Directions

Techniques for wirelessly examining neuronal activity are rapidly adapting. The major limitation to characterizing and quantifying sleep in lower vertebrates and other model organisms is in the inability to reliably measure neural activity in a freely behaving animal. Automated video tracking paired with machine learning algorithms is being applied to other model organisms to noninvasively measure sleep-wake states. These tools paired with in vivo fluorescence imaging provide the means for appropriately characterizing sleep in zebrafish using high-throughput, unbiased metrics. As the precision of techniques for spatiotemporal control of gene expression adapts, their application in zebrafish paired with the benefits of high-throughput phenotyping will position zebrafish as a prime resource for rapidly examining mechanistic control of behavior in vertebrates.

References

1. Teame T, et al. The use of zebrafish (Danio rerio) as biomedical models. Anim Front. 2019;9:68–77. https://doi.org/10.1093/af/vfz020.
2. Hurd MW, Debruyne J, Straume M, Cahill GM. Circadian rhythms of locomotor activity in zebrafish. Physiol Behav. 1998;65:465–72. https://doi.org/10.1016/s0031-9384(98)00183-8.
3. Rafferty SA, Quinn TA. A beginner's guide to understanding and implementing the genetic modification of zebrafish. Prog Biophys Mol Biol. 2018;138:3–19. https://doi.org/10.1016/j.pbiomolbio.2018.07.005.
4. Howe K, et al. The zebrafish reference genome sequence and its relationship to the human genome. Nature. 2013;496:498–503. https://doi.org/10.1038/nature12111.
5. Seth A, Stemple DL, Barroso I. The emerging use of zebrafish to model metabolic disease. Dis Model Mech. 2013;6:1080–8. https://doi.org/10.1242/dmm.011346.
6. Chiu CN, Prober DA. Regulation of zebrafish sleep and arousal states: current and prospective approaches. Front Neural Circuits. 2013;7:58. https://doi.org/10.3389/fncir.2013.00058.
7. Rihel J, Prober DA, Schier AF. Monitoring sleep and arousal in zebrafish. Methods Cell Biol. 2010;100:281–94. https://doi.org/10.1016/B978-0-12-384892-5.00011-6.
8. Sorribes A, et al. The ontogeny of sleep-wake cycles in zebrafish: a comparison to humans. Front Neural Circuits. 2013;7:178. https://doi.org/10.3389/fncir.2013.00178.

9. Lieschke GJ, Currie PD. Animal models of human disease: zebrafish swim into view. Nat Rev Genet. 2007;8:353–67. https://doi.org/10.1038/nrg2091.
10. Vascotto SG, Beckham Y, Kelly GM. The zebrafish's swim to fame as an experimental model in biology. Biochem Cell Biol. 1997;75:479–85.
11. Kimmel CB, Ballard WW, Kimmel SR, Ullmann B, Schilling TF. Stages of embryonic development of the zebrafish. Dev Dyn. 1995;203:253–310. https://doi.org/10.1002/aja.1002030302.
12. Lawrence C. Advances in zebrafish husbandry and management. Methods Cell Biol. 2011;104:429–51. https://doi.org/10.1016/B978-0-12-374814-0.00023-9.
13. Abdollahpour H, Falahatkar B, Lawrence C. The effect of photoperiod on growth and spawning performance of zebrafish, Danio rerio. Aquacult Rep. 2020;17:100295. https://doi.org/10.1016/j.aqrep.2020.100295.
14. Adatto I, Lawrence C, Thompson M, Zon LI. A new system for the rapid collection of large numbers of developmentally staged zebrafish embryos. PLoS One. 2011;6:e21715. https://doi.org/10.1371/journal.pone.0021715.
15. Rieger S, Wang F, Sagasti A. Time-lapse imaging of neural development: zebrafish lead the way into the fourth dimension. Genesis. 2011;49:534–45. https://doi.org/10.1002/dvg.20729.
16. White RM, et al. Transparent adult zebrafish as a tool for in vivo transplantation analysis. Cell Stem Cell. 2008;2:183–9. https://doi.org/10.1016/j.stem.2007.11.002.
17. Antinucci P, Hindges R. A crystal-clear zebrafish for in vivo imaging. Sci Rep. 2016;6:29490. https://doi.org/10.1038/srep29490.
18. Kunst M, et al. A cellular-resolution atlas of the larval zebrafish brain. Neuron. 2019;103:21–38 e25. https://doi.org/10.1016/j.neuron.2019.04.034.
19. Hildebrand DGC, et al. Whole-brain serial-section electron microscopy in larval zebrafish. Nature. 2017;545:345–9. https://doi.org/10.1038/nature22356.
20. Baier H, Scott EK. Genetic and optical targeting of neural circuits and behavior--zebrafish in the spotlight. Curr Opin Neurobiol. 2009;19:553–60. https://doi.org/10.1016/j.conb.2009.08.001.
21. Friedrich RW, Genoud C, Wanner AA. Analyzing the structure and function of neuronal circuits in zebrafish. Front Neural Circuits. 2013;7:71. https://doi.org/10.3389/fncir.2013.00071.
22. Orger MB, de Polavieja GG. Zebrafish behavior: opportunities and challenges. Annu Rev Neurosci. 2017;40:125–47. https://doi.org/10.1146/annurev-neuro-071714-033857.
23. Cong L, et al. Rapid whole brain imaging of neural activity in freely behaving larval zebrafish (Danio rerio). elife. 2017;6 https://doi.org/10.7554/eLife.28158.
24. Ahrens MB, et al. Brain-wide neuronal dynamics during motor adaptation in zebrafish. Nature. 2012;485:471–7. https://doi.org/10.1038/nature11057.
25. Chen X, et al. Brain-wide organization of neuronal activity and convergent sensorimotor transformations in Larval zebrafish. Neuron. 2018;100:876–890 e875. https://doi.org/10.1016/j.neuron.2018.09.042.
26. Wanner AA, Vishwanathan A. Methods for mapping neuronal activity to synaptic connectivity: lessons from larval zebrafish. Front Neural Circuits. 2018;12:89. https://doi.org/10.3389/fncir.2018.00089.
27. Portugues R, Feierstein CE, Engert F, Orger MB. Whole-brain activity maps reveal stereotyped, distributed networks for visuomotor behavior. Neuron. 2014;81:1328–43. https://doi.org/10.1016/j.neuron.2014.01.019.
28. Dal Maschio M, Donovan JC, Helmbrecht TO, Baier H. Linking neurons to network function and behavior by two-photon holographic optogenetics and volumetric imaging. Neuron. 2017;94:774–789 e775. https://doi.org/10.1016/j.neuron.2017.04.034.
29. Kettleborough RN, et al. A systematic genome-wide analysis of zebrafish protein-coding gene function. Nature. 2013;496:494–7. https://doi.org/10.1038/nature11992.
30. Rihel J, Schier AF. Behavioral screening for neuroactive drugs in zebrafish. Dev Neurobiol. 2012;72:373–85. https://doi.org/10.1002/dneu.20910.

31. Rihel J, et al. Zebrafish behavioral profiling links drugs to biological targets and rest/wake regulation. Science. 2010;327:348–51. https://doi.org/10.1126/science.1183090.
32. Collins JE, White S, Searle SM, Stemple DL. Incorporating RNA-seq data into the zebrafish Ensembl genebuild. Genome Res. 2012;22:2067–78. https://doi.org/10.1101/gr.137901.112.
33. Postlethwait J, Amores A, Force A, Yan Y-L. In: William Detrich H, Westerfield M, Leonard IZ, editors. Methods in cell biology, vol. 60. Academic Press; 1998. p. 149–63.
34. Meyer A, Schartl M. Gene and genome duplications in vertebrates: the one-to-four (-to-eight in fish) rule and the evolution of novel gene functions. Curr Opin Cell Biol. 1999;11:699–704. https://doi.org/10.1016/s0955-0674(99)00039-3.
35. Taylor JS, Braasch I, Frickey T, Meyer A, Van de Peer Y. Genome duplication, a trait shared by 22000 species of ray-finned fish. Genome Res. 2003;13:382–90. https://doi.org/10.1101/gr.640303.
36. Köster R, Sassen W. A molecular toolbox for genetic manipulation of zebrafish. Advances in Genomics and Genetics. 2015;5:151. https://doi.org/10.2147/AGG.S57585.
37. Huang P, Zhu Z, Lin S, Zhang B. Reverse genetic approaches in zebrafish. J Genet Genomics. 2012;39:421–33. https://doi.org/10.1016/j.jgg.2012.07.004.
38. Varshney GK, Sood R, Burgess SM. Understanding and editing the zebrafish genome. Adv Genet. 2015;92:1–52. https://doi.org/10.1016/bs.adgen.2015.09.002.
39. Bedell VM, et al. In vivo genome editing using a high-efficiency TALEN system. Nature. 2012;491:114–8. https://doi.org/10.1038/nature11537.
40. Hsu PD, Lander ES, Zhang F. Development and applications of CRISPR-Cas9 for genome engineering. Cell. 2014;157:1262–78. https://doi.org/10.1016/j.cell.2014.05.010.
41. Jinek M, et al. A programmable dual-RNA-guided DNA endonuclease in adaptive bacterial immunity. Science. 2012;337:816–21. https://doi.org/10.1126/science.1225829.
42. Hendricks JC, Sehgal A, Pack AI. The need for a simple animal model to understand sleep. Prog Neurobiol. 2000;61:339–51. https://doi.org/10.1016/s0301-0082(99)00048-9.
43. Hartse KM., in Handbook of Clinical Neurology, Vol. 98 (eds. P Montagna, S Chokroverty) 97–109 (Elsevier, 2011).
44. Allada R, Siegel JM. Unearthing the phylogenetic roots of sleep. Curr Biol. 2008;18:R670–9. https://doi.org/10.1016/j.cub.2008.06.033.
45. Cirelli C, Tononi G. Is sleep essential? PLoS Biol. 2008;6:e216. https://doi.org/10.1371/journal.pbio.0060216.
46. Medeiros SLS, et al. Cyclic alternation of quiet and active sleep states in the octopus. iScience. 2021;24:102223. https://doi.org/10.1016/j.isci.2021.102223.
47. Anafi RC, Kayser MS, Raizen DM. Exploring phylogeny to find the function of sleep. Nat Rev Neurosci. 2019;20:109–16. https://doi.org/10.1038/s41583-018-0098-9.
48. Yang Q, Pando BF, Dong G, Golden SS, van Oudenaarden A. Circadian gating of the cell cycle revealed in single cyanobacterial cells. Science. 2010;327:1522–6. https://doi.org/10.1126/science.1181759.
49. Oikonomou G, Prober DA. Attacking sleep from a new angle: contributions from zebrafish. Curr Opin Neurobiol. 2017;44:80–8. https://doi.org/10.1016/j.conb.2017.03.009.
50. Zhdanova IV. Sleep and its regulation in zebrafish. Rev Neurosci. 2011;22:27–36. https://doi.org/10.1515/RNS.2011.005.
51. Yokogawa T, et al. Characterization of sleep in zebrafish and insomnia in hypocretin receptor mutants. PLoS Biol. 2007;5:e277. https://doi.org/10.1371/journal.pbio.0050277.
52. Appelbaum L, et al. Sleep-wake regulation and hypocretin-melatonin interaction in zebrafish. Proc Natl Acad Sci USA. 2009;106:21942–7. https://doi.org/10.1073/pnas.906637106.
53. Leung LC, et al. Neural signatures of sleep in zebrafish. Nature. 2019;571:198–204. https://doi.org/10.1038/s41586-019-1336-7.
54. Zhdanova IV, Wang SY, Leclair OU, Danilova NP. Melatonin promotes sleep-like state in zebrafish. Brain Res. 2001;903:263–8. https://doi.org/10.1016/s0006-8993(01)02444-1.
55. Gandhi AV, Mosser EA, Oikonomou G, Prober DA. Melatonin is required for the circadian regulation of sleep. Neuron. 2015;85:1193–9. https://doi.org/10.1016/j.neuron.2015.02.016.

56. Harding EC, Franks NP, Wisden W. The temperature dependence of sleep. Front Neurosci. 2019;13:336. https://doi.org/10.3389/fnins.2019.00336.
57. Jhaveri KA, Trammell RA, Toth LA. Effect of environmental temperature on sleep, locomotor activity, core body temperature and immune responses of C57BL/6J mice. Brain Behav Immun. 2007;21:975–87. https://doi.org/10.1016/j.bbi.2007.03.007.
58. Singh C, Rihel J, Prober DA. Neuropeptide Y regulates sleep by modulating noradrenergic signaling. Curr Biol. 2017;27:3796–3811 e3795. https://doi.org/10.1016/j.cub.2017.11.018.
59. Ji RR, Baba H, Brenner GJ, Woolf CJ. Nociceptive-specific activation of ERK in spinal neurons contributes to pain hypersensitivity. Nat Neurosci. 1999;2:1114–9. https://doi.org/10.1038/16040.
60. Randlett O, et al. Whole-brain activity mapping onto a zebrafish brain atlas. Nat Methods. 2015;12:1039–46. https://doi.org/10.1038/nmeth.3581.
61. Prober DA, Rihel J, Onah AA, Sung RJ, Schier AF. Hypocretin/orexin overexpression induces an insomnia-like phenotype in zebrafish. J Neurosci. 2006;26:13400–10. https://doi.org/10.1523/JNEUROSCI.4332-06.2006.
62. Kaneko M, Cahill GM. Light-dependent development of circadian gene expression in transgenic zebrafish. PLoS Biol. 2005;3:e34. https://doi.org/10.1371/journal.pbio.0030034.
63. Hurd MW, Cahill GM. Entraining signals initiate behavioral circadian rhythmicity in larval zebrafish. J Biol Rhythm. 2002;17:307–14. https://doi.org/10.1177/074873002129002618.
64. Kazimi N, Cahill GM. Development of a circadian melatonin rhythm in embryonic zebrafish. Brain Res Dev Brain Res. 1999;117:47–52. https://doi.org/10.1016/s0165-3806(99)00096-6.
65. Cahill GM, Hurd MW, Batchelor MM. Circadian rhythmicity in the locomotor activity of larval zebrafish. Neuroreport. 1998;9:3445–9. https://doi.org/10.1097/00001756-199810260-00020.
66. Tamai TK, Young LC, Whitmore D. Light signaling to the zebrafish circadian clock by Cryptochrome 1a. Proc Natl Acad Sci. 2007;104:14712–7. https://doi.org/10.1073/pnas.0704588104.
67. Delaunay F, Thisse C, Marchand O, Laudet V, Thisse B. An inherited functional circadian clock in zebrafish embryos. Science. 2000;289:297–300. https://doi.org/10.1126/science.289.5477.297.
68. Whitmore D, Foulkes NS, Strahle U, Sassone-Corsi P. Zebrafish clock rhythmic expression reveals independent peripheral circadian oscillators. Nat Neurosci. 1998;1:701–7. https://doi.org/10.1038/3703.
69. Cahill GM. Clock mechanisms in zebrafish. Cell Tissue Res. 2002;309:27–34. https://doi.org/10.1007/s00441-002-0570-7.
70. Vatine G, Vallone D, Gothilf Y, Foulkes NS. It's time to swim! Zebrafish and the circadian clock. FEBS Lett. 2011;585:1485–94. https://doi.org/10.1016/j.febslet.2011.04.007.
71. Dekens MP, Whitmore D. Autonomous onset of the circadian clock in the zebrafish embryo. EMBO J. 2008;27:2757–65. https://doi.org/10.1038/emboj.2008.183.
72. Pando MP, Pinchak AB, Cermakian N, Sassone-Corsi P. A cell-based system that recapitulates the dynamic light-dependent regulation of the vertebrate clock. Proc Natl Acad Sci USA. 2001;98:10178–83. https://doi.org/10.1073/pnas.181228598.
73. Kobayashi Y, et al. Molecular analysis of zebrafish photolyase/cryptochrome family: two types of cryptochromes present in zebrafish. Genes Cells. 2000;5:725–38. https://doi.org/10.1046/j.1365-2443.2000.00364.x.
74. Cermakian N, Whitmore D, Foulkes NS, Sassone-Corsi P. Asynchronous oscillations of two zebrafish CLOCK partners reveal differential clock control and function. Proc Natl Acad Sci USA. 2000;97:4339–44. https://doi.org/10.1073/pnas.97.8.4339.
75. Mosser EA, et al. Identification of pathways that regulate circadian rhythms using a larval zebrafish small molecule screen. Sci Rep. 2019;9:12405. https://doi.org/10.1038/s41598-019-48914-7.
76. Whitmore D, Foulkes NS, Sassone-Corsi P. Light acts directly on organs and cells in culture to set the vertebrate circadian clock. Nature. 2000;404:87–91. https://doi.org/10.1038/35003589.

77. Carr AJ, et al. Light reaches the very heart of the zebrafish clock. Chronobiol Int. 2006;23:91–100. https://doi.org/10.1080/07420520500464395.
78. Steindal IAF, Whitmore D. Zebrafish circadian clock entrainment and the importance of broad spectral light sensitivity. Front Physiol. 2020;11:1002. https://doi.org/10.3389/fphys.2020.01002.
79. Fernandes AM, et al. Deep brain photoreceptors control light-seeking behavior in zebrafish larvae. Curr Biol. 2012;22:2042–7. https://doi.org/10.1016/j.cub.2012.08.016.
80. Moore HA, Whitmore D. Circadian rhythmicity and light sensitivity of the zebrafish brain. PLoS One. 2014;9:e86176. https://doi.org/10.1371/journal.pone.0086176.
81. Tamai TK, Carr AJ, Whitmore D. Zebrafish circadian clocks: cells that see light. Biochem Soc Trans. 2005;33:962–6. https://doi.org/10.1042/BST20050962.
82. Hartmann S, et al. Zebrafish larvae show negative phototaxis to near-infrared light. PLoS One. 2018;13:e0207264. https://doi.org/10.1371/journal.pone.0207264.
83. Lin Q, Jesuthasan S. Masking of a circadian behavior in larval zebrafish involves the thalamo-habenula pathway. Sci Rep. 2017;7:4104. https://doi.org/10.1038/s41598-017-04205-7.
84. Keene AC, Duboue ER. The origins and evolution of sleep. J Exp Biol. 2018;221 https://doi.org/10.1242/jeb.159533.
85. Doyle S, Menaker M. Circadian photoreception in vertebrates. Cold Spring Harb Symp Quant Biol. 2007;72:499–508. https://doi.org/10.1101/sqb.2007.72.003.
86. Arendt J. Melatonin and the pineal gland: influence on mammalian seasonal and circadian physiology. Rev Reprod. 1998;3:13–22. https://doi.org/10.1530/ror.0.0030013.
87. Chen S, et al. Light-dependent regulation of sleep and wake states by Prokineticin 2 in zebrafish. Neuron. 2017;95:153–168 e156. https://doi.org/10.1016/j.neuron.2017.06.001.
88. Ben-Moshe Livne Z, et al. Genetically blocking the zebrafish pineal clock affects circadian behavior. PLoS Genet. 2016;12:e1006445. https://doi.org/10.1371/journal.pgen.1006445.
89. Underwood H. The pineal and melatonin: regulators of circadian function in lower vertebrates. Experientia. 1990;46:120–8. https://doi.org/10.1007/BF01955437.
90. Noche RR, Lu PN, Goldstein-Kral L, Glasgow E, Liang JO. Circadian rhythms in the pineal organ persist in zebrafish larvae that lack ventral brain. BMC Neurosci. 2011;12:7. https://doi.org/10.1186/1471-2202-12-7.
91. Panula P, et al. Modulatory neurotransmitter systems and behavior: towards zebrafish models of neurodegenerative diseases. Zebrafish. 2006;3:235–47. https://doi.org/10.1089/zeb.2006.3.235.
92. Inutsuka A, Yamanaka A. The physiological role of orexin/hypocretin neurons in the regulation of sleep/wakefulness and neuroendocrine functions. Front Endocrinol (Lausanne). 2013;4:18. https://doi.org/10.3389/fendo.2013.00018.
93. Ohno K, Sakurai T. Orexin neuronal circuitry: role in the regulation of sleep and wakefulness. Front Neuroendocrinol. 2008;29:70–87. https://doi.org/10.1016/j.yfrne.2007.08.001.
94. Peyron C, et al. Neurons containing hypocretin (orexin) project to multiple neuronal systems. J Neurosci. 1998;18:9996–10015. https://doi.org/10.1523/JNEUROSCI.18-23-09996.1998.
95. de Lecea L, et al. The hypocretins: hypothalamus-specific peptides with neuroexcitatory activity. Proc Natl Acad Sci USA. 1998;95:322–7. https://doi.org/10.1073/pnas.95.1.322.
96. de Lecea L, Sutcliffe JG. The hypocretins and sleep. FEBS J. 2005;272:5675–88. https://doi.org/10.1111/j.1742-4658.2005.04981.x.
97. Siegel JM. Hypocretin (orexin): role in normal behavior and neuropathology. Annu Rev Psychol. 2004;55:125–48. https://doi.org/10.1146/annurev.psych.55.090902.141545.
98. Sakurai T. The neural circuit of orexin (hypocretin): maintaining sleep and wakefulness. Nat Rev Neurosci. 2007;8:171–81. https://doi.org/10.1038/nrn2092.
99. Chemelli RM, et al. Narcolepsy in orexin knockout mice: molecular genetics of sleep regulation. Cell. 1999;98:437–51. https://doi.org/10.1016/s0092-8674(00)81973-x.
100. Lin L, et al. The sleep disorder canine narcolepsy is caused by a mutation in the hypocretin (orexin) receptor 2 gene. Cell. 1999;98:365–76. https://doi.org/10.1016/s0092-8674(00)81965-0.

101. Willie JT, et al. Distinct narcolepsy syndromes in Orexin receptor-2 and Orexin null mice: molecular genetic dissection of Non-REM and REM sleep regulatory processes. Neuron. 2003;38:715–30. https://doi.org/10.1016/s0896-6273(03)00330-1.
102. Hara J, et al. Genetic ablation of orexin neurons in mice results in narcolepsy, hypophagia, and obesity. Neuron. 2001;30:345–54. https://doi.org/10.1016/s0896-6273(01)00293-8.
103. Peyron C, et al. A mutation in a case of early onset narcolepsy and a generalized absence of hypocretin peptides in human narcoleptic brains. Nat Med. 2000;6:991–7. https://doi.org/10.1038/79690.
104. Thannickal TC, et al. Reduced number of hypocretin neurons in human narcolepsy. Neuron. 2000;27:469–74. https://doi.org/10.1016/s0896-6273(00)00058-1.
105. Elbaz I, Levitas-Djerbi T, Appelbaum L. The hypocretin/orexin neuronal networks in zebrafish. Curr Top Behav Neurosci. 2017;33:75–92. https://doi.org/10.1007/7854_2016_59.
106. Elbaz I, Yelin-Bekerman L, Nicenboim J, Vatine G, Appelbaum L. Genetic ablation of hypocretin neurons alters behavioral state transitions in zebrafish. J Neurosci. 2012;32:12961–72. https://doi.org/10.1523/JNEUROSCI.1284-12.2012.
107. Kaslin J, Nystedt JM, Ostergard M, Peitsaro N, Panula P. The orexin/hypocretin system in zebrafish is connected to the aminergic and cholinergic systems. J Neurosci. 2004;24:2678–89. https://doi.org/10.1523/JNEUROSCI.4908-03.2004.
108. Faraco JH, et al. Regulation of hypocretin (orexin) expression in embryonic zebrafish. J Biol Chem. 2006;281:29753–61. https://doi.org/10.1074/jbc.M605811200.
109. Naumann EA, Kampff AR, Prober DA, Schier AF, Engert F. Monitoring neural activity with bioluminescence during natural behavior. Nat Neurosci. 2010;13:513–20. https://doi.org/10.1038/nn.2518.
110. Yelin-Bekerman L, et al. Hypocretin neuron-specific transcriptome profiling identifies the sleep modulator Kcnh4a. elife. 2015;4:e08638. https://doi.org/10.7554/eLife.08638.
111. Cipolla-Neto J, Amaral FGD. Melatonin as a hormone: new physiological and clinical insights. Endocr Rev. 2018;39:990–1028. https://doi.org/10.1210/er.2018-00084.
112. Lima-Cabello E, et al. A review of the melatonin functions in zebrafish physiology. J Pineal Res. 2014;57:1–9. https://doi.org/10.1111/jpi.12149.
113. Falcon J, Besseau L, Sauzet S, Boeuf G. Melatonin effects on the hypothalamo-pituitary axis in fish. Trends Endocrinol Metab. 2007;18:81–8. https://doi.org/10.1016/j.tem.2007.01.002.
114. Begay V, Falcon J, Cahill GM, Klein DC, Coon SL. Transcripts encoding two melatonin synthesis enzymes in the teleost pineal organ: circadian regulation in pike and zebrafish, but not in trout. Endocrinology. 1998;139:905–12. https://doi.org/10.1210/endo.139.3.5790.
115. Hastings MH, Vance G, Maywood E. Some reflections on the phylogeny and function of the pineal. Experientia. 1989;45:903–9.
116. Underwood H, Goldman BD. Vertebrate circadian and photoperiodic systems: role of the pineal gland and melatonin. J Biol Rhythm. 1987;2:279–315. https://doi.org/10.1177/074873048700200404.
117. Zachmann A, Ali MA, Falc6n J., in Rhythms in fishes (ed MA Ali) 149–165 (Springer US, 1992).
118. Ekstrzm P, Meissl H. The pineal organ of teleost fishes. Rev Fish Biol Fish. 1997;7:199–284. https://doi.org/10.1023/A:1018483627058.
119. Ben-Moshe Z, et al. The light-induced transcriptome of the zebrafish pineal gland reveals complex regulation of the circadian clockwork by light. Nucleic Acids Res. 2014;42:3750–67. https://doi.org/10.1093/nar/gkt1359.
120. Reppert SM, Weaver DR, Cassone VM, Godson C, Kolakowski LF Jr. Melatonin receptors are for the birds: molecular analysis of two receptor subtypes differentially expressed in chick brain. Neuron. 1995;15:1003–15. https://doi.org/10.1016/0896-6273(95)90090-x.
121. Lynch HJ, Wurtman RJ, Moskowitz MA, Archer MC, Ho MH. Daily rhythm in human urinary melatonin. Science. 1975;187:169–71. https://doi.org/10.1126/science.1167425.

122. Akerstedt T, Froberg JE, Friberg Y, Wetterberg L. Melatonin excretion, body temperature and subjective arousal during 64 hours of sleep deprivation. Psychoneuroendocrinology. 1979;4:219–25. https://doi.org/10.1016/0306-4530(79)90005-2.
123. Tzischinsky O, Shlitner A, Lavie P. The association between the nocturnal sleep gate and nocturnal onset of urinary 6-sulfatoxymelatonin. J Biol Rhythm. 1993;8:199–209. https://doi.org/10.1177/074873049300800303.
124. Zhdanova IV, Wurtman RJ, Morabito C, Piotrovska VR, Lynch HJ. Effects of low oral doses of melatonin, given 2-4 hours before habitual bedtime, on sleep in normal young humans. Sleep. 1996;19:423–31. https://doi.org/10.1093/sleep/19.5.423.
125. Dollins AB, Zhdanova IV, Wurtman RJ, Lynch HJ, Deng MH. Effect of inducing nocturnal serum melatonin concentrations in daytime on sleep, mood, body temperature, and performance. Proc Natl Acad Sci USA. 1994;91:1824–8. https://doi.org/10.1073/pnas.91.5.1824.
126. Stone BM, Turner C, Mills SL, Nicholson AN. Hypnotic activity of melatonin. Sleep. 2000;23:663–9.
127. Ebihara S, Marks T, Hudson DJ, Menaker M. Genetic control of melatonin synthesis in the pineal gland of the mouse. Science. 1986;231:491–3. https://doi.org/10.1126/science.3941912.
128. Joiner WJ. Unraveling the evolutionary determinants of sleep. Curr Biol. 2016;26:R1073–87. https://doi.org/10.1016/j.cub.2016.08.068.
129. Panula P, et al. The comparative neuroanatomy and neurochemistry of zebrafish CNS systems of relevance to human neuropsychiatric diseases. Neurobiol Dis. 2010;40:46–57. https://doi.org/10.1016/j.nbd.2010.05.010.
130. Berman JR, Skariah G, Maro GS, Mignot E, Mourrain P. Characterization of two melanin-concentrating hormone genes in zebrafish reveals evolutionary and physiological links with the mammalian MCH system. J Comp Neurol. 2009;517:695–710. https://doi.org/10.1002/cne.22171.
131. Peyron C, Sapin E, Leger L, Luppi PH, Fort P. Role of the melanin-concentrating hormone neuropeptide in sleep regulation. Peptides. 2009;30:2052–9. https://doi.org/10.1016/j.peptides.2009.07.022.
132. Oikonomou G, et al. The serotonergic raphe promote sleep in zebrafish and mice. Neuron. 2019;103:686–701 e688. https://doi.org/10.1016/j.neuron.2019.05.038.
133. Candy J, Collet C. Two tyrosine hydroxylase genes in teleosts. Biochim Biophys Acta. 2005;1727:35–44. https://doi.org/10.1016/j.bbaexp.2004.11.005.
134. Sehgal A, Mignot E. Genetics of sleep and sleep disorders. Cell. 2011;146:194–207. https://doi.org/10.1016/j.cell.2011.07.004.
135. Renier C, et al. Genomic and functional conservation of sedative-hypnotic targets in the zebrafish. Pharmacogenet Genomics. 2007;17:237–53. https://doi.org/10.1097/FPC.0b013e3280119d62.
136. Sundvik M, et al. The histaminergic system regulates wakefulness and orexin/hypocretin neuron development via histamine receptor H1 in zebrafish. FASEB J. 2011;25:4338–47. https://doi.org/10.1096/fj.11-188268.
137. Chen A, Singh C, Oikonomou G, Prober DA. Genetic analysis of histamine signaling in larval zebrafish sleep. eNeuro. 2017;4 https://doi.org/10.1523/ENEURO.0286-16.2017.
138. Panula P, Chen Y-C, Baronio D, Lewis S, Sundvik M. In: Yanai K, Passani MB, editors. The functional roles of histamine receptors. Springer International Publishing; 2022. p. 291–302.
139. Ruuskanen JO, Peitsaro N, Kaslin JV, Panula P, Scheinin M. Expression and function of alpha-adrenoceptors in zebrafish: drug effects, mRNA and receptor distributions. J Neurochem. 2005;94:1559–69. https://doi.org/10.1111/j.1471-4159.2005.03305.x.
140. Ruuskanen JO, et al. Conserved structural, pharmacological and functional properties among the three human and five zebrafish alpha 2-adrenoceptors. Br J Pharmacol. 2005;144:165–77. https://doi.org/10.1038/sj.bjp.0706057.

141. Singh A, Subhashini N, Sharma S, Mallick BN. Involvement of the α1-adrenoceptor in sleep–waking and sleep loss-induced anxiety behavior in zebrafish. Neuroscience. 2013;245:136–47. https://doi.org/10.1016/j.neuroscience.2013.04.026.
142. Singh C, Oikonomou G, Prober DA. Norepinephrine is required to promote wakefulness and for hypocretin-induced arousal in zebrafish. elife. 2015;4:e07000. https://doi.org/10.7554/eLife.07000.
143. Sigurgeirsson B, þorsteinsson H, Arnardóttir H, Jóhannesdóttir I þ, Karlsson KÆ. Effects of modafinil on sleep–wake cycles in Larval zebrafish. Zebrafish. 2011;8:133–40. https://doi.org/10.1089/zeb.2011.0708.
144. Borbely AA. A two process model of sleep regulation. Hum Neurobiol. 1982;1:195–204.
145. Borbely AA, Achermann P. Sleep homeostasis and models of sleep regulation. J Biol Rhythm. 1999;14:557–68. https://doi.org/10.1177/074873099129000894.
146. Tononi G, Cirelli C. Sleep and synaptic down-selection. Eur J Neurosci. 2020;51:413–21. https://doi.org/10.1111/ejn.14335.
147. Tononi G, Cirelli C. Sleep and the price of plasticity: from synaptic and cellular homeostasis to memory consolidation and integration. Neuron. 2014;81:12–34. https://doi.org/10.1016/j.neuron.2013.12.025.
148. Appelbaum L, et al. Circadian and homeostatic regulation of structural synaptic plasticity in hypocretin neurons. Neuron. 2010;68:87–98. https://doi.org/10.1016/j.neuron.2010.09.006.
149. Bjorness TE, et al. An adenosine-mediated glial-neuronal circuit for homeostatic sleep. J Neurosci. 2016;36:3709–21. https://doi.org/10.1523/JNEUROSCI.3906-15.2016.
150. Diering GH, et al. Homer1a drives homeostatic scaling-down of excitatory synapses during sleep. Science. 2017;355:511–5. https://doi.org/10.1126/science.aai8355.
151. Porkka-Heiskanen T, et al. Adenosine: a mediator of the sleep-inducing effects of prolonged wakefulness. Science. 1997;276:1265–8. https://doi.org/10.1126/science.276.5316.1265.
152. Scharf MT, Naidoo N, Zimmerman JE, Pack AI. The energy hypothesis of sleep revisited. Prog Neurobiol. 2008;86:264–80. https://doi.org/10.1016/j.pneurobio.2008.08.003.
153. Landolt HP, et al. Caffeine attenuates waking and sleep electroencephalographic markers of sleep homeostasis in humans. Neuropsychopharmacology. 2004;29:1933–9. https://doi.org/10.1038/sj.npp.1300526.
154. Boehmler W, et al. Identification of zebrafish A2 adenosine receptors and expression in developing embryos. Gene Expr Patterns. 2009;9:144–51. https://doi.org/10.1016/j.gep.2008.11.006.
155. Clayman CL, Connaughton VP. Neurochemical and behavioral consequences of ethanol and/or caffeine exposure: effects in zebrafish and rodents. Curr Neuropharmacol. 2022;20:560–78. https://doi.org/10.2174/1570159X19666211111142027.
156. Reichert S, Pavon Arocas O, Rihel J. The neuropeptide Galanin is required for homeostatic rebound sleep following increased neuronal activity. Neuron. 2019;104:370–384 e375. https://doi.org/10.1016/j.neuron.2019.08.010.
157. Bellesi M, Riedner BA, Garcia-Molina GN, Cirelli C, Tononi G. Enhancement of sleep slow waves: underlying mechanisms and practical consequences. Front Syst Neurosci. 2014;8:208. https://doi.org/10.3389/fnsys.2014.00208.
158. Finelli LA, Borbely AA, Achermann P. Functional topography of the human nonREM sleep electroencephalogram. Eur J Neurosci. 2001;13:2282–90. https://doi.org/10.1046/j.0953-816x.2001.01597.x.
159. Huber R, Ghilardi MF, Massimini M, Tononi G. Local sleep and learning. Nature. 2004;430:78–81. https://doi.org/10.1038/nature02663.
160. Werth E, Achermann P, Borbely AA. Fronto-occipital EEG power gradients in human sleep. J Sleep Res. 1997;6:102–12. https://doi.org/10.1046/j.1365-2869.1997.d01-36.x.
161. Krueger JM, Obal F. A neuronal group theory of sleep function. J Sleep Res. 1993;2:63–9. https://doi.org/10.1111/j.1365-2869.1993.tb00064.x.
162. Krueger JM, et al. Sleep as a fundamental property of neuronal assemblies. Nat Rev Neurosci. 2008;9:910–9. https://doi.org/10.1038/nrn2521.

163. Krueger JM, Nguyen JT, Dykstra-Aiello CJ, Taishi P. Local sleep. Sleep Med Rev. 2019;43:14–21. https://doi.org/10.1016/j.smrv.2018.10.001.
164. Roy S, Krueger JM, Rector DM, Wan Y. A network model for activity-dependent sleep regulation. J Theor Biol. 2008;253:462–8. https://doi.org/10.1016/j.jtbi.2008.03.033.
165. Krueger JM, Obal F Jr. Sleep function. Front Biosci. 2003;8:d511–9. https://doi.org/10.2741/1031.
166. Donlea JM. Neuronal and molecular mechanisms of sleep homeostasis. Curr Opin Insect Sci. 2017;24:51–7. https://doi.org/10.1016/j.cois.2017.09.008.
167. Allada R, Cirelli C, Sehgal A. Molecular mechanisms of sleep homeostasis in flies and mammals. Cold Spring Harb Perspect Biol. 2017;9 https://doi.org/10.1101/cshperspect.a027730.
168. Alam MA, Kumar S, McGinty D, Alam MN, Szymusiak R. Neuronal activity in the preoptic hypothalamus during sleep deprivation and recovery sleep. J Neurophysiol. 2014;111:287–99. https://doi.org/10.1152/jn.00504.2013.
169. Szymusiak R, Alam N, Steininger TL, McGinty D. Sleep-waking discharge patterns of ventrolateral preoptic/anterior hypothalamic neurons in rats. Brain Res. 1998;803:178–88. https://doi.org/10.1016/s0006-8993(98)00631-3.
170. McGinty D, Szymusiak R. Hypothalamic regulation of sleep and arousal. Front Biosci. 2003;8:s1074–83. https://doi.org/10.2741/1159.
171. Arrigoni E, Fuller PM. The sleep-promoting ventrolateral preoptic nucleus: what have we learned over the past 25 years? Int J Mol Sci. 2022;23 https://doi.org/10.3390/ijms23062905.
172. Kroeger D, et al. Galanin neurons in the ventrolateral preoptic area promote sleep and heat loss in mice. Nat Commun. 2018;9:4129. https://doi.org/10.1038/s41467-018-06590-7.
173. Gaus SE, Strecker RE, Tate BA, Parker RA, Saper CB. Ventrolateral preoptic nucleus contains sleep-active, galaninergic neurons in multiple mammalian species. Neuroscience. 2002;115:285–94. https://doi.org/10.1016/s0306-4522(02)00308-1.
174. Guo X, et al. RNA-seq analysis of galaninergic neurons from ventrolateral preoptic nucleus identifies expression changes between sleep and wake. BMC Genomics. 2020;21:633. https://doi.org/10.1186/s12864-020-07050-7.
175. Campbell SS, Tobler I. Animal sleep: a review of sleep duration across phylogeny. Neurosci Biobehav Rev. 1984;8:269–300. https://doi.org/10.1016/0149-7634(84)90054-x.
176. Pack AI, et al. Novel method for high-throughput phenotyping of sleep in mice. Physiol Genomics. 2007;28:232–8. https://doi.org/10.1152/physiolgenomics.00139.2006.
177. Cirelli C. The genetic and molecular regulation of sleep: from fruit flies to humans. Nat Rev Neurosci. 2009;10:549–60. https://doi.org/10.1038/nrn2683.
178. Zimmerman JE, Naidoo N, Raizen DM, Pack AI. Conservation of sleep: insights from non-mammalian model systems. Trends Neurosci. 2008;31:371–6. https://doi.org/10.1016/j.tins.2008.05.001.
179. Lee DA, Oikonomou G, Prober DA. Large-scale analysis of sleep in zebrafish. Bio Protoc. 2022;12:e4313. https://doi.org/10.21769/BioProtoc.4313.
180. Aho V, et al. Homeostatic response to sleep/rest deprivation by constant water flow in larval zebrafish in both dark and light conditions. J Sleep Res. 2017;26:394–400. https://doi.org/10.1111/jsr.12508.
181. Zimmerman JE, Raizen DM, Maycock MH, Maislin G, Pack AI. A video method to study Drosophila sleep. Sleep. 2008;31:1587–98. https://doi.org/10.1093/sleep/31.11.1587.
182. Mueller T, Dong Z, Berberoglu MA, Guo S. The dorsal pallium in zebrafish, Danio rerio (Cyprinidae, Teleostei). Brain Res. 2011;1381:95–105. https://doi.org/10.1016/j.brainres.2010.12.089.
183. Freeman J, et al. Mapping brain activity at scale with cluster computing. Nat Methods. 2014;11:941–50. https://doi.org/10.1038/nmeth.3041.
184. Cancedda L, et al. Patterned vision causes CRE-mediated gene expression in the visual cortex through PKA and ERK. J Neurosci. 2003;23:7012–20. https://doi.org/10.1523/JNEUROSCI.23-18-07012.2003.

185. Hussain A, et al. High-affinity olfactory receptor for the death-associated odor cadaverine. Proc Natl Acad Sci USA. 2013;110:19579–84. https://doi.org/10.1073/pnas.1318596110.
186. Ashlin TG, Blunsom NJ, Ghosh M, Cockcroft S, Rihel J. Pitpnc1a regulates zebrafish sleep and wake behavior through modulation of insulin-like growth factor signaling. Cell Rep. 2018;24:1389–96. https://doi.org/10.1016/j.celrep.2018.07.012.
187. Sigurgeirsson B, et al. Sleep-wake dynamics under extended light and extended dark conditions in adult zebrafish. Behav Brain Res. 2013;256:377–90. https://doi.org/10.1016/j.bbr.2013.08.032.
188. Talbot WS, Hopkins N. Zebrafish mutations and functional analysis of the vertebrate genome. Genes Dev. 2000;14:755–62.
189. Sung YH, et al. Highly efficient gene knockout in mice and zebrafish with RNA-guided endonucleases. Genome Res. 2014;24:125–31. https://doi.org/10.1101/gr.163394.113.
190. Baier H, et al. Genetic dissection of the retinotectal projection. Development. 1996;123:415–25. https://doi.org/10.1242/dev.123.1.415.
191. Mullins MC, Hammerschmidt M, Haffter P, Nusslein-Volhard C. Large-scale mutagenesis in the zebrafish: in search of genes controlling development in a vertebrate. Curr Biol. 1994;4:189–202. https://doi.org/10.1016/s0960-9822(00)00048-8.
192. Nüsslein-Volhard C. The zebrafish issue of development. Development. 2012;139:4099–103. https://doi.org/10.1242/dev.085217.
193. Driever W, et al. A genetic screen for mutations affecting embryogenesis in zebrafish. Development. 1996;123:37–46. https://doi.org/10.1242/dev.123.1.37.
194. Schier AF, et al. Mutations affecting the development of the embryonic zebrafish brain. Development. 1996;123:165–78. https://doi.org/10.1242/dev.123.1.165.
195. Moens CB, Donn TM, Wolf-Saxon ER, Ma TP. Reverse genetics in zebrafish by TILLING. Brief Funct Genomic Proteomic. 2008;7:454–9. https://doi.org/10.1093/bfgp/eln046.
196. Rosen JN, Sweeney MF, Mably JD. Microinjection of zebrafish embryos to analyze gene function. J Vis Exp. 2009; https://doi.org/10.3791/1115.
197. Nasevicius A, Ekker SC. Effective targeted gene 'knockdown' in zebrafish. Nat Genet. 2000;26:216–20. https://doi.org/10.1038/79951.
198. Holm A, et al. The evolutionarily conserved miRNA-137 targets the neuropeptide hypocretin/orexin and modulates the wake to sleep ratio. Proc Natl Acad Sci USA. 2022;119:e2112225119. https://doi.org/10.1073/pnas.2112225119.
199. Srdanovic S, et al. Transient knock-down of kcna2 reduces sleep in larval zebrafish. Behav Brain Res. 2017;326:13–21. https://doi.org/10.1016/j.bbr.2017.02.026.
200. Cirelli C, et al. Reduced sleep in Drosophila Shaker mutants. Nature. 2005;434:1087–92. https://doi.org/10.1038/nature03486.
201. Koh K, et al. Identification of SLEEPLESS, a sleep-promoting factor. Science. 2008;321:372–6. https://doi.org/10.1126/science.1155942.
202. Douglas CL, et al. Sleep in Kcna2 knockout mice. BMC Biol. 2007;5:42. https://doi.org/10.1186/1741-7007-5-42.
203. He Y, et al. The transcriptional repressor DEC2 regulates sleep length in mammals. Science. 2009;325:866–70. https://doi.org/10.1126/science.1174443.
204. Baier PC, et al. Mice lacking the circadian modulators SHARP1 and SHARP2 display altered sleep and mixed state endophenotypes of psychiatric disorders. PLoS One. 2014;9:e110310. https://doi.org/10.1371/journal.pone.0110310.
205. Gerety SS, Wilkinson DG. Morpholino artifacts provide pitfalls and reveal a novel role for pro-apoptotic genes in hindbrain boundary development. Dev Biol. 2011;350:279–89. https://doi.org/10.1016/j.ydbio.2010.11.030.
206. Law SH, Sargent TD. The serine-threonine protein kinase PAK4 is dispensable in zebrafish: identification of a morpholino-generated pseudophenotype. PLoS One. 2014;9:e100268. https://doi.org/10.1371/journal.pone.0100268.

207. Bill BR, Petzold AM, Clark KJ, Schimmenti LA, Ekker SC. A primer for morpholino use in zebrafish. Zebrafish. 2009;6:69–77. https://doi.org/10.1089/zeb.2008.0555.
208. Perrimon N, Ni JQ, Perkins L. In vivo RNAi: today and tomorrow. Cold Spring Harb Perspect Biol. 2010;2:a003640. https://doi.org/10.1101/cshperspect.a003640.
209. Kelly A, Hurlstone AF. The use of RNAi technologies for gene knockdown in zebrafish. Brief Funct Genomics. 2011;10:189–96. https://doi.org/10.1093/bfgp/elr014.
210. Dong M, et al. Heritable and lineage-specific gene knockdown in zebrafish embryo. PLoS One. 2009;4:e6125. https://doi.org/10.1371/journal.pone.0006125.
211. Halpern ME, et al. Gal4/UAS transgenic tools and their application to zebrafish. Zebrafish. 2008;5:97–110. https://doi.org/10.1089/zeb.2008.0530.
212. Shoji W, Sato-Maeda M. Application of heat shock promoter in transgenic zebrafish. Develop Growth Differ. 2008;50:401–6. https://doi.org/10.1111/j.1440-169X.2008.01038.x.
213. Lee DA, et al. Genetic and neuronal regulation of sleep by neuropeptide VF. elife. 2017;6 https://doi.org/10.7554/eLife.25727.
214. Doyon Y, et al. Heritable targeted gene disruption in zebrafish using designed zinc-finger nucleases. Nat Biotechnol. 2008;26:702–8. https://doi.org/10.1038/nbt1409.
215. Meng X, Noyes MB, Zhu LJ, Lawson ND, Wolfe SA. Targeted gene inactivation in zebrafish using engineered zinc-finger nucleases. Nat Biotechnol. 2008;26:695–701. https://doi.org/10.1038/nbt1398.
216. Berridge CW, Schmeichel BE, Espana RA. Noradrenergic modulation of wakefulness/arousal. Sleep Med Rev. 2012;16:187–97. https://doi.org/10.1016/j.smrv.2011.12.003.
217. Chen S, et al. A large-scale in vivo analysis reveals that TALENs are significantly more mutagenic than ZFNs generated using context-dependent assembly. Nucleic Acids Res. 2013;41:2769–78. https://doi.org/10.1093/nar/gks1356.
218. Lee DA, et al. Evolutionarily conserved regulation of sleep by epidermal growth factor receptor signaling. Sci Adv. 2019;5:eaax4249. https://doi.org/10.1126/sciadv.aax4249.
219. Jao LE, Wente SR, Chen W. Efficient multiplex biallelic zebrafish genome editing using a CRISPR nuclease system. Proc Natl Acad Sci USA. 2013;110:13904–9. https://doi.org/10.1073/pnas.1308335110.
220. Hwang WY, et al. Efficient genome editing in zebrafish using a CRISPR-Cas system. Nat Biotechnol. 2013;31:227–9. https://doi.org/10.1038/nbt.2501.
221. Chang N, et al. Genome editing with RNA-guided Cas9 nuclease in zebrafish embryos. Cell Res. 2013;23:465–72. https://doi.org/10.1038/cr.2013.45.
222. Sorlien EL, Witucki MA, Ogas J. Efficient production and identification of CRISPR/Cas9-generated gene knockouts in the model system Danio rerio. J Vis Exp. 2018; https://doi.org/10.3791/56969.
223. Kroll F, et al. A simple and effective F0 knockout method for rapid screening of behaviour and other complex phenotypes. elife. 2021;10 https://doi.org/10.7554/eLife.59683.
224. Palermo J, et al. Variant-to-gene-mapping followed by cross-species genetic screening identifies GPI-anchor biosynthesis as novel regulator of sleep. bioRxiv. 2021:2021.2012.2019.472248, https://doi.org/10.1101/2021.12.19.472248.
225. Zimmerman AJ, et al. Zebrafish screen of high-confidence effector genes at insomnia GWAS loci implicates conserved regulators of sleep-wake behaviors. bioRxiv. 2021:2022.2010.2005.511011. https://doi.org/10.1101/2022.10.05.511011.
226. Thyme SB, et al. Phenotypic landscape of schizophrenia-associated genes defines candidates and their shared functions. Cell. 2019;177:478–491 e420. https://doi.org/10.1016/j.cell.2019.01.048.
227. Tran S, Prober DA. Validation of candidate sleep disorder risk genes using zebrafish. Front Mol Neurosci. 2022;15:873520. https://doi.org/10.3389/fnmol.2022.873520.
228. Del Bene F, Wyart C. Optogenetics: a new enlightenment age for zebrafish neurobiology. Dev Neurobiol. 2012;72:404–14. https://doi.org/10.1002/dneu.20914.
229. Antinucci P, et al. A calibrated optogenetic toolbox of stable zebrafish opsin lines. elife. 2020;9 https://doi.org/10.7554/eLife.54937.

230. Burgess HA, Granato M. Modulation of locomotor activity in larval zebrafish during light adaptation. J Exp Biol. 2007;210:2526–39. https://doi.org/10.1242/jeb.003939.
231. Chen S, Chiu CN, McArthur KL, Fetcho JR, Prober DA. TRP channel mediated neuronal activation and ablation in freely behaving zebrafish. Nat Methods. 2016;13:147–50. https://doi.org/10.1038/nmeth.3691.
232. Chiu CN, et al. A zebrafish genetic screen identifies Neuromedin U as a regulator of sleep/wake states. Neuron. 2016;89:842–56. https://doi.org/10.1016/j.neuron.2016.01.007.
233. Michel MC. Neuropeptide Y and related peptides. Berlin Heidelberg: Springer; 2012.
234. Sohn JW, Elmquist JK, Williams KW. Neuronal circuits that regulate feeding behavior and metabolism. Trends Neurosci. 2013;36:504–12. https://doi.org/10.1016/j.tins.2013.05.003.
235. Yokobori E, et al. Neuropeptide Y stimulates food intake in the zebrafish, Danio rerio. J Neuroendocrinol. 2012;24:766–73. https://doi.org/10.1111/j.1365-2826.2012.02281.x.
236. Clift D, Richendrfer H, Thorn RJ, Colwill RM, Creton R. High-throughput analysis of behavior in zebrafish larvae: effects of feeding. Zebrafish. 2014;11:455–61. https://doi.org/10.1089/zeb.2014.0989.
237. Zada D, Bronshtein I, Lerer-Goldshtein T, Garini Y, Appelbaum L. Sleep increases chromosome dynamics to enable reduction of accumulating DNA damage in single neurons. Nat Commun. 2019;10:895. https://doi.org/10.1038/s41467-019-08806-w.
238. Zada D, et al. Parp1 promotes sleep, which enhances DNA repair in neurons. Mol Cell. 2021;81:4979–4993 e4977. https://doi.org/10.1016/j.molcel.2021.10.026.
239. Wang H, et al. Genome-wide association analysis of self-reported daytime sleepiness identifies 42 loci that suggest biological subtypes. Nat Commun. 2019;10:3503. https://doi.org/10.1038/s41467-019-11456-7.
240. Van Buskirk C, Sternberg PW. Epidermal growth factor signaling induces behavioral quiescence in Caenorhabditis elegans. Nat Neurosci. 2007;10:1300–7. https://doi.org/10.1038/nn1981.
241. Konietzka J, et al. Epidermal growth factor signaling promotes sleep through a combined series and parallel neural circuit. Curr Biol. 2020;30:1–16 e13. https://doi.org/10.1016/j.cub.2019.10.048.
242. Foltenyi K, Greenspan RJ, Newport JW. Activation of EGFR and ERK by rhomboid signaling regulates the consolidation and maintenance of sleep in Drosophila. Nat Neurosci. 2007;10:1160–7. https://doi.org/10.1038/nn1957.
243. Donlea JM, Ramanan N, Shaw PJ. Use-dependent plasticity in clock neurons regulates sleep need in Drosophila. Science. 2009;324:105–8. https://doi.org/10.1126/science.1166657.
244. Harbison ST, McCoy LJ, Mackay TF. Genome-wide association study of sleep in Drosophila melanogaster. BMC Genomics. 2013;14:281. https://doi.org/10.1186/1471-2164-14-281.
245. Harbison ST, Serrano Negron YL, Hansen NF, Lobell AS. Selection for long and short sleep duration in Drosophila melanogaster reveals the complex genetic network underlying natural variation in sleep. PLoS Genet. 2017;13:e1007098. https://doi.org/10.1371/journal.pgen.1007098.
246. Levitas-Djerbi T, Appelbaum L. Modeling sleep and neuropsychiatric disorders in zebrafish. Curr Opin Neurobiol. 2017;44:89–93. https://doi.org/10.1016/j.conb.2017.02.017.
247. Doldur-Balli F, et al. Synaptic dysfunction connects autism spectrum disorder and sleep disturbances: a perspective from studies in model organisms. Sleep Med Rev. 2022;62:101595. https://doi.org/10.1016/j.smrv.2022.101595.
248. Robea MA, et al. Preliminary results regarding sleep in a zebrafish model of autism spectrum disorder. Brain Sci. 2021;11 https://doi.org/10.3390/brainsci11050556.
249. Ozcan GG, Lim S, Leighton P, Allison WT, Rihel J. Sleep is bi-directionally modified by amyloid beta oligomers. elife. 2020;9 https://doi.org/10.7554/eLife.53995.
250. Shenoy A, Banerjee M, Upadhya A, Bagwe-Parab S, Kaur G. The brilliance of the zebrafish model: perception on behavior and Alzheimer's disease. Front Behav Neurosci. 2022;16:861155. https://doi.org/10.3389/fnbeh.2022.861155.
251. Lyons DA, Talbot WS. Glial cell development and function in zebrafish. Cold Spring Harb Perspect Biol. 2014;7:a020586. https://doi.org/10.1101/cshperspect.a020586.

Evolved Loss of Sleep and Circadian Rhythms in Cavefish

Alex C. Keene, Erik R. Duboue, Nicholas S. Foulkes, and Cristiano Bertolucci

Abstract Sleep and locomotor rhythms are nearly universal behaviors that are critical for the function of diverse biological processes. While the timing and duration of sleep vary dramatically throughout the animal kingdom, little is known about the genetic or evolutionary basis underlying species-specific differences in these behaviors. Growing evidence suggests differences in sleep duration and timing are influenced by the ecological environment and evolutionary history. Cave environments insulated from the effects of the day-night cycle provide a unique opportunity to investigate the function and evolution of sleep and circadian clocks. Over 200 species of cavefish have been identified throughout the world, allowing for the investigation of trait evolution in a relatively constant environment. The Mexican cavefish and Somalian cavefish, in particular, have emerged as highly useful laboratory models to investigate the mechanistic and ecological basis for the evolution of circadian rhythms. Furthermore, the Mexican cavefish has provided new insight into the evolution and functional significance of sleep. These species display loss of both circadian processes and light detection under very different environmental conditions and timescales. Here, we review recent findings that identify roles for sensory processing, neuromodulation, and DNA repair pathways in the evolution of sleep and circadian rhythms. Furthermore, we highlight the implementation of gene-editing and transgenic tools that allow for mechanistic studies of the neural basis for the evolution of sleep and circadian rhythms. These studies have potential to shed light on the fundamental functions of sleep-wake regulation and the genetic basis for interspecies trait diversity throughout the animal kingdom.

A. C. Keene (✉)
Department of Biology, Texas A&M University, College Station, TX, USA

E. R. Duboue
Department of Biology, Florida Atlantic University, Jupiter, FL, USA

N. S. Foulkes
Institute for Chemical and Biological Systems, Karlsruhe Institute of Technology, Eggenstein-Leopoldshafen, Germany

C. Bertolucci
Department of Life Sciences and Biotechnology, University of Ferrara, Ferrara, Italy

P. Gehrman et al. (eds.), *Genetics of Sleep and Sleep Disorders*,
https://doi.org/10.1007/978-3-031-62723-1_5

Keywords Evolution · Hypocretin/orexin · Astyanax · Genetic variation

Introduction

Sleep and circadian rhythms are nearly ubiquitous, complex behaviors displayed by most organisms ranging from "primitive" species such as jellyfish and nematodes to humans [1–3]. Species that inhabit diverse environments synchronize the timing of sleep and activity with environmental cues including light and food availability [4, 5]. The circadian clock is a highly conserved, ubiquitous timing mechanism which temporally coordinates most aspects of behavior and physiology in a species-specific fashion, including the timing of sleep [6, 7]. The timing and duration of sleep varies dramatically between species, with numerous species including armadillos and opossums sleeping over 18 h each day and others including elephants and giraffes sleeping only a few hours each day [2, 8]. While the functional relevance of these fundamental differences is poorly understood, species-specific differences in sleep and circadian behavior are both thought to have been shaped by the environment and ecological history. Understanding how the cellular and ecological mechanisms underly the evolution of sleep and circadian behavior should provide insight into the relationship between these traits and their environment.

A central question is how sleep and circadian rhythms are functionally and evolutionarily related. Cavefish provide a unique opportunity to investigate how sleep and clocks evolve in response to extreme environmental perturbation and to uncover physiological, developmental, and behavioral processes regulated by sleep and the circadian clock [1, 4]. Worldwide, over 200 different species of cavefish have been identified, encompassing diverse taxonomic orders [9, 10]. While the subterrestrial aquatic environments inhabited by cavefish differ dramatically, they largely share a number of features including perpetual darkness, reduced predation, and restricted food availability [10, 11]. As a likely consequence, most cave organisms studied to date possess a number of common morphological as well as behavioral traits. For example, most cavefish species exhibit eye loss and albinism, which have been studied since the time of Darwin [12, 13]. However, much less is known about the evolution of behavior in cave environments. Here, we describe recent findings that have documented sleep loss and disrupted circadian activity rhythms in various species of cavefish. In turn, we discuss how these behavioral phenotypes may have been shaped by very different cave environments and how they may help us to elucidate the functional relevance of sleep and its evolution in response to a changing environment.

Investigation of Sleep and Circadian Rhythms in Fish Species

Diverse fish have been widely used as models for investigating various aspects of behavior. However, it is the zebrafish, *Danio rerio*, that has emerged as a widely used model system for studying the genetic and neural basis of vertebrate behavior [14]. Together with its repertoire of genetic tools, this species displays robust circadian and sleep behaviors, providing a basis for detailed mechanistic studies of these traits [15–17]. Typically, experiments in this species use larvae because of their small size, transparency, and amenability to high-throughput behavioral screening [18]. Zebrafish are diurnal with robust activity patterns regulated by a central circadian clock that is highly homologous to that in mammals [19, 20]. Furthermore, clocks in zebrafish peripheral tissues and even cell lines are entrained by direct exposure to light making this species an attractive model for studying generally how light entrains circadian clocks [21].

In animal species ranging from jellyfish through to humans, prolonged periods of immobility associated with a reduced arousal threshold are typically used to define sleep [1, 22, 23]. These behavioral criteria for sleep are robust in zebrafish during the night and have been exploited as a behavioral metric that defines sleep [24, 25]. In zebrafish adults and larvae, 6 s and 1 s of immobility respectively have been used to define sleep since these periods are associated with other conserved behavioral metrics including species-specific posture and reduced arousal threshold [24, 26]. While the field of sleep research in fish and other emerging genetic models has historically focused on behavior as a readout of sleep, recent studies in zebrafish have applied brain imaging to identify neurophysiological signatures of sleep and wakefulness using genetically encoded Ca^{2+} sensors to image the activity of broad brain regions in larval zebrafish and identified sleep stages analogous to slow-wave and paradoxical sleep [27, 28]. Therefore, the neurophysiological signatures of sleep appear to be highly conserved in fish, and functional polysomnography can be applied to complement behavioral analysis in the future.

A strength of fish is their amenability to genetic and pharmacological manipulation. Studies of sleep and circadian behavior in zebrafish have paved the way for genetic and pharmacological screens that have identified novel regulators of these behaviors, many of which are conserved in mammalian systems [29]. Perhaps less appreciated, the approaches developed in zebrafish can be readily applied to other fish models that have unique strengths including evolutionary adaptations resulting from defined ecological constraints and naturally occurring variation in sleep or circadian behavior. The advantages of cavefish as an evolutionary model system make them a powerful comparative model to transfer approaches initially developed in zebrafish and to systematically investigate how sleep and circadian traits have evolved in extreme environments (Fig. 1).

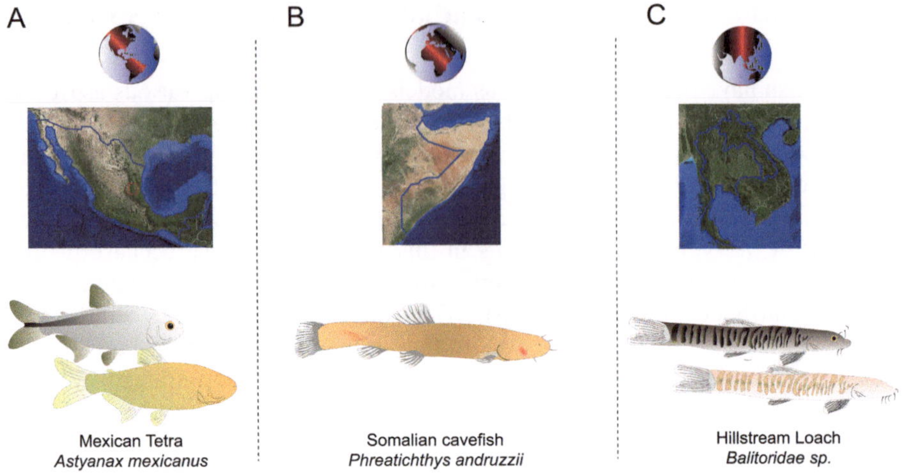

Fig. 1 Geographic location of cavefish used in sleep and circadian research. *Astyanax mexicanus* from northeast Mexico and *Phreatichthys andruzzii* from Somalia are commonly used fish models to study alternations in sleep-rest regulation. Other cavefish models, including Balitoridae from Southeast Asia, appear to have converged on many shared traits

Diversity of Cavefish and Cave Habitats

The majority of identified cavefish species exhibit the classic troglomorphic traits of eye loss and reduced pigmentation [10, 11]. These species originate from diverse fish lineages including carp, perch, and Characidae [9], providing powerful evidence for convergent evolution of troglomorphic phenotypes. Cavefish are encountered throughout the world with over 80 cavefish species resident in China alone, most of which belong to the cyprinid family [9, 10]. Estimates of the evolutionary divergence of cavefish from surface-dwelling ancestors differ dramatically between cavefish populations ranging from millions of years to as little as 10,000 years [30–32]. The lower end of this range suggests the potential for relatively rapid evolution upon encountering a novel environment [33]. This naturally occurring rapid evolution provides the basis for mapping the genetic basis of a diversity of traits and the ecological factors that contribute to selection.

The ability to compare the evolution of traits in diverse cavefish populations with the associated cave ecology and to compare cavefish with extant related fish that inhabit nearby rivers or lakes provide powerful models for studying the mechanisms underlying evolution in these extreme environments. Despite the presence of cavefish throughout the world, laboratory-based studies of cavefish, particularly for sleep and circadian rhythms, have focused nearly exclusively on two species of cavefish, the Somalian cavefish, *Phreatichthys andruzzii*, and Mexican cavefish, *Astyanax mexicanus*. Both species have been studied in a laboratory setting for decades, can readily reproduce in captivity, and, as a consequence, have emerged as leading models for investigating the evolution of sleep and circadian rhythms [34,

35]. Importantly, these two species evolved under very different environmental conditions, over dramatically different timescales, and consequently their troglomorphic adaptations differ considerably. Nevertheless, their convergence on dysregulated circadian rhythms provides an opportunity to investigate the mechanisms by which clocks are dysregulated and the biological adaptations that appear to compensate for a dysregulated clock.

A. mexicanus were first characterized in the 1930s and have been established as a leading model to investigate trait evolution because of its well-defined evolutionary history and the availability of multiple, independently derived populations [13, 36, 37]. At least 30 independent cave populations have been identified, nearly all of them displaying convergence to the classic troglomorphic traits of reduced pigmentation and eye size. Estimates for the point of divergence between cave forms and surface ancestors range from 10,000 years to 1–2 million years, likely arising from different molecular and modeling methodology [31, 38, 39]. While the most up-to-date methodology appears to identify a middle ground of ~200 k years, the disparate timelines highlight the challenges of identifying evolutionary divergence, and the need for additional efforts to define evolutionary history, even in widely studied fish models [39, 40].

Importantly, cave populations of *A. mexicanus* are interfertile with one another, as well as with extant surface fish, allowing for hybridization and genomic mapping experiments in a laboratory setting [41]. The convergence of multiple independently evolved cavefish populations for many traits has been classically used to identify quantitative trait loci (QTL), and in some cases individual genes, that underlie morphological differences [32, 42, 43]. More recently, these approaches have been applied to study differences in sleep and locomotor behavior. While the initial QTL experiments were likely underpowered and did not identify strong candidate regulators of sleep-wake regulation [44], recent advances in genome sequencing and QTL mapping highlight the potential of this approach [43, 45]. In addition, targeted analysis of candidate regulators has identified dysregulation of core-circadian gene expression and altered regulation of the hypothalamic neuropeptide *hypocretin/orexin* as a contributing to the regulation of sleep and locomotor rhythms [46, 47].

P. andruzzii is a cyprinid species resident in central Somalia, in bodies of water locked inside phreatic rock formations beneath the desert. It originated from normal-eyed ancestors that lived in surface water when Somalia had a wetter climate. As a consequence of the desertification that was driven by increases in Northeast African aridity [48], 2.8 million years ago, the *P. andruzzii* ancestors that originally colonized the cave environment remained isolated in the dark of the phreatic waters. This complete isolation was reinforced by the disappearance of surface populations, unlike some other species that maintain some level of gene flow with surface-dwelling populations. Most likely as a result of this extended isolation period, the troglomorphic phenotypes of *P. andruzzii* are [11, 49] among the most extreme observed in all cavefish species. For this reason, it has attracted particular attention as a model for studying how a range of phenotypes, affecting the circadian clock and the photoreactivation DNA repair system, have arisen during a prolonged period of evolution in constant darkness [4, 50, 51].

While the morphology of many cavefish species has been well documented, differences in their behavior and physiology have received considerably less attention, in part because these studies often require laboratory rearing. This is particularly apparent for the investigation of sleep and circadian rhythms, which requires multiple-day recordings that are often not possible outside of the laboratory environment. To our knowledge, circadian locomotor activity and sleep-wake regulation has only been characterized in *A. mexicanus* and *P. andruzzii* as well as a third species of cavefish, the hillstream loach, Balitoridae, that originates in Southeast Asia [46, 50, 52, 53]. In many cases, the use of these species to study sleep is challenging because there is a lack of closely related surface species, limiting an ability to interpret phenotypes. This is less important for circadian rhythms that are set by a 24-h cycle, allowing for the broader use of cavefish. Comparisons between these cavefish species and related surface-dwelling fish have provided insight into the evolution of these behavioral traits.

There is enormous diversity in the ecology of caves inhabited by different species of cavefish and even between individual populations of *A. mexicanus* [10]. These ecological differences are likely to represent key factors shaping differential evolution of behavioral traits and the genetic mechanisms through which these traits evolve. In the case of the Somalian cavefish *P. andruzzii*, we still lack a detailed understanding of its habitat due to the dangerous local political situation which prevents researchers from gaining access. However, information obtained when the fish were originally collected by visiting Italian scientists between 1968 and 1982 suggests that these fish inhabit narrow bodies of water lying within extensive and complex phreatic aquifers which underlie the Somalian desert [54]. Details of food availability and factors such as temperature and water quality are difficult to predict without careful field studies that are limited by access at this time. In addition, insight into the ecological phenomena that drive evolution is limited even with a detailed understanding of the present climate, because the historical ecology that drives evolution is almost always poorly understood. Interestingly, the fish were originally collected in natural wells within this region, which are interconnected with the complex phreatic system, suggesting that the fish are able to contact the surface. Furthermore, the fish were only observed in these wells during the night, an observation that already hints at some degree of circadian rhythmicity in *P. andruzzii* behavior in its natural habitat. These observations highlight the need to investigate fish under standardized laboratory conditions, as well as in their natural environment.

More is known about the caves inhabited by *A. mexicanus,* due in large part to a series of expeditions led by Robert Mitchell in the 1960s–1970s [49]. Remarkable differences have been identified in the geology and ecosystem of individual caves. For example, the commonly studied Pachón cave is a perched pool, while the Soyate cave is a pit cave over 800 m below the surface [11]. In addition, most of the caves are inhabited by "pure cavefish," while at least three caves, Rio Subterraneo, Chica, and Caballe Moro, contain surface-cave hybrid fish, and genomic analysis suggests varying degrees of admixture with surface fish populations. While it is suggested that nutrient levels are reduced across the caves, seasonal flooding appears to be a major contributor to most caves, likely resulting in annual changes in water chemistry and food availability [49]. Another interesting feature of the environment of certain caves inhabited by *A. mexicanus* is the presence of roosting sites for bats. Each morning these bats fly back to the cave following a night of feeding, and they produce a significant amount of guano that falls into the cave water and therefore serves as a potential food source for the resident cavefish. However, other caves apparently lack this zeitgeber cues that are commonly used to establish behavioral and physiological sleep-activity rhythms (Ref). Therefore, the differences in factors that include the presence of bat colonies, seasonal flooding that may wash nutrients into the caves, time of divergence from surface fish populations, and admixture rates all provide factors that may lead to differences between individual cave populations [39]. The constancy of temperature, light, and food availability is likely to influence the requirements for circadian clocks and provide an opportunity to examine how changes in selective pressure shape circadian clocks and the molecular basis of clock evolution [10, 49].

Altered Circadian Rhythms in Cavefish

From a purely conceptual point of view, there would seem little value in conserving a light-regulated, 24-h timing mechanism during evolution in a perpetually dark, relatively constant environment. One could therefore simplistically predict that the circadian clock mechanism would be lost in blind cavefish in a manner similar to eyes and pigmentation. Energy conservation has been proposed as the major driver of active selection for eye regression in cavefish adapted to an energy-deprived environments [55, 56], and it is possible that this extends to behavioral regulation. Considering the energy constraints present in the cave environment, the loss of the circadian clock mechanism could also be a strategy to reduce metabolic consumption [55]. Alternatively, conservation of the circadian clock might indicate that it did actually confer some selective advantage in the cave environment and may even be entrained by non-photic cues indicative of the time of day but originating outside of the cave. It could also represent more subtle cues, such as changes in temperature or water chemistry, that have yet to be detected. If procuring food is a major challenge within the hypogean environment, maintaining a working circadian clock entrained by food availability could confer a selective advantage. Thus, the unique ecology associated with each cave environment could result in individual species of cavefish experiencing very different selective constraints in terms of their circadian clock function. Initial results characterizing circadian rhythms in blind cavefish point to a complex interplay between the circadian clock, ecology, and evolution.

Circadian locomotor rhythms are ultimately driven by a core-transcriptional feedback loop that includes the transcriptional activators CLOCK and BMAL and transcriptional repressors PER and CRY that are highly conserved from fruit flies to mammals. This cell-autonomous system for maintaining 24-h rhythms is expressed throughout the brain and in cell types throughout the body. The clock contributes to the regulation of most aspects of cellular and organismal physiology, and using different genomic techniques it has been determined that a significant fraction of the genes is rhythmically expressed, depending on the cell type or tissue and environmental conditions [57, 58]. In addition, posttranscriptional regulation of core-clock components likely plays a key role in the regulation of circadian rhythms though this is less well understood, particularly for fish models [59, 60]. Most animals tested, including zebrafish and surface fish populations of *A. mexicanus*, and African cichlids display robust circadian locomotor rhythms, suggesting circadian rhythms are present in commonly used fish models [52, 61, 62].

In the case of *A. mexicanus* and *P. andruzzii* cavefish, several lines of evidence point to aberrations in clock-regulated behavior [46, 50]. A striking feature of *P. andruzzii* is that light is unable to entrain its circadian clock as measured both at the behavioral level and rhythmic clock gene expression in peripheral tissues [50]. Circadian cycling of gene expression can also be studied in cell culture. In serum-shocked *P. andruzzii*-derived cell lines, infradian rhythms of clock gene expression are observed. Loss-of-function mutations in a subset of opsin genes as well as mutations affecting the light-regulated *period2* gene have been linked with this blind, infradian clock phenotype. Specifically, point mutations lead to C-terminally truncated Opn4m2 and TMT opsins which lack the chromophore binding site [50], and the *period2* gene transcript is aberrantly spliced resulting in translation of a cytoplasmic protein [63]. In contrast, regular feeding has been shown to elicit circadian rhythms of locomotor activity in this species, suggesting food-entrainable, but not light-entrainable, locomotor rhythms have been maintained [50]. This points to genetically distinct food and light-entrainable clock mechanisms which have been differentially affected during *P. andruzzii* evolution. Consequently, this suggests that a robust, feeding-entrainable oscillator may confer a selective advantage in the Somalian cave environment. The original anecdotal observations that adult fish could only be collected in wells during the night support this notion.

Dampened circadian rhythms under both light-dark and free-running laboratory conditions have been reported in the blind cave forms of *A. mexicanus* [46]. Interestingly, the pattern of clock gene expression has been reported to resemble that typically encountered under constant lighting conditions [46, 64]. Under artificial light-dark cycle conditions, the molecular clocks of *A. mexicanus* cavefish populations retain their light entrainability, though these rhythms are absent in constant darkness revealing a lack of a free-running clock [46]. Interestingly, a first investigation of *A. mexicanus* has revealed the entrainment of locomotor activity both in blind and sighted morphs exposed to 24-h feeding cycles, with a pattern of synchronization very similar to that shown previously in *P. andruzzii* [50, 52]. It is tempting to speculate that rhythmicity in food availability potentially resulting from the daily bat migrations into certain caves might therefore serve as a non-photic zeitgeber for circadian clock entrainment. However, assays of clock gene expression in wild-caught fish collected from their natural cave environments at different timepoints during the day-night cycle failed to reveal any significant circadian rhythmicity, arguing against an entraining effect of this regular food availability in these particular caves [46].

Recently, transcriptomic analysis has been used to investigate whether the loss of molecular locomotor rhythms has occurred through shared genetic processes across multiple, independently evolved populations of cavefish. Comparing whole-body gene expression levels between cavefish and three different populations of cavefish under dark-dark conditions revealed robust circadian gene expression in surface fish, while rhythms were largely absent in core-clock genes across all three cavefish populations [65]. The genes that remained cycling in cavefish populations largely differed, offering further support for the notion that clock dysregulation has independent origins across cavefish population. Nevertheless, there were many identified genes involved in circadian regulation that were dysregulated across all three cave populations but not surface populations. For example, *arylalkylamine N-acetyltransferase 2* (*aanat2*), a pineal gland enzyme that is critical for melatonin synthesis, was downregulated across all cavefish populations tested. Deletion of *aanat2* in surface fish increased nighttime activity, abolishing circadian locomotor rhythms and inducing a cavefish-like phenotype [65]. These findings build on longstanding evidence in zebrafish that melatonin is a critical regulator of sleep and circadian rhythms [25, 66, 67] and suggest that naturally occurring clock dysregulation in cavefish can be used to identify novel genes involved in circadian regulation.

The adaptive basis of loss of circadian rhythmicity is unclear. From one extreme point of view, the presence of a functional circadian clock that entrains locomotor rhythms may provide no selective advantage to these species of cavefish and therefore is in the process of being lost. The presence of multiple loss-of-function gene mutations in the core-clock genes of *P. andruzzii* in contrast to more "subtle," regulatory changes encountered in *A. mexicanus* may simply reflect the longer period of isolation of the Somalian species. An alternative point of view postulates that the loss of a light-entrainable molecular clock provides a mechanism for energy conservation. Supporting this notion, a global reduction in metabolic rate observed in *A. mexicanus* cavefish raises the possibility that loss of circadian cycling in metabolic rate provides a mechanism for energy conservation [55]. In addition, the loss of circadian locomotor rhythms may allow for reduced sleep and therefore an increased foraging time, providing a selective advantage in a nutrient-poor environment [1]. However, the loss of circadian locomotor activity is not always associated with loss of metabolic rhythms. The North American cavefish *Amblyopsis* exhibits loss of circadian locomotor activity while maintaining metabolic rhythms, revealing that these two processes can be biologically and evolutionarily decoupled [68]. These findings raise the possibility of tissue-specific selective pressure on clock regulation. Therefore, the convergence on dampened molecular and behavioral rhythms in different species of cavefish raises the possibility that this trait is adaptive, but the functional significance of these evolved changes has not been determined.

An alternative approach to examining the adaptive value of circadian regulation is to examine the association between behavioral and genomic changes across caves with defined ecological differences. As noted above, this requires a better understanding of the ecology within the caves themselves. In addition, in an ideal situation, this would involve measuring behavior in its natural context. While a number of studies have examined behavior within the caves [69, 70], it is particularly challenging for sleep behavior because of requirements for long-term monitoring that is often challenging in caves that are difficult to access. Recently, a study compared behavior and morphological adaptations across individual pools of the Chica cave that houses surface-cave hybrids. The diversity in both behavioral and morphological phenotypes allows for inferences about how local environment is shaping trait evolution. While the behaviors were not measured in the wild, fish were transported into a lab and measured under standard conditions, representing the first behavioral measurements of circadian locomotor activity in wild-caught *A. mexicanus* [71]. Analysis revealed numerous cavefish phenotypes including eye loss and short sleep associated with the same pool, supporting the notion that these traits coevolve. Whole-genome sequencing and hybridization analysis were applied in behaviorally phenotyped fish to infer genetic pathways that may be involved in the morphological changes. This analysis revealed circadian pathways to be highlight enriched for genetic divergence. Furthermore, a mutation identified in the circadian transcriptional repressor *Cryptochrome 1A* is conserved in many other cavefish populations, suggesting deep evolutionary conservation of clock-related changes [71]. These findings provide support for the hypothesis that loss of clock function provides critical adaptive value in cave-dwelling fish, though further studies are required to identify the physiological consequences of reduced clock function and its adaptive value.

It appears that the molecular basis for the loss of circadian rhythms differs fundamentally between Somalian and Mexican cavefish. While loss of circadian function in Mexican cavefish appears to largely derive from alterations in the regulation of clock gene transcription, mutations affecting the coding sequences of the *period2* core-clock gene as well as opsin photoreceptor genes are present in Somalian cavefish [63]. The robust differences in circadian behavior between surface fish and cave forms of *A. mexicanus*, combined with the ability to generate hybrids, have allowed for QTL mapping of circadian traits. Multiple studies have examined QTL loci associated with circadian behavior and locomotor activity using surface-cave hybrid populations. One study identified six linkage groups, and another identified multiple QTLs that associate with global changes in locomotor activity [44, 64]. However, the identification of the causal genes within these linkage groups represents a significant bottleneck for subsequent analysis. A sequenced genome and the identification of selection signatures throughout the genome may allow for the identification of relevant genes that lie within linkage groups [39, 43, 72]. This, combined with the recent application of CRISPR gene editing in *A. mexicanus*, should enable the testing of the role played by individual genes identified through QTL analysis of this complex behavior.

Sleep Regulation in Cavefish

A primary function of the clock is to regulate an animal's rest-activity patterns. As a consequence, the timing of sleep is largely dependent on a functioning clock [73]. While classic models suggest the timing of sleep is controlled by a mechanism that is distinct from those conferring sleep homeostasis and duration, growing evidence suggests these processes are likely to be more interrelated than previously appreciated [73–75]. Further, while circadian rhythms tend to be approximately 24 h in length, sleep duration is highly variable across the animal kingdom, presenting a significant challenge to define and quantify sleep using comparative models [1, 76]. In species throughout the animal kingdom, shared behavioral signatures of sleep, including elevated arousal threshold, stereotyped posture, and homeostatic recovery following deprivation, appear to be highly conserved and can be used to characterize sleep behavior in the absence of physiological measurements [23]. These have been used to characterize sleep across a wide array of mammalian and nonmammalian species and to reveal key aspects of sleep evolution and function [1, 23, 76]. However, the need to define sleep and its variability has significantly hampered progress in understanding the evolution of sleep. For comparative approaches, inferring changes in sleep require either comparisons between closely related species or large-scale phylogenetic analysis. Therefore, most of the research to date investigating the evolution of sleep in cavefish has focused on *A. mexicanus.*

There is broad agreement that ecology and evolutionary history potently shape sleep differences across species and disparate populations within a species [1, 77]. While the loss of circadian rhythms in response to the perpetual darkness of caves is well known, the change in ecology from river to cave environments also appears to profoundly impact rest-activity across multiple species of cave animals [53] [52, 53]. Sleep per se has only been characterized in *A. mexicanus* where at least four cavefish have reduced sleep, suggesting that sleep loss has independently evolved multiple times in evolution [52, 78].

This reduction in the duration of sleep in cavefish populations is already evident early in development (3–4 dpf) and persists throughout adulthood [44, 47], raising the possibility that genetic factors regulating sleep in early development are stable throughout adulthood (Fig. 2). The study of sleep in *A. mexicanus* populations, like in other fish models, has focused on behavioral quiescence as a measure of sleep. In larvae, gentle tapping was applied during a quiescent period to define the length of inactivity used to define sleep [52]. Since the response to vibration differs substantially in adult fish, a mild electric current was applied to measure arousal threshold at this life stage [78]. At both stages, arousal threshold was elevated following approximately 1 min of immobility, and this duration of quiescence was used as a measure of sleep [52, 78]. While this provides general guidance for behavioral analysis, it is critical to revisit this definition of sleep as animal tracking and brain imaging approaches improve, providing potential for more accurate quantification of sleep.

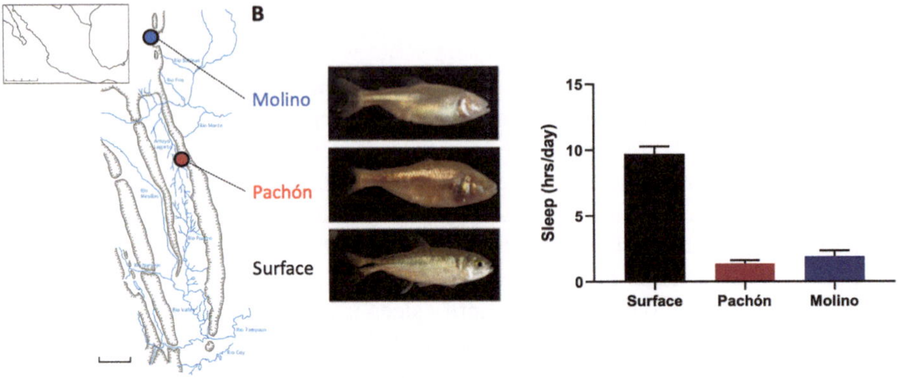

Fig. 2 The evolution of sleep loss in Mexican cavefish. (**a**) Map of the Sierra de El Abra region in NE Mexico. Surface fish (bottom) inhabit rivers within the same region as Molino (blue) and Pachòn (red) cave populations. (**b**) All six cavefish populations tested to date have converged on sleep loss

An additional measure of sleep is homeostatic rebound following deprivation. While the nature of sleep debt remains poorly understood, the ability of animals to detect sleep loss and compensate by increasing sleep is wide-ranging and a central behavioral feature of sleep [79]. Zebrafish have long been known to exhibit rebound sleep following deprivation by gentle tapping or electrical stimulation, confirming the presence of this sleep characteristic in fish [25, 26]. In *A. mexicanus*, significantly less is known about this sleep characteristic than behavioral quiescence and elevated arousal threshold. One study sleep deprived juvenile *A. mexicanus* by mechanical shaking over a 10-h period during the night and measured sleep during the subsequent day [80]. Both surface fish and three cavefish populations increased daytime sleep following nighttime deprivation, suggesting that sleep homeostasis is intact in cavefish despite the robust reduction in sleep duration. The genes that are differentially expressed following sleep deprivation are largely unique in each *A. mexicanus* population, raising the possibility that the mechanism of homeostasis differs across surface fish and each population of cavefish. The evolutionary reasons for maintaining sleep homeostasis despite loss of sleep and the biological underpinnings require further study. However, these results suggest that sleep may be critical for cavefish function and that the reduced sleep in cavefish may represent more efficient sleep rather than a reduced need for sleep.

While sleep has yet to be defined in other populations of cavefish, the shared behavioral and morphological traits of distantly evolved cavefish suggest a likelihood that sleep will be reduced compared to related surface populations. In four different populations of *balitorid* cave loaches, rest periods appear to be dramatically reduced compared to surface species, yet circadian regulation of activity rhythms persists [53]. These findings implicate independent evolution of sleep and circadian systems in multiple species and raise the possibility that these behaviors may be under distinct evolutionary pressure. Sleep was not defined in each of these species, for example, by measuring how elevated arousal threshold associates with behavioral quiescence, and therefore the effects on sleep itself cannot be inferred. Furthermore, the study was limited by the relatively small number of animals tested, given the limited availability of the species that were under investigation. However, these findings do provide a framework for studying rest duration across additional cavefish species. Therefore, future work in cavefish promises to shed important new light on the evolution of sleep regulation.

Mechanisms of Sleep Loss

Sleep is largely believed to be a phenomenon controlled by the brain, though there is increasing evidence that sensory mechanisms and peripheral endocrine cues regulate sleep [75, 81]. In zebrafish, a combination of genetic manipulations of brain function and behavioral screening have been applied to identify the neural mechanisms of sleep regulation, and many of these processes are conserved with those in mammals (28391131; 28414966). The convergence on sleep loss in *A. mexicanus* provides the opportunity to examine the biological basis of sleep loss and possibly identify novel conserved mechanisms regulating sleep. To this end, many of the approaches first developed to study the neural and genetic mechanisms of sleep regulation in zebrafish have been transferred to *A. mexicanus* and used to define differences between surface and cave populations that contribute to the evolution of sleep loss.

The gating of sensory information is thought to be critical for animals to enter a sleep-like state, and neural mechanisms have been identified in species ranging from *C. elegans* to mammals that filter incoming sensory information during sleep [82, 83]. Cavefish have evolved a rapid expansion of many nonvisual sensory systems, raising the possibility that changes in sensory processing contribute to sleep loss [84]. In *A. mexicanus*, one particularly prominent sensory change is an expansion of the lateral line, the mechanosensory system that detects water flow in fish [85]. The lateral line is a critical regulator of vibration attraction behavior that is thought to be associated with foraging [85]. This system has also been found to suppress sleep in cavefish from the Pachón population. Pharmacological ablation of the lateral line restores surface fish-like sleep to Pachón cavefish, suggesting that expanded mechanosensation underlies sleep loss [86]. In addition, lateral line ablation reduces orexin/hypocretin levels, revealing a sensory-hypothalamic axis that drives sleep loss. These findings reveal a role for the mechanosensory system in regulation of behavior, as well as transcriptional regulation in wake-promoting neurons in the hypothalamus. This is particularly interesting because hypothalamic HCRT neurons in mammals are responsive to numerous sensory modalities revealing a conserved mechanism of HCRT regulation [87].

It is possible that the neural processes regulating sleep differ across cavefish populations or that populations have independently evolved shared mechanisms of sleep loss. Ablation of the lateral line does not affect sleep in other populations of cavefish suggesting distinct mechanisms underlie sleep loss in different *A. mexicanus* populations [47]. It is possible that this reflects unique ecological parameters within the Pachón cave, though reasons for these differences across *A. mexicanus* populations are unknown. Therefore, future mechanistic studies exploring sleep loss in different populations of *A. mexicanus* and additional cavefish species such as *P. andruzzii* which exhibits more severe troglomorphic phenotypes, may provide important insight into the many ways that sensory systems may impact sleep can be altered.

Pharmacological Regulation of Sleep

In humans, disorders related to sleep loss are highly pleiotropic; however, the conservation of sleep has led to fish being used as a model system to examine the mechanisms underlying sleep loss [88]. To date, all mechanistic studies on sleep regulation in cavefish have focused on the Pachón cave population of *A. mexicanus*, likely because this population breeds readily in captivity. Initial studies examining the mechanism of sleep loss have largely focused on restoring sleep loss to Pachón cavefish. One way to approach this is to screen for pharmacological agents that modulate sleep. Larval fish are naturally well suited for this because they absorb drugs readily and can be housed in small amounts of water, so making them particularly well suited for high-throughput screening [29]. An initial targeted screen tested the effects of neurotransmitter receptor antagonists on sleep in surface and cavefish. The six drugs tested did not affect sleep in surface fish, while the NMDA-receptor antagonist MK-801 and the β-blocker propranolol restored sleep to cavefish [89]. The effects of propranolol were particularly striking because sleep was restored to surface fish levels, suggesting that an elevated adrenergic system contributes to sleep loss in cavefish. This was followed up with tests of more selective antagonists that indicated the β1 receptor-selective antagonist increases sleep [89]. Later, mutations in the human β1 receptor were found to result in short sleep behavior, and the mutation also impacted sleep in mice [90]. Therefore, the study of pharmacological restoration of sleep in cavefish identified highly conserved receptors that modulate sleep.

The use of pharmacological approaches to identify regulators of sleep has been extended to known regulators of sleep. For example, the wake-promoting neuropeptide was found to be upregulated in cavefish and was functionally validated by testing sleep in fish exposed to hypocretin receptor antagonists. Multiple different hypocretin antagonists increased sleep in cavefish, without affecting sleep in surface fish [47]. Pharmacological agents can also be used to identify brain regions associated with altered sleep. For example, a study examined neural activity, measured by immunostaining for phosphorylated ERK, in the brains of fish treated with propranolol or the hypocretin receptor antagonist EMPA. While both drugs induced sleep in cavefish, they had altered the activity of different populations of neurons across the brain. However, there were shared areas. For example, both drugs reduced activity in at least one population of cavefish in the preoptic hypothalamus [91]. With the increasing availability of resources and genetic tools for mapping brain activity, the identification of drugs that differentially impact sleep in surface and cavefish provides an approach to identify how different modulations of sleep uniquely impact neural activity.

Identification of Genes Contributing to Evolved Sleep Differences in Cavefish

Over recent decades, genetically amenable models including *C. elegans,* zebrafish, and *Drosophila* have been used to identify genes that regulate sleep. While a lack of genetic tools has presented a major challenge in *A. mexicanus*, the recent application of CRISPR-based gene editing allows for the study of how single genes regulate behavior. Like many new models, a starting point has been to examine genes that are both differentially regulated and known to modulate sleep in other animal systems. For example, hypocretin is upregulated in cavefish, and inhibition of HCRT using morpholinos, or genetically silencing a subset of hypocretin neurons using the GAL4/UAS system, restores sleep to cavefish [47]. In an alternative approach, gene expression profiling was used to identify genes that were differentially regulated across the circadian cycle. Genes with impaired transcriptional cycling were then functionally validated by targeted knockdown using CRISPR in surface fish. This approach validated a role for the melatonin-synthesis enzyme *aanat2* and the circadian clock output gene *RORca* in time of day-specific regulation of sleep [65]. While these approaches provide proof of principle for the use of genetic approaches to identify sleep regulators, they only scratch the surface of the systems potential. In both zebrafish and cavefish, CRISPR-based mutagenesis is highly efficient, allowing injected fish to be screened for sleep without the need to generate stable lines [42, 65, 92, 93]. Indeed, this approach has been used in zebrafish and medaka to screen for genetic regulators of behavior [92]. The application of high-throughput CRISPR screening in cavefish will likely lead to the identification of novel regulators of sleep that have not been identified in other animal models.

A unique advantage of studying sleep in *A. mexicanus* is the ability to generate surface x cave hybrid fish and to map genetic variation associated with phenotypic traits. The use of quantitative trait loci (QTL) mapping has been broadly applied to identify loci associated with albinism, eye loss, foraging, schooling, and many other behaviors [94]. The power of this approach has been aided by the available genome sequences for surface fish and Pachón cavefish [43, 45]. An initial study applying QTL analysis to the study of sleep examined adult Pachón x cave hybrids. The study was able to identify significant QTL for locomotor activity but not sleep [78]. This is presumably due to increased variability associated with sleep and highlights the need for higher-powered studies. Another study measured sleep in 20-day-old juvenile hybrids [95]. While sequencing was not performed, sleep was reduced in albino hybrids compared to pigmented hybrids. In Pachón cavefish, albinism is caused by mutation of a single gene, *Oculocutaneous albinism 2*, that is required for melanin synthesis [95, 96]. Interestingly, *oca2* shares a pathway with catecholamine biosynthesis, and levels of the wake-promoting neurotransmitter norepinephrine are elevated in cavefish [96]. Indeed, surface fish with CRISPR-based mutation of *oca2* sleep less than control counterparts, confirming a role for *Oca2* in sleep regulation [95]. Together, these studies highlight the ability of trait-association studies and QTL mapping to identify regulators of sleep and the potential of higher-powered studies combined with functional validation of candidate genes.

Neural Correlates of Sleep Loss

The sleep differences between surface and cavefish can also be used to map neuroanatomical differences associated with sleep. *A. mexicanus* cavefish have undergone significant changes in brain structure, including a reduction in the optic tectum and expansion of the hypothalamus, likely to compensate for changes in the sensory and metabolic demands of the cave environment [97, 98]). Both zebrafish and *A. mexicanus* larvae are amenable to whole-brain imaging in intact fish [91, 99–101]. In zebrafish, many groups have created a computationally derived brain atlas that is derived from the merger of many (sometimes 100's) of brains, resulting in a standard neuroanatomical atlas with brain regions and subclasses of neurons identified [99–101]. The generation of brain atlases using whole-brain imaging has the potential to define anatomical differences between surface fish and cavefish across the entire brain. An early integration of a brain atlas was developed using a combination to total ERK expression, a broadly expressed protein used as an anatomical label, and labeling with more selectively expressed neuropeptides [91]. The brain regions were then manually segmented resulting in comparisons of 30 different brain regions between surface fish and three different populations of cavefish. Perhaps, unsurprisingly, broad changes were observed in anatomical structure, both between surface and cave populations and between different cave populations. Tyrosine hydroxylase-expressing catecholamine neurons (TH+) were increased in the hypothalamus of all three cavefish populations. However, in the medial octavolateralis nucleus, a brain region associated with mechanosensory processing, the number of TH+ neurons was only increased in two of the three cavefish populations examined. In addition, the atlas was combined with markers of neural activity, such as phosphorylated ERK to define how activity changes during sleep, feeding, and drug treatment [91]. Therefore, this atlas provides structural and functional insight into the differences between the brains of surface fish and cavefish.

The reliance on hand segmentation of brain regions represents a major limitation of the initial version of the *A. mexicanus* brain atlas, because it limits the number of brain regions that can be realistically studied. To overcome these limitations, a zebrafish analysis pipeline has been applied that allows for automated segmentation of ~200 brain regions [102]. This unprecedented comparison of neuroanatomical changes across the entire brain resulted in the identification of a contraction of the cavefish dorsal diencephalon compared to surface fish. The pipeline was also effective at measuring volume and shape across the brain of surface x cave hybrid fish [103]. Analysis of F2 hybrids was then used to examine the relationship between individual brain regions. While this atlas has not been applied to the study of sleep, the combination of automated high-resolution brain anatomical mapping in hybrid fish with behavioral analysis and QTL analysis has the great potential to reveal the relationship between genetic variation, neuroanatomical variability, and sleep. Therefore, the application of neuroimaging tools in a comparative model can address fundamental questions about biological variability that are inaccessible in classic genetic models.

The Functional Significance of Reducing Sleep in the Cave Environment

One fundamental question in cavefish biology concerns whether there may be some benefit associated with reduced sleep in cave habitats [104]. An extreme point of view posits that sleep, or other behavioral traits lost in cavefish, confer no selective advantage in this extreme environment and so the mechanisms which maintain this function are progressively being lost in cavefish. Alternatively, it is also possible that a reduction of the time dedicated to sleep may actually enhance survival in caves since it may extend the time that can be dedicated to foraging, an argument also used to account for loss of circadian clock function in cavefish (ref). Given the limited food supply, reduced sleep would then be predicted to enhance survival by providing the fish more time to forage (ref). Furthermore, it is possible that sleep loss arises from changes in predation. One hypothesis of sleep regulation suggests sleep occurs due to adaptive inactivity allowing for predator avoidance [105]. Accordingly, the lack of predators in the cave environment may relax selection on sleep systems. While these questions are difficult to address experimentally, the presence of sleep loss in many different *A. mexicanus* cave populations provides the possibility to examine the relationship between sleep loss and differing ecological factors in each cave.

One recently proposed function of sleep relates to enabling periods of DNA repair in neurons to compensate for DNA damage incurred during extended periods of elevated synaptic activity and cellular respiration [106]. Interestingly, within this context, DNA repair systems have also been subject to change during the evolution of blind cavefish. Specifically, the photoreactivation system which repairs UV-induced DNA photoproducts is modified significantly in both *A. mexicanus* and *P. andruzzii*. This repair system consists of three DNA photolyase enzymes (6–4, CPD, and DASH) which are close relatives of the cryptochromes and which harness visible light as energy to enable catalysis of the repair reaction. In the case of *A. mexicanus*, basal expression levels of CPD photolyase are elevated, and consistently photoreactivation repair is enhanced in cave forms compared with surface fish

[46]. Curiously, in *P. andruzzii*, the opposite seems to be the case, and there is a complete loss of photoreactivation function. This results in part from truncation mutations in the 6–4 and DASH photolyase proteins which are therefore rendered cytoplasmic. Differently, the CPD photolyase gene carries no truncation mutations and is a predominantly nuclear protein. However, *P. andruzzii* cells also lack light, UV, and ROS-induced photolyase gene transcription which results from loss of function of the D-box enhancer promoter element [51]. Consequently, the remaining CPD photolyase activity is not upregulated upon exposure to sunlight or oxidative stress, and so photoreactivation function under artificial lighting conditions is greatly attenuated.

Of course, in their natural habitat exposure to sunlight is limited, so the degree of UV-induced DNA damage to contend with is likely reduced and there is an absence of visible light to drive DNA repair. Therefore, photoreactivation in cavefish is simply not possible or even relevant. However, the upregulation of CPD photolyase expression in *A. mexicanus* as well as the conservation of the CPD photolyase coding sequence in *P. andruzzii* has led to speculation that this particular photolyase enzyme might possess an additional light-independent DNA repair function that targets other types of DNA damage encountered in the cave environment. Thus, the types of DNA damage encountered in each cave environment may shape the evolution of DNA repair systems of the resident cavefish species. The potential relevance of these modified DNA repair systems for the changes in sleep observed in cavefish represents a fascinating area for future investigation.

Linking Evolution with Ecology in Cavefish

Mounting evidence points to similarities but also to significant differences in the evolution of circadian rhythms and sleep between different species of cavefish. This is most likely the consequence of species-specific or population-specific differences in the time of isolation from the sunlit surface environment as well as the very different cave environments which they inhabit. Therefore, it is increasingly clear that a comprehensive understanding of the ecology of each species is essential for understanding the mechanisms shaping the evolution of these complex phenotypes in cave environments. This is a challenging goal since it requires detailed experiments in the field. Long-term behavior experiments are technically challenging, and few have been performed in cave species [69, 70]. Many cave sites are difficult to access due to geographic or political impediments. In addition, the nature of studying sleep and circadian rhythms often requires long-term recordings which are difficult in the cave environment. One approach has been to compare circadian gene expression in fin clips from wild-caught fish collected in caves and nearby rivers [46]. These studies, as predicted by laboratory studies, reveal reduced cycling gene expression in cavefish. Unlike circadian rhythms, much less is known about molecular markers that code for sleep drive, and therefore the study of sleep in cave populations will inevitably require behavioral analysis. Nevertheless, the ability to combine insight gained from laboratory-based studies together with the results of studies of animals in the wild and the detailed characterization of cave environments promises to provide a unique level of insight into how these extreme environments can shape the evolution of complex behavioral phenotypes.

Future Directions

The evolution of fundamental differences in the circadian timing system between surface fish and cavefish provides the opportunity to identify novel molecular and physiological regulators of sleep and circadian rhythms. Recently, whole-brain imaging in zebrafish harboring a transgene for GCaMP has been applied to identify neural correlates of sleep that appear to be highly conserved with those in mammals. Slow-wave sleep is a primary indicator for sleep in mammals and is present within the dorsal pallium of zebrafish, a structure analogous to the mammalian cortex [28]. These findings provide a physiological measure of sleep and the opportunity to examine the neural activity associated with sleep differences observed in cavefish. Recently, the tol2 transgenesis system commonly used in zebrafish has been applied to cavefish [93, 107, 108], and GCaMP imaging has been used to compare differences in neural activity within visual processing centers (ref). The availability of this methodology and transgenic fish allows for analysis to extend beyond behavioral analysis to the changes in neural activity that accompany sleep loss in cave populations of *A. mexicanus*. In addition, these tools could be applied to studies to examine circadian regulation of neural activity.

Circadian rhythms and sleep are highly pleiotropic, and it is likely that many more genes and molecular pathways govern the evolution of these behaviors than have presently been discovered in Somalian and Mexican cavefish. In *A. mexicanus*, QTL mapping has been widely used to identify genomic regions associated with trait variance, but there are few examples of genes that have been confirmed to underlie trait evolution. While *A. mexicanus* has the potential to identify novel regulators of sleep and circadian rhythms, and how genetic variation contributes to interspecies differences, progress has been limited by the throughput of assays and methodology for validating hits. Recently, sleep loss was identified in *A. mexicanus* at 3 days post-hatching, allowing for increased throughput of behavioral analysis [109]. Larval zebrafish have been used for high-throughput pharmacological and mutagenesis screens [29, 110], and the application of these systems for GWAS experiments should increase the statistical power of this approach. In addition, TALEN and CRISPR gene-editing approaches have been effectively used in cavefish to validate loss-of-function mutations that cause albinism [42, 108, 111]. Beyond identifying novel genes involved in sleep and circadian regulation, *A. mexicanus* is well poised to examine how genetic variation encodes for differences in these traits. Targeted mutagenesis has been effectively applied in zebrafish to engineer specific mutations [112–114], and these approaches could be transferred to *A. mexicanus* to induce population-specific genetic mutations. Rather than deleting genes, this will require swapping alleles between populations and evaluating the effects on behavior. The combination of these advances provides opportunity for the use of *A. mexicanus* to identify novel regulations of sleep and circadian rhythms.

Studies on sleep and circadian rhythms in cavefish may also provide important new insight into the relationship between these traits and pathology in humans, with circadian dysfunction and short sleep being associated with a host of diseases including cardiac disease, diabetes, and obesity [81]. Interestingly, genetic mutations in the *A. mexicanus* insulin receptor render them insensitive to insulin, largely phenocopying type 2 diabetes [115]. Furthermore, multiple populations of cavefish are hyperphagic and obese and have fatty livers [116]. Along with sleep and circadian changes, cavefish exhibit many genetic features associated with human disease, yet there are no reports of adverse health or differences in lifespan between surface fish and cavefish. Therefore, it is possible that cavefish may have evolved resilience to traits that include sleep loss, diabetes, and obesity, raising the potential to use *A. mexicanus* as a model to understand resistance to sleep- and circadian-associated disorders.

Taken together, cavefish represent powerful models for investigating the fundamental mechanisms underlying sleep and circadian regulation. The diversity of cavefish throughout the world has the potential to reveal both shared and independent mechanisms regulating clock function. Relating these changes to particular features of the cave environments which they inhabit will provide novel insight into the ecological factors that drive trait evolution.

References

1. Keene AC, Duboue ER. The origins and evolution of sleep. J Exp Biol. 2018;221. https://doi.org/10.1242/jeb.159533.
2. Siegel JM. Clues to the functions of mammalian sleep. Nature. 2005;437:1264–71.
3. Joiner WJ. Unraveling the evolutionary determinants of sleep. Curr Biol. 2016;26:R1073–87.
4. Foulkes NS, Whitmore D, Vallone D, Bertolucci C. Studying the evolution of the vertebrate circadian clock: the power of fish as comparative models. Adv Genet. 2016;95:1–30.
5. Bhadra U, Thakkar N, Das P, Pal Bhadra M. Evolution of circadian rhythms: from bacteria to human. Sleep Med. 2017;35:49–61.
6. Mohawk JA, Green CB, Takahashi JS. Central and peripheral circadian clocks in mammals. Annu Rev Neurosci. 2012;35:445–62.
7. Hardin PE, Panda S. Circadian timekeeping and output mechanisms in animals. Curr Opin Neurobiol. 2013;23:724–31.
8. Capellini I, Barton RA, McNamara P, Preston BT, Nunn CL. Phylogenetic analysis of the ecology and evolution of mammalian sleep. Evolution (N Y). 2008;62(1764–1776):1764.
9. Borowsky R. Cavefishes. Curr Biol. 2018;28:R60–4.
10. Culver DC, Pipan T. The biology of caves and other subterranean habitats. Oxford: Oxford University Press; 2009.
11. Keene A, Yoshizawa M, McGaugh S. Biology and evolution of the Mexican cavefish. 1st ed. New York: Academic; 2015.
12. Jeffery WR. Regressive evolution in *Astyanax* cavefish. Annu Rev Genet. 2009;43:25–47.
13. Gross JB. The complex origin of Astyanax cavefish. BMC Evol Biol. 2012;12. https://doi.org/10.1186/1471-2148-12-105.

14. Orger MB, de Polavieja GG. Zebrafish behavior: opportunities and challenges. Annu Rev Neurosci. 2017;40:125–47.
15. Kawakami K, Asakawa K, Hibi M, Itoh M, Muto A, Wada H. Gal4 driver transgenic zebrafish: powerful tools to study developmental biology, organogenesis, and neuroscience. Adv Genet. 2016;95:65–87.
16. Chiu CN, Prober DA. Regulation of zebrafish sleep and arousal states: current and prospective approaches. Front Neural Circuits. 2013;7:58.
17. Elbaz I, Foulkes NS, Gothilf Y, Appelbaum L. Circadian clocks, rhythmic synaptic plasticity and the sleep-wake cycle in zebrafish. Front Neural Circuits. 2013;7. https://doi.org/10.3389/fncir.2013.00009.
18. Rihel J, Schier AF. Sites of action of sleep and wake drugs: insights from model organisms. Curr Opin Neurobiol. 2013;23:831–40.
19. Frøland Steindal I, Whitmore D. Circadian clocks in fish—what have we learned so far? Biology (Basel). 2019;8:E17.
20. Idda ML, Bertolucci C, Vallone D, Gothilf Y, Sánchez-Vázquez FJ, Foulkes NS. Circadian clocks: lessons from fish. Prog Brain Res. 2012;199:41–57.
21. Sassone-Corsi P, Whitmore D, Foulkes NS. Light acts directly on organs and cells in culture to set the vertebrate circadian clock. Nature. 2000;404:87–91.
22. Nath RD, Bedbrook CN, Abrams MJ, Basinger T, Bois JS, Prober DA, Sternberg PW, Gradinaru V, Goentoro L. The Jellyfish Cassiopea exhibits a sleep-like state. Curr Biol. 2017;27:2984–2990.e3.
23. Campbell SS, Tobler I. Animal sleep: a review of sleep duration across phylogeny. Neurosci Biobehav Rev. 1984;8:269–300.
24. Prober DA, Rihel J, Onah AA, Sung R-J, Schier AF. Hypocretin/orexin overexpression induces an insomnia-like phenotype in zebrafish. J Neurosci. 2006;26:13400–10.
25. Zhdanova IV, Wang SY, Leclair OU, Danilova NP. Melatonin promotes sleep-like state in zebrafish. Brain Res. 2001;903:263–8.
26. Yokogawa T, Marin W, Faraco J, Pézeron G, Appelbaum L, Zhang J, Rosa F, Mourrain P, Mignot E. Characterization of sleep in zebrafish and insomnia in hypocretin receptor mutants. PLoS Biol. 2007;5:2379–97.
27. Leung LC, Wang GX, Mourrain P. Imaging zebrafish neural circuitry from whole brain to synapse. Front Neural Circuits. 2013;7. https://doi.org/10.3389/fncir.2013.00076.
28. Leung LC, Wang GX, Madelaine R, Skariah G, Kawakami K, Deisseroth K, Urban AE, Mourrain P. Neural signatures of sleep in zebrafish. Nature. 2019;571:198–204.
29. Rihel J, Prober DA, Arvanites A, Lam K, Zimmerman S, Jang S, Haggarty SJ, Kokel D, Rubin LL, Peterson RT, Schier AF. Zebrafish behavioral profiling links drugs to biological targets and rest/wake regulation. Science. 2010;327:348–51.
30. Behrmann-Godel J, Nolte AW, Kreiselmaier J, Berka R, Freyhof J. The first European cave fish. Curr Biol. 2017;27:257–8.
31. Fumey J, Hinaux H, Noirot C, Thermes C, Rétaux S, Casane D. Evidence for late Pleistocene origin of Astyanax mexicanus cavefish. BMC Evol Biol. 2018;18:43.
32. Bradic M, Teotónio H, Borowsky RL. The population genomics of repeated evolution in the blind cavefish astyanax mexicanus. Mol Biol Evol. 2013;30:2383–400.
33. Behrmann-Godel J, Nolte AW, Kreiselmaier J, Berka R, Freyhof J. The first European cave fish. Curr Biol. 2017;27:R257–8.
34. Dezfuli BS, Capuano S, Magosso S, Giari L, Berti R. The lateral line system in larvae of the blind cyprinid cavefish, Phreatichthys andruzzii. Anat Rec. 2009;292:423–30.
35. Borowsky R. Handling Astyanax mexicanus eggs and fry. Cold Spring Harb Protoc. 2008;3. https://doi.org/10.1101/pdb.prot5093.
36. Breder CM, Rasquin P. Comparative studies in the light sensitivity of blind characins from a series of Mexican caves. Bull Amer Mus Nat Hist. 1947;89:325–51.
37. Breder C. Descriptive ecology of La Cueva Chica, with especial reference to the blind fish. Anoptichthys Zoologica. 1942;27:7–15.

38. Bradic M, Beerli P, Garcia-de Leon FJ, Esquivel-Bobadilla S, Borowsky RL. Gene flow and population structure in the Mexican blind cavefish complex (Astyanax mexicanus). BMC Evol Biol. 2012;12. https://doi.org/10.1186/1471-2148-12-9.
39. Herman A, Brandvain Y, Weagley J, Jeffery WR, Keene AC, Kono TJY, Bilandžija H, Borowsky R, Espinasa L, O'Quin K, Ornelas-García CP, Yoshizawa M, Carlson B, Maldonado E, Gross JB, Cartwright RA, Rohner N, Warren WC, McGaugh SE. The role of gene flow in rapid and repeated evolution of cave-related traits in Mexican tetra, Astyanax mexicanus. Mol Ecol. 2018;22:4397–416.
40. Moran R, Richards E, Ornelas-Garcia C, Gross J, Gross A, Donny J, Keene AC, Kowalko J, Rohner N, McGaugh SE. Selection-driven trait loss in independently evolved cavefish populations. BioRxiv. 2022:518185.
41. Sadoglu P. Mendelian inheritance in the hybrids between the Mexican blind cave fishes and their overground ancestor. Verh Dtsch Zool Ges Graz. 1957:423–39.
42. Klaassen H, Wang Y, Adamski K, Rohner N, Kowalko JE. CRISPR mutagenesis confirms the role of oca2 in melanin pigmentation in Astyanax mexicanus. Dev Biol. 2018;441:313–8.
43. McGaugh SE, Gross JB, Aken B, Blin M, Borowsky R, Chalopin D, Hinaux H, Jeffery WR, Keene A, Ma L, Minx P, Murphy D, O'Quin KE, Rétaux S, Rohner N, Searle SMJ, Stahl BA, Tabin C, Volff J-N, Yoshizawa M, Warren WC. The cavefish genome reveals candidate genes for eye loss. Nat Commun. 2014;5. https://doi.org/10.1038/ncomms6307.
44. Yoshizawa M, Robinson BG, Duboué ER, Masek P, Jaggard JB, O'Quin KE, Borowsky RL, Jeffery WR, Keene AC. Distinct genetic architecture underlies the emergence of sleep loss and prey-seeking behavior in the Mexican cavefish. BMC Biol. 2015;13. https://doi.org/10.1186/s12915-015-0119-3.
45. Warren WC, Boggs TE, Borowsky R, Carlson BM, Ferrufino E, Gross JB, Hillier L, Hu Z, Keene AC, Kenzior A, Kowalko JE, Tomlinson C, Kremitzki M, Lemieux ME, Graves-Lindsay T, McGaugh SE, Miller JT, Mommersteeg MTM, Moran RL, Peuß R, Rice ES, Riddle MR, Sifuentes-Romero I, Stanhope BA, Tabin CJ, Thakur S, Yamamoto Y, Rohner N. A chromosome-level genome of Astyanax mexicanus surface fish for comparing population-specific genetic differences contributing to trait evolution. Nat Commun. 2021;12:1447.
46. Beale A, Guibal C, Tamai TK, Klotz L, Cowen S, Peyric E, Reynoso VH, Yamamoto Y, Whitmore D. Circadian rhythms in Mexican blind cavefish Astyanax mexicanus in the lab and in the field. Nat Commun. 2013;4:2769. https://doi.org/10.1038/ncomms3769.
47. Jaggard JB, Stahl BA, Lloyd E, Prober DA, Duboue ER, Keene AC. Hypocretin underlies the evolution of sleep loss in the Mexican cavefish. Elife. 2018;7. https://doi.org/10.7554/eLife.32637.
48. de Menocal PB. Plio-Pleistocene African climate. Science. 1995;270:53–9.
49. Mitchell RW, Russell WH, Elliott WR. Mexican eyeless characin fishes, genus Astyanax: environment, distribution, and evolution. Lubbock: Texas Tech Press; 1977.
50. Cavallari N, Frigato E, Vallone D, Fröhlich N, Lopez-Olmeda JF, Foà A, Berti R, Sánchez-Vázquez FJ, Bertolucci C, Foulkes NS. A blind circadian clock in cavefish reveals that opsins mediate peripheral clock photoreception. PLoS Biol. 2011;9:e1001142. https://doi.org/10.1371/journal.pbio.1001142.
51. Zhao H, di Mauro G, Lungu-Mitea S, Negrini P, Guarino AM, Frigato E, Braunbeck T, Ma H, Lamparter T, Vallone D, Bertolucci C, Foulkes NS. Modulation of DNA repair systems in blind cavefish during evolution in constant darkness. Curr Biol. 2018;28:3229–43.
52. Duboué ER, Keene AC, Borowsky RL. Evolutionary convergence on sleep loss in cavefish populations. Curr Biol. 2011;21:671–6.
53. Duboué ER, Borowsky RL. Altered rest-activity patterns evolve via circadian independent mechanisms in cave adapted balitorid loaches. PLoS One. 2012;7:e30868.
54. Berti R, Messana G. Subterranean fishes of Africa. In: Biology of subterranean fishes. Science Publishers; 2010.
55. Moran D, Softley R, Warrant EJ. Eyeless Mexican cavefish save energy by eliminating the circadian rhythm in metabolism. PLoS One. 2014;9:e107877.

56. Jeffery WR. Adaptive evolution of eye degeneration in the Mexican blind cavefish. J Hered. 2005;96:185–96.
57. Patke A, Young MW, Axelrod S. Molecular mechanisms and physiological importance of circadian rhythms. Nat Rev Mol Cell Biol. 2020;21:67–84.
58. Pilorz V, Helfrich-Förster C, Oster H. The role of the circadian clock system in physiology. Pflugers Arch. 2018;470:227–39.
59. Parnell AA, de Nobrega AK, Lyons LC. Translating around the clock: multi-level regulation of post-transcriptional processes by the circadian clock. Cell Signal. 2021;80:109904.
60. Anna G, Kannan NN. Post-transcriptional modulators and mediators of the circadian clock. Chronobiol Int. 2021;38:1244–61.
61. Cahill GM, Hurd MW, Batchelor MM. Circadian rhythmicity in the locomotor activity of larval zebrafish. Neuroreport. 1998;9:3445–9.
62. Lloyd E, Chhouk B, Conith AJ, Keene AC, Albertson RC. Diversity in rest-activity patterns among Lake Malawi cichlid fishes suggests a novel axis of habitat partitioning. J Exp Biol. 2021;224. https://doi.org/10.1242/jeb.242186.
63. Ceinos RM, Frigato E, Pagano C, Fröhlich N, Negrini P, Cavallari N, Vallone D, Fuselli S, Bertolucci C, Foulkes NS. Mutations in blind cavefish target the light-regulated circadian clock gene, period 2. Sci Rep. 2018;8:8754.
64. Carlson BM, Gross JB. Characterization and comparison of activity profiles exhibited by the cave and surface morphotypes of the blind Mexican tetra, Astyanax mexicanus. Comp Biochem Physiol C Toxicol Pharmacol. 2018;208:114–29.
65. Mack KL, Jaggard JB, Persons JL, Roback EY, Passow CN, Stanhope BA, Ferrufino E, Tsuchiya D, Smith SE, Slaughter BD, Kowalko J, Rohner N, Keene AC, McGaugh SE. Repeated evolution of circadian clock dysregulation in cavefish populations. PLoS Genet. 2021;17:1–32.
66. Lima-Cabello E, Díaz-Casado ME, Guerrero JA, Otalora BB, Escames G, López LC, Reiter RJ, Acuña-Castroviejo D. A review of the melatonin functions in zebrafish physiology. J Pineal Res. 2014;57:1–9.
67. Appelbaum L, Wang GX, Maro GS, Mori R, Tovin A, Marin W, Yokogawa T, Kawakami K, Smith SJ, Gothilf Y, Mignot E, Mourrain P. Sleep-wake regulation and hypocretin-melatonin interaction in zebrafish. Proc Natl Acad Sci USA. 2009;106:21942–7.
68. Niemiller ML, Poulson TL. Subterranean fishes of North America: amblyopsidae. In: Biology of Subterranean fishes; 2010. p. 169–280.
69. Hyacinthe C, Attia J, Rétaux S. Evolution of acoustic communication in blind cavefish. Nat Commun. 2019;10:4231.
70. Bibliowicz J, Alié A, Espinasa L, Yoshizawa M, Blin M, Hinaux H, Legendre L, Père S, Rétaux S. Differences in chemosensory response between eyed and eyeless Astyanax mexicanus of the Rio Subterráneo cave. EvoDevo. 2013;4:25.
71. Moran RL, Jaggard JB, Roback EY, Kenzior A, Rohner N, Kowalko JE, Ornelas-García CP, McGaugh SE, Keene AC. Hybridization underlies localized trait evolution in cavefish. iScience. 2022;25:103778.
72. Warren W, Boggs T, Borowsk R, Carlson B, Ferrufino E, Gross J, Hu Z, Keene A, Kenzior A, Kowalko J, Tomlinon C, Kremitzki M, Lemieux M, Graves-Lindsay T, McGaugh S, Miller J, Mommersteeg M, Moran R, Peus R, Rice E, Riddle M, Sifuentes-Romero I, Stahope B, Tabin C, Thankur S, Yoshiyuki Y, Rohner N. A chromosome level genome of Astyanax mexicanus surface fish for comparing population-specific genetic differences contributing to trait evolution. bioRxiv. 2020:189654.
73. Borbély AA. A two process model of sleep regulation. Hum Neurobiol. 1982;1:195–204.
74. Saper CB, Fuller PM, Pedersen NP, Lu J, Scammell TE. Sleep state switching. Neuron. 2010;68:1023–42.
75. Shafer OT, Keene AC. The regulation of drosophila sleep. Curr Biol. 2021;31:R38–49.
76. Allada R, Siegel JM. Unearthing the phylogenetic roots of sleep. Curr Biol. 2008;18:R670–9.
77. McNamara P, Barton R, Nunn C. Evolution of sleep: phylogenetic and functional perspectives; 2009. http://www.lavoisier.fr/notice/frTWOL6K6ASRW3AO.html

78. Yoshizawa M, Robinson BG, Duboué ER, Masek P, Jaggard JBJ, O'Quin KEK, Borowsky RLR, Jeffery WRW, Keene ACA, Duboue E, Masek P, Jaggard JBJ, O'Quin KEK, Borowsky RLR, Jeffery WRW, Keene ACA. Distinct genetic architecture underlies the emergence of sleep loss and prey-seeking behavior in the Mexican cavefish. BMC Biol. 2015;20:15.
79. Allada R, Cirelli C, Sehgal A. Molecular mechanisms of sleep homeostasis in flies and mammals. Cold Spring Harb Perspect Biol. 2017;9. https://doi.org/10.1101/cshperspect.a027730.
80. McGaugh SE, Passow CN, Jaggard JB, Stahl BA, Keene AC. Unique transcriptional signatures of sleep loss across independently evolved cavefish populations. J Exp Zool B Mol Dev Evol. 2020:22949. https://doi.org/10.1002/jez.b.22949.
81. Arble DM, Bass J, Behn CD, Butler MP, Challet E, Czeisler C, Depner CM, Elmquist J, Franken P, Grandner MA, Hanlon EC, Keene AC, Joyner MJ, Karatsoreos I, Kern PA, Klein S, Morris CJ, Pack AI, Panda S, Ptacek LJ, Punjabi NM, Sassone-Corsi P, Scheer FA, Saxena R, Seaquest ER, Thimgan MS, van Cauter E, Wright KP. Impact of sleep and circadian disruption on energy balance and diabetes: a summary of workshop discussions. Sleep. 2015;38:1849–60.
82. Scammell TE, Arrigoni E, Lipton JO. Neural circuitry of wakefulness and sleep. Neuron. 2017;93:747–65.
83. Cho JY, Sternberg PW. Multilevel modulation of a sensory motor circuit during C. elegans sleep and arousal. Cell. 2014;156:249–60.
84. The evolution of sensory adaptation in Astyanax mexicanus. https://books.google.com/books?hl=en&lr=&id=YDBOBQAAQBAJ&oi=fnd&pg=PA247&dq=The+evolution+of+sensory+adaptation+in+Astyanax+mexicanus&ots=NZHaciJWhi&sig=fsJZYyC8kfAQd-MVxzBLo6kjqs#v=onepage&q=The%20evolution%20of%20sensory%20adaptation%20in%20Astyanax%20mexicanus&f=false
85. Yoshizawa M, Gorički Š, Soares D, Jeffery WR. Evolution of a behavioral shift mediated by superficial neuromasts helps cavefish find food in darkness. Curr Biol. 2010;20:1631–6.
86. Jaggard J, Robinson BG, Stahl BA, Oh I, Masek P, Yoshizawa M, Keene AC. The lateral line confers evolutionarily derived sleep loss in the Mexican cavefish. J Exp Biol. 2017;220. https://doi.org/10.1242/jeb.145128.
87. Mileykovskiy BY, Kiyashchenko LI, Siegel JM. Behavioral correlates of activity in identified hypocretin/orexin neurons. Neuron. 2005;46:787–98.
88. Keene AC, Appelbaum L. Sleep in fish models. In: Handbook of behavioral neuroscience; 2019.
89. Duboué ERER, Borowsky RLRL, Keene ACAC. β-adrenergic signaling regulates evolutionarily derived sleep loss in the mexican cavefish. Brain Behav Evol. 2012;80:233–43.
90. Shi G, Xing L, Wu D, Bhattacharyya BJ, Jones CR, McMahon T, Chong SYC, Chen JA, Coppola G, Geschwind D, Krystal A, Ptáček LJ, Fu YH. A rare mutation of β1-adrenergic receptor affects sleep/wake behaviors. Neuron. 2019;103:1044–1055.e7.
91. Jaggard JB, Lloyd E, Yuiska A, Patch A, Fily Y, Kowalko JE, Appelbaum L, Duboue ER, Keene AC. Cavefish brain atlases reveal functional and anatomical convergence across independently evolved populations. Sci Adv. 2020;6. https://doi.org/10.1126/sciadv.aba3126.
92. Kroll F, Powell GT, Ghosh M, Gestri G, Antinucci P, Hearn TJ, Tunbak H, Lim S, Dennis HW, Fernandez JM, Whitmore D, Dreosti E, Wilson SW, Hoffman EJ, Rihel J. A simple and effective F0 knockout method for rapid screening of behaviour and other complex phenotypes. Elife. 2021;10. https://doi.org/10.7554/eLife.59683.
93. Stahl BA, Peuß R, McDole B, Kenzior A, Jaggard JB, Gaudenz K, Krishnan J, McGaugh SE, Duboue ER, Keene AC, Rohner N. Stable transgenesis in Astyanax mexicanus using the Tol2 transposase system. Dev Dyn. 2019;248:679–87.
94. Kowalko J. Utilizing the blind cavefish Astyanax mexicanus to understand the genetic basis of behavioral evolution. J Exp Biol. 2020;223. https://doi.org/10.1242/JEB.208835.
95. O'Gorman M, Thakur S, Imrie G, Moran RL, Choy S, Sifuentes-Romero I, Bilandžija H, Renner KJ, Duboué E, Rohner N, McGaugh SE, Keene AC, Kowalko JE. Pleiotropic function of the oca2 gene underlies the evolution of sleep loss and albinism in cavefish. Curr Biol. 2021;31:3694–3701.e4.

96. Bilandžija H, Ma L, Parkhurst A, Jeffery WR. A potential benefit of albinism in Astyanax cavefish: downregulation of the oca2 gene increases tyrosine and catecholamine levels as an alternative to melanin synthesis. PLoS One. 2013;8:e80823.
97. Soares D, Yamamoto Y, Strickler AG, Jeffery WR. The lens has a specific influence on optic nerve and tectum development in the blind cavefish Astyanax. Dev Neurosci. 2004;26:308–17.
98. Alié A, Devos L, Torres-Paz J, Prunier L, Boulet F, Blin M, Elipot Y, Retaux S, Alie A, Devos L, Torres-Pas J, Prunier L, Boulet F, Blin M, Elipot Y, Reteaux S. Developmental evolution of the forebrain in cavefish: from natural variations in neuropeptides to behavior. Elife. 2018;7. https://doi.org/10.7554/eLife.32808.
99. Randlett O, Wee CL, Naumann EA, Nnaemeka O, Schoppik D, Fitzgerald JE, Portugues R, Lacoste AMB, Riegler C, Engert F, Schier AF. Whole-brain activity mapping onto a zebrafish brain atlas. Nat Methods. 2015;12:1–12.
100. Ullmann JFP, Cowin G, Kurniawan ND, Collin SP. A three-dimensional digital atlas of the zebrafish brain. NeuroImage. 2010;51:76–82.
101. Kunst M, Laurell E, Mokayes N, Kramer A, Kubo F, Fernandes AM, Förster D, Dal Maschio M, Baier H. A cellular-resolution atlas of the larval zebrafish brain. Neuron. 2019. https://doi.org/10.1016/j.neuron.2019.04.034.
102. Gupta T, Marquart GD, Horstick EJ, Tabor KM, Pajevic S, Burgess HA. Morphometric analysis and neuroanatomical mapping of the zebrafish brain. Methods. 2018;150:49–62.
103. Kozol R, Conith A, Yuiska A, Cree-Newman A, Tolentio B, Banesh K, Paz A, Lloyd E, Kowalko J. A brain-wide analysis maps structural evolution to distinct anatomical modules. Elife. 12:e80777.
104. Duboué ER, Keene AC, Investigating the evolution of sleep in the Mexican cavefish; 2015.
105. Hartmann EL. The functions of sleep. Yale University Press; 1973.
106. Zada D, Bronshtein I, Lerer-Goldshtein T, Garini Y, Appelbaum L. Sleep increases chromosome dynamics to enable reduction of accumulating DNA damage in single neurons. Nat Commun. 2019;10:895.
107. Elipot Y, Legendre L, Père S, Sohm F, Rétaux S. Astyanax transgenesis and husbandry: how cavefish enters the laboratory. Zebrafish. 2014;11:291–9.
108. Stahl BA, Jaggard JB, Chin JSR, Kowalko JE, Keene AC, Duboué ER. Manipulation of gene function in Mexican cavefish. J Vis Exp. 2019. https://doi.org/10.3791/59093.
109. Jaggard JB, Stahl BA, Lloyd E, Prober DA, Duboue ER, Keene AC. Hypocretin underlies the evolution of sleep loss in the Mexican cavefish. elife. 2017; https://doi.org/10.1101/122903.
110. Liu J, Merkle FT, Gandhi AV, Gagnon JA, Woods IG, Chiu CN, Shimogori T, Schier AF, Prober DA. Evolutionarily conserved regulation of hypocretin neuron specification by Lhx9. Development (Cambridge). 2015;142:1113–24.
111. Ma L, Jeffery WR, Essner JJ, Kowalko JE. Genome editing using TALENs in blind Mexican cavefish, Astyanax mexicanus. PLoS One. 2015;10. https://doi.org/10.1371/journal.pone.0119370.
112. Bedell VM, Wang Y, Campbell JM, Poshusta TL, Starker CG, Krug RG, Tan W, Penheiter SG, Ma AC, Leung AYH, Fahrenkrug SC, Carlson DF, Voytas DF, Clark KJ, Essner JJ, Ekker SC. In vivo genome editing using a high-efficiency TALEN system. Nature. 2012;491:114–8.
113. Hwang WY, Fu Y, Reyon D, Maeder ML, Tsai SQ, Sander JD, Peterson RT, Yeh JR, Joung JK. Efficient genome editing in zebrafish using a CRISPR-Cas system. Nat Biotechnol. 2013;31:227–9.
114. Chang N, Sun C, Gao L, Zhu D, Xu X, Zhu X, Xiong J-W, Xi JJ. Genome editing with RNA-guided Cas9 nuclease in zebrafish embryos. Cell Res. 2013;23:465–72.
115. Riddle MR, Aspiras AC, Gaudenz K, Peuß R, Sung JY, Martineau B, Peavey M, Box AC, Tabin JA, McGaugh S, Borowsky R, Tabin CJ, Rohner N. Insulin resistance in cavefish as an adaptation to a nutrient-limited environment. Nature. 2018;555:647–51.
116. Aspiras A, Rohner N, Marineau B, Borowsky R, Tabin J. Melanocortin 4 receptor mutations contribute to the adaptation of cavefish to nutrient-poor conditions. Proc Natl Acad Sci. 2015;112:9688–73.

The Might of Mice in Advancing Sleep Genetics

Sigrid Carlen Veasey

Abstract Forward and reverse genetic approaches in simpler animal models, including zebrafish, flies, and worms, have greatly accelerated sleep gene discovery. Yet, the mouse remains as an indispensable and complimentary model in sleep genetics. Murine models are used to confirm and specify further the role(s) of a gene identified in simpler animal models plays in mammalian sleep. As we now appreciate the importance of specific sleep state-specific electrographic waveforms in the consolidation of memories, mouse models play critical roles in characterizing gene effects on electrographic phenomena in sleep. At the same time, newer strategies in forward genetics in the mouse have identified both dominant and recessive genes influencing sleep. To date, diverse families of genes, from ion channel subunit genes to genes regulating circadian rhythms, metabolism, and inflammation, have been shown to influence sleep in the mouse. Human genome wide association studies have identified genetic loci associated with specific sleep disorders, and the mouse has been indispensable in establishing the roles of putative sleep genes identified in human studies. Armed with novel genetic tools and approaches coupled with advancements in our understanding of the role of sleep in memory and brain health, we are well-poised to more fully realize the might of the mouse in advancing the genetics of sleep.

Keywords Mouse · Non-rapid eye movement sleep · Forward genetics · Microarray · Memory

---

S. C. Veasey (✉)
Chronobiology and Sleep Institute, Department of Medicine, Perelman School of Medicine, University of Pennsylvania, Philadelphia, PA, USA
e-mail: veasey@pennmedicine.upenn.edu

P. Gehrman et al. (eds.), *Genetics of Sleep and Sleep Disorders*,
https://doi.org/10.1007/978-3-031-62723-1_6

Introduction

Although amounts of sleep accrued across 24 h vary with species, all animal species studied, thus far, require sleep for cognition, health, and in many cases survival, yet fundamental questions regarding sleep remain unanswered. To date we do not understand fully why animals require sleep; what the molecular occurrences are that regulate sleep timing and state changes; and what sleep accomplishes at the cellular level and for the brain and body. Genetics has been used to advance many fundamental biological questions, and, thus, it would seem genetics could be applied to advance sleep science. One of the earliest quests in sleep genetics began over 50 years ago when Valatz and Bugat examined electrographic sleep in six distinct inbred mouse strains [1]. They determined that sleep state duration and timing varied significantly across the inbred strains, and more so than within a given strain, thereby supporting genetic influences in sleep [1]. Using F1 hybrid mice and back crosses, they identified that specific waveforms in NREM sleep and REM sleep are genetically determined, with each sleep state seeming genetically independent of the other, as ratios of the states and waveform differences varied across inbred strains [1]. These strain-dependent findings strongly support the potential of using genetics to answer the mysteries of sleep.

Genetic research has delivered tremendous insight into many fundamental biological processes, including circadian rhythms. Indeed, genetics has played a central role in elucidating the core molecular mechanisms of the 24-h cellular clock [2–4]. A highly fruitful paradigm was set up to examine the timing of rest and activity across constant darkness by measuring fly movement (via infrared beam splitting) in individual tubes to measure the circadian period. A comparable high throughput system, however, does not currently exist for the many intricacies of sleep staging, timing, depth, and recovery. Sleep is composed of various behavioral states defined by electrographic activity patterns (described below). Measuring and characterizing this behavioral state-dependent electrographic activity, at present, requires surgical implantation of electrodes and an elaborate sleep scoring analysis system. There are several additional challenges with trying to identify key genes in sleep. While circadian rhythms are relatively robust, responding primarily to the timing of ambient light, and to a lesser extent to several other signals, sleep behaviors may be profoundly influenced by numerous conditions that may indirectly affect sleep, e.g., iron deficiency resulting in restless leg syndrome, diet, infection, generalized anxiety, or chronic pain. Additionally, alterations in the production, uptake or metabolism of sleep regulatory neurotransmitters/neuromodulators, or alterations in post-synaptic signal transduction pathways, synapses, connectivity may alter sleep, but likely influence other brain functions, as well. These challenges do not render sleep genetics impossible; rather phenotyping should be comprehensive to fully understand how a specific gene influences sleep. For example, a short sleep mutation may be beneficial if sleep need is also less, but detrimental is sleep need is not met with short sleep. The mouse model is well suited to addressing these issues, as recovery sleep and wake and cognitive studies may help address benefit and/or harm of a specific mutation. This chapter summarizes key discoveries made our understanding of some of the functions of sleep and then incorporates these advancements into a review of using both forward and reverse genetic tactics in the mouse to advance sleep neurobiology. The distinct advantages, disadvantages, and limitations with each of the described techniques are discussed, and strategies are proposed to integrate use of murine genetics with genetics in both simpler animal models and humans to best advance our understanding of sleep neurobiology.

The Expanding Phenotype of Sleep

Despite its seemingly inactive state, sleep can be considered a complex behavior, one comprised of multiple distinct behavioral states, each with its own electrographic signature, and specific central and peripheral effects on the sleeper. Appreciation of this complexity guides the focus of genetics of the state and enable us to understand what questions may be answered with genetics approaches.

Like simpler animal models, behavioral sleep in the mouse may be defined as a reversible state of behavioral quiescence and reduced responsiveness to the environment that is regulated primarily by the circadian rhythm and sleep homeostasis [5]. Simply put, there are specific times within a 24-h period that are more conducive to sleep (circadian influences on sleep/wake), and the longer one is awake, the greater the drive to sleep is (sleep homeostasis). It is important to recognize that behavioral quiescence is only a proxy for sleep. If a gene impacts muscle function or physical activity in sleep, a screen in the simpler mutants may not accurately estimate sleep. In the mouse, as in the human, this behavioral quiescence and reduced responsiveness is comprised of a repeated patterned series of stages defined by distinctive combinations of various electroencephalographic (EEG) and electromyographic (EMG) waveforms, which may then be used to not only confirm sleep but also distinguish behavioral sleep states. The waveforms have the advantage of confirming or refuting sleep across behavioral quiescence and periods of physical activity. EEG waveforms are measured with either cranial surface or intracranial electrodes, while surface or intramuscular electrodes are used to measure EMG activity.

The two distinct behavioral states of sleep are non-rapid-eye movement sleep (NREMS) and rapid-eye-movement sleep (REMS). NREMS is defined electrographically by an overall slowing in the EEG with the combinatorial occurrence of neocortical slow oscillations (0.5–1 Hz), delta waves (1–4 Hz), spindles (9–16 Hz), and hippocampal sharp-wave ripples (SWR, waveforms consisting of a large negative wave followed by a brief series of very fast waves, 100–200 Hz) [6, 7]. NREMS may be further divided into states of intermittent slow wave activity (light NREMS) and periods of predominant delta waveform activity (slow wave sleep). In contrast, REMS is defined by postural muscle atonia, coincident with a predominance of fast desynchronized EEG waveforms and/or theta oscillations (5–8 Hz), either with phasic activity, e.g., bursts of eye movements, muscle twitches, and ponto-geniculate occipital waves (phasic REMS), or without phasic activity (tonic REMS) [7]. In mammals, sleep follows strikingly consistent patterns beginning with an entry into lighter stages of NREMS before deep NREMS (slow wave sleep), where slow oscillations and delta waves may predominate. In healthy mammals, NREMS precedes REMS, and the spectral power of slow wave sleep is at its greatest in the beginning of the night and then exponentially declines across sleep [6, 8]. There is a progressive shift across major sleep periods from longer, more pronounced slow wave sleep periods in the first third of the night to minimal slow wave activity, and more REMS later in sleep. When mammals are deprived of sleep, the slow wave activity is even more pronounced in the beginning of recovery sleep. Slow wave activity in response to sleep loss is considered a measure of sleep homeostasis.

These sleep state-specific EEG waveforms are more than simply biomarkers of specific sleep states. Research over the past decade has identified essential roles of theta waves, spindles, sharp wave ripples, and slow wave activity in the consolidation of memory and transfer of learned engrams from the hippocampus to the cortex and consolidation in the cortex. Early research by Wilson and McNaughton advanced an active role for sleep in memory consolidation in the hippocampus when they discovered that pairs of neurons co-activated in the hippocampus in a learning trial were reactivated together in NREMS and that this reactivation occurred temporally with SWR events [9]. A subsequent series of studies established that NREMS is essential for hippocampal memory consolidation. Loss of sleep in mice within a key window after hippocampal-dependent contextual learning prevents consolidation of that information [10]. Notably, cued memory, which is not hippocampal dependent, is unaffected by sleep [10]. Similarly, in humans hippocampal-dependent declarative and procedural (motor) memories benefit from sleep within a specific post-learning window, as measured by task performance, while hippocampal-independent memory tasks are not influenced by post-encoding sleep [11].

The mechanisms underlying sleep state-dependent hippocampal memory consolidation are now more completely understood and involve a convergence of network activity by three NREMS waveforms: hippocampal SWR's, neocortical slow oscillations, and thalamic spindle activity (see Fig. 1). During NREMS, CA3 neurons that were activated during encoding are reactivated and produce a large depolarization (the sharp wave component) at the stratum radiatum dendrites of encoding-paired CA1 neurons and a brief (50–100 msec) but rapid excitation (100–200 Hz), the ripple component) at the CA1 somata. Acutely suppressing the ripple component of a SWR prevents the consolidation of hippocampal-dependent spatial memory without disrupting NREMS [12, 13], demonstrating the importance of this particular waveform in sleep for consolidation of memory. Concurrently, in the neocortex during NREMS, recently encoded declarative memories are selectively reactivated and strengthened through communication of locally generated slow oscillations [14]. Spindles are also essential for memory, where suppression of spindles in sleep prevents memory consolidation [15, 16]. The stronger the encoding, the greater the spectral power of network slow oscillations and the greater the improvement in task performance after sleep [17]. The hyperpolarizing down-states and depolarizing up-states of neocortical slow oscillations also allow synchronization between the locally encoded network in the neocortex and relevant thalamic

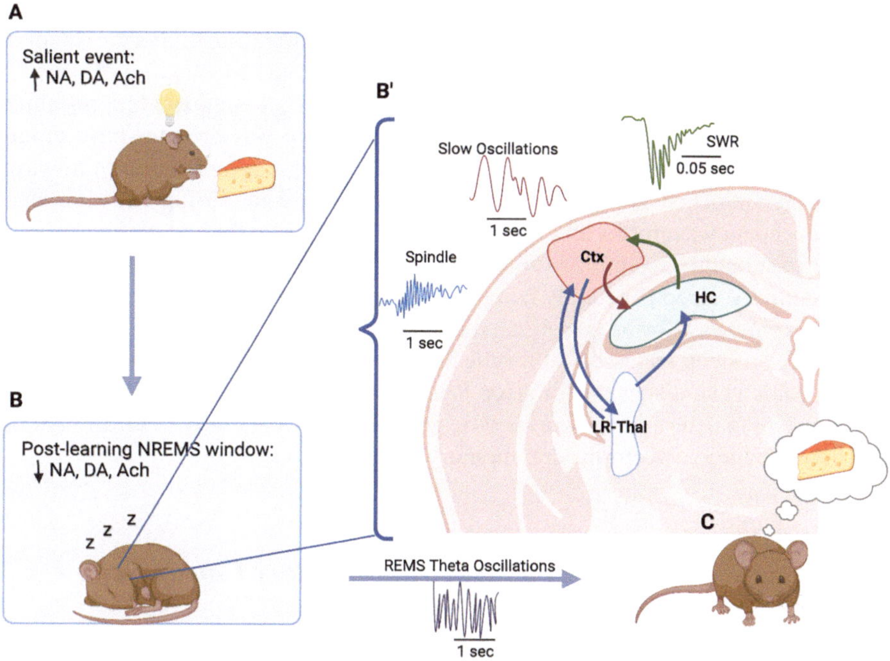

Fig. 1 Sleep state-dependent waveforms are essential for memory consolidation. Not all events are consolidated into lasting memories. Those most salient become more robust and lasting memories. (**a**) In this example a laboratory mouse has never seen chesse before and finds it quite tasty. Several wake neuromodulators are released in the hippocampus and cortex including noradrenaline (NA), dopamine (DA), and acetylcholine (Ach). (**b**) In a 3–4 h window after experiencing chesse for the first time, NREM sleep will result in lowering NA, DA and Ach levels in the forebrain and will serve to initiate a series of waveforms in various brain regions necessary for memory consolidation and long-term storage. These waveforms include in NREM sleep, thalamocortical spindles and slow oscillations, and in the hippocampus, sharp wave ripples. In REM sleep theta waves predominate in the hippocampus. (**c**) Memory of cheese will now persist

and hippocampal connections that, in turn, enable the cooccurrence of spindles in the thalamus and SWR in the hippocampus [18]. The arrival of spindles in the neocortex during the depolarizing up-state allows the transfer of reactivated memory information to the related neocortical network by promoting Ca^{2+} influx into neocortical network neurons and Ca^{2+} channel-dependent long-term potentiation [19].

There is evidence that REMS also serves (at least in part) to consolidate memories. REMS increases in density and phasic activity after learning in mice and in humans [20], and the magnitude of the increase in REMS after learning predicts performance of the learned task [20]. Two recent studies show critical roles for REM sleep in memory consolidation. The first study took advantage of the observation that hippocampal theta waves require GABAergic input from the medial septum of the basal forebrain [21]. Silencing medial septum GABAergic neurons optogenetically in REMS reduced theta activity in the hippocampus and impaired hippocampal dependent memory consolidation, without interfering with REMS [22]. Pertinent to the above discussion on the role of NREMS in hippocampal memory consolidation, silencing the GABAergic input in NREMS had no effect [22]. This suggests that there is an additive benefit to REMS for contextual hippocampal memory consolidation. Notably, theta waves and sharp wave ripples have unique CAMKII neuronal Ca^{2+} transient signatures [23], suggesting that variance in waveforms may influence synaptic responses. This strong relationship between sleep state-dependent waveforms and memory means that genetic differences in a specific waveform should be examined for effects not only on sleep but memory as well.

In summary, sleep is a highly complex behavior, where a pattern of very specific electrographic waveforms is generated in a sleep state-dependent fashion and these waveforms in sleep are essential for the consolidation of memory. This information should guide gene discovery for sleep neurobiology and as genes are found, it will be important to determine how a specific gene influences not only sleep amount but the electrographic waveforms and memory consolidation.

Sleep-Dependent Synaptic Plasticity in Memory Consolidation

Synaptic plasticity appears important for long-term memory consolidation [24, 25], and sleep appears to influence synaptic plasticity. In humans, paired associative stimulation of a peripheral nerve and relevant cortical region can induce LTP or LTD depending on the stimulus interval, and the density of regional spindles in sleep following the stimulation predicts the development of either LTP or LTD [26]. In contrast, measures to induce LTD through reduced use of a circuit reduce local slow wave activity [26]. Motor task learning induces branch-specific spine formation in mice, as measured with two-photon imaging of the motor cortex across learning and consolidation [27]. If, however, mice are deprived of sleep across the critical window for NREMS consolidation, the task-responsive branch visualized does not show task-responsive spine enlargement and learning is impaired, suggesting that sleep is essential for not only memory but the related structural plasticity [27]. A recent series of studies has defined key roles for synaptic plasticity initially in the hippocampus and then in the neocortex to underlie sleep-dependent consolidation of memory [28]. Structural plasticity events require physical modification of the actin infrastructure in synaptic spines [29]. Depending on post-translational modifications and concentration of available cofilin, cofilin will stabilize or destabilize its local actin infrastructure [30]. Using a novel optogenetic technique to silence cofilin and prevent LTP in specific populations of neurons, Goto et al. recently identified a clear three-step plasticity process for a hippocampal-dependent contextual memory task [28]. Specifically, they identified an initial local hippocampal LTP within an hour of learning, then during the ensuing sleep, a network LTP develops in the hippocampus without LTP in the neocortex. On the following day after sleep, a neocortical LTP (anterior cingulate) develops [28]. While all three events are critical for memory consolidation, both the hippocampal and the neocortex plasticity events require sleep, such that disruption of LTP outside of sleep does not impair consolidation.

REMS also plays an important role in synaptic plasticity. Visualizing task-responsive motor cortex branches with two-photon imaging in mice, synaptic spines can be monitored across motor task learning and across sleep. Spines formed in response to learning a new motor task increase in size across REMS [31]. If mice are deprived of REMS, the spines will not increase in size after learning and their long-term stability (at 4 days) is also jeopardized. In contrast, depriving the mice of NREMS neither influenced spines in the first day nor long-term [31]. Moreover, REMS-deprived mice do not perform as well 24 h after learning the motor task, compared to both rested control mice and NREMS deprived mice. Thus, REMS plays important roles in synapse strengthening and stability. Because interference with either NREMS or REMS can impair consolidation of memories with a hippocampal dependent step, it is most likely that the two stages play complementary roles where REMS as the later sleep state plays a role in refining or reinforcing NREMS plasticity and circuits.

Sleep Is Also Necessary for Synaptic Normalization

Wakefulness imparts a constant barrage of sensory information, which should generate an enormous number of synapses every day. In the adult, however, the total number of spines is relatively stable, and many individual spines are stable for months, if not longer [32]; thus, numerous dendritic spines formed across wakefulness must be eliminated prior to the ensuing period of wakefulness. Strikingly, although NREMS and REM strengthen some new synapses, both states contribute to synapse pruning or elimination. Tononi and Cirelli have hypothesized that sleep serves to prune sufficient synapses to correct synaptic homeostasis prior to the next wake period [33]. In support, De Vivo et al. used serial electron microscopy to create three-dimensional images of layer II motor cortex synapses to assess the effects of sleep and wake on synapse characteristics [34]. They found that while sleep did not change the number or density of synapses, the overall axon spine interface (ASI) per synapse, a measure of synaptic strength, was reduced in sleeping mice relative to mice kept awake. This was, however, only true for medium to small ASI's, as large ASI's (stronger synapses) did not change with sleep. A similar picture of increased ASI was observed in the hippocampus in adolescent mice after 6–8 h of waking [35]. REMS deprivation after learning a motor task reduces the pruning of recently formed synapses, and interestingly, NREMS deprivation did not influence pruning of the recently formed spines [36]. Elimination of spines during REMS facilitates new motor task learning the following day.

This rich but exceptionally complex phenotype requires that when a gene is found, characterization of the effect of the gene on electrographic sleep should be complemented with a description of effects of the gene on learning and memory, especially for hippocampal-dependent processes and synaptic homeostasis to more fully understand the gene's role in these integrated processes. At the same time, the advancements in sleep neurobiology provide clues for putative sleep genes.

Genetic Tactics to Advance Sleep Neurobiology in Mice

Unbiased Approaches for Sleep Genetics

Forward genetics is an unbiased approach with the capability to explore many phenotypes simultaneously to discover genes that influence distinct measurable phenotypes. One of the first approaches to forward genetics was the use of quantitative trait loci (QTL). QTL studies have been used in mice to discover genes involved in many complex behaviors and have been implemented to discover sleep QTL's. Two strains of mice with distinct sleep state-dependent EEG activity are BALB/cBy and C57BL/6By, where their crosses are termed CXB [37]. A set of recombinant inbred mice was generated, from which several QTL's were identified for effects on NREM and REM sleep, including one for the amount of REM sleep in a 24-h cycle [38]. Using CXB recombinant inbred (RI) mice in another study, a loci with an LOD score of >30 on chromosome 5 was identified for theta frequency in REM sleep. A gene mapping to this region is *short chain acyl-coenzyme A dehydrogenase* (*SCAD*). SCAD is essential for fatty acid oxidation and humans with SCAD deficiency have lethargy and developmental delay, yet sleep has not been characterized. Using a second RI scheme with C57BL/6 J and DBA/2 J (BXD), a homeostatic sleep genetic effect was observed. The QTL was named delta-power-sleep 1, (DPS1) [39]. Homer1a is a candidate gene for this locus [40]. QTL is most valuable for identifying genes with very larger trait differences across strains and requires large numbers of mice to be phenotyped for study of the one trait. For this reason, newer techniques to analyze multiple traits at the same time have been developed.

A newly developed strategy to identify novel sleep genes has been developed using diversity outbred mice [41]. Large populations of diversity outbred mice have been derived from the major inbred laboratory mouse strains, through use of randomized outbreeding across multiple generations. Using these collections of mice, QTL can be identified for a specific sleep phenotype, where many of the QTL regions involve only a few genes. In a study phenotyping sleep activity (infrared beam splitting) in over 300 mice, a linkage peak was identified to associate to short sleep, for a loci with the gene for dihydropyrimidine dehydrogenease *(Dpyd)*, which breaks down thymine and uracil nucleotides. Similarly, a *Dyp*d knockout mouse showed reduced less sleep in the dark (active) period. Intriguingly, this is the first mutant sleep gene that appears to specifically affect sleep in that in a larger behavioral/cognitive screen, the mutant mice appeared normal. Overall, this approach seems quite promising and could unveil more novel sleep genes when performed on a far larger scale.

In mice, a third approach that has been implemented is random mutagenesis by administering *N*-ethyl-*N*-nitrosurea (ENU) to adult male mice to then examine mutations in progeny [42]. ENU induces random point mutations in the germline, and then mutagenized males are bred to wild-type females one to two generations, and progeny are assessed for the measurable phenotype to determine heritability, hopefully a phenotype accessible to a high-throughput screening technique [43]. Because the mutation may involve a gene involved in transcriptional regulation, the mutation can result in an increase or decrease of a specific gene product or the mutation directly alters the gene product (resulting in a gain of function or loss of function mutation). Mice with genetically tractable abnormal phenotypes are then be sequenced to identify the gene of interest for the altered phenotype.

In a remarkable tour de force, the above strategy was used to generate over 8000 F1 progeny of ENU mutagenized mice that were then remarkably implanted with EEG and EMG recording electrodes, and sleep was then recorded and analyzed to detect outliers [44]. Variables measured were 24-h times spent in wake, NREMS and REMS, and bout durations of each state, the presence of muscle atonia in REMS and the slow wave activity after 4 h of total sleep deprivation by gentle handling. The research team focused this initial look on mutants deviating more than three standard deviations from the mean. Two mutants were identified. One mutant revealed a prolonged sleep time. Whole exome sequencing identified a single nucleotide substitution in the salt-inducible kinase-3 (*SIK3*) gene. SIK3 is a protein kinase, that regulates energy availability. The team went on to demonstrate in *SIK3* mutants a normal circadian period and normal wake-promoting behaviors and that the mutant mice had increase sleep need. In support, the mice showed greater slow wave activity in recovery sleep after sleep loss. They went on to show that sleep deprivation prompts phosphorylation of SIK3. Interestingly, the team went on to confirm a similar phenotype for the mutation in the fly homolog [44]. A second mutation in a sodium leak channel gene (*Nalcn*) was found to reduce the amount of REMS but manifest normal theta activity in REMS and normal nuchal muscle atonia. This mutation increases excitability of neurons, but how this translates into reduced REMS will require further study. A third mutant with a sleep phenotype was a mutation in the voltage-dependent calcium channel, CACNA1a, which weakly reduced wakefulness, which was also associated with ataxic gait [45].

Because sleep parameters are generally normally distributed, a thousand mutants must be screened to find three mutants (value beyond three standard deviations) for a given trait, and then perhaps only a fraction will show heritability. The research group that reported the above forward genetics screen for sleep mutants has now screened over 10,000 mice and has found 3 genes that influence sleep. Notably, Sik3 has come up in the screen in four independent pedigrees.

More recently, an approach with forward genetics was taken to find recessive mutations after ENU mutagenesis, and this strategy was used to identify sleep mutants [46]. To accomplish this, large groups of G3 mice generated from the founder G1 male and multiple offspring females to generate over a hundred pedigrees. The group used an activity assay as a proxy for sleep and identified several mice with less activity in the lights-on (inactive) phase. Ultimately, they identified the mutation in VAMP2, a neuronal N-ethylmaleimide-sensitive factor attachment protein receptor (SNARE) complex, that is essential for neurotransmitter exocytosis (release). The electrographic phenotype revealed one hour less NREMS and one hour less REMS, with longer bouts of wake. Interestingly, there were no cognitive deficits associated with the observed reductions in sleep. This study serves to emphasize an important point about forward genetic studies. The search for recessive mutations will likely yield a very different set of genes. For example, VAMP2 absence is perinatal lethal, while the haploinsufficient VAMP2 mouse has a relatively mild phenotype.

There are several caveats with using mutagenetic forward genetic approaches to understand the genetics of sleep. First it is labor-intensive and costly, given the numbers of mice assays and the phenotyping involved. Second, the traditional mutagenesis screening algorithm as performed by Funato et al., looks for autosomal dominant genes with mutant phenotypes. A significant VAMP2 mutation would be missed in the Funato et al., screen [44], because it is lethal. Thus, genes that serve multiple roles including one or more essential to life yet also influencing sleep will be missed by the traditional autosomal dominant screens. However, this does not mean forwards screens for autosomal dominant sleep genes should not be performed, on the contrary, as several large effect sleep genes have been found in human studies to be autosomal dominant [47, 48]. Thus, it makes sense to continue to screen for autosomal dominant mutations, while building mechanisms to efficiently identify recessive genes. One additional concern with all random mutagenesis approaches, is that by chance larger genes with more sites to mutate are more likely to be mutated and thus signal and receptor proteins and other large proteins may be found with greater frequency.

The International Knockout Mouse Consortium has been developing mouse lines each with a knockout of a protein-coding gene, with the intent of developing strains for all murine proteins. The International Mouse Phenotyping Consortium is charged with developing phenotypes for each of the genes knock out. Fortunately, sleep behavior (piezoelectric cage floor monitors to detect movements/wakefulness) has been incorporated into the phenotyping to provide values for sleep and wake timing and breathing movements for the 24 hr period, the light and dark cycles. To date, 572 genes have been tested, and 18 homozygous knockout strains have been identified with abnormal sleep phenotypes, suggesting that 3% of genes affect sleep/wake activities. Only three gene knockouts (*PPP1r9b*, a protein phosphatase; *CCl26*, a chemokine ligand, and *Chn1*, a GTPase-activating signal transduction neuronal protein) influenced total (24-h) sleep time. It is interesting that the genes identified, thus far, are very different genes than those picked up in the mutagenesis screens. This screen does not include a homeostatic component and does not distinguish sleep/wake states (NREM sleep versus REM sleep). However, this approach has the capacity to provide a basic rest/activity pattern and timing phenotype for a most comprehensive set of genes, including many types of genes that would be missed by the other forward genetic approaches. A major challenge for this screen is that the variances for the selected sleep wake parameters are quite high.

Sleep Neurobiology Advancements Using Reverse Genetics

Reverse genetics is used to determine the role that a gene of interest plays in a particular phenotype. In the study of sleep genetics, reverse genetics in mice can be used to define the effect of putative sleep genes, identified in fly, worm, or zebrafish forward mutagenesis screens, on electrographic sleep amounts, quality, and timing in the mouse. Reverse genetics in the mouse is also used to define the role of candidate genes suggested in a human GWAS study, and where pathways and circuits have been identified in the mouse to influence sleep, reverse genetics can test a host of genes (neurotransmitters, receptors, signal transduction genes) that may influence the circuit and sleep/wake behaviors. Typically, the role of the gene of interest is tested by deleting the gene prior to development and examining the effects of the knockout on sleep. Identifying a gene with a profound influence on sleep has been the Holy Grail for reverse genetics studies of sleep in mice. Here, we provide some of the most successful examples of using reverse genetics in mice to understand a gene (as summarized in Fig. 2). We then provide a synopsis of sleep responses for specific genes in each of the categories of sleep genes assessed and discuss how the information has informed sleep neurobiology.

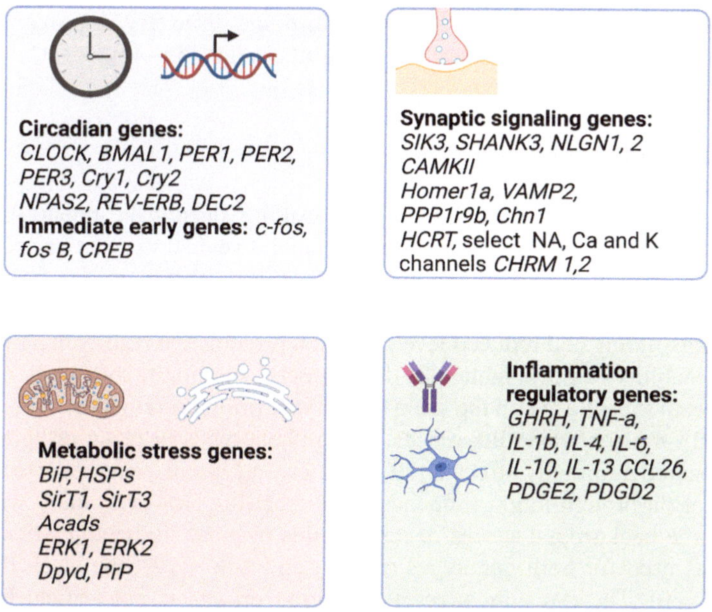

Fig. 2 Functional categories of identified sleep genes. Categories (bold) and genes below were identified in forward and reverse genetics screens to have effects on sleep timing, duration and/or waveforms. In each caterogiry, genes have been identified to have diverse and at times opposing effects on behavioral state timing or duration, sleep homeostatic response and electroencephalographic waveform spectral power

A Gene for Narcolepsy

An illustrative example of the power of the mouse in understanding the role in sleep served by a gene was the identification of orexin (also known as hypocretin) as a key neurotransmitter affected in narcolepsy. From studies in humans with narcolepsy and based on studies of specific dog lines within breeds with increased prevalence of narcolepsy [49], it was evident that narcolepsy was genetic, yet the cause was not known. Two labs independently identified a novel hypothalamic neuropeptide thought to play a role in feeding (named by one lab hypocretin and named orexin by the other) [50, 51]. Immunohistochemical studies showed diffuse orexinergic projections throughout the cortices and brainstem, suggesting that orexin may play a role in a more widespread behavior than feeding [52]. A knock-out mouse targeting (*prepro-orexin*, the gene for the precursor protein for orexin-A and -B) was generated and studied for effects on feeding and activity [53]. The mutant mice evidenced periods of inactivity, particularly in the lights off (active) period. To look for epileptiform activity, the mice were implanted with EEG and EMG electrodes and sleep wake was recorded [53]. The sleep/wake phenotype in orexin-deficient recapitulated narcolepsy with short wake, sleep bouts, a reduction of the diurnal ratio of sleep amount across the lights on and off periods, the occurrence of sleep onset REMS periods and cataplexy (loss of muscle tone in waking). Brains of humans previously diagnosed with narcolepsy were then examined, confirming the loss of orexin in the lateral hypothalamus [54]. In turn, the discovery of orexin has facilitated the elucidation of novel sleep and arousal circuits [55–57].

Fatal Familial Insomnia Genetics

Another sleep disorder in humans with a striking sleep/wake phenotype is fatal familial insomnia, a rare and devastating degenerative disorder that has been identified in just over 40 families, although sporadic cases have also been reported. A mutation (D178N/M129) in the prion protein gene (*PrP*) underlies the disorder and induces misfolding and reduced levels of prion protein, and results in a rapidly progressive inability to sleep, metabolic disturbances, autonomic dysfunction, and ultimately death [58]. Based on the sleep phenotype, prion protein knockout mice were examined for sleep/wake differences. Sleep phenotypes in prion protein deficient mice appeared minimally affected [59]. Shorter wake bouts were observed, suggesting that prion protein might influence wake stability. However, there were differences in the background strains of mice in this original study, and when a similar strain was used for both genotypes (mutant and wild type), the sleep differences resolved [60]. These results suggest that prion protein is not critical for normal sleep/wake activity. Intriguingly another mutation in the prion protein D178/V127 is observed in the neurodegenerative process, Creutzfeldt-Jakob disease (CJD). Sleep disruption is also common in this prion disorder, manifesting as fragmented sleep, loss of spindles, and markedly reduced REM sleep [61]. Mice with this mutation do show greater sleep disruption and loss of REM sleep [62]. Spindle activity was not quantified in the mutant mice. Nonetheless, this work shows very nicely that absence of a gene can have minimal effects on phenotype, in this case the sleep phenotype, while a single specific mutation in gene can have profound effects.

Sleep Genetics: Generating Hypotheses with Microarrays

Several genetics approaches have been employed to generate possible sleep genes that could then be tested using reverse genetics. Early exploratory approaches to identify novel genes influencing sleep included subtractive hybridization, differential display, and microarrays. More recently, RNA sequencing and translating ribosome affinity purification (TRAP) sequencing in rats and/or mice across extended wakefulness and sleep have been used to identify potential sleep genes [63]. Mice exposed to sleep deprivation show alterations in hippocampal post-synaptic densities and hippocampal dependent memory impairments, suggesting that the hippocampus is susceptible to sleep loss [10, 64]. Microarray analysis of hippocampal tissue in mice with or without sleep deprivation has identified sleep state-dependent changes in chaperone, metabolic, immediate early genes, and synaptic scaffolding genes, consistent with the hypothesis that sleep, in part, promotes protein synthesis [65]. These findings are reasonably consistent with sleep genes identified in microarray work in other labs [66]. TRAP-sequencing provides the active translatome response, in this case to sleep loss, and this technique was recently used to determine genes being translated in response to sleep deprivation in CAMKII hippocampal neurons [67]. Considerably fewer genes were identified for sleep deprivation effects on translation [67], yet the functional categories of genes up- and downregulated are similar in microarray and TRAP studies. Upregulated pathways included protein folding, and downregulated genes included ion channels and circadian genes. A major challenge for sleep is that because the durations of each behavioral state of sleep are so brief, it is not possible at present to assess gene changes across spontaneous behavioral states (e.g., waking to NREM sleep). A second challenge is that gene responses to sleep loss may vary across brain regions and specific cell types. More recently single-cell RNA sequencing in the mouse brain was performed to compare gene response differences across cell types, using in situ hybridization in the relevant tissue to validate some genes [68]. Twelve hours of sleep deprivation in the mouse alters the transcriptome for neurons, astrocytes, and microglia, and notably, the profiles of gene responses are distinct for each cell type, suggesting that sleep loss affects each cell type in unique ways. Surprisingly, the unfolded protein response was noted primarily in endothelial cells and not in neurons [68]. Parallel proteomics and phosphoproteomic studies showed neurons to have greater proteomic changes than astrocytes [68]. The biological function pathways impacted by sleep loss in astrocytes included synapse size and activity organization, cytoskeletal infrastructure organization, ion transport upregulation, and fatty acid degradation in astrocytes, while the neuronal responses included synaptic vesicle transport regulation, neurotransmitter synthesis, and glucose metabolism [68], pathways consistent with synaptic and energy homeostasis hypotheses of sleep [69, 70], yet not delivering as of yet, novel sleep genes.

Circadian Genes and Sleep Behavior: Bi-directional Relationships

Sleep neurobiologists have generally considered sleep to be regulated by two independent processes, sleep need, or homeostasis, and the circadian rhythm. However, recent studies provide evidence that the two processes are inextricably linked at the cellular level, as sleep influences core clock genes, and alterations in core clock genes influence sleep. Cellular circadian rhythm is regulated by a transcriptional feedback loop. CLOCK and its heterodimer, BMAL1, activate transcription of E-box-containing genes, period (*PER*) and cryptochrome (*CRY*) 1 and 2 [71]. When PER and CRY proteins reach sufficient concentrations and dimerize, they enter the nucleus to suppress BMAL1:CLOCK transcriptional activities [72]. Most of the reverse genetic studies of clock gene effects on sleep involve whole animal knock out or knock down prior to development. Thus, it is not at all clear which effects are direct clock gene effects on sleep and which are secondary to either transcription of another gene or gene set or secondary to alterations in circuit development and the health of neural tissue. Nonetheless, BMAL1 knockout mice show increased 24-hr sleep and increased fragmentation of sleep (greater transitions to wake) [73]. NPAS2 is another binding partner of BMAL1 and when deleted mice show reduced NREM sleep across the dark period [74]. This finding suggests that distinct gene sets downstream of CLOCK/BMAL1 and NPAS2/BMAL determine circadian gene effects on sleep. Mice lacking CRY1 or CRY2 protein have a normal sleep phenotype, while the double knockouts show increased sleep in the dark period (the opposite of NPAS2 deletion) and altered sleep homeostasis. Less consistent effects on sleep are observed for Per 1 and 2-deficient mice [75, 76]. Collectively these findings support the concept that the clock itself may regulate circadian sleep timing without influencing sleep quantities, while specific circadian transcriptional factors may have distinct effects on gene sets that then influence sleep/wake activity. REV-ERB-α and -β negatively regulate the transcription of BMAL1, CLOCK, and NPAS2. REV-ERB-α and -β are perhaps the more interesting of the potential sleep-clock genes in that REV-ERB-α deletion does not affect basal sleep wake activity but reduces sleep homeostasis [77], and REV-ERB-β deletion results in increased NREM and REM sleep in the dark period, and delta power across the entire 24-h cycle is reduced, without influencing homeostasis [78]. Interestingly, REV-ERB-β deletion results in increased BMAL1, CLOCK, NPAS2, Per1, and Per2 transcription, but also results in marked changes in transcription of genes that promote sleep, including Cacna1a, Gabra5, and Kcnma5 [78]. To advance the role of circadian genes in sleep regulation, studies are now needed to induce knockdowns in the adult animal (avoid confounds of development) and in various cell groups for each of these genes (to refine mechanisms and understand when and where specific circadian genes influence sleep).

Cellular Stress in Response to Prolonged Wakefulness

Several of the above gene exploration studies in mice have demonstrated that sleep loss upregulates genes involved in protein folding and handling, supporting a role for sleep in proteostasis. Cirelli and Tononi were first to note the heat shock gene response to sleep deprivation, by showing increases in several chaperones in response to short-term sleep deprivation [79]. The increase in heat shock genes was confirmed by a second group of researchers and expanded to show changes in multiple brain regions and delayed recovery [80]. The basal forebrain cholinergic neurons contribute to sleep/wake control, including homeostatic regulation of sleep [81–83]. Cholinergic neuron gene responses to sleep deprivation was performed by capturing single neurons from mice expressing green fluorescent protein (GFP) by way of the choline acetyltransferase promoter [84]. These basal forebrain cholinergic neurons were assessed for gene differences across various times of wakefulness or circadian time points. As expected, circadian genes differed for time points of lights on and lights off. Genes upregulated across sleep deprivation were largely protein chaperone and other folding genes. This stands in contrast with the cortical single cell sequencing that found chaperone changes only in endothelial cells, but cholinergic neurons are such a small fraction of cells in the single cell sequence experiment, a change specific to the cholinergic cells would be obscured. Whether these proteostasis genes change to promote sleep, or in response to altered brain temperature in wakefulness, relative to sleep, or in response to stress has not been elucidated. Grp78, or BiP, is the major protein folding chaperone in the endoplasmic reticulum. Absence of Grp78 is embryonically lethal, but mice with one copy can survive into adulthood. Mice with Grp78 heterozygosity show fragmented sleep and more pronounced fragmentation of sleep across aging [85]. Thus, this major chaperone protein does not influence sleep until advanced age. The effects of BiP haploinsufficiency on sleep homeostasis have not been examined. Sirtuins are deacetylases that play key roles in neuroprotection and in countering cellular senescence and oxidative stress. Reductions in Sirtuin Type 1 (SirT1) and Sirtuin Type 3 (SirT3) both influence sleep. SirT1 and SirT3 loss result in increased sleepiness, reduced theta waveform activity, reduced homeostasis yet only SirT1 affects total sleep times (increasing both NREM and REM sleep time/24 h) and have profound effects on the health of wake-activated neurons including noradrenergic locus coeruleus and orexinergic neurons [86, 87].

Synaptic Genes and Sleep

Electroencephalographic (EEG) waveforms, including spindles, theta, sharp wave ripples, and slow wave activity, are essential for memory consolidation in sleep and for the restorative properties of sleep. Thus, it is unsurprising that many of the genes associated with changes in one or more of the sleep-related EEG waveform signals are ion channels genes. It is, however, surprising that these ion channel genes also have effects on specific sleep state durations/24 hr or across either the light or dark phase. A related category of genes that influence behavioral state are genes regulating synaptic activity.

Voltage-dependent calcium channels (VDCC) regulate the influx of calcium into cells, including neurons, and thus in part determine the overall excitability of neurons and influence signal transduction and other molecular pathways [88]. Three types of VDCC expressed on neurons are the N-, P/Q- and R-type, where the N-type VDCC's regulate glutamate release and thereby excitability [89]. *Cav2.2* encodes a subunit that is specific to the N-type VDCC. Mice lacking the *Cav2.2* gene were assessed for sleep/wake activity [90]. Locomotor activity increased in the dark period in *Cav2.2* mice. Total time/24 h spent in each state did not vary but mice lacking had longer NREM sleep and wake bouts, suggesting that the presence of *Cav2.2* promotes state change. As noted previously, because many proteins, particular ion channel proteins may influence development in a transgenic mouse, which might then influence sleep, we cannot conclude that alterations in *Cav2.2* directly influence sleep architecture. A related channel, *Cav2.3,* has been shown critical for corticothalamic rhythmicity and burst activity [91]. Reverse genetic studies in mice lacking *Cav2.3* found markedly fragmented wake (wake instability) and reduced wake for 24 h, predominantly secondary to changes in the dark period [92]. Wakefulness and NREM sleep both had increased EEG slowing, and a greater homeostatic response to sleep loss, suggesting that this N-type Calcium channel is critical for optimal wakefulness. *Cav2.1* has also been explored for effects on behavioral state and these mice have less NREM sleep and less of a waking response to caffeine [93]. T-type calcium channels (*Cav 3.2* and *Cav 3.2*) have also been shown to have effects on sleep and sleep waveforms. A missense mutation in schizophrenia for the *Cav 3.3* channel in mice results in reduced sleep spindles [94]. Similarly transgenic deletion of *Cav3.3* results in abolition of spindles [94]. Interestingly, despite the pronounced effect on spindles, NREM sleep time was largely unaffected, suggesting that spindles do not influence NREM sleep time. Mice lacking both *Cav3.2* and *Cav3.3* were examined and found to have in addition to poor spindle activity, reduced wake, and increased slow wave activity [95]. A VDCC in the thalamus is the T-type channel, *Cav3.1,* which has been shown through reverse genetics to be essential for NREM sleep slow wave activity and slow wave activity generation [96]. An ambitious reverse genetics study of 21 different mouse strains with deletions in one of the VDCC, or calcium-dependent potassium channels, all result in less sleep (potentially less sleep need) to varying degrees [97].

Voltage-dependent potassium mutations that have profound effects on sleep wake timing have been identified in *drosophila* [98], prompting reverse genetics for potassium channel genes in mice. Transgenic depletion of the voltage-dependent potassium channel subunit, *Kv1.2*, results in increased waking and reduced NREM sleep without affecting REM sleep [99].

Sodium channel gene mutations have been associated with alterations in sleep [100, 101], but because the mutations increase epilepsy, it is difficult to parse out primary from secondary sleep effects. Ca^{2+}–dependent small-conductance (SK)-type K^+ channel 2 (SK2) deficiency does reduce spectral power at delta, theta and spindle frequencies, and fragments NREM sleep, suggesting this gene has, to date, the most extensive effects on sleep and electrographic activity in sleep [102]. Intriguingly, overexpressing SK2 consolidates sleep and wakefulness and increases the arousal threshold [103].

In contrast to the pronounced effects of orexin deficiency on sleep, deficiencies in other wake-promoting neuromodulators have rather small effects on sleep. However, a double knockout of cholinergic muscarinic receptors, *CHRM1* and *CHRM 3*, results in abolition of REM sleep [104]. Forward genetics identified a genetic line with a mutation in a vesicular protein gene, VAMP2, with reduced sleep [46]. Mice with this mutation were then shown to have reduced sleep and lower amplitude delta and theta amplitudes in sleep [46].

Several mutations in synaptic scaffolding proteins impact sleep wake architecture. *Homer1A* is a post-synaptic scaffolding protein, and its absence impairs sleep homeostasis. Loss of *SHANK3*, another post-synaptic scaffolding protein, results in 1-h less sleep in the dark period. Neuroligins (*NLGN1* and *NLGN2*) have also been shown to influence sleep in a manner remarkably similar to *SHANK3* [105, 106].

Post-synaptic kinases have also been shown using reverse genetics to have roles in behavioral state timing. Adenosine kinase haploinsufficiency results in greater wake (1 h more) and less pronounced homeostatic response to sleep loss [107]. Sik3 was discussed above in the forward genetic section, but a mutation in a single phosphorylation site (Serine 551) which is a protein kinase A site results in pronounced hypersomnia with 2 h more NREM sleep in the dark period and increased slow wave activity in both waking and NREM sleep [108]. When this mutation is induced specifically in neurons in mice at 2–3 weeks of age (after brain development), a similar phenotype is seen with increased sleep [109]. Transgenic knock out of extracellular signal-regulated kinases (ERK1 or ERK2) results in a small increase in wakefulness [110].

Inflammatory Genes and Sleep

A bidirectional feed forward relationship was identified early on for sleep loss and muramyl peptide, a component of the bacterial cell wall with immune modulatory functions [111–113]. Most of the earlier studies were performed in rats and rabbits, and later confirmed to have similar effects in mice. Cytokines tumor necrosis factor-a (TNF-a), interleukin-1b (IL-1b), and growth hormone releasing hormone (GHRH and prostaglandin D2) have all been shown to reduce wakefulness and increase NREM sleep [114–116]. In contrast, cytokines IL-4, IL-10, and IL-13 all suppress NREM sleep [117]. Intriguingly, pro-inflammatory cytokines can suppress expression of core clock genes in at least the hypothalamus and periphery [118, 119].

Future Directions

As diversely as animal species have evolved on this planet, they all have in common the need for sleep. It is evident from studies, not only in mice but all animals, that genetics contributes significantly to sleep architecture, sleep time, sleep and wake quality and sleep intensity. Sleep is essential across development for proper brain connectivity and function and survival. In this chapter, we have raised several challenges in studying the genetics of sleep. The loss of a gene or a significant mutation in a gene that is essential for sleep is likely to lead to embryonic lethality, such that the above forward and reverse genetic approaches will not find and be able to characterize such a gene. There is a need to develop high throughput methods to induce mutations in the adult to avoid developmental confounds. The above studies highlight other challenges to identifying important sleep genes. Forward genetic screens will create more mutations in larger genes, so that the smallest of genes will be harder to find in a screen. For genes playing opposing roles in different brain regions or brain and periphery, a global knockout may obscure local genetic effects. It may now be important to re-think our phenotype of sleep to focus on restorative sleep, the functions of sleep: memory consolidation, brain health, and synaptic homeostasis. These are all areas in which murine studies will contribute to advance the genetics of sleep. Human GWAS will likely continue to provide candidate genes for specific sleep phenotypes, particularly those for which the subjective report on quality of sleep is important (e.g., insomnia). The mouse provides an ideal model with which to explore the role of GWAS candidate genes. If we expand these phenotypes to include gene effects on electrographic sleep state-dependent waveforms, sleep depth (arousability thresholds), and hippocampal memory studies across sleep loss and recovery, we will utilize the full might of the mouse in sleep genetics.

References

1. Valatx JL, Bugat R. Genetic factors as determinants of the waking-sleep cycle in the mouse (author's transl). Brain Res. 1974;69(2):315–30. https://doi.org/10.1016/0006-8993(74)90009-2. Epub 1974/04/05. PubMed PMID: 4362814
2. Konopka RJ, Benzer S. Clock mutants of Drosophila melanogaster. Proc Natl Acad Sci USA. 1971;68(9):2112–6. https://doi.org/10.1073/pnas.68.9.2112. Epub 1971/09/01. PubMed PMID: 5002428; PMCID: PMC389363
3. Bargiello TA, Young MW. Molecular genetics of a biological clock in Drosophila. Proc Natl Acad Sci USA. 1984;81(7):2142–6. https://doi.org/10.1073/pnas.81.7.2142. Epub 1984/04/01. PubMed PMID: 16593450; PMCID: PMC345453
4. Sehgal A, Price JL, Man B, Young MW. Loss of circadian behavioral rhythms and per RNA oscillations in the Drosophila mutant timeless. Science. 1994;263(5153):1603–6. https://doi.org/10.1126/science.8128246. Epub 1994/03/18. PubMed PMID: 8128246
5. Campbell SS, Tobler I. Animal sleep: a review of sleep duration across phylogeny. Neurosci Biobehav Rev. 1984;8(3):269–300. https://doi.org/10.1016/0149-7634(84)90054-x. Epub 1984/01/01. PubMed PMID: 6504414

6. Borbely AA, Baumann F, Brandeis D, Strauch I, Lehmann D. Sleep deprivation: effect on sleep stages and EEG power density in man. Electroencephalogr Clin Neurophysiol. 1981;51(5):483–95. Epub 1981/05/01. PubMed PMID: 6165548
7. Mendelson WB, Vaughn WJ, Walsh MJ, Wyatt RJ. A signal analysis approach to rat sleep scoring instrumentation. Waking Sleeping. 1980;4(1):1–8. Epub 1980/01/01. PubMed PMID: 7395193
8. Borbely AA. A two process model of sleep regulation. Hum Neurobiol. 1982;1(3):195–204. Epub 1982/01/01. PubMed PMID: 7185792
9. Wilson MA, McNaughton BL. Reactivation of hippocampal ensemble memories during sleep. Science. 1994;265(5172):676–9. https://doi.org/10.1126/science.8036517. Epub 1994/07/29. PubMed PMID: 8036517
10. Graves LA, Heller EA, Pack AI, Abel T. Sleep deprivation selectively impairs memory consolidation for contextual fear conditioning. Learn Mem. 2003;10(3):168–76. https://doi.org/10.1101/lm.48803. 10/3/168 [pii]. Epub 2003/05/30. PubMed PMID: 12773581; PMCID: 202307
11. van der Helm E, Gujar N, Nishida M, Walker MP. Sleep-dependent facilitation of episodic memory details. PLoS One. 2011;6(11):e27421. https://doi.org/10.1371/journal.pone.0027421. Epub 2011/11/25. PubMed PMID: 22114672; PMCID: PMC3219667
12. Ego-Stengel V, Wilson MA. Disruption of ripple-associated hippocampal activity during rest impairs spatial learning in the rat. Hippocampus. 2010;20(1):1–10. https://doi.org/10.1002/hipo.20707. Epub 2009/10/10. PubMed PMID: 19816984; PMCID: PMC2801761
13. Girardeau G, Benchenane K, Wiener SI, Buzsaki G, Zugaro MB. Selective suppression of hippocampal ripples impairs spatial memory. Nat Neurosci. 2009;12(10):1222–3. https://doi.org/10.1038/nn.2384. Epub 2009/09/15. PubMed PMID: 19749750
14. Diekelmann S, Born J. The memory function of sleep. Nat Rev Neurosci. 2010;11(2):114–26. https://doi.org/10.1038/nrn2762. Epub 2010/01/05. PubMed PMID: 20046194
15. Swift KM, Gross BA, Frazer MA, Bauer DS, Clark KJD, Vazey EM, Aston-Jones G, Li Y, Pickering AE, Sara SJ, Poe GR. Abnormal locus coeruleus sleep activity alters sleep signatures of memory consolidation and impairs place cell stability and spatial memory. Curr Biol. 2018;28(22):3599–609.e4. https://doi.org/10.1016/j.cub.2018.09.054. Epub 2018/11/06. PubMed PMID: 30393040; PMCID: PMC7556718
16. Kjaerby C, Andersen M, Hauglund N, Untiet V, Dall C, Sigurdsson B, Ding F, Feng J, Li Y, Weikop P, Hirase H, Nedergaard M. Memory-enhancing properties of sleep depend on the oscillatory amplitude of norepinephrine. Nat Neurosci. 2022;25(8):1059–70. https://doi.org/10.1038/s41593-022-01102-9. Epub 2022/07/08. PubMed PMID: 35798980
17. Huber R, Ghilardi MF, Massimini M, Tononi G. Local sleep and learning. Nature. 2004;430(6995):78–81. https://doi.org/10.1038/nature02663. Epub 2004/06/09. PubMed PMID: 15184907
18. Siapas AG, Wilson MA. Coordinated interactions between hippocampal ripples and cortical spindles during slow-wave sleep. Neuron. 1998;21(5):1123–8. https://doi.org/10.1016/s0896-6273(00)80629-7. Epub 1998/12/18. PubMed PMID: 9856467
19. Rosanova M, Ulrich D. Pattern-specific associative long-term potentiation induced by a sleep spindle-related spike train. J Neurosci. 2005;25(41):9398–405. https://doi.org/10.1523/JNEUROSCI.2149-05.2005. Epub 2005/10/14. PubMed PMID: 16221848; PMCID: PMC6725710
20. Smith C, Lapp L. Increases in number of REMS and REM density in humans following an intensive learning period. Sleep. 1991;14(4):325–30. https://doi.org/10.1093/sleep/14.4.325. Epub 1991/08/01. PubMed PMID: 1947596
21. Simon AP, Poindessous-Jazat F, Dutar P, Epelbaum J, Bassant MH. Firing properties of anatomically identified neurons in the medial septum of anesthetized and unanesthetized restrained rats. J Neurosci. 2006;26(35):9038–46. https://doi.org/10.1523/JNEUROSCI.1401-06.2006. Epub 2006/09/01. PubMed PMID: 16943562; PMCID: PMC6675331

22. Boyce R, Glasgow SD, Williams S, Adamantidis A. Causal evidence for the role of REM sleep theta rhythm in contextual memory consolidation. Science. 2016;352(6287):812–6. https://doi.org/10.1126/science.aad5252. Epub 2016/05/14. PubMed PMID: 27174984
23. Li P, Geng X, Jiang H, Caccavano A, Vicini S, Wu JY. Measuring sharp waves and oscillatory population activity with the genetically encoded calcium indicator GCaMP6f. Front Cell Neurosci. 2019;13:274. https://doi.org/10.3389/fncel.2019.00274. Epub 2019/07/06. PubMed PMID: 31275115; PMCID: PMC6593119
24. Tsien JZ, Huerta PT, Tonegawa S. The essential role of hippocampal CA1 NMDA receptor-dependent synaptic plasticity in spatial memory. Cell. 1996;87(7):1327–38. https://doi.org/10.1016/s0092-8674(00)81827-9. Epub 1996/12/27. PubMed PMID: 8980238
25. Frankland PW, O'Brien C, Ohno M, Kirkwood A, Silva AJ. Alpha-CaMKII-dependent plasticity in the cortex is required for permanent memory. Nature. 2001;411(6835):309–13. https://doi.org/10.1038/35077089. Epub 2001/05/18. PubMed PMID: 11357133
26. Bergmann TO, Molle M, Marshall L, Kaya-Yildiz L, Born J, Roman Siebner H. A local signature of LTP- and LTD-like plasticity in human NREM sleep. Eur J Neurosci. 2008;27(9):2241–9. https://doi.org/10.1111/j.1460-9568.2008.06178.x. Epub 2008/05/01. PubMed PMID: 18445215
27. Yang G, Lai CS, Cichon J, Ma L, Li W, Gan WB. Sleep promotes branch-specific formation of dendritic spines after learning. Science. 2014;344(6188):1173–8. https://doi.org/10.1126/science.1249098. Epub 2014/06/07. PubMed PMID: 24904169; PMCID: PMC4447313
28. Goto A, Bota A, Miya K, Wang J, Tsukamoto S, Jiang X, Hirai D, Murayama M, Matsuda T, McHugh TJ, Nagai T, Hayashi Y. Stepwise synaptic plasticity events drive the early phase of memory consolidation. Science. 2021;374(6569):857–63. https://doi.org/10.1126/science.abj9195. Epub 2021/11/12. PubMed PMID: 34762472
29. Okamoto K, Nagai T, Miyawaki A, Hayashi Y. Rapid and persistent modulation of actin dynamics regulates postsynaptic reorganization underlying bidirectional plasticity. Nat Neurosci. 2004;7(10):1104–12. https://doi.org/10.1038/nn1311. Epub 2004/09/14. PubMed PMID: 15361876
30. Bosch M, Castro J, Saneyoshi T, Matsuno H, Sur M, Hayashi Y. Structural and molecular remodeling of dendritic spine substructures during long-term potentiation. Neuron. 2014;82(2):444–59. https://doi.org/10.1016/j.neuron.2014.03.021. Epub 2014/04/20. PubMed PMID: 24742465; PMCID: PMC4281348
31. Li W, Ma L, Yang G, Gan WB. REM sleep selectively prunes and maintains new synapses in development and learning. Nat Neurosci. 2017;20(3):427–37. https://doi.org/10.1038/nn.4479. Epub 2017/01/17. PubMed PMID: 28092659; PMCID: PMC5535798
32. Grutzendler J, Kasthuri N, Gan WB. Long-term dendritic spine stability in the adult cortex. Nature. 2002;420(6917):812–6. https://doi.org/10.1038/nature01276. Epub 2002/12/20. PubMed PMID: 12490949
33. Tononi G, Cirelli C. Sleep and synaptic homeostasis: a hypothesis. Brain Res Bull. 2003;62(2):143–50. https://doi.org/10.1016/j.brainresbull.2003.09.004. Epub 2003/11/26. PubMed PMID: 14638388
34. de Vivo L, Bellesi M, Marshall W, Bushong EA, Ellisman MH, Tononi G, Cirelli C. Ultrastructural evidence for synaptic scaling across the wake/sleep cycle. Science. 2017;355(6324):507–10. https://doi.org/10.1126/science.aah5982. Epub 2017/02/06. PubMed PMID: 28154076; PMCID: PMC5313037
35. Spano GM, Banningh SW, Marshall W, de Vivo L, Bellesi M, Loschky SS, Tononi G, Cirelli C. Sleep deprivation by exposure to novel objects increases synapse density and axon-spine interface in the Hippocampal CA1 region of adolescent mice. J Neurosci. 2019;39(34):6613–25. https://doi.org/10.1523/JNEUROSCI.0380-19.2019. Epub 2019/07/03. PubMed PMID: 31263066; PMCID: PMC6703893
36. Bushey D, Tononi G, Cirelli C. Sleep and synaptic homeostasis: structural evidence in Drosophila. Science. 2011;332(6037):1576–81. https://doi.org/10.1126/science.1202839. Epub 2011/06/28. PubMed PMID: 21700878; PMCID: PMC3128387

37. Tafti M, Franken P, Kitahama K, Malafosse A, Jouvet M, Valatx JL. Localization of candidate genomic regions influencing paradoxical sleep in mice. Neuroreport. 1997;8(17):3755–8. https://doi.org/10.1097/00001756-199712010-00019. Epub 1998/01/14. PubMed PMID: 9427364
38. Toth LA, Williams RW. A quantitative genetic analysis of slow-wave sleep and rapid-eye movement sleep in CXB recombinant inbred mice. Behav Genet. 1999;29(5):329–37. https://doi.org/10.1023/a:1021609917126. Epub 2000/04/15. PubMed PMID: 10765561
39. Franken P, Chollet D, Tafti M. The homeostatic regulation of sleep need is under genetic control. J Neurosci. 2001;21(8):2610–21. https://doi.org/10.1523/JNEUROSCI.21-08-02610.2001. 21/8/2610 [pii]. Epub 2001/04/18. PubMed PMID: 11306614
40. Mackiewicz M, Paigen B, Naidoo N, Pack AI. Analysis of the QTL for sleep homeostasis in mice: Homer1a is a likely candidate. Physiol Genomics. 2008;33(1):91–9. https://doi.org/10.1152/physiolgenomics.00189.2007. Epub 2008/01/04. PubMed PMID: 18171722
41. Keenan BT, Galante RJ, Lian J, Zhang L, Guo X, Veatch OJ, Chesler EJ, O'Brien WT, Svenson KL, Churchill GA, Pack AI. The dihydropyrimidine dehydrogenase gene contributes to heritable differences in sleep in mice. Curr Biol. 2021;31(23):5238–48.e7. https://doi.org/10.1016/j.cub.2021.09.049. Epub 2021/10/16. PubMed PMID: 34653361; PMCID: PMC8665053
42. Nolan PM, Peters J, Strivens M, Rogers D, Hagan J, Spurr N, Gray IC, Vizor L, Brooker D, Whitehill E, Washbourne R, Hough T, Greenaway S, Hewitt M, Liu X, McCormack S, Pickford K, Selley R, Wells C, Tymowska-Lalanne Z, Roby P, Glenister P, Thornton C, Thaung C, Stevenson JA, Arkell R, Mburu P, Hardisty R, Kiernan A, Erven A, Steel KP, Voegeling S, Guenet JL, Nickols C, Sadri R, Nasse M, Isaacs A, Davies K, Browne M, Fisher EM, Martin J, Rastan S, Brown SD, Hunter J. A systematic, genome-wide, phenotype-driven mutagenesis programme for gene function studies in the mouse. Nat Genet. 2000;25(4):440–3. https://doi.org/10.1038/78140. Epub 2000/08/10. PubMed PMID: 10932191
43. Hrabe de Angelis MH, Flaswinkel H, Fuchs H, Rathkolb B, Soewarto D, Marschall S, Heffner S, Pargent W, Wuensch K, Jung M, Reis A, Richter T, Alessandrini F, Jakob T, Fuchs E, Kolb H, Kremmer E, Schaeble K, Rollinski B, Roscher A, Peters C, Meitinger T, Strom T, Steckler T, Holsboer F, Klopstock T, Gekeler F, Schindewolf C, Jung T, Avraham K, Behrendt H, Ring J, Zimmer A, Schughart K, Pfeffer K, Wolf E, Balling R. Genome-wide, large-scale production of mutant mice by ENU mutagenesis. Nat Genet. 2000;25(4):444–7. https://doi.org/10.1038/78146. Epub 2000/08/10. PubMed PMID: 10932192
44. Funato H, Miyoshi C, Fujiyama T, Kanda T, Sato M, Wang Z, Ma J, Nakane S, Tomita J, Ikkyu A, Kakizaki M, Hotta-Hirashima N, Kanno S, Komiya H, Asano F, Honda T, Kim SJ, Harano K, Muramoto H, Yonezawa T, Mizuno S, Miyazaki S, Connor L, Kumar V, Miura I, Suzuki T, Watanabe A, Abe M, Sugiyama F, Takahashi S, Sakimura K, Hayashi Y, Liu Q, Kume K, Wakana S, Takahashi JS, Yanagisawa M. Forward-genetics analysis of sleep in randomly mutagenized mice. Nature. 2016;539(7629):378–83. https://doi.org/10.1038/nature20142. Epub 2016/11/05. PubMed PMID: 27806374; PMCID: PMC6076225
45. Miyoshi C, Kim SJ, Ezaki T, Ikkyu A, Hotta-Hirashima N, Kanno S, Kakizaki M, Yamada M, Wakana S, Yanagisawa M, Funato H. Methodology and theoretical basis of forward genetic screening for sleep/wakefulness in mice. Proc Natl Acad Sci USA. 2019;116(32):16062–7. https://doi.org/10.1073/pnas.1906774116. Epub 2019/07/25. PubMed PMID: 31337678; PMCID: PMC6689935
46. Banks GT, Guillaumin MCC, Heise I, Lau P, Yin M, Bourbia N, Aguilar C, Bowl MR, Esapa C, Brown LA, Hasan S, Tagliatti E, Nicholson E, Bains RS, Wells S, Vyazovskiy VV, Volynski K, Peirson SN, Nolan PM. Forward genetics identifies a novel sleep mutant with sleep state inertia and REM sleep deficits. Sci Adv. 2020;6(33):eabb3567. https://doi.org/10.1126/sciadv.abb3567. Epub 2020/08/28. PubMed PMID: 32851175; PMCID: PMC7423362
47. Toh KL, Jones CR, He Y, Eide EJ, Hinz WA, Virshup DM, Ptacek LJ, Fu YH. An hPer2 phosphorylation site mutation in familial advanced sleep phase syndrome. Science.

2001;291(5506):1040–3. https://doi.org/10.1126/science.1057499. Epub 2001/03/10. PubMed PMID: 11232563

48. Xu Y, Padiath QS, Shapiro RE, Jones CR, Wu SC, Saigoh N, Saigoh K, Ptacek LJ, Fu YH. Functional consequences of a CKIdelta mutation causing familial advanced sleep phase syndrome. Nature. 2005;434(7033):640–4. https://doi.org/10.1038/nature03453. Epub 2005/04/01. PubMed PMID: 15800623
49. Baker TL, Foutz AS, McNerney V, Mitler MM, Dement WC. Canine model of narcolepsy: genetic and developmental determinants. Exp Neurol. 1982;75(3):729–42. https://doi.org/10.1016/0014-4886(82)90038-3. Epub 1982/03/01. PubMed PMID: 7199479
50. Sakurai T, Amemiya A, Ishii M, Matsuzaki I, Chemelli RM, Tanaka H, Williams SC, Richarson JA, Kozlowski GP, Wilson S, Arch JR, Buckingham RE, Haynes AC, Carr SA, Annan RS, DE MN, Liu WS, Terrett JA, Elshourbagy NA, Bergsma DJ, Yanagisawa M. Orexins and orexin receptors: a family of hypothalamic neuropeptides and G protein-coupled receptors that regulate feeding behavior. Cell. 1998;92(5):1 page following 696. Epub 1998/04/04. PubMed PMID: 9527442
51. de Lecea L, Kilduff TS, Peyron C, Gao X, Foye PE, Danielson PE, Fukuhara C, Battenberg EL, Gautvik VT, Bartlett FS 2nd, Frankel WN, van den Pol AN, Bloom FE, Gautvik KM, Sutcliffe JG. The hypocretins: hypothalamus-specific peptides with neuroexcitatory activity. Proc Natl Acad Sci U S A. 1998;95(1):322–7. Epub 1998/02/21. PubMed PMID: 9419374; PMCID: 18213
52. Peyron C, Tighe DK, van den Pol AN, de Lecea L, Heller HC, Sutcliffe JG, Kilduff TS. Neurons containing hypocretin (orexin) project to multiple neuronal systems. J Neurosci. 1998;18(23):9996–10015. Epub 1998/11/21. PubMed PMID: 9822755; PMCID: PMC6793310
53. Chemelli RM, Willie JT, Sinton CM, Elmquist JK, Scammell T, Lee C, Richardson JA, Williams SC, Xiong Y, Kisanuki Y, Fitch TE, Nakazato M, Hammer RE, Saper CB, Yanagisawa M. Narcolepsy in orexin knockout mice: molecular genetics of sleep regulation. Cell. 1999;98(4):437–51. https://doi.org/10.1016/s0092-8674(00)81973-x. Epub 1999/09/11. PubMed PMID: 10481909
54. Thannickal TC, Moore RY, Nienhuis R, Ramanathan L, Gulyani S, Aldrich M, Cornford M, Siegel JM. Reduced number of hypocretin neurons in human narcolepsy. Neuron. 2000;27(3):469–74. https://doi.org/10.1016/s0896-6273(00)00058-1. Epub 2000/10/31. PubMed PMID: 11055430; PMCID: PMC8760623
55. De Luca R, Nardone S, Grace KP, Venner A, Cristofolini M, Bandaru SS, Sohn LT, Kong D, Mochizuki T, Viberti B, Zhu L, Zito A, Scammell TE, Saper CB, Lowell BB, Fuller PM, Arrigoni E. Orexin neurons inhibit sleep to promote arousal. Nat Commun. 2022;13(1):4163. https://doi.org/10.1038/s41467-022-31591-y. Epub 2022/07/20. PubMed PMID: 35851580; PMCID: PMC9293990
56. Carter ME, Brill J, Bonnavion P, Huguenard JR, Huerta R, de Lecea L. Mechanism for Hypocretin-mediated sleep-to-wake transitions. Proc Natl Acad Sci USA. 2012;109(39):E2635–44. https://doi.org/10.1073/pnas.1202526109. 1202526109 [pii] Epub 2012/09/08. PubMed PMID: 22955882; PMCID: 3465396
57. Carter ME, Yizhar O, Chikahisa S, Nguyen H, Adamantidis A, Nishino S, Deisseroth K, de Lecea L. Tuning arousal with optogenetic modulation of locus coeruleus neurons. Nat Neurosci. 2010;13(12):1526–33. https://doi.org/10.1038/nn.2682. nn.2682 [pii] Epub 2010/11/03. PubMed PMID: 21037585; PMCID: 3174240
58. Medori R, Tritschler HJ, Leblanc A, Villare F, Manetto V, Chen HY, Xue R, Leal S, Montagna P, Cortelli P, et al. Fatal familial insomnia, a prion disease with a mutation at codon 178 of the prion protein gene. N Engl J Med. 1992;326(7):444–9. https://doi.org/10.1056/NEJM199202133260704. Epub 1992/02/13. PubMed PMID: 1346338; PMCID: PMC6151859
59. Tobler I, Deboer T, Fischer M. Sleep and sleep regulation in normal and prion protein-deficient mice. J Neurosci. 1997;17(5):1869–79. https://doi.org/10.1523/

JNEUROSCI.17-05-01869.1997. Epub 1997/03/01. PubMed PMID: 9030645; PMCID: PMC6573394

60. Huber R, Deboer T, Tobler I. Sleep deprivation in prion protein deficient mice sleep deprivation in prion protein deficient mice and control mice: genotype dependent regional rebound. Neuroreport. 2002;13(1):1–4. https://doi.org/10.1097/00001756-200201210-00005. Epub 2002/04/02. PubMed PMID: 11924867
61. Landolt HP, Glatzel M, Blattler T, Achermann P, Roth C, Mathis J, Weis J, Tobler I, Aguzzi A, Bassetti CL. Sleep-wake disturbances in sporadic Creutzfeldt-Jakob disease. Neurology. 2006;66(9):1418–24. https://doi.org/10.1212/01.wnl.0000210445.16135.56. Epub 2006/05/10. PubMed PMID: 16682677
62. Dossena S, Imeri L, Mangieri M, Garofoli A, Ferrari L, Senatore A, Restelli E, Balducci C, Fiordaliso F, Salio M, Bianchi S, Fioriti L, Morbin M, Pincherle A, Marcon G, Villani F, Carli M, Tagliavini F, Forloni G, Chiesa R. Mutant prion protein expression causes motor and memory deficits and abnormal sleep patterns in a transgenic mouse model. Neuron. 2008;60(4):598–609. https://doi.org/10.1016/j.neuron.2008.09.008. Epub 2008/11/29. PubMed PMID: 19038218
63. Mackiewicz M, Zimmerman JE, Shockley KR, Churchill GA, Pack AI. What are microarrays teaching us about sleep? Trends Mol Med. 2009;15(2):79–87. https://doi.org/10.1016/j.molmed.2008.12.002. Epub 2009/01/24. PubMed PMID: 19162550; PMCID: PMC2942088
64. Havekes R, Park AJ, Tudor JC, Luczak VG, Hansen RT, Ferri SL, Bruinenberg VM, Poplawski SG, Day JP, Aton SJ, Radwanska K, Meerlo P, Houslay MD, Baillie GS, Abel T. Sleep deprivation causes memory deficits by negatively impacting neuronal connectivity in hippocampal area CA1. elife. 2016;5:e13424. https://doi.org/10.7554/eLife.13424. Epub 2016/08/24. PubMed PMID: 27549340; PMCID: PMC4996653
65. Vecsey CG, Peixoto L, Choi JH, Wimmer M, Jaganath D, Hernandez PJ, Blackwell J, Meda K, Park AJ, Hannenhalli S, Abel T. Genomic analysis of sleep deprivation reveals translational regulation in the hippocampus. Physiol Genomics. 2012;44(20):981–91. https://doi.org/10.1152/physiolgenomics.00084.2012. Epub 2012/08/30. PubMed PMID: 22930738; PMCID: PMC3472468
66. Wang H, Liu Y, Briesemann M, Yan J. Computational analysis of gene regulation in animal sleep deprivation. Physiol Genomics. 2010;42(3):427–36. https://doi.org/10.1152/physiolgenomics.00205.2009. Epub 2010/05/27. PubMed PMID: 20501693
67. Lyons LC, Chatterjee S, Vanrobaeys Y, Gaine ME, Abel T. Translational changes induced by acute sleep deprivation uncovered by TRAP-Seq. Mol. Brain. 2020;13(1):165. https://doi.org/10.1186/s13041-020-00702-5. Epub 2020/12/05. PubMed PMID: 33272296; PMCID: PMC7713217
68. Jha PK, Valekunja UK, Ray S, Nollet M, Reddy AB. Single-cell transcriptomics and cell-specific proteomics reveals molecular signatures of sleep. Commun Biol. 2022;5(1):846. https://doi.org/10.1038/s42003-022-03800-3. Epub 2022/08/20. PubMed PMID: 35986171; PMCID: PMC9391396
69. Tononi G, Cirelli C. Sleep and the price of plasticity: from synaptic and cellular homeostasis to memory consolidation and integration. Neuron. 2014;81(1):12–34. https://doi.org/10.1016/j.neuron.2013.12.025. Epub 2014/01/15. PubMed PMID: 24411729; PMCID: PMC3921176
70. Benington JH, Heller HC. Restoration of brain energy metabolism as the function of sleep. Prog Neurobiol. 1995;45(4):347–60. https://doi.org/10.1016/0301-0082(94)00057-o. Epub 1995/03/01. PubMed PMID: 7624482
71. Lowrey PL, Takahashi JS. Mammalian circadian biology: elucidating genome-wide levels of temporal organization. Annu Rev Genomics Hum Genet. 2004;5:407–41. https://doi.org/10.1146/annurev.genom.5.061903.175925. Epub 2004/10/16. PubMed PMID: 15485355; PMCID: PMC3770722

72. Mehra A, Baker CL, Loros JJ, Dunlap JC. Post-translational modifications in circadian rhythms. Trends Biochem Sci. 2009;34(10):483–90. https://doi.org/10.1016/j.tibs.2009.06.006. Epub 2009/09/11. PubMed PMID: 19740663; PMCID: PMC2765057
73. Laposky A, Easton A, Dugovic C, Walisser J, Bradfield C, Turek F. Deletion of the mammalian circadian clock gene BMAL1/Mop3 alters baseline sleep architecture and the response to sleep deprivation. Sleep. 2005;28(4):395–409. https://doi.org/10.1093/sleep/28.4.395. Epub 2005/09/21. PubMed PMID: 16171284
74. Franken P, Dudley CA, Estill SJ, Barakat M, Thomason R, O'Hara BF, McKnight SL. NPAS2 as a transcriptional regulator of non-rapid eye movement sleep: genotype and sex interactions. Proc Natl Acad Sci U S A. 2006;103(18):7118–23. https://doi.org/10.1073/pnas.0602006103. Epub 2006/04/26. PubMed PMID: 16636276; PMCID: PMC1459027
75. Kopp C, Albrecht U, Zheng B, Tobler I. Homeostatic sleep regulation is preserved in mPer1 and mPer2 mutant mice. Eur J Neurosci. 2002;16(6):1099–106. https://doi.org/10.1046/j.1460-9568.2002.02156.x. Epub 2002/10/18. PubMed PMID: 12383239
76. Shiromani PJ, Xu M, Winston EM, Shiromani SN, Gerashchenko D, Weaver DR. Sleep rhythmicity and homeostasis in mice with targeted disruption of mPeriod genes. Am J Physiol Regul Integr Comp Physiol. 2004;287(1):R47–57. https://doi.org/10.1152/ajpregu.00138.2004. Epub 2004/03/20. PubMed PMID: 15031135
77. Mang GM, La Spada F, Emmenegger Y, Chappuis S, Ripperger JA, Albrecht U, Franken P. Altered Sleep Homeostasis in Rev-erbalpha Knockout Mice. Sleep. 2016;39(3):589–601. https://doi.org/10.5665/sleep.5534. Epub 2015/11/14. PubMed PMID: 26564124; PMCID: PMC4763348
78. Amador A, Kamenecka TM, Solt LA, Burris TP. REV-ERBbeta is required to maintain normal wakefulness and the wake-inducing effect of dual REV-ERB agonist SR9009. Biochem Pharmacol. 2018;150:1–8. https://doi.org/10.1016/j.bcp.2018.01.009. Epub 2018/01/23. PubMed PMID: 29355503
79. Cirelli C, Tononi G. Gene expression in the brain across the sleep-waking cycle. Brain Res. 2000;885(2):303–21. https://doi.org/10.1016/s0006-8993(00)03008-0. Epub 2000/12/05. PubMed PMID: 11102586
80. Terao A, Steininger TL, Hyder K, Apte-Deshpande A, Ding J, Rishipathak D, Davis RW, Heller HC, Kilduff TS. Differential increase in the expression of heat shock protein family members during sleep deprivation and during sleep. Neuroscience. 2003;116(1):187–200. https://doi.org/10.1016/s0306-4522(02)00695-4. Epub 2003/01/22. PubMed PMID: 12535952
81. Anaclet C, Pedersen NP, Ferrari LL, Venner A, Bass CE, Arrigoni E, Fuller PM. Basal forebrain control of wakefulness and cortical rhythms. Nat Commun. 2015;6:8744. https://doi.org/10.1038/ncomms9744. Epub 2015/11/04. PubMed PMID: 26524973; PMCID: PMC4659943
82. Kalinchuk AV, Porkka-Heiskanen T, McCarley RW, Basheer R. Cholinergic neurons of the basal forebrain mediate biochemical and electrophysiological mechanisms underlying sleep homeostasis. Eur J Neurosci. 2015;41(2):182–95. https://doi.org/10.1111/ejn.12766. Epub 2014/11/06. PubMed PMID: 25369989; PMCID: PMC4460789
83. Xu M, Chung S, Zhang S, Zhong P, Ma C, Chang WC, Weissbourd B, Sakai N, Luo L, Nishino S, Dan Y. Basal forebrain circuit for sleep-wake control. Nat Neurosci. 2015;18(11):1641–7. https://doi.org/10.1038/nn.4143. Epub 2015/10/13. PubMed PMID: 26457552; PMCID: PMC5776144
84. Nikonova EV, Gilliland JD, Tanis KQ, Podtelezhnikov AA, Rigby AM, Galante RJ, Finney EM, Stone DJ, Renger JJ, Pack AI, Winrow CJ. Transcriptional profiling of cholinergic neurons from basal forebrain identifies changes in expression of genes between sleep and wake. Sleep. 2017;40(6):zsx059. https://doi.org/10.1093/sleep/zsx059. Epub 2017/04/19. PubMed PMID: 28419375; PMCID: PMC6075396
85. Naidoo N, Zhu J, Galante RJ, Lian J, Strus E, Lee A, Keenan BT, Pack AI. Reduction of the molecular chaperone binding immunoglobulin protein (BiP) accentuates the effect of aging

on sleep-wake behavior. Neurobiol Aging. 2018;69:10–25. https://doi.org/10.1016/j.neurobiolaging.2018.04.011. Epub 2018/05/31. PubMed PMID: 29843048; PMCID: PMC6064380
86. Zhang J, Zhu Y, Zhan G, Fenik P, Panossian L, Wang MM, Reid S, Lai D, Davis JG, Baur JA, Veasey S. Extended wakefulness: compromised metabolics in and degeneration of locus ceruleus neurons. J Neurosci. 2014;34(12):4418–31. https://doi.org/10.1523/JNEUROSCI.5025-12.2014. PubMed PMID: 24647961; PMCID: 3960479
87. Panossian L, Fenik P, Zhu Y, Zhan G, McBurney MW, Veasey S. SIRT1 regulation of wakefulness and senescence-like phenotype in wake neurons. J Neurosci. 2011;31(11):4025–36. https://doi.org/10.1523/JNEUROSCI.5166-10.2011. PubMed PMID: 21411645; PMCID: 3065120
88. Diverse-Pierluissi M, Goldsmith PK, Dunlap K. Transmitter-mediated inhibition of N-type calcium channels in sensory neurons involves multiple GTP-binding proteins and subunits. Neuron. 1995;14(1):191–200. https://doi.org/10.1016/0896-6273(95)90254-6. Epub 1995/01/01. PubMed PMID: 7826637
89. Ertel EA, Campbell KP, Harpold MM, Hofmann F, Mori Y, Perez-Reyes E, Schwartz A, Snutch TP, Tanabe T, Birnbaumer L, Tsien RW, Catterall WA. Nomenclature of voltage-gated calcium channels. Neuron. 2000;25(3):533–5. https://doi.org/10.1016/s0896-6273(00)81057-0. Epub 2000/04/25. PubMed PMID: 10774722
90. Beuckmann CT, Sinton CM, Miyamoto N, Ino M, Yanagisawa M. N-type calcium channel alpha1B subunit (Cav2.2) knock-out mice display hyperactivity and vigilance state differences. J Neurosci. 2003;23(17):6793–7. https://doi.org/10.1523/JNEUROSCI.23-17-06793.2003. Epub 2003/08/02. PubMed PMID: 12890773; PMCID: PMC6740709
91. Zaman T, Lee K, Park C, Paydar A, Choi JH, Cheong E, Lee CJ, Shin HS. Cav2.3 channels are critical for oscillatory burst discharges in the reticular thalamus and absence epilepsy. Neuron. 2011;70(1):95–108. https://doi.org/10.1016/j.neuron.2011.02.042. Epub 2011/04/13. PubMed PMID: 21482359
92. Siwek ME, Muller R, Henseler C, Broich K, Papazoglou A, Weiergraber M. The CaV2.3 R-type voltage-gated Ca2+ channel in mouse sleep architecture. Sleep. 2014;37(5):881–92. https://doi.org/10.5665/sleep.3652. Epub 2014/05/03. PubMed PMID: 24790266; PMCID: PMC3985108
93. Deboer T, van Diepen HC, Ferrari MD, Van den Maagdenberg AM, Meijer JH. Reduced sleep and low adenosinergic sensitivity in cacna1a R192Q mutant mice. Sleep. 2013;36(1):127–36. https://doi.org/10.5665/sleep.2316. Epub 2013/01/05. PubMed PMID: 23288979; PMCID: PMC3524534
94. Ghoshal A, Uygun DS, Yang L, McNally JM, Lopez-Huerta VG, Arias-Garcia MA, Baez-Nieto D, Allen A, Fitzgerald M, Choi S, Zhang Q, Hope JM, Yan K, Mao X, Nicholson TB, Imaizumi K, Fu Z, Feng G, Brown RE, Strecker RE, Purcell SM, Pan JQ. Effects of a patient-derived de novo coding alteration of CACNA1I in mice connect a schizophrenia risk gene with sleep spindle deficits. Transl. Psychiatry. 2020;10(1):29. https://doi.org/10.1038/s41398-020-0685-1. Epub 2020/02/19. PubMed PMID: 32066662; PMCID: PMC7026444
95. Pellegrini C, Lecci S, Luthi A, Astori S. Suppression of Sleep Spindle Rhythmogenesis in Mice with Deletion of CaV3.2 and CaV3.3 T-type Ca(2+) Channels. Sleep. 2016;39(4):875–85. https://doi.org/10.5665/sleep.5646. Epub 2015/11/28. PubMed PMID: 26612388; PMCID: PMC4791621
96. Lee J, Kim D, Shin HS. Lack of delta waves and sleep disturbances during non-rapid eye movement sleep in mice lacking alpha1G-subunit of T-type calcium channels. Proc Natl Acad Sci U S A. 2004;101(52):18195–9. https://doi.org/10.1073/pnas.0408089101. Epub 2004/12/17. PubMed PMID: 15601764; PMCID: PMC539778
97. Tatsuki F, Sunagawa GA, Shi S, Susaki EA, Yukinaga H, Perrin D, Sumiyama K, Ukai-Tadenuma M, Fujishima H, Ohno R, Tone D, Ode KL, Matsumoto K, Ueda HR. Involvement of Ca(2+)-dependent hyperpolarization in sleep duration in mammals. Neuron. 2016;90(1):70–85. https://doi.org/10.1016/j.neuron.2016.02.032. Epub 2016/03/22. PubMed PMID: 26996081

98. Cirelli C, Bushey D, Hill S, Huber R, Kreber R, Ganetzky B, Tononi G. Reduced sleep in Drosophila Shaker mutants. Nature. 2005;434(7037):1087–92. https://doi.org/10.1038/nature03486.nature03486. [pii] Epub 2005/04/29. PubMed PMID: 15858564
99. Douglas CL, Vyazovskiy V, Southard T, Chiu SY, Messing A, Tononi G, Cirelli C. Sleep in Kcna2 knockout mice. BMC Biol. 2007;5:42. https://doi.org/10.1186/1741-7007-5-42. Epub 2007/10/11. PubMed PMID: 17925011; PMCID: PMC2151933
100. Chen C, Westenbroek RE, Xu X, Edwards CA, Sorenson DR, Chen Y, McEwen DP, O'Malley HA, Bharucha V, Meadows LS, Knudsen GA, Vilaythong A, Noebels JL, Saunders TL, Scheuer T, Shrager P, Catterall WA, Isom LL. Mice lacking sodium channel beta1 subunits display defects in neuronal excitability, sodium channel expression, and nodal architecture. J Neurosci. 2004;24(16):4030–42. https://doi.org/10.1523/JNEUROSCI.4139-03.2004. Epub 2004/04/23. PubMed PMID: 15102918; PMCID: PMC6729427
101. Papale LA, Makinson CD, Christopher Ehlen J, Tufik S, Decker MJ, Paul KN, Escayg A. Altered sleep regulation in a mouse model of SCN1A-derived genetic epilepsy with febrile seizures plus (GEFS+). Epilepsia. 2013;54(4):625–34. https://doi.org/10.1111/epi.12060. Epub 2013/01/15. PubMed PMID: 23311867; PMCID: PMC3703918
102. Cueni L, Canepari M, Lujan R, Emmenegger Y, Watanabe M, Bond CT, Franken P, Adelman JP, Luthi A. T-type Ca2+ channels, SK2 channels and SERCAs gate sleep-related oscillations in thalamic dendrites. Nat Neurosci. 2008;11(6):683–92. https://doi.org/10.1038/nn.2124. Epub 2008/05/20. PubMed PMID: 18488023
103. Wimmer RD, Astori S, Bond CT, Rovo Z, Chatton JY, Adelman JP, Franken P, Luthi A. Sustaining sleep spindles through enhanced SK2-channel activity consolidates sleep and elevates arousal threshold. J Neurosci. 2012;32(40):13917–28. https://doi.org/10.1523/JNEUROSCI.2313-12.2012. Epub 2012/10/05. PubMed PMID: 23035101; PMCID: PMC3658620
104. Niwa Y, Kanda GN, Yamada RG, Shi S, Sunagawa GA, Ukai-Tadenuma M, Fujishima H, Matsumoto N, Masumoto KH, Nagano M, Kasukawa T, Galloway J, Perrin D, Shigeyoshi Y, Ukai H, Kiyonari H, Sumiyama K, Ueda HR. Muscarinic acetylcholine receptors Chrm1 and Chrm3 are essential for REM sleep. Cell Rep. 2018;24(9):2231–47.e7. https://doi.org/10.1016/j.celrep.2018.07.082. Epub 2018/08/30. PubMed PMID: 30157420
105. Seok BS, Cao F, Belanger-Nelson E, Provost C, Gibbs S, Jia Z, Mongrain V. The effect of Neuroligin-2 absence on sleep architecture and electroencephalographic activity in mice. Mol. Brain. 2018;11(1):52. https://doi.org/10.1186/s13041-018-0394-3. Epub 2018/09/21. PubMed PMID: 30231918; PMCID: PMC6146600
106. El Helou J, Belanger-Nelson E, Freyburger M, Dorsaz S, Curie T, La Spada F, Gaudreault PO, Beaumont E, Pouliot P, Lesage F, Frank MG, Franken P, Mongrain V. Neuroligin-1 links neuronal activity to sleep-wake regulation. Proc Natl Acad Sci U S A. 2013;110(24):9974–9. https://doi.org/10.1073/pnas.1221381110. Epub 2013/05/30. PubMed PMID: 23716671; PMCID: PMC3683757
107. Palchykova S, Winsky-Sommerer R, Shen HY, Boison D, Gerling A, Tobler I. Manipulation of adenosine kinase affects sleep regulation in mice. J Neurosci. 2010;30(39):13157–65. https://doi.org/10.1523/JNEUROSCI.1359-10.2010. Epub 2010/10/01. PubMed PMID: 20881134; PMCID: PMC2950004
108. Honda T, Fujiyama T, Miyoshi C, Ikkyu A, Hotta-Hirashima N, Kanno S, Mizuno S, Sugiyama F, Takahashi S, Funato H, Yanagisawa M. A single phosphorylation site of SIK3 regulates daily sleep amounts and sleep need in mice. Proc Natl Acad Sci USA. 2018;115(41):10458–63. https://doi.org/10.1073/pnas.1810823115. Epub 2018/09/27. PubMed PMID: 30254177; PMCID: PMC6187192
109. Iwasaki K, Fujiyama T, Nakata S, Park M, Miyoshi C, Hotta-Hirashima N, Ikkyu A, Kakizaki M, Sugiyama F, Mizuno S, Abe M, Sakimura K, Takahashi S, Funato H, Yanagisawa M. Induction of mutant Sik3(sleepy) allele in neurons in late infancy increases sleep need. J Neurosci. 2021;41(12):2733–46. https://doi.org/10.1523/JNEUROSCI.1004-20.2020. Epub 2021/02/10. PubMed PMID: 33558433; PMCID: PMC8018738

110. Mikhail C, Vaucher A, Jimenez S, Tafti M. ERK signaling pathway regulates sleep duration through activity-induced gene expression during wakefulness. Sci Signal. 2017;10(463):eaai9219. https://doi.org/10.1126/scisignal.aai9219. Epub 2017/01/26. PubMed PMID: 28119463
111. Krueger JM, Obal FJ, Fang J, Kubota T, Taishi P. The role of cytokines in physiological sleep regulation. Ann N Y Acad Sci. 2001;933:211–21. Epub 2002/05/10. PubMed PMID: 12000022
112. Opp MR, Krueger JM. Anti-interleukin-1 beta reduces sleep and sleep rebound after sleep deprivation in rats. Am J Phys. 1994;266(3 Pt 2):R688–95. Epub 1994/03/01. PubMed PMID: 8160860
113. Churchill L, Rector DM, Yasuda K, Fix C, Rojas MJ, Yasuda T, Krueger JM. Tumor necrosis factor alpha: activity dependent expression and promotion of cortical column sleep in rats. Neuroscience. 2008;156(1):71–80. https://doi.org/10.1016/j.neuroscience.2008.06.066. S0306–4522(08)00986-X [pii] Epub 2008/08/13. PubMed PMID: 18694809; PMCID: 2654198
114. Kapas L, Hong L, Cady AB, Opp MR, Postlethwaite AE, Seyer JM, Krueger JM. Somnogenic, pyrogenic, and anorectic activities of tumor necrosis factor-alpha and TNF-alpha fragments. Am J Phys. 1992;263(3 Pt 2):R708–15. Epub 1992/09/01. PubMed PMID: 1357984
115. Opp MR, Postlethwaite AE, Seyer JM, Krueger JM. Interleukin 1 receptor antagonist blocks somnogenic and pyrogenic responses to an interleukin 1 fragment. Proc Natl Acad Sci U S A. 1992;89(9):3726–30. Epub 1992/05/01. PubMed PMID: 1533282; PMCID: 525563
116. Ingiosi AM, Opp MR, Krueger JM. Sleep and immune function: glial contributions and consequences of aging. Curr Opin Neurobiol. 2013;23(5):806–11. https://doi.org/10.1016/j.conb.2013.02.003. S0959-4388(13)00051-2 [pii] Epub 2013/03/05. PubMed PMID: 23452941; PMCID: 3695049
117. Kubota T, Fang J, Kushikata T, Krueger JM. Interleukin-13 and transforming growth factor-beta1 inhibit spontaneous sleep in rabbits. Am J Physiol Regul Integr Comp Physiol. 2000;279(3):R786–92. https://doi.org/10.1152/ajpregu.2000.279.3.R786. Epub 2000/08/24. PubMed PMID: 10956235
118. Cavadini G, Petrzilka S, Kohler P, Jud C, Tobler I, Birchler T, Fontana A. TNF-alpha suppresses the expression of clock genes by interfering with E-box-mediated transcription. Proc Natl Acad Sci U S A. 2007;104(31):12843–8. https://doi.org/10.1073/pnas.0701466104. Epub 2007/07/25. PubMed PMID: 17646651; PMCID: PMC1937554
119. Yamamura Y, Yano I, Kudo T, Shibata S. Time-dependent inhibitory effect of lipopolysaccharide injection on Per1 and Per2 gene expression in the mouse heart and liver. Chronobiol Int. 2010;27(2):213–32. https://doi.org/10.3109/07420521003769111. Epub 2010/04/08. PubMed PMID: 20370466

Cnidarian Sleep

Diana A. Francis and Michael J. Abrams

Abstract Perhaps nothing is stronger evidence of the importance of sleep than its conservation across animals. Even cnidarians, gelatinous aquatic animals such as the jellyfish Cassiopea and the polyp Hydra exhibit the hallmark features of sleep observed in animals with centralized nervous systems: reversible quiescence, increased latency to arousal, and homeostatic regulation. Despite lacking a brain, cnidarians possess muscles controlled by neural networks and neurotransmitter systems, including GABA, dopamine, acetylcholine, and melatonin. Further, conserved sleep regulators, including the cGMP-dependent protein kinase 1, PRKG1, and GABA, in Hydra, and a nicotinic cholinergic receptor alpha subunit, CHRNAL-E, in Cassiopea, emphasize that not only is the behavioral state conserved, but so too are components of the regulatory system. We also discuss possible functions of sleep in Cnidaria, including roles in cell proliferation and the homeostasis of neurological pacemaker activity. Overall, investigations in cnidarian sleep have shed light on fundamental aspects of sleep regulation and underscore the importance of studying these organisms to gain insight into the evolutionary origins and basic mechanisms of sleep.

Keywords Cnidaria · Conservation · Regulation · Function · Evolution

Cnidarians and Behavior

Cnidarians are gelatinous aquatic animals that occupy a unique evolutionary position in animal phylogeny as members of the sister clade to Bilateria (Fig. 1a). Based of molecular clock and fossil data, Cnidarians have existed for over 700 million

D. A. Francis
Berkeley, CA, USA

M. J. Abrams (✉)
Albany, CA, USA
e-mail: mjabrams@berkeley.edu

P. Gehrman et al. (eds.), *Genetics of Sleep and Sleep Disorders*,
https://doi.org/10.1007/978-3-031-62723-1_7

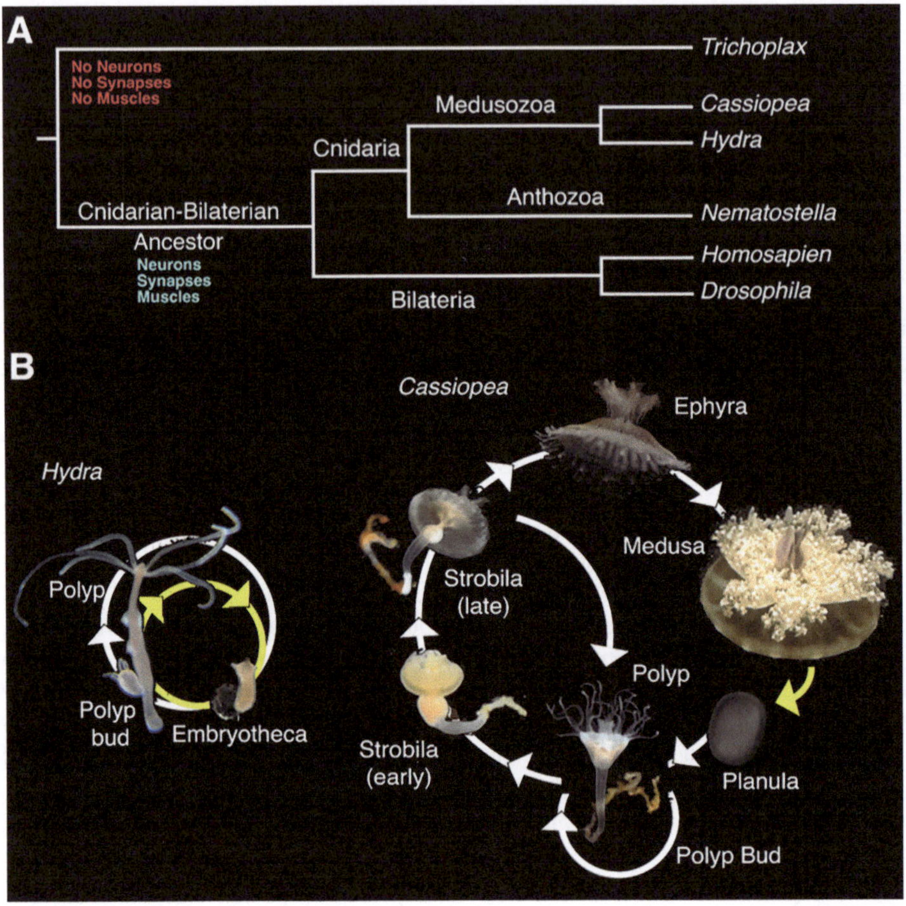

Fig. 1 Nervous system evolution and basic Cnidarian life cycles. (**a**) Conserved neurons, synapses, and muscle emerged in animals within the Cnidarian-Bilaterian ancestor [15]. (**b**) Cnidarian life cycles are often dimorphic, with one from being sexually reproductive, while the other is capable of asexual reproduction

years [1] and are considered the simplest animals to have true tissues, a digestive cavity, and a nervous system. Their body plans generally consist of a simple sac-like structure with a central mouth surrounded by tentacles armed with specialized stinging cells called cnidocytes, which are used for capturing prey and defense. These represent the simplest animals where sleep has been defined at the behavioral and neural levels.

The cnidarian life cycle most often involves two distinct forms: a sessile polyp and a free-swimming medusa (Fig. 1b). The polyp form is typically attached to a substrate and reproduces asexually by budding, while the medusa form is free-swimming and reproduces sexually. In some species, the medusa form is absent, and some polyps can also reproduce sexually. While there are some examples of symmetry breaking in Cnidaria [2], cnidarians are canonically considered radially symmetrical, with no brain, blood, and only a hydrostatic skeleton.

Cnidarins are the earliest animals to have evolved bilaterian-like neurons (numerous conserved components) that synapse to muscles to facilitate behavioral motor functions [3] (Fig. 1a). However, rather than a brain that controls behavior of the entire animal, Cnidarians have a decentralized nervous system. However, between species, there are varying degrees of regional condensation and specialization. For example, Polyps like *Hydra* have a diffuse nerve net, but with higher neural density around the mouth and foot, while true jellyfish, like *Cassiopea*, have highly organized ganglia that perceive motion and light and facilitate directed movement [4]. While it is not understood how whole-body phenomena like sleep arise in the absence of a centralized nervous system, we must admit that sleep likely evolved first in animals without a brain. Perhaps we can consider these animals to have some intermediate combination of structural and functional specialization without a true central nervous system.

Neurons appear to likely be derived, somewhat confusingly, from different cells depending on the Cnidarian. *Hydra* neurons, and nematocysts, begin as multipotent cells, while non-hydrozoan cnidarians rely on epithelial cells to generate these cell types [5]. Unlike bilaterians, the main muscle cell type of cnidarians is epitheliomuscular cells, specialized epithelial cells containing smooth myofilaments, that form the building blocks of the two body layers, the epidermis and gastrodermis [6, 7]. These muscles allow for basic movements, locomotion, and enable the various behaviors of Cnidarians.

Cnidarians display several core animal behaviors. (1) A clear feeding behavior that is typically characterized by the use of nematocyst, which immobilizes or kills the prey, and an increase in muscle contractions, that bring prey toward the mouth. The most well characterized is the glutathione-induced feeding response of *Hydra* [8]. (2) While jellyfish are not generally known for their courtship behaviors, some species have been observed engaging in various types of displays and interactions prior to mating. One cubozoa species, *Copula sivickisi*, has an intricate courtship dance in which the male will use his tentacles to grab a female's tentacles and drag her around for a while before pulling her in close to him [9]. The male then produces and releases a spermatophore (sperm packet) and uses his tentacles to pass it to one of the female's tentacles. The pair splits, and the female eats the sperm packet [9]. (3) Medusae also display an obvious escape response that can be triggered mechanically [10]. Cnidarians also exhibit a range of responses to environmental stimuli. For example, many species of jellyfish and coral are sensitive to changes in water temperature, salinity, and pH, and will alter their behavior in response to these changes [10]. Cnidarians also exhibit phototaxis, which has been linked to neuropeptides [10].

Behavioral control has been studied in *Hydra*, in which whole-body muscle activity is mapped using genetically encoded fluorescent calcium indicators. Seven distinct spatiotemporal activity patterns have been identified that include global, local, and propagating activity [11]. In jellyfish, pulsing activity is initiated by pacemaker cells within ganglia that are radially spaced around the bell margin [12].

Because the pacemaker neurons form a neural circuit that maintains rhythmic activity in the absence of a rhythmic input, they are considered Central Pattern Generators (CPGs). CPG activity has been studied in the context of epilepsy and parasomnia in humans, and imbalances between arousal and sleep forces can lead to variable motor outputs determined by factors including the involvement and thresholds of specific brain areas and neural circuits, and CPGs [13]. To study how the distributed ganglia in a decentralized neural architecture control pulsing and are affected by sleep or sleep-deprivation, a high-speed imaging and analysis system was developed to locate the region of muscle-contraction initiation, and by proxy, the ganglion that initiated the contraction [14]. Together, the phylogenetic position, relatively simple tissues and body plan, and the accessibility of neurological behavioral assessment, makes Cnidaria a powerful system for studying the fundamental control and function of sleep.

Conservation of Neural Regulatory Pathways

There is evidence from experimental approaches that both fast (acetylcholine, glutamate, GABA, glycine) and slow (serotonin, dopamine, octopamine, histamine, and melatonin) transmitters, as well as neuropeptides, including the well characterized RFamide, are involved in cnidarian neurotransmission [15]. Indeed, it appears likely that the majority of the canonical low-molecular-weight transmitters were recruited to neural signaling in the bilaterian–cnidarian ancestor [16] (Figs. 1a and 2). Experiments utilizing immunohistochemical, pharmacological, behavioral, electrophysiological, and genetic evidence approaches are discussed below.

There is strong biochemical evidence for the presence of receptors to glutamate, GABA, and glycine in cnidarians. Receptor binding studies on *Hydra vulgaris* show (1) glutamate has a single receptor class in membranes and plays an inhibitory role in regulating the glutathione-induced feeding response [15], (2) GABA receptors have biochemical and pharmacological properties similar to those of mammalian ionotropic GABA receptors [17, 18], and (3) glycine receptors have been demonstrated and characterized in membrane preparations [19].

Only moderate progress has been made towards biochemical analysis of the cholinergic system. Several early attempts by a number of workers to identify acetylcholine itself or its metabolic enzymes have been unsuccessful, although recently acetylcholinesterase activity was quantified in *Actinia equina* and *Clytia hemisphaerica* [20, 21]. Importantly, in *Nematostella vectensis*, isolated nematocytes were screened for responses to well-conserved neurotransmitters and found only acetylcholine elicited a significant response; acetylcholine-evoked outward currents were abolished by nicotinic acetylcholine receptor antagonists and recapitulated by nicotine [22]. Acetylcholinesterase activity was also found in the tentacles, indicating that nematocytes use cholinergic signaling to regulate K+ currents, similar to how efferent cholinergic innervation of vertebrate hair cells modulates auditory responses [23].

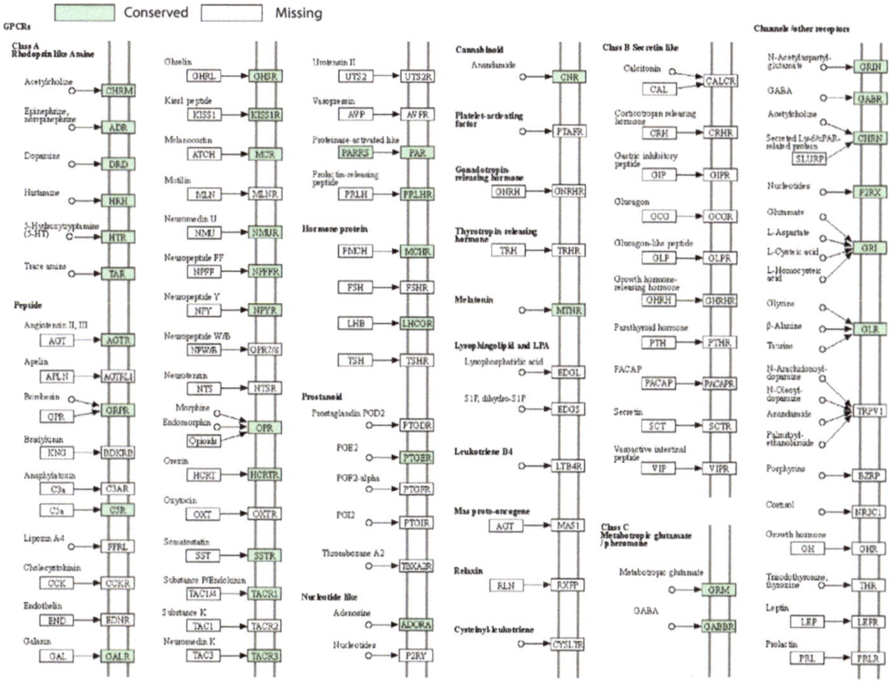

Fig. 2 Numerous neuroactive components are conserved in Cnidaria. KEGG pathways showing presence (tan-box)/absence (white-box) of neuroactive ligand-receptor interactions

It is difficult to say how generalizable these findings are across Cnidaria. Few species have been investigated within each class, and for each question, very little comparative analysis has been accomplished. For this reason, it is important to consider each finding on its own and how it may relate to our broader understanding of animal nervous systems.

Gene expression offers a more comprehensive view and suggests many of the core neuroactive ligand-receptor interactions are present in *Cassiopea* [24] (Fig. 2), and broadly in cnidarians [15]. Fast neurotransmitter related genes has been recently identified in the genome of the sea anemone, *Nematostella vectensis*, but results indicate that many are not directly involved in vertebrate-like fast neurotransmission [25]; however, cholinergic receptors are expressed in both in neural and non-neural regions [26].

Finally, melatonin, one of the phylogenetically oldest and most ubiquitous pleiotropic agents in nature (acting as an antioxidant [27], immunoenhancing [28] and neuromodulator [29]), is also found in cnidarians. Its levels appear to correspond to sexual maturation, and gamete-bearing colonial masses, and importantly, in association with neurons wrapped around gametes. Further, the distribution of melatonin receptors strongly implies its association with the neural network [30].

Behavioral Characterization of Sleep in Cnidaria

Sleep across animals is defined by three behavioral criteria: reversible quiescence, homeostasis, and increased latency or threshold to arousal [31]. There is a general understanding that sleep behavior in cephalized animals has two systems of regulation: (1) the circadian system, which controls the timing and cyclical rhythms of sleep, and (2) the homeostatic system, which drives the onset and maintains the duration of sleep [32]. Evidence suggests that the same two systems also work in tandem to regulate sleep in some cnidarians, including the upside-down jellyfish *Cassiopea xamachana*. First, *Cassiopea* express several of the core circadian machinery, including *cryptochromes* and *clock* [24] (Fig. 3a), and their pulsing activity that oscillates according to a typical circadian 12:12 light:dark (LD) cycle (Fig. 3b) is maintained in free-running constant dark conditions [29]. Second, *Cassiopea* undergoes a compensatory period of rebound sleep following mechanical or light stimulation during their sleep phase, a rebound which reflects an underlying homeostatic sleep pressure that increases with prolonged wakefulness [31, 33]. Though the two-pronged framework of regulatory processes generally applies to cephalized animals, the circadian clock is not required for sleep behavior, as demonstrated in mice and *Drosophila* circadian clock mutants [34, 35]. Moreover, sleep in adult *C. elegans*, rather than aligning with circadian rhythms, occurs during distinct developmental periods and following incidence of cellular stress or satiety [36]. Similarly, evidence indicates that sleep in *Hydra* is under homeostatic but not circadian control; *Hydra* displays a homeostatic rebound following nighttime

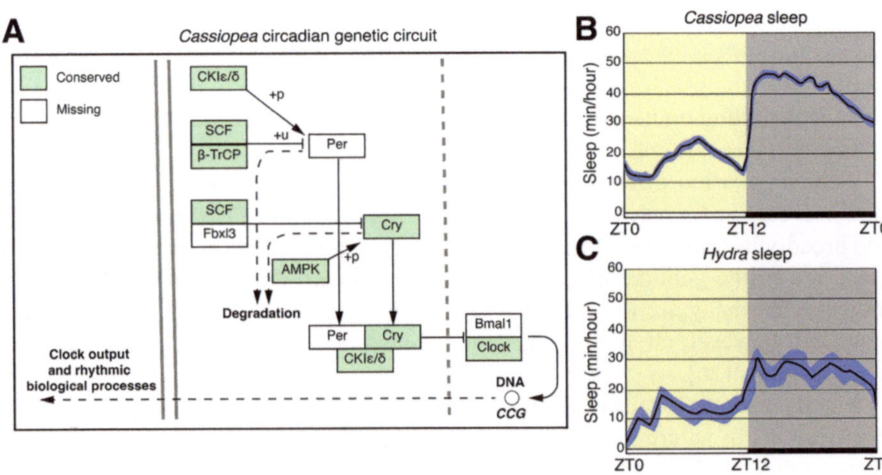

Fig. 3 Circadian clock genes in *Cassiopea*, and the sleep cycle of *Cassiopea* and *Hydra*. (**a**) A condensed KEGG pathway showing presence (tan-box)/absence (white-box) of circadian rhythm. The daily sleep cycle of (**b**) *Cassiopea* and (**c**) *Hydra* in 12:12 LD conditions. *Cassiopea* continue to have circadian sleep in DD free-running conditions; however, Hydra show ultradian rhythmicity, indicating the lack of circadian regulation of Hydra sleep

stimulation, but while wild-type sleep oscillations are present during the normal LD cycle (Fig. 3c), it is lost in DD and constant light conditions (LL) [37]. Interestingly, in lieu of wild-type 12:12 sleep-wake cycling, *Hydra* in DD and LL displays 4-h cycling of sleep states, which is more pronounced in DD [37]. These 4-h ultradian rhythms are thought to be either ancestral precursors or intermediaries to circadian rhythms, but it is possible that they reflect an evolutionary loss of function.

Sleep in cnidarians is behaviorally characterized by three key characteristics that are also observed in animals with centralized nervous systems [14, 24, 33, 37]. Sleep is defined by (1) quiescent periods that are rapidly reversible, during which there is, (2) increased latency (or threshold) to arousal, and if deprived of quiescence, animals experience a compensatory low activity period due to (3) homeostatic regulation.

Of these three metrics, quiescence is the most well-documented in cnidarians thus far, largely because cnidarian sleep is a newly emerging subfield and quiescence is relatively easy to determine with only observation-based methods. In some venomous box jellyfish (*Chironex fleckeri*, *Carybdea marsupialis*, and *Carybdea rastonii*), no quiescence has been observed, but other box jellyfish species, *Tripedalia cystophora and Copula sivickisi,* undergo 12-h quiescent periods [38]. During these periods, *T. cystophora* localizes to the ocean floor and reduces its pulse rate, whereas *C. sivickisi* attaches itself to a substrate, encloses its tentacles inside of its bell, and becomes motionless. Interestingly, though, while *T. cystophora* pulses actively during the day and becomes quiescent at night, *C. sivickisi* exhibits daytime quiescence and nighttime activation. These opposing patterns of quiescence could be indicative of an evolutionary switch to nocturnality in *C. sivickisi*, but the molecular mechanisms of these box jellyfish have yet to be studied, and it is unclear whether such cycling is under circadian control.

Quiescent periods are not necessarily consistent with sleep states, however, as they may instead reflect other behavioral states like dormancy. In corals, quiescent periods are seasonal; the northern star coral *Astrangia poculata* enters winter quiescent periods induced by cold thermal stress and characterized by a retraction of its tentacles as well as a reduction in its feeding behavior and colony growth [39]. However, the corals do not respond to mechanical stimulation, and this lack of rapid reversibility indicates that they do not enter a sleep state in winter but rather a dormancy state that may differ in its function and underlying machinery. Notably, *Cassiopea* and *Hydra* each experience an afternoon siesta and a 12-h nighttime quiescent period that are both rapidly reversible. *Cassiopea* significantly slows its pulsing frequency and *Hydra* only enters low mobility "housekeeping" states during these periods, but they increase their activity level upon introduction of light, food, and mechanical stimuli. This rapid reversibility is consistent with a state change transition from sleep to wake.

Even when reversible, quiescent episodes could be indicative of lower-activity wakefulness rather than sleep. The second behavioral metric, latency to arousal, or the delayed responsiveness to stimuli, differentiates quiescence in sleep from quiescence in wakefulness. Essentially, when stimulated, sleeping animals take longer to react than awake animals. During the nighttime quiescent periods of both *Cassiopea* and *Hydra*, there is an extended interval between sensory stimuli and responsiveness, consistent with sleep. This latency to arousal has also been shown to be rapidly reversible in *Cassiopea*: when subjected to two rounds of mechanical perturbation in the dark [33], *Cassiopea* is faster to pulse following the second round than it is following the first. In *Hydra*, animals in the quiescent state for longer than 20 min showed a significant delay in their responses to a light pulse [37].

Recently, in *Cassiopea*, discrete periods of sleep were characterized using latency-to-arousal tests [14, 40–44] in two steps: (1) a non-aroused activity level was determined, and then (2) the response time to an arousing stimulus was measured [14]. Cassiopea have a strong pause in pulsation when lights are turned off, perhaps akin to the hyper-arousal freeze response of other animals [45]. Using this approach, it was found that *Cassiopea* tend to take longer to freeze in response to light-dark changes, and therefore have an increased latency-to-arousal when pulsing slowly (below their median pulse rate). If the light-dark test is repeated within 30 s, animals respond quickly, indicating the animals are aroused from the light-dark test, and is further evidence of reversibility of quiescence in *Cassiopea*. Evidence from *Hydra* and *Cassiopea* shows they have discrete sleep bouts that can be characterized using similar methods as have been applied to bilaterians.

It may be good to take a moment now to step back and just consider the enormity of these findings. Sleep, a behavior so complex we are only now beginning to understand its regulation and function, is so important for animal life, that the first animals to have evolved neurons, synapses, and muscle (Fig. 1a), likely used conserved neurotransmitters (Fig. 2) to generate a state that cycles similarly across animals (Fig. 3) and can be detected using the same behavioral tests.

Conserved Regulation of Sleep in Cnidaria

Though the mechanisms of sleep regulation in cnidarians are largely uncharacterized, several well-documented sleep modulators from cephalized animals are known to be conserved. One is melatonin, the critical vertebrate sleep hormone, administration of which induces concentration-dependent sleep states in *Cassiopea* and *Hydra* [33, 37], suggesting melatonin has a conserved role as a sleep-promoting or sleep-timing factor. It may also interact with the cnidarian circadian clock, as work in *Nematostella* shows that melatonin plays a role in their circadian behavior. In this case, there is cyclic expression of this hormone under diel conditions, light-dependent oscillations in genes in the melatonin synthesis pathway, and melatonin initiating expression of circadian clock genes [46]. These animals have melatonin receptors [30], and melatonin synthesis genes [30, 46, 47], indicating that melatonin signaling to regulate sleep may be deeply conserved. Further, melatonin localization and melatonin receptors expression patterns in nematostella, an Anthozoan, implicates ancient roles in both the reproductive and nervous systems [30].

The cGMP-dependent protein kinase 1 (PRKG1) is another sleep promoting factor that is conserved from cnidarians through mammals. For example, flies with high levels PRKG1 do not display deficits in short-term memory following 12 h of sleep deprivation, that are otherwise present in wildtype flies [48]. PRKG1 in worms regulates sleep behavior and functions in sensory neurons to promote worm sleep [49]. PRKG1 mutant mice showed a strikingly altered distribution of sleep and wakefulness over the 24 h of a day as well as reductions in rapid-eye-movement sleep (REMS) duration and in non-REM sleep (NREMS) consolidation, and their ability to sustain waking episodes was compromised [50]. PRKG1 was upregulated in the hypostome (head/mouth) of sleep-deprived *Hydra*. As further evidence of this theory, pharmacological activation of PRKG1 promotes sleep in *Hydra*, whereas a PRKG1 inhibitor has a concentration-dependent arousing effect [37]. Together, this points towards PRKG1 as an ancestral sleep drive gene.

The GABAergic system has also been implicated in cnidarian sleep. Incubation in the inhibitory neurotransmitter GABA increases daytime sleep in *Hydra* [37], consistent with a wide body of evidence designating GABAergic pathways as sleep-promoting [51]. Additionally, sleep deprivation drove a decrease in expression of one GABAergic receptor in *Cassiopea* [24].

The cholinergic system is also connected to sleep regulation in a variety of animal taxa. The *Drosophila* mutant Redeye (*rye*), which experiences a severe reduction in sleep both during the day and at night, has a polymorphism in a nicotinic acetylcholine receptor alpha subunit (nAChRa) [52]. In *Cassiopea*, many components of the vertebrate cholinergic synapse are present (Fig. 4). One nAChRa named *chrnal-E*, which bears appreciable similarity to Chrna7 in both sequence and structure (Fig. 5a), was one of the most highly differentially expressed gene within a set of sleep-deprived animals [23]. nAChRs composed of Chrna7 subunits is highly permeable to calcium [53, 54], which would affect membrane potential, and the excitability of neurons within the ganglia, possibly including the pacemaker neurons that initiate muscle contractions.

Silencing *chrnal-E* by RNA interference resulted in a rebound-like phenotype: they sleep significantly more during the midday-siesta and less at night compared to the control group (Fig. 5b). As such, it is reasonable to assume that *chrnal-E* could have an ancestral role in the homeostatic sleep drive, but further assays must be done to assess the function of *chrnal-E* in circadian regulation. That being said, given that Drosophila *rye* homozygotes exhibit wild-type cycling of the circadian clock protein PERIOD and that most *rye* mutants are rhythmic, it is likely that *chrnal-E*, like *rye*, has little involvement in circadian rhythms.

The function of *chrnal-E* in the sleep drive of *Cassiopea*, while probable, remains undefined. One possibility is that *chrnal-E* serves as a tracker of sleep debt, a hypothesis supported by the cycling of RYE protein levels in wild-type *Drosophila*. RYE peaks transiently at the onset of a sleep period and declines with prolonged sleep. Notably, instead of increasing gradually during wake, RYE levels remain low throughout the course of the day. This suggests that *rye* activation is dependent on the accumulation of a different sleep drive component, and *chrnal-E* may similarly track the concentration of other factors. It is also possible that *chrnal-E* signals to the GABAergic system since sleep-promoting cholinergic fibers in the mammalian brain activate GABAergic neurons.

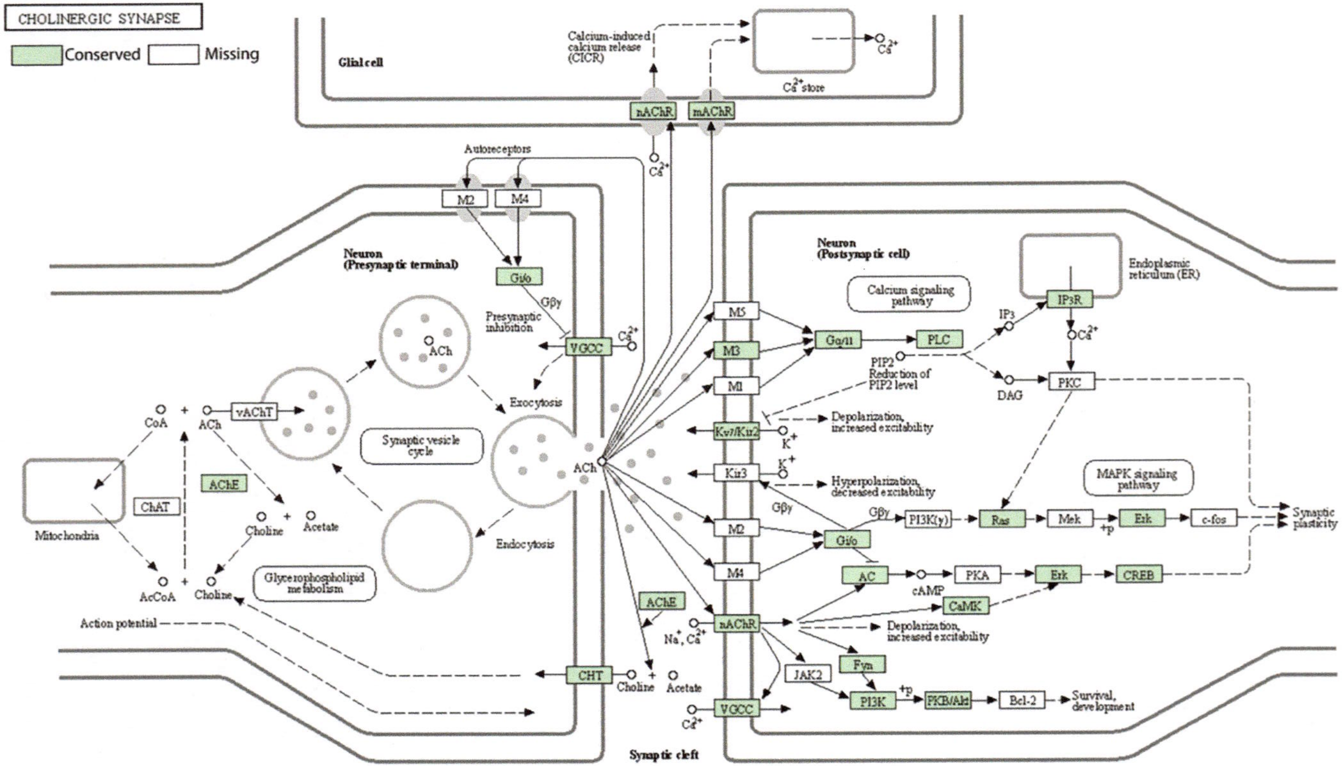

Fig. 4 Many components of the cholinergic synapse are conserved in Cassiopea. KEGG pathways showing presence (tan-box)/absence (white-box) of the cholinergic synapse

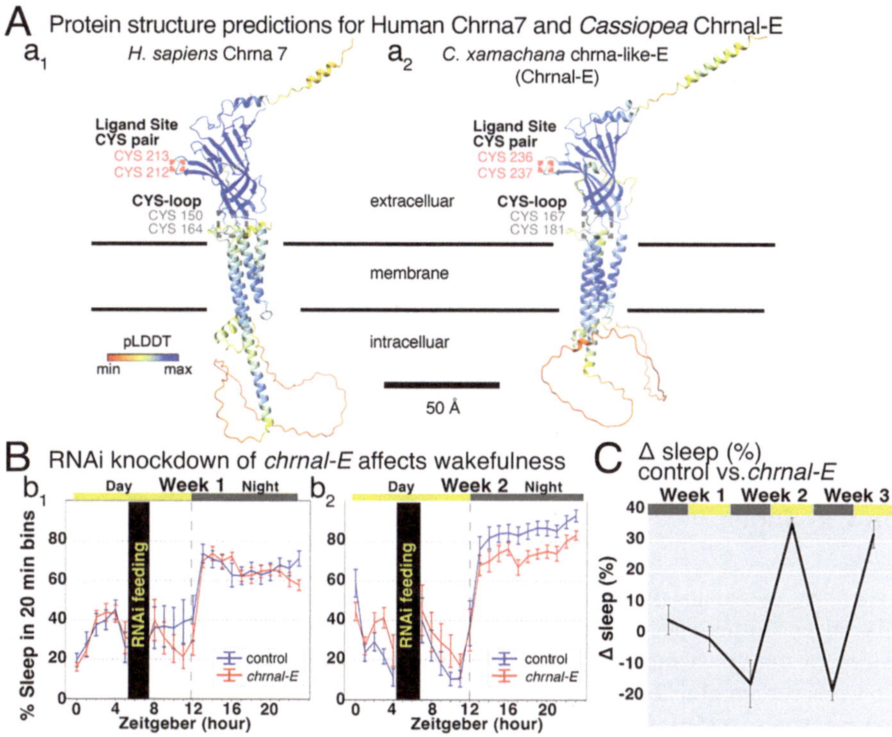

Fig. 5 A nAChRa-like subunit (Chrnal-*E*) that is upregulated in Cassiopea during sleep-deprivation shares domain and sequence similarity to Human nAChRa (Chrna7). (**a**) Alpha fold protein structure predictions for (**a₁**) Human Chrna-7, and (**a₂**) *C. xamachana* chrma-like-E (Chrnal-E), with labeled Cys-Cys ligand site and CYS-loop indicating alpha-subunit identity. Extracellular, membrane, and intracellular regions demarcated. pLDDT color legend indicates structure confidence. (**b**) Animals were fed RNAi their pulsing behavior was recorded. Blue lines are average control behavior and the red lines are the average *chrnal-E* knockdown behavior. A black box labeled RNAi feeding designates the period. Error bars are SEM. (**c**) The Δ sleep (%) between control and *chrnal-E*, error bars are 95% confidence interval for Percent Effect

There exists a key difference between the *rye* and *chrnal-E* loss of function assays. Unlike Drosophila *rye* mutants, *chrnal-E*-depleted *Cassiopea* do not exhibit a global reduction of sleep in both daytime and nighttime hours and instead shift their sleep-wake cycles as if responding to sleep deprivation (Fig. 5c). Such a discrepancy could be explained by differences in experimental technique; the *chrnal-E* assay is a knockdown, not a knockout, so presumably, low levels of CHRNAL-E could accumulate throughout the night and induce daytime sleep.

While evidence indicates that the molecular underpinnings of cnidarian sleep are largely conserved, some factors differ in function between cnidarians and later-diverging phyla. Dopamine is arousing in cephalized animals, yet in *Hydra*, dopamine and its precursors promote sleep [37]. On the other hand, orinthine, the non-proteinogenic amino acid involved in the urea cycle, is sufficient for daytime sleep promotion in *Drosophila* but suppresses sleep in *Hydra* [37]. This functional flipping of the ancestral dopaminergic and orinthine pathways in sleep is thought to have occurred in the development of the CNS, perhaps as a consequence of circuitry rewiring.

Together, it is profound that conserved components of neural regulatory systems are frequently involved in regulating sleep even in the simplest animals. While the exact mechanisms are not generally conserved, this is not surprising as in many cases genes involved in the regulatory circuit in vertebrates are not present in sequenced cnidarians. Only through continued investigation of sleep regulation in cnidarians will we understand the extent of mechanistic conservation, though it is already clear that evolution does not invent neural regulatory systems just to discard them.

Sleep Regulates Central Pattern Generator Plasticity in Cassiopea

There are two major theories for how global brain states arise to generate behaviors like sleep. Either specialized regions control the switch between wakefulness and sleep (a top-down mechanism) [55]; or neural networks have an emergent bias towards certain global states that are affected by local regulatory circuits (a bottom-up mechanism) [56]. From work in humans and model animal systems, there is a growing appreciation that sleep is regulated globally, regionally, and locally by intersecting mechanisms [57]. However, sleep plays roles that appear specific to the brain, including synaptic homeostasis [58], neurotransmitter regulation [59], cellular repair [60], memory consolidation [61], and neural plasticity [62, 63]. While sleep across animals plays many roles, would any of these functions of sleep be relevant to an animal without a brain?

Though *Cassiopea*, as with all cnidarians, do not have a centralized nervous system, they do have ganglia that are radially spaced around the bell margin (Fig. 6a). As was mentioned in the first section, the ganglia contain pacemaker neurons as part of Central Pattern Generators (CPGs) that initiate muscle contractions, and in recording setups using infrared light and high-speed cameras (Fig. 6b), the point of muscle contraction initiation can be detected for every pulse. Tracking the position of muscle contraction initiation acts as proxy for local ganglionic CPG activity in scyphozoans [14]. Plotting the position of contraction initiating ganglia on the y-axis, and time on the x-axis (Fig. 6c1) generates actograms to visualize ganglionic CPG activity within the network. From the actograms, we can see that a subset of ganglia leads activity for long periods, but that they exchange over time. A systemic analysis revealed that sleep regulates the homeostasis of ganglionic CPG activity.

Analysis of ganglionic CPG activity in control, sleep deprived, and recovery conditions yielded strong evidence for sleep in *Cassiopea* playing a key role in

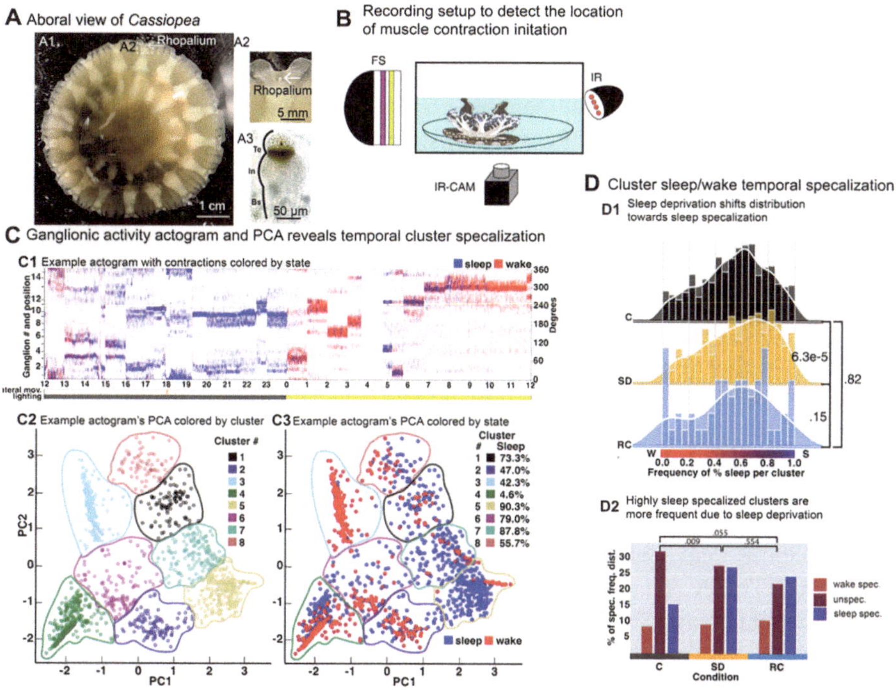

Fig. 6 Sleep is required to maintain plasticity in the CPG network of *Cassiopea*. (**a**) *Cassiopea* pulse in place, aboral side down, facilitating long-term recordings. (**a1**) Aboral view of *Cassiopea*, with a labeled rhopalium, each of which contains a CPG. (**a2**) Rhopalia are positioned radially around the bell margin. (**a3**) Rhopalia are comprised of three segments, a terminal (te), intermediate (in) and basal (bs). (**b**) Ganglia tracker setup and computer vision pipeline. (**b**) Cassiopea cannot escape, or maneuver such that their bell margins are obscured, and are maintained in full-spectrum (FS) light, on a 12:12 cycle, fed every third day, and are recorded in IR with a 120 fps camera (CAM).(**c**) Example actograms from a single animal 24-h recording, and the associated PCAs. (**c1**) Using ganglia activity 1-min bins, and k-means clustering. Colors and numbers indicate cluster. (**c3**) The actograms were then re-colored using the cluster color associated with each 1-min bin for. (**d**) PCA sleep/wake and night/day specialization analysis of $n = 14$ control animals with 385 clusters, $n = 14$ SD animals with 223 clusters, and $n = 5$ RC animals and 54 clusters. (**d1**) Distributions of the specialization % for each cluster, separated by condition, visualized using matplotlib Gaussian kernel density estimator. *Y*-axis is normalized density, and *X*-axis is the frequency of ($\mathbf{d_1}$) % sleep; 1 is maximum sleep, 0 is maximum wake biased, and 0.5 is unbiased. *p*-value determined by Kolmogorov-Smirnov test. Categorical counts, normalized to total count, for each state and condition, given as the percent of the cluster bias. (**d2**) Sleep biased (0.8–1.0), unbiased (0.4–0.6), and wake biased (0.0–0.2). *p*-values determined by 2 × 3 chi-square

regulating neural network plasticity. Patterns of CPG usage within 24-h actograms were processed using Principle Component Analysis (PCA), a technique that transforms a dataset to give each point a coordinate that describes its variability in fewer dimensions. Each point in a PCA is the linear transformation of 1-min bins of actogram data, using the activity of each CPG as its features (Fig. 6c2). The average sleep activity of the bin can be overlaid each actogram's PCA to ascertain the percentage of bins in each cluster that occurred during sleep or wake, allowing for the calculation of the temporal sleep bias of each cluster, representing an active CPG network (Fig. 6c3). In control conditions, clusters tend to be unbiased (near the center of the distribution), meaning certain CPG networks active during sleep are also active during wake-CPG networks, therefore, change homeostatically over time. In other systems, it has been shown that neural networks are constantly regulated by activity-dependent processes. These processes include changes in the transmission efficacy of existing synapses or changes in the circuit connectivity by the formation and deletion of synapses [64]. Changes in CPG networks in *Cassiopea* likely involve transmission efficacy plasticity which encompasses adaptive changes in presynaptic, extracellular, or postsynaptic molecules that lead to plasticity of the neural network without involving modification of synaptic structure.

Remarkably, the CPG networks in sleep deprived animals become far more sleep biased than those in control conditions (Fig. 6d), seen by the shift to the right in the sleep cluster distribution peak. Therefore, clusters active during rebound sleep tend not to be active during wake. In the days following sleep deprivation, the distribution appears to return to a distribution similar to control (Fig. 6d), indicating a level of reversibility, and a pressure to return to a more homeostatically exchanging network state. The PCA and cluster bias analysis across conditions supports a role for sleep in maintaining CPG plasticity.

Here it is interesting to consider a well-studied CPG paradigm, the sinoatrial node (SAN). Underlying the adaptability of heart rate is SAN plasticity, and a major pathway for acute regulation of SAN function is through altered gating of ion channels important for SAN cell automaticity, either through ligand binding or post-translational modification, another mechanism of SAN regulation is through changes in expression of ion channels, which have been documented as an important cause of altered SAN function and heart rate in the setting of chronic stimuli (e.g., disease, aging, long-term exercise training) [65]. Similar mechanisms of plasticity may also be present in the Interstitial Cells of Cajal, that act as gut CPGs [66]. Sleep deprivation in *Cassiopea* may be causing a similar phenomenon in the CPG network as chronic stimuli does in the SAN or in ICCs, causing changes in the plasticity of the CPGs.

Indeed, it is conceivable that neurons in the ganglia monitor byproducts of prolonged wakefulness, such as changes in the strength of afferent synapses [67], the release of adenosine [68, 69], or rising levels of unfolded proteins or reactive oxygen [70]. Once a sensor reports a deviation from the set point that calls for corrective action, the homeostat must compare a physiological variable representing the sleep balance of the organism to a reference representing the homeostatic set point. As was mentioned previously in this chapter, the sleep-deprivation induced overexpression of *chrnal-E* could lead to higher calcium membrane permeability, a sensible feature for a homeostat. Perhaps like the fly dorsal fan shaped body neurons [71], action potentials generated by pacemaker neurons in the ganglia would be a cellular mechanism for making such a comparison in *Cassiopea*. Perhaps the ability for certain ganglia to generate action potentials, and lead activity during sleep, varies as a function of sleep need, a potentially interesting avenue for future research.

Because of the relatively simple nervous system architecture, few cell types, and ganglionic redundancy, it is possible to perform experiments in *Cassiopea* and possibly *Hydra* and other cnidarians as well, that are not possible in traditional invertebrate or vertebrate systems, in the effort to address the original question of how sleep arises in the nervous system. For example, highly sleep-active CPG can be detected, amputated, and CPG networks can be reassessed to reveal how sleep is regulated when the sleep-leaders are removed. At the same time, action potentials from amputated CPGs can be detected using suction electrodes, and action potential frequencies can be tracked in the light and dark, to determine if highly sleep- or wake-active CPGs maintain their bias outside of the CPG network. These perturbations, and others, can be combined with pharmacological interventions, to study the how the CPG networks function to generate animal behavior. These approaches may yield insights into how the wake and sleep states arise and reveal mechanisms of bottom-up or top-down sleep regulation.

Possible Functions of Sleep in Cnidaria

When considering the function of sleep, it is important to first consider the possible overlap between circadian rhythms and sleep. Circadian clocks have evolved multiple times during eukaryotic evolution, allowing organisms, including plants, fungi, and animals, to coordinate their physiology and behavior with predictable environmental rhythms [72, 73]. These molecular clocks help organize numerous metabolic and behavioral processes [74]. Cnidarians, the sister phylum to the bilaterians, also possess most of the circadian genes present in bilaterians. The diurnal transcription profiles of many of these genes in the sea anemone *Nematostella vectensis* and the stony coral *Acropora millepora* suggest that the conserved topology of the animal circadian clock may pre-date the bilaterian-cnidarian split [75–78]. Indeed, the circadian rhythm genes *Clock* and ARNT appear in the Ctenophore, *M. leidyi* [79], placing these genes at the root of the animal tree. However, results from the sponge *A. queenslandica* indicate (1) the ancestral metazoan circadian network was simpler than the eumetazoan network, or (2) that this sponge has lost some components, as has occurred in some other animals such as *Hydra* [37]. So, when considering the function of sleep, it is important consider it as separate from circadian regulated functions.

Compared to those with circadian rhythms, fewer organisms sleep, and by considering what is different about them may reveal sleep's basic function. To detect sleep in an organism, as mentioned previously, it must have a quiescent state that increases its latency or threshold to arousal; further this quiescence must be homeostatically regulated. This definition would likely need to be re-interpreted to define sleep in organisms lacking neurons. We, therefore, propose that sleep in its most basic form arose to facilitate neural activity. It will be fascinating to learn if cntephores sleep (we would predict they do), supporting the idea that neurons and sleep phylogentically appear concurrently. The finding that *Hydra* lost circadian genes but retained sleep, while sponges likely lost both systems, is further evidence that the functions of circadian rhythms and sleep are separate, and neurally focused. Conversely, while sleep is drastically reduced in some animals that maintain nervous system [80], they have become highly optimized to specific environments and have radically altered physiology that may compensate [81].

The finding that sleep-deprivation drives a shift towards sleep bias of the ganglia network implies that sleep promotes the unbiased state through increased plasticity. It is interesting to consider the benefits of a more plastic network as it is difficult to imagine that it would facilitate learning or memory in *Cassiopea.* However, *Cassiopea* live in an environment with predators, intermittent prey availability, and breeding cycles [82], so perhaps the unbiased system is better primed to respond to a changing environment. The plasticity could allow the animal to rest some ganglia without affecting behavior much, which may be like how some birds and ocean mammals have unihemispheric sleep [83], allowing half of the brain to sleep while the other maintains necessary activity, or another phenomenon, local sleep, that has been recorded in mammals, in which brain regions appear asleep even while the animal remains conscious [83]. Perhaps there are deleterious effects of prolonged activity being concentrated in one set of ganglia, and the exchange of ganglia allows the inactive ones to recover from wakefulness.

As mentioned previously, *Cassiopea* and *Hydra* undergo a homeostatic rebound following deprivation of nighttime quiescence. Given that sleep deprivation is detected by the homeostat, it follows that tissue-specific changes correlated with sleep deprivation may be directly or indirectly associated with the homeostatic sleep drive. Accordingly, sleep deprivation suppresses cell proliferation in *Drosophila* [84] and rats [85], and a similar trend exists in *Hydra* [37]. Proliferation reduces along the entire body plan of *Hydra* subjected to mechanical or pharmacological sleep deprivation. Based on evidence from *Hydra*, the natural inference is that cell proliferation may be downregulated by the sleep drive, and sleep-dependent proliferation likely serves as an ancestral and evolutionarily conserved sleep function. Cell proliferation is one of the most fundamental biological processes, and for it to be regulated by sleep in even the most basic animals, emphasizes the deep physiological need for sleep.

Lastly, because cnidarians are transparent or semi-transparent, they are amenable to genetically encoded calcium imaging, which has already been used to great effect in *Hydra*. As discussed previously, using this approach, seven distinct spatiotemporal activity patterns have been identified that either originate or spread in different neural regions [11]. It would be every interesting to determine if the activity patterns that drive whole body muscle contractions in *Hydra* show different calcium patterns between sleep and wake. Similar analysis may be achievable in *Clytia*, as they have already been shown to have functionally subdivided diffuse nerve nets that have spatially localized subassemblies whose synchronous activation controls directional food transfer from the tentacles to the mouth [86]. Further, neural cell type markers are being identified in *Hydra* [87], and other cnidarians [88], including *Clytia* [89], and mapping the connectome to understand the neural circuits that are involved in cnidarian behavior is underway. As progress is made, we will begin to comprehend the extent to which different whole animal behaviors involve different neural types and circuits, and how sleep is organized without a centralized nervous system. This will inform our overall understanding of how sleep is controlled, and perhaps identify the simplest circuit requirements for sleep.

References

1. Erwin DH. Early metazoan life: divergence, environment and ecology. Philos Trans R Soc B Biol Sci. 2015;370(1684):20150036.
2. Finnerty JR, Pang K, Burton P, Paulson D, Martindale MQ. Origins of bilateral symmetry: Hox and dpp expression in a sea anemone. Science. 2004;304(5675):1335–7.
3. Watanabe H, Fujisawa T, Holstein TW. Cnidarians and the evolutionary origin of the nervous system. Develop Growth Differ. 2009;51(3):167–83.
4. Bosch TC, Klimovich A, Domazet-Lošo T, Gründer S, Holstein TW, Jékely G, Miller DJ, Murillo-Rincon AP, Rentzsch F, Richards GS, Schröder K. Back to the basics: cnidarians start to fire. Trends Neurosci. 2017;40(2):92–105.
5. Gold DA, Lau CLF, Fuong H, Kao G, Hartenstein V, Jacobs DK. Mechanisms of cnidocyte development in the moon jellyfish Aurelia. Evol Dev. 2019;21(2):72–81.
6. Leclère L, Röttinger E. Diversity of cnidarian muscles: function, anatomy, development and regeneration. Front Cell Dev Biol. 2017;4:157.
7. Brusca RC, Brusca GJ. Invertebrates. 2nd ed. Sunderland: Sinauer Associates; 2003.
8. Loomis WF. Glutathione control of the specific feeding reactions of hydra. Ann N Y Acad Sci. 1955;62(9):211–27.
9. Lewis C, Long TA. Courtship and reproduction in Carybdea sivickisi (Cnidaria: Cubozoa). Mar Biol. 2005;147:477–83.
10. Mackie GO. Fast pathways and escape behavior in Cnidaria. In: Eaton RC, editor. Neural mechanisms of startle behavior. Boston: Springer; 1984.
11. Szymanski JR, Yuste R. Mapping the whole-body muscle activity of Hydra vulgaris. Curr Biol. 2019;29(11):1807–17.
12. Pallasdies F, Goedeke S, Braun W, Memmesheimer RM. From single neurons to behavior in the jellyfish Aurelia aurita. elife. 2019;8:e50084.
13. Mutti C, Bernabè G, Barozzi N, Ciliento R, Trippi I, Pedrazzi G, Azzi N, Parrino L. Commonalities and differences in NREM parasomnias and sleep-related epilepsy: is there a continuum between the two conditions? Front Neurol. 2020;11:600026.
14. Abrams MJ, Zhang L, von Emster K, Lee BH, Zeigler H, Jain T, Jafri A, Chen Z, Harland RM. Sleep is required for neural network plasticity in the jellyfish Cassiopea. Curr Biol. 2023, in revision
15. Kass-Simon G, Pierobon P. Cnidarian chemical neurotransmission, an updated overview. Comp Biochem Physiol A Mol Integr Physiol. 2007;146(1):9–25.
16. Moroz LL, Kohn AB. Unbiased view of synaptic and neuronal gene complement in ctenophores: are there pan-neuronal and pan-synaptic genes across metazoa? Integr Comp Biol. 2015;55(6):1028–49.
17. Pierobon P, Concas A, Santoro G, Marino G, Minei R, Pannaccione A, Mostallino MC, Biggio G. Biochemical and functional identification of GABA receptors in Hydra vulgaris. Life Sci. 1995;56(18):1485–97.
18. Concas A, Pierobon P, Mostallino MC, Marino G, Minei R, Biggio G. Modulation of γ-aminobutyric acid (GABA) receptors and the feeding response by neurosteroids in Hydra vulgaris. Neuroscience. 1998;85(3):979–88.
19. Pierobon P, Minei R, Porcu P, Sogliano C, Tino A, Marino G, Biggio G, Concas A. Putative glycine receptors in Hydra: a biochemical and behavioural study. Eur J Neurosci. 2001;14(10):1659–66.

20. Denker E, Chatonnet A, Rabet N. Acetylcholinesterase activity in Clytia hemisphaerica (Cnidaria). Chem Biol Interact. 2008;175(1–3):125–8.
21. Talesa V, Romani R, Rosi G, Giovannini E. Presence of an acetylcholinesterase in the cnidarian Actinia equina (Anthozoa: Actiniaria) and of a thiocholine ester-hydrolyzing esterase in the sponge Spongia officinalis (Demospongiae: Keratosa). J Exp Zool. 1996;276(2):102–11.
22. Weir K, Dupre C, van Giesen L, Lee AS, Bellono NW. A molecular filter for the cnidarian stinging response. elife. 2020;9:e57578.
23. Elgoyhen AB, Katz E. The efferent medial olivocochlear-hair cell synapse. J Physiol Paris. 2012;106(1–2):47–56.
24. Abrams MJ, Francis AD, Chen H, Zeigler H, Hsu YC, Ohdera A, Harland RM. Conserved cholinergic regulation of sleep in the jellyfish *Cassiopea*. PNAS. 2023, in revision
25. Oren M, Brikner I, Appelbaum L, Levy O. Fast neurotransmission related genes are expressed in non nervous endoderm in the sea anemone Nematostella vectensis. PLoS One. 2014;9(4):e93832.
26. Faltine-Gonzalez DZ, Layden MJ. Characterization of nAChRs in Nematostella vectensis supports neuronal and non-neuronal roles in the cnidarian–bilaterian common ancestor. EvoDevo. 2019;10(1):1–18.
27. Reiter RJ, Tan DX, Osuna C, Gitto E. Actions of melatonin in the reduction of oxidative stress: a review. J Biomed Sci. 2000;7(6):444–58.
28. Srinivasan V, Maestroni GJ, Cardinali DP, Esquifino AI, Perumal SP, Miller SC. Melatonin, immune function and aging. Immun Ageing. 2005;2:1–10.
29. Gandhi AV, Mosser EA, Oikonomou G, Prober DA. Melatonin is required for the circadian regulation of sleep. Neuron. 2015;85(6):1193–9.
30. Roopin M, Levy O. Melatonin distribution reveals clues to its biological significance in basal metazoans. PLoS One. 2012;7(12):e52266.
31. Joiner WJ. Unraveling the evolutionary determinants of sleep. Curr Biol. 2016;26(20):R1073–87.
32. Borbély AA. A two-process model of sleep regulation. Hum Neurobiol. 1982;1:195–204.
33. Nath RD, Bedbrook CN, Abrams MJ, Basinger T, Bois JS, Prober DA, Sternberg PW, Gradinaru V, Goentoro L. The jellyfish Cassiopea exhibits a sleep-like state. Curr Biol. 2017;27(19):2984–90.
34. Naylor E, Bergmann BM, Krauski K, Zee PC, Takahashi JS, Vitaterna MH, Turek FW. The circadian clock mutation alters sleep homeostasis in the mouse. J Neurosci. 2000;20(21):8138–43.
35. Hendricks JC, Lu S, Kume K, Yin JCP, Yang Z, Sehgal A. Gender dimorphism in the role of cycle (BMAL1) in rest, rest regulation, and longevity in Drosophila melanogaster. J Biol Rhythm. 2003;18(1):12–25.
36. Trojanowski NF, Raizen DM. Call it worm sleep. Trends Neurosci. 2016;39(2):54–62.
37. Kanaya HJ, Park S, Kim JH, Kusumi J, Krenenou S, Sawatari E, Sato A, Lee J, Bang H, Kobayakawa Y, Lim C. A sleep-like state in Hydra unravels conserved sleep mechanisms during the evolutionary development of the central nervous system. Sci Adv. 2020;6(41):eabb9415.
38. Garm A, Bielecki J, Petie R, Nilsson DE. Opposite patterns of diurnal activity in the box jellyfish Tripedalia cystophora and Copula sivickisi. Biol Bull. 2012;222(1):35–45.
39. Grace S. Winter quiescence, growth rate, and the release from competition in the temperate scleractinian coral Astrangia poculata (Ellis & Solander 1786). Northeast Nat. 2017;24(sp7):B119–34.
40. Campbell SS, Tobler I. Animal sleep: a review of sleep duration across phylogeny. Neurosci Biobehav Rev. 1984;8(3):269–300.
41. Greenspan RJ, Tononi G, Cirelli C, Shaw PJ. Sleep and the fruit fly. Trends Neurosci. 2001;24(3):142–5.
42. Zhdanova IV, Wang SY, Leclair OU, Danilova NP. Melatonin promotes sleep-like state in zebrafish. Brain Res. 2001;903(1–2):263–8.
43. Singh K, Chao MY, Somers GA, Komatsu H, Corkins ME, Larkins-Ford J, Tucey T, Dionne HM, Walsh MB, Beaumont EK, Hart DP. C. elegans Notch signaling regulates adult chemosensory response and larval molting quiescence. Curr Biol. 2011;21(10):825–34.

44. Singh K, Ju JY, Walsh MB, DiIorio MA, Hart AC. Deep conservation of genes required for both Drosophila melanogaster and Caenorhabditis elegans sleep includes a role for dopaminergic signaling. Sleep. 2014;37(9):1439–51.
45. Roelofs K. Freeze for action: neurobiological mechanisms in animal and human freezing. Philo Trans R Soc B Biol Sci. 2017;372(1718):20160206.
46. Peres R, Reitzel AM, Passamaneck Y, Afeche SC, Cipolla-Neto J, Marques AC, Martindale MQ. Developmental and light-entrained expression of melatonin and its relationship to the circadian clock in the sea anemone Nematostella vectensis. EvoDevo. 2014;5(1):1–18.
47. Anctil M. Chemical transmission in the sea anemone Nematostella vectensis: a genomic perspective. Comp Biochem Physiol Part D Genomics Proteomics. 2009;4(4):268–89.
48. Donlea J, Leahy A, Thimgan MS, Suzuki Y, Hughson BN, Sokolowski MB, Shaw PJ. Foraging alters resilience/vulnerability to sleep disruption and starvation in Drosophila. Proc Natl Acad Sci. 2012;109(7):2613–8.
49. Raizen DM, Zimmerman JE, Maycock MH, Ta UD, You YJ, Sundaram MV, Pack AI. Lethargus is a Caenorhabditis elegans sleep-like state. Nature. 2008;451(7178):569–72.
50. Langmesser S, Franken P, Feil S, Emmenegger Y, Albrecht U, Feil R. cGMP-dependent protein kinase type I is implicated in the regulation of the timing and quality of sleep and wakefulness. PLoS One. 2009;4(1):e4238.
51. Lancel M. Role of GABAA receptors in the regulation of sleep: initial sleep responses to peripherally administered modulators and agonists. Sleep. 1999;22(1):33–42.
52. Shi M, Yue Z, Kuryatov A, Lindstrom JM, Sehgal A. Identification of Redeye, a new sleep-regulating protein whose expression is modulated by sleep amount. Elife 2014;3:e01473.
53. Shen JX, Yakel JL. Nicotinic acetylcholine receptor-mediated calcium signaling in the nervous system. Acta Pharmacol Sin. 2009;30(6):673–80.
54. Berridge MJ, Bootman MD, Roderick HL. Calcium signalling: dynamics, homeostasis and remodelling. Nat Rev Mol Cell Biol. 2003;4(7):517–29.
55. Saper CB, Fuller PM, Pedersen NP, Lu J, Scammell TE. Sleep state switching. Neuron. 2010;68(6):1023–42.
56. Nichols AL, Eichler T, Latham R, Zimmer M. A global brain state underlies C. elegans sleep behavior. Science. 2017;356(6344):eaam6851.
57. Zielinski MR, McKenna JT, McCarley RW. Functions and mechanisms of sleep. AIMS Neurosci. 2016;3(1):67.
58. Tononi G, Cirelli C. Sleep function and synaptic homeostasis. Sleep Med Rev. 2006;10(1):49–62.
59. Shaw PJ, Cirelli C, Greenspan RJ, Tononi G. Correlates of sleep and waking in Drosophila melanogaster. Science. 2000;287(5459):1834–7.
60. Eugene AR, Masiak J. The neuroprotective aspects of sleep. MEDtube Sci. 2015;3(1):35.
61. Walker MP, Stickgold R. Sleep-dependent learning and memory consolidation. Neuron. 2004;44(1):121–33.
62. Tononi G, Cirelli C. Sleep and the price of plasticity: from synaptic and cellular homeostasis to memory consolidation and integration. Neuron. 2014;81(1):12–34.
63. Frank MG, Cantera R. Sleep, clocks, and synaptic plasticity. Trends Neurosci. 2014;37(9):491–501.
64. Von Bernhardi R, Bernhardi LEV, Eugenín J. What is neural plasticity? In: The plastic brain. Adv Exp Med Biol. 2017;1015:1–15. https://doi.org/10.1007/978-3-319-62817-2_1
65. Mohler PJ, Hund TJ. Novel pathways for regulation of sinoatrial node plasticity and heart rate. Circ Res. 2017;121(9):1027–8.
66. Ördög T, Baldo M, Danko R, Sanders KM. Plasticity of electrical pacemaking by interstitial cells of Cajal and gastric dysrhythmias in W/Wv mutant mice. Gastroenterology. 2002;123(6):2028–40.
67. Tononi G, Cirelli C. Sleep and synaptic homeostasis: a hypothesis. Brain Res Bull. 2003;62(2):143–50.
68. Brown RE, Basheer R, McKenna JT, Strecker RE, McCarley RW. Control of sleep and wakefulness. Physiol Rev. 2012;92(3):1087–187.

69. Porkka-Heiskanen T, Strecker RE, Thakkar M, Bjørkum AA, Greene RW, McCarley RW. Adenosine: a mediator of the sleep-inducing effects of prolonged wakefulness. Science. 1997;276(5316):1265–8.
70. Shaw PJ, Tononi G, Greenspan RJ, Robinson DF. Stress response genes protect against lethal effects of sleep deprivation in Drosophila. Nature. 2002;417(6886):287–91.
71. Donlea JM, Pimentel D, Miesenböck G. Neuronal machinery of sleep homeostasis in Drosophila. Neuron. 2014;81(4):860–72.
72. Bell-Pedersen D, Cassone VM, Earnest DJ, Golden SS, Hardin PE, Thomas TL, Zoran MJ. Circadian rhythms from multiple oscillators: lessons from diverse organisms. Nat Rev Genet. 2005;6(7):544–56.
73. Takahashi JS. Transcriptional architecture of the mammalian circadian clock. Nat Rev Genet. 2017;18(3):164–79.
74. Huang W, Ramsey KM, Marcheva B, Bass J. Circadian rhythms, sleep, and metabolism. J Clin Invest. 2011;121(6):2133–41.
75. Vize PD. Transcriptome analysis of the circadian regulatory network in the coral Acropora millepora. Biol Bull. 2009;216(2):131–7.
76. Reitzel AM, Behrendt L, Tarrant AM. Light entrained rhythmic gene expression in the sea anemone Nematostella vectensis: the evolution of the animal circadian clock. PLoS One. 2010;5(9):e12805.
77. Reitzel AM, Tarrant AM, Levy O. Circadian clocks in the cnidaria: environmental entrainment, molecular regulation, and organismal outputs. Integr Comp Biol. 2013;53(1):118–30.
78. Brady AK, Snyder KA, Vize PD. Circadian cycles of gene expression in the coral, Acropora millepora. PLoS One. 2011;6(9):e25072.
79. Jindrich K, Roper KE, Lemon S, Degnan BM, Reitzel AM, Degnan SM. Origin of the animal circadian clock: diurnal and light-entrained gene expression in the sponge Amphimedon queenslandica. Front Mar Sci. 2017;4:327.
80. Jaggard JB, Stahl BA, Lloyd E, Prober DA, Duboue ER, Keene AC. Hypocretin underlies the evolution of sleep loss in the Mexican cavefish. elife. 2018;7:e32637.
81. Mascetti GG. Unihemispheric sleep and asymmetrical sleep: behavioral, neurophysiological, and functional perspectives. Nat Sci Sleep. 2016;8:221–38.
82. Ohdera AH, Abrams MJ, Ames CL, Baker DM, Suescún-Bolívar LP, Collins AG, Freeman CJ, Gamero-Mora E, Goulet TL, Hofmann DK, Jaimes-Becerra A. Upside-down but headed in the right direction: review of the highly versatile Cassiopea xamachana system. Front Ecol Evol. 2018;6:35.
83. Vyazovskiy VV, Olcese U, Hanlon EC, Nir Y, Cirelli C, Tononi G. Local sleep in awake rats. Nature. 2011;472(7344):443–7.
84. Szuperak M, Churgin MA, Borja AJ, Raizen DM, Fang-Yen C, Kayser MS. A sleep state in Drosophila larvae required for neural stem cell proliferation. elife. 2018;7:e33220.
85. Guzmán-Marín R, Suntsova N, Stewart DR, Gong H, Szymusiak R, McGinty D. Sleep deprivation reduces proliferation of cells in the dentate gyrus of the hippocampus in rats. J Physiol. 2003;549(2):563–71.
86. Weissbourd B, Momose T, Nair A, Kennedy A, Hunt B, Anderson DJ. A genetically tractable jellyfish model for systems and evolutionary neuroscience. Cell. 2021;184(24):5854–68.
87. Badhiwala KN, Primack AS, Juliano CE, Robinson JT. Multiple neuronal networks coordinate Hydra mechanosensory behavior. elife. 2021;10:e64108.
88. Sprecher SG. Neural cell type diversity in cnidaria. Front Neurosci. 2022;16:909400.
89. Chari T, Weissbourd B, Gehring J, Ferraioli A, Leclère L, Herl M, Gao F, Chevalier S, Copley RR, Houliston E, Anderson DJ. Whole-animal multiplexed single-cell RNA-seq reveals transcriptional shifts across Clytia medusa cell types. Sci Adv. 2021;7(48):eabh1683.

Part II
Human Sleep and Sleep Disorders

Approaches to Studying Human Genetics

Olivia J. Veatch

Abstract It is well-established that genetic factors influence sleep-related traits and chronotype—which is the behavioral manifestation of the genetically encoded molecular oscillator known as the circadian clock (Gehrman et al. Psychiatr Clin North Am 38:667–681, 2015; Rosensweig and Green Eur J Neurosci 51:139–165, 2020; Veatch et al. Lancet Neurol 16:158–170, 2017). Advances in molecular and computational analysis methods, along with the evolution of biobanks in recent years, have provided an explosion of knowledge into the underlying genetic mechanisms contributing to variability in sleep/wake behavior as well as expression of sleep disorders (Cade et al. Genome Med 13:136, 2021; Dashti et al. Nat Commun 10:1100, 2019a; Jasen et al. Nat Genet 51:394–403, 2019; Lane et al. Nat Genet 49:274–281, 2017; Wang et al. Nat Commun 10:3503, 2019). However, several technical (and ethical) challenges remain that must be considered when undertaking human genetic studies. These include the need for a solid understanding of the assumptions being made with the various approaches implemented and the limitations of these approaches that impede the translation of findings. This chapter will offer a broad overview of approaches currently applied in the field of genetic epidemiology as well as background information that should inform future studies aimed at deciphering the complicated genetic architecture underlying expression of sleep and circadian traits in humans.

Keywords Phenotype definitions · Genetic epidemiology · Linkage analysis · Association studies · Ancestry

---

O. J. Veatch (✉)
Department of Psychiatry and Behavioral Sciences, University of Kansas Medical Center, Kansas City, Kansas, USA
e-mail: oveatch@kumc.edu

P. Gehrman et al. (eds.), *Genetics of Sleep and Sleep Disorders*,
https://doi.org/10.1007/978-3-031-62723-1_8

Defining Phenotypes

One would be remiss to begin a chapter describing approaches to genetic analyses of human traits without first highlighting the importance of phenotyping. It is unlikely that simplistic phenotype-genotype relationships exist where one altered gene has a singular and replicable causal effect on expression of a complex human trait. When planning an analysis, it is crucial to contemplate which phenotype definitions are more suitable to studying genetic effects. A phenotype is broadly defined as an observable characteristic of an organism. In genetic studies of sleep and sleep disorders, this could be the presence or absence of some behavior, like evening chronotype, or of a diagnostically defined disorder, like obstructive sleep apnea. Defining the phenotype that will be evaluated for a statistical relationship with genetic variation is central to the study design. This impacts both the interpretation of results as well as the power of the study—or the probability of detecting a true effect given it exists. Some general concepts to ponder when defining the phenotype to evaluate in a genetic analysis include (1) most diagnoses are based on, and influenced by, multiple biological traits, (2) many phenotypes are more appropriately modeled as quantitative versus discrete traits, and (3) as phenotype definitions become more reliant on dynamic human behavior, the influence of genotype becomes less direct. Another important point of consideration is that ultimately an assumption is being made that the phenotype being tested for associations with genetic factors is the expression of underlying biology. To help ensure this assumption is met, it is useful to conduct analyses that provide evidence supporting this expectation.

Determining Genetic Contributions to Phenotypes

There are many approaches available to help determine if a phenotype has a genetic component. This includes assessing if a trait or disorder aggregates in families. In essence, an investigator with access to phenotype information from related individuals, that spans multiple generations of a pedigree, can ascertain if there is a family history of a particular trait that is statistically significant. Other useful methods to assess a statistically significant family history include calculating the attributable fraction, or determining if a quantitative trait measurement correlates among related individuals [28]. Calculating the familial recurrence risk can also inform an investigator if evidence for genetic influences exist, by testing if there is increased risk for some trait in relatives when compared to unrelated individuals [38–40]. Heritability estimates can also be calculated to help determine the amount of variability in a trait that can be attributed to genetics when compared to other factors like the environment [50]. While the afore mentioned approaches may provide support for the hypothesis that a phenotype is influenced by genomic variation, methods to assess trait segregation in families do not completely remove effects that may be attributable to common environment. Twin and adoption studies can help tease apart genetic versus environmental influences on sleep-related traits [33, 52]; however, to ultimately determine if phenotypes are genetic, it is beneficial to conduct studies aimed at identifying specific genetic variants that are linked or associated with the trait.

Identifying Trait-Associated Genetic Factors

Genetic studies in humans have historically focused on generating sequencing or genotyping data using DNA samples extracted from biospecimens to identify markers that vary in a sample population and have statistical evidence for a relationship with a phenotypic trait that also varies in the population. These studies arguably fall under two broad categories of linkage and association analyses. Linkage analysis of a trait evaluates if any specific chromosomal position, or loci, where a genetic marker (e.g., Single Nucleotide Variant, Copy Number Variant) is present co-segregates with expression of the trait within families—usually tracking the trait and variant to determine if it transmits from parents to offspring. Association studies can be conducted in either families or unrelated individuals to determine if genetic markers are more frequently observed in individuals with some measurable expression of a phenotype compared to individuals with distinct or absent expression of the phenotype. Linkage and association studies can be used to evaluate both dichotomous (case-control) and quantitative (QTL) trait definitions. Most studies either focus on evaluating statistical relationships between traits and genetic markers in specific genes already hypothesized to influence trait expression, or all available markers across the human genome which allows investigators to generate hypotheses regarding genes of interest to further study.

Candidate Gene and Genome-Wide Interrogations

Prior to the generation of a reference human genome and advent of genome-wide technologies for genotyping and sequencing, most human genetic studies focused on candidate genes with a priori knowledge of a biological relationship to the phenotype of interest. For example, genetic analyses of human sleep often focused on genes encoding proteins that had evidence for influencing sleep/wake cycles (e.g., circadian rhythm regulation, orexinergic/hypocretinergic transmissions, melatonergic transmissions) [16, 18]. While incorporating biological knowledge into genetic studies is important, candidate gene approaches are potentially biased as these rely on preselecting genomic regions to investigate. This limits discovery of novel factors influencing phenotype expression. Genome-wide interrogations have offered the opportunity to advance the understanding of genetic mechanisms underlying sleep by providing additional evidence for previously implicated mechanisms as well as identify novel mechanisms. These methods may offer a less biased evaluation of the human genome compared to candidate gene approaches, but they are not without limitations. These include financial constraints, as technologies used to generate whole-genome genotyping and sequencing data are relatively expensive compared to targeted genotyping/sequencing. In addition, the cost to store these data once generated can be a limiting factor. Furthermore, to be well-powered, genome-wide analyses require large datasets that include information collected on thousands of individuals. High quality, accurate phenotypes can be difficult to obtain in these large samples. Results from genome-wide analyses can also be difficult to interpret as an association between a genetic marker and a trait can reflect a direct relationship, or an indirect relationship that is detected because there is allelic association between the marker and another region of the genome, often referred to as linkage disequilibrium (LD).

Influences of Genetic Ancestry

Notably, patterns of LD can vary among individuals from different ancestral backgrounds which can result in population stratification confounding associations between a genotype and the trait of interest [22]. Variability in allele frequencies and LD patterns among human ancestral populations can also result in lack of generalizability of findings across populations with diverse genetic backgrounds [20, 37]. When considering the impact ancestry can have on identification and interpretation of genetic markers associated with human phenotypes, it is also important to recognize that most genetically informed definitions of ancestry are global—using proportions averaged across the entire genome—and based on comparisons with genomes collected from a subset of individuals sampled from specific geographic areas [14, 17, 24, 42]. In other words, ancestry estimates for any individual included in an analysis dataset are only as accurate as the reference population that is used to infer this information. This also impacts the accuracy of predicting genetic markers that are not directly genotyped or sequenced via imputation, as these predictions are based on what is known about LD patterns in the ancestral reference populations [35]. Ancestry is also important to keep in mind when interpreting the significance of results from genome-wide association studies (GWAS). Of note, the *p*-value threshold currently considered genome-wide significant ($p = 5.0 \times 10^{-8}$) is based on a Bonferroni correction using the estimated number of independent common variants observed across European ancestral genomes [4, 13, 25]. A number of additional features of the human genome further complicate interpretation of results from statistical genetic analyses; however, many approaches have been, and are continually being, developed to help identify biologically and clinically relevant trait-associated genetic factors.

Approaches to Aid with Interpretation of Results

Among the other features that can affect interpretation of results from GWAS is steadily accumulating evidence indicating assignment of disease-associated variants to the genes in which they reside, or those genes encoded in the closest genomic regions, may not be the most accurate way to predict functional relationships [5, 44]. Indeed, the human genome contains a number of long-range and trans-regulatory elements that can be predicted by leveraging data available via RNA-sequencing experiments aimed at identifying expression quantitative trait loci [15, 27, 31, 36, 54] and chromosome confirmation capture techniques (e.g., Hi-C, CHIP-Seq) aimed at predicting open regions of chromatin and topologically associating domains [32, 47]. Once trait-associated variants have been assigned to the most appropriate effector genes, further analyses can then be conducted to help translate results into biological and clinical knowledge.

For instance, individual effects of most genomic variants identified in human genetic studies on expression of phenotypic traits are modest [34]. These smaller effects can easily be overlooked in a typical approach to analysis of genome-wide data, which aims to identify the most significantly associated individual common variants or rare variants binned into the same biologically relevant genomic region. In reality, genes often work as complex interacting networks, and it is expected that genomic variation has cumulative effects on human health. Polygenic risk scores which calculate the estimated effect of many genetic variants associated with a phenotype [7] have been useful to exploring whether the accumulation of genetic risk factors underlies variable expressivity of sleep-related traits [10, 19]. Pathway-based analysis is another approach that considers the combined effects of multiple genomic variants functioning together [6, 8, 53]. These analyses offer the potential to help overcome phenotypic and genetic heterogeneity, implicate entire genetic mechanisms (e.g., biochemical pathways, biological processes, cellular/molecular mechanisms, and protein-protein interaction networks), and give biological insights that would otherwise be imperceptible [1, 23, 30]. Ultimately, pathway-based approaches are powerful methods to help overcome issues arising due to underlying genetic heterogeneity that offer the opportunity to expand the utility of genome-wide data and allow for more biologically relevant interpretations of results from these studies.

In addition, most complex traits—like sleep and sleep disorders—exhibit variable expressivity and many individuals diagnosed with a particular sleep disorder (e.g., insomnia, obstructive sleep apnea) will also present with a broad array of co-occurring conditions. It is becoming apparent that pleiotropic genetic effects are abundant in human genomes indicating the same genetic mechanisms influence risk for multiple clinically distinct disorders [43, 49]. Various approaches exist that can help detect and characterize pleiotropic effects connecting seemingly distinct complex disorders. These include LD score regression which examines the relationship between results from GWAS conducted on separate phenotypes, while incorporating LD structure, to determine if genetic risk factors for two distinct traits are correlated [2]. Another approach for detecting pleiotropy includes the Phenome-wide Association Study (PheWAS) which tests if specific genetic markers are associated with multiple conditions and traits [12]. PheWAS requires access to large datasets where extensive phenotyping data are available for each individual, such as electronic health record-derived data [48]. Phenome-wide analyses provide evidence of pleiotropy which can present in many forms. This includes when variation in one gene influences multiple traits, or when multiple traits cannot be directly connected to one gene. For example, it is possible a genetic variant will influence expression of one phenotype that subsequently results in expression of a second phenotype [45]. A computational approach that can help tease apart different pleiotropic effects is Mendelian randomization. This method tests the hypothesis that one trait is an intermediate phenotype and causes another trait. To infer causality using this approach, a genetic variant must be strongly associated with the intermediate phenotype of interest but not the outcome or any other confounders [11]. These and other post hoc analysis methods offer powerful tools useful for interpreting results from genetic studies. These efforts in turn will further the understanding of how genetic data can move the field towards the eventual goal of genomic driven medical care for individuals with sleep and circadian disturbances.

Conclusions

It is difficult to offer in a single chapter a thorough understanding of the many factors that should be considered when designing human genetic studies. For the reader who is intrigued and interested in performing a deeper dive into this literature, there are excellent resources available that offer more details on the various concepts presented here [21, 46]. While genomic technologies and analysis methods are rapidly advancing, many basic concepts necessary to conduct a powerful and informative genetic analysis will continue to remain valid. It should be noted that human genetic studies of sleep conducted thus far have often been required to focus on phenotype definitions that relied on potentially inaccurate self-reported traits in order to gain power through large sample sizes. To fully elucidate genetic factors contributing to sleep and sleep disorders, future work that is focused on determining more optimal ways to define these phenotypes would be beneficial. For implicated genes and genomic regions, in vivo and in vitro follow-up studies are also necessary to better understand which genes are being implicated via genome-wide analyses and the functional effects of genomic variation associated with human sleep-related traits.

References

1. Askland K, Read C, Moore J. Pathways-based analyses of whole-genome association study data in bipolar disorder reveal genes mediating ion channel activity and synaptic neurotransmission. Hum Genet. 2009;125(1):63–79. https://doi.org/10.1007/s00439-008-0600-y.
2. Bulik-Sullivan BK, Loh PR, Finucane HK, Ripke S, Yang J, Schizophrenia Working Group of the Psychiatric Genomics C, Patterson N, Daly MJ, Price AL, Neale BM. LD Score regression distinguishes confounding from polygenicity in genome-wide association studies. Nat Genet. 2015;47(3):291–5. https://doi.org/10.1038/ng.3211.
3. Cade BE, Lee J, Sofer T, Wang H, Zhang M, Chen H, Gharib SA, Gottlieb DJ, Guo X, Lane JM, Liang J, Lin X, Mei H, Patel SR, Purcell SM, Saxena R, Shah NA, Evans DS, Hanis CL, et al. Whole-genome association analyses of sleep-disordered breathing phenotypes in the NHLBI TOPMed program. Genome Med. 2021;13(1):136. https://doi.org/10.1186/s13073-021-00917-8.
4. Chen Z, Boehnke M, Wen X, Mukherjee B. Revisiting the genome-wide significance threshold for common variant GWAS. G3 (Bethesda). 2021;11(2) https://doi.org/10.1093/g3journal/jkaa056.
5. Chesi A, Wagley Y, Johnson ME, Manduchi E, Su C, Lu S, Leonard ME, Hodge KM, Pippin JA, Hankenson KD, Wells AD, Grant SFA. Genome-scale Capture C promoter interactions implicate effector genes at GWAS loci for bone mineral density. Nat Commun. 2019;10(1):1260. https://doi.org/10.1038/s41467-019-09302-x.
6. Chimusa ER, Dalvie S, Dandara C, Wonkam A, Mazandu GK. Post genome-wide association analysis: dissecting computational pathway/network-based approaches. Brief Bioinform. 2019;20(2):690–700. https://doi.org/10.1093/bib/bby035.
7. Choi SW, Mak TS, O'Reilly PF. Tutorial: a guide to performing polygenic risk score analyses. Nat Protoc. 2020;15(9):2759–72. https://doi.org/10.1038/s41596-020-0353-1.
8. Cirillo E, Parnell LD, Evelo CT. A review of pathway-based analysis tools that visualize genetic variants. Front Genet. 2017;8:174. https://doi.org/10.3389/fgene.2017.00174.
9. Dashti HS, Jones SE, Wood AR, Lane JM, van Hees VT, Wang H, Rhodes JA, Song Y, Patel K, Anderson SG, Beaumont RN, Bechtold DA, Bowden J, Cade BE, Garaulet M, Kyle SD, Little MA, Loudon AS, Luik AI, et al. Genome-wide association study identifies genetic loci for self-reported habitual sleep duration supported by accelerometer-derived estimates. Nat Commun. 2019a;10(1):1100. https://doi.org/10.1038/s41467-019-08917-4.
10. Dashti HS, Redline S, Saxena R. Polygenic risk score identifies associations between sleep duration and diseases determined from an electronic medical record biobank. Sleep. 2019b;42(3) https://doi.org/10.1093/sleep/zsy247.
11. Davey Smith G, Hemani G. Mendelian randomization: genetic anchors for causal inference in epidemiological studies. Hum Mol Genet. 2014;23(R1):R89–98. https://doi.org/10.1093/hmg/ddu328.
12. Denny JC, Ritchie MD, Basford MA, Pulley JM, Bastarache L, Brown-Gentry K, Wang D, Masys DR, Roden DM, Crawford DC. PheWAS: demonstrating the feasibility of a phenome-wide scan to discover gene-disease associations. Bioinformatics. 2010;26(9):1205–10. https://doi.org/10.1093/bioinformatics/btq126.
13. Fadista J, Manning AK, Florez JC, Groop L. The (in)famous GWAS P-value threshold revisited and updated for low-frequency variants. Eur J Hum Genet. 2016;24(8):1202–5. https://doi.org/10.1038/ejhg.2015.269.
14. Fairley S, Lowy-Gallego E, Perry E, Flicek P. The International Genome Sample Resource (IGSR) collection of open human genomic variation resources. Nucleic Acids Res. 2020;48(D1):D941–7. https://doi.org/10.1093/nar/gkz836.
15. Gamazon ER, Wheeler HE, Shah KP, Mozaffari SV, Aquino-Michaels K, Carroll RJ, Eyler AE, Denny JC, Consortium GT, Nicolae DL, Cox NJ, Im HK. A gene-based association method for mapping traits using reference transcriptome data. Nat Genet. 2015;47(9):1091–8. https://doi.org/10.1038/ng.3367.

16. Gehrman PR, Keenan BT, Byrne EM, Pack AI. Genetics of sleep disorders. Psychiatr Clin North Am. 2015;38(4):667–81. https://doi.org/10.1016/j.psc.2015.07.004.
17. GenomeAsia KC. The GenomeAsia 100K project enables genetic discoveries across Asia. Nature. 2019;576(7785):106–11. https://doi.org/10.1038/s41586-019-1793-z.
18. Goel N. Genetic markers of sleep and sleepiness. Sleep Med Clin. 2017;12(3):289–99. https://doi.org/10.1016/j.jsmc.2017.03.005.
19. Goodman MO, Cade BE, Shah NA, Huang T, Dashti HS, Saxena R, Rutter MK, Libby P, Sofer T, Redline S. Pathway-specific polygenic risk scores identify obstructive sleep apnea-related pathways differentially moderating genetic susceptibility to coronary artery disease. Circ Genom Precis Med. 2022;15(5):e003535. https://doi.org/10.1161/CIRCGEN.121.003535.
20. Greene CS, Penrod NM, Williams SM, Moore JH. Failure to replicate a genetic association may provide important clues about genetic architecture. PLoS One. 2009;4(6):e5639. https://doi.org/10.1371/journal.pone.0005639.
21. Haines JLP-VMA. Genetic analysis of complex diseases. 2nd ed. Wiley-Liss; 2006.
22. Hellwege JN, Keaton JM, Giri A, Gao X, Velez Edwards DR, Edwards TL. Population stratification in genetic association studies. Curr Protoc Hum Genet. 2017;95:1–22. https://doi.org/10.1002/cphg.48.
23. Holmans P, Green EK, Pahwa JS, Ferreira MA, Purcell SM, Sklar P, Wellcome Trust Case-Control, C, Owen MJ, O'Donovan MC, Craddock N. Gene ontology analysis of GWA study data sets provides insights into the biology of bipolar disorder. Am J Hum Genet. 2009;85(1):13–24. https://doi.org/10.1016/j.ajhg.2009.05.011.
24. International HapMap C. The international HapMap project. Nature. 2003;426(6968):789–96. https://doi.org/10.1038/nature02168.
25. International HapMap C. A haplotype map of the human genome. Nature. 2005;437(7063):1299–320. https://doi.org/10.1038/nature04226.
26. Jansen PR, Watanabe K, Stringer S, Skene N, Bryois J, Hammerschlag AR, de Leeuw CA, Benjamins JS, Munoz-Manchado AB, Nagel M, Savage JE, Tiemeier H, White T, Me Research T, Tung JY, Hinds DA, Vacic V, Wang X, Sullivan PF, et al. Genome-wide analysis of insomnia in 1,331,010 individuals identifies new risk loci and functional pathways. Nat Genet. 2019;51(3):394–403. https://doi.org/10.1038/s41588-018-0333-3.
27. Jhamb D, Magid-Slav M, Hurle MR, Agarwal P. Pathway analysis of GWAS loci identifies novel drug targets and repurposing opportunities. Drug Discov Today. 2019;24(6):1232–6. https://doi.org/10.1016/j.drudis.2019.03.024.
28. Khoury MJ, Flanders WD. Nontraditional epidemiologic approaches in the analysis of gene-environment interaction: case-control studies with no controls! Am J Epidemiol. 1996;144(3):207–13. https://doi.org/10.1093/oxfordjournals.aje.a008915.
29. Lane JM, Liang J, Vlasac I, Anderson SG, Bechtold DA, Bowden J, Emsley R, Gill S, Little MA, Luik AI, Loudon A, Scheer FA, Purcell SM, Kyle SD, Lawlor DA, Zhu X, Redline S, Ray DW, Rutter MK, Saxena R. Genome-wide association analyses of sleep disturbance traits identify new loci and highlight shared genetics with neuropsychiatric and metabolic traits. Nat Genet. 2017;49(2):274–81. https://doi.org/10.1038/ng.3749.
30. Lesnick TG, Papapetropoulos S, Mash DC, Ffrench-Mullen J, Shehadeh L, de Andrade M, Henley JR, Rocca WA, Ahlskog JE, Maraganore DM. A genomic pathway approach to a complex disease: axon guidance and Parkinson disease. PLoS Genet. 2007;3(6):e98. https://doi.org/10.1371/journal.pgen.0030098.
31. Li B, Verma SS, Veturi YC, Verma A, Bradford Y, Haas DW, Ritchie MD. Evaluation of PrediXcan for prioritizing GWAS associations and predicting gene expression. Pac Symp Biocomput. 2018a;23:448–59. https://www.ncbi.nlm.nih.gov/pubmed/29218904
32. Li T, Jia L, Cao Y, Chen Q, Li C. OCEAN-C: mapping hubs of open chromatin interactions across the genome reveals gene regulatory networks. Genome Biol. 2018b;19(1):54. https://doi.org/10.1186/s13059-018-1430-4.

33. Madrid-Valero JJ, Rubio-Aparicio M, Gregory AM, Sanchez-Meca J, Ordonana JR. The heritability of insomnia: systematic review and meta-analysis of twin studies. Sleep Med Rev. 2021;58:101437. https://doi.org/10.1016/j.smrv.2021.101437.
34. Marigorta UM, Rodriguez JA, Gibson G, Navarro A. Replicability and prediction: lessons and challenges from GWAS. Trends Genet. 2018;34(7):504–17. https://doi.org/10.1016/j.tig.2018.03.005.
35. Naj AC. Genotype imputation in genome-wide association studies. Curr Protoc Hum Genet. 2019;102(1):e84. https://doi.org/10.1002/cphg.84.
36. Pasman JA, Verweij KJH, Gerring Z, Stringer S, Sanchez-Roige S, Treur JL, Abdellaoui A, Nivard MG, Baselmans BML, Ong JS, Ip HF, van der Zee MD, Bartels M, Day FR, Fontanillas P, Elson SL, Me Research T, de Wit H, Davis LK, et al. GWAS of lifetime cannabis use reveals new risk loci, genetic overlap with psychiatric traits, and a causal influence of schizophrenia. Nat Neurosci. 2018;21(9):1161–70. https://doi.org/10.1038/s41593-018-0206-1.
37. Popejoy AB, Fullerton SM. Genomics is failing on diversity. Nature. 2016;538(7624):161–4. https://doi.org/10.1038/538161a.
38. Risch N. Linkage strategies for genetically complex traits. I. Multilocus models. Am J Hum Genet. 1990a;46(2):222–8. https://www.ncbi.nlm.nih.gov/pubmed/2301392
39. Risch N. Linkage strategies for genetically complex traits. II. The power of affected relative pairs. Am J Hum Genet. 1990b;46(2):229–41. https://www.ncbi.nlm.nih.gov/pubmed/2301393
40. Risch N. Linkage strategies for genetically complex traits. III. The effect of marker polymorphism on analysis of affected relative pairs. Am J Hum Genet. 1990c;46(2):242–53. https://www.ncbi.nlm.nih.gov/pubmed/2301394
41. Rosensweig C, Green CB. Periodicity, repression, and the molecular architecture of the mammalian circadian clock. Eur J Neurosci. 2020;51(1):139–65. https://doi.org/10.1111/ejn.14254.
42. Sirugo G, Tishkoff SA, Williams SM. The quagmire of race, genetic ancestry, and health disparities. J Clin Invest. 2021;131(11) https://doi.org/10.1172/JCI150255.
43. Sivakumaran S, Agakov F, Theodoratou E, Prendergast JG, Zgaga L, Manolio T, Rudan I, McKeigue P, Wilson JF, Campbell H. Abundant pleiotropy in human complex diseases and traits. Am J Hum Genet. 2011;89(5):607–18. https://doi.org/10.1016/j.ajhg.2011.10.004.
44. Smemo S, Tena JJ, Kim KH, Gamazon ER, Sakabe NJ, Gomez-Marin C, Aneas I, Credidio FL, Sobreira DR, Wasserman NF, Lee JH, Puviindran V, Tam D, Shen M, Son JE, Vakili NA, Sung HK, Naranjo S, Acemel RD, et al. Obesity-associated variants within FTO form long-range functional connections with IRX3 [Research Support, N.I.H., Extramural Research Support, Non-U.S. Gov't]. Nature. 2014;507(7492):371–5. https://doi.org/10.1038/nature13138.
45. Tyler AL, Crawford DC, Pendergrass SA. The detection and characterization of pleiotropy: discovery, progress, and promise. Brief Bioinform. 2016;17(1):13–22. https://doi.org/10.1093/bib/bbv050.
46. Uffelmann E, Huang QQ, Munung NS, de Vries J, Okada Y, Martin AR, Martin HC, Lappalainen T, Posthuma D. Genome-wide association studies. Nat Rev Methods Primers. 2021;1(1):59. https://doi.org/10.1038/s43586-021-00056-9.
47. Ulianov SV, Khrameeva EE, Gavrilov AA, Flyamer IM, Kos P, Mikhaleva EA, Penin AA, Logacheva MD, Imakaev MV, Chertovich A, Gelfand MS, Shevelyov YY, Razin SV. Active chromatin and transcription play a key role in chromosome partitioning into topologically associating domains. Genome Res. 2016;26(1):70–84. https://doi.org/10.1101/gr.196006.115.
48. Veatch OJ, Bauer CR, Keenan BT, Josyula NS, Mazzotti DR, Bagai K, Malow BA, Robishaw JD, Pack AI, Pendergrass SA. Characterization of genetic and phenotypic heterogeneity of obstructive sleep apnea using electronic health records. BMC Med Genet. 2020;13(1):105. https://doi.org/10.1186/s12920-020-00755-4.
49. Veatch OJ, Keenan BT, Gehrman PR, Malow BA, Pack AI. Pleiotropic genetic effects influencing sleep and neurological disorders. Lancet Neurol. 2017;16(2):158–70. https://doi.org/10.1016/S1474-4422(16)30339-8.
50. Visscher PM, Hill WG, Wray NR. Heritability in the genomics era—concepts and misconceptions. Nat Rev. Genet. 2008;9(4):255–66. https://doi.org/10.1038/nrg2322.

51. Wang H, Lane JM, Jones SE, Dashti HS, Ollila HM, Wood AR, van Hees VT, Brumpton B, Winsvold BS, Kantojarvi K, Palviainen T, Cade BE, Sofer T, Song Y, Patel K, Anderson SG, Bechtold DA, Bowden J, Emsley R, et al. Genome-wide association analysis of self-reported daytime sleepiness identifies 42 loci that suggest biological subtypes. Nat Commun. 2019;10(1):3503. https://doi.org/10.1038/s41467-019-11456-7.
52. Wang Z, Yang W, Li X, Qi X, Pan KY, Xu W. Association of sleep duration, napping, and sleep patterns with risk of cardiovascular diseases: a nationwide twin study. J Am Heart Assoc. 2022;11(15):e025969. https://doi.org/10.1161/JAHA.122.025969.
53. White MJ, Yaspan BL, Veatch OJ, Goddard P, Risse-Adams OS, Contreras MG. Strategies for pathway analysis using GWAS and WGS data. Curr Protoc Hum Genet. 2019;100(1):e79. https://doi.org/10.1002/cphg.79.
54. Wu C, Pan W. Integrating eQTL data with GWAS summary statistics in pathway-based analysis with application to schizophrenia. Genet Epidemiol. 2018;42(3):303–16. https://doi.org/10.1002/gepi.22110.

Mendelian Randomization of Sleep and Circadian Traits

Shilpa Sonti and Struan F. Grant

Abstract Maintaining proper circadian rhythm and good sleep quality is essential for proper physical, mental, and behavioral performance. The sleep-wake cycle represents one of the most conserved circadian rhythms; however, its molecular mechanisms are not fully understood. Interruption of the circadian rhythm is known to have a major impact on sleep quality and can result in several short- and long-term health complications. The causal association relationship of disruption of sleep and circadian patterns with disease states has been ambiguous. Several observational studies and clinical trials have attempted to establish causal relationships, but the presence of confounding factors has made drawing conclusions challenging. The discovery of a genetic component in the pathogenesis of circadian and sleep patterns has opened up a whole new perspective in understanding the pathophysiology of circadian disruption and sleep disturbances with respect to their role in disease presentation. The advent of large prospective studies, including the establishing of biobanks, has made genome wide association studies (GWAS) possible and enabled the identification of genetic variants statistically significantly associated with circadian rhythms and sleep traits. A major objective downstream of the discovery of these genetic variants has been the investigation of a possible causal relationship

S. Sonti (✉)
Center for Spatial and Functional Genomics, Children's Hospital of Philadelphia, Philadelphia, PA, USA
e-mail: sontis@chop.edu

S. F. Grant
Center for Spatial and Functional Genomics, Children's Hospital of Philadelphia, Philadelphia, PA, USA

Department of Genetics, University of Pennsylvania, Philadelphia, PA, USA

Institute for Diabetes, Obesity and Metabolism, Perelman School of Medicine, University of Pennsylvania, Philadelphia, PA, USA

Department of Pediatrics, The University of Pennsylvania Perelman School of Medicine, Philadelphia, PA, USA

Divisions of Human Genetics and Endocrinology & Diabetes, Children's Hospital of Philadelphia, Philadelphia, PA, USA

P. Gehrman et al. (eds.), *Genetics of Sleep and Sleep Disorders*,
https://doi.org/10.1007/978-3-031-62723-1_9

between circadian and sleep disturbances with health complications. Mendelian randomization (MR) is an analytical approach that utilizes these genetic variants as instruments to establish causal effects between circadian and sleep disturbances and disease outcomes. Given that genetic variants are randomly inherited at birth, MR allows the bypassing of confounding biases commonly experienced in traditional analyses in order to establish reliable causal conclusions. In this chapter, we highlight the concept and importance of MR and illustrate the many causal relationships identified by MR analyses between sleep and circadian traits and various health/disease outcomes.

Keywords Mendelian randomization · Confounding factors · Genetic variations · Genome wide association studies (GWAS) · Causal associations

Introduction

Circadian rhythms are natural 24-h cycle processes that primarily respond to light and dark. These rhythms are essential for maintaining physical and mental wellbeing in most living organisms. Circadian rhythms are maintained by the suprachiasmatic nucleus (SCN; master clock) in the hypothalamus of the human brain. The SCN receives input directly from the eyes and regulates the expression of respective genes (biological clocks) across all cells to regulate the circadian rhythm [1]. Circadian rhythms influence several important functions, such as hormone release, eating habits, sleep-wake cycle, and body temperature. However, the most noticeable effect of circadian rhythm is on both sleep patterns and sleep-wake cycle. Sleep is therefore undoubtedly essential for all living organisms [2]. Circadian genes that encode for proteins that influence sensations of sleepiness in the night and wakefulness in the morning (activated in response to the amount of light during night and day respectively) determine sleep patterns in humans. When circadian rhythm is disrupted, the duration of sleep changes and the quality of sleep decreases. Prolonged disruption of this pattern can result in health complications such as depression and obesity. On the other hand, serious health complications such as Alzheimer's or Parkinson's disease can alter sleep patterns which in turn disrupts the circadian rhythm and ultimately alters other physiological functions therefore causing further aggravation. This illustrates the complex association between sleep patterns, circadian rhythms, and adverse health effects. Therefore, it is necessary for us to characterize the true causality of these associations in order to administer the correct course of treatment for a given condition to avoid aggravating the disorder further.

Observational studies are often conducted to study the associations between exposure elements (such as environmental, chemical and pathogen factors, or genetic predisposition) and disease outcome (such as circadian, sleep disturbances, or other disease state). The conclusions of these studies have been widely unreliable, given that when the same exposure-outcome intervention (or hypothesis) was tested via a randomized clinical trial (RCT), the association previously identified was absent.

Randomized clinical trials (RCTs) randomly assign participants into two groups for comparison to assess the causal relationship between exposure and outcome. Despite being expensive and time consuming, RCTs are relatively informative given they employ strong statistical measures and are resourceful in identifying significant association trends. However, the conclusions derived by RCTs may not truly pinpoint causality as an increased risk of a health outcome in the presence of an exposure doesn't necessarily imply a causal relationship between the exposure and outcome. Indeed, there could be a number of intermediate variables, termed confounding variables (or confounding factors), dictating the nature of the relationship between an effect and outcome [3].

The confounding factor is associated (causally or non-causally) with the exposure and affects the outcome independent of the exposure. For example, cross-sectional and meta-analyses of prospective studies have revealed that obstructive sleep apnea (OSA) is associated with an increased risk of lung cancer [4, 5]. While these studies suggest a possible association between OSA and lung cancer, they do not necessarily imply that OSA causes esophageal cancer. There are other factors that these studies might not have considered while drawing conclusions, including most obviously smoking. Epidemiological studies implicate the association of smoking with OSA [6, 7] as well as lung cancer [8]. Hence, in this example, smoking becomes a confounding factor—it is associated with OSA (exposure) and it causes lung cancer (outcome); it causes lung cancer independent of OSA incidence thus distorting the causal association between poor oral health and esophageal cancer.

Fortunately confounding factors can be controlled, and adjustments for these factors can be made at the design stage of epidemiological studies as well as during analysis [9]. By making the comparative groups similar with respect to the confounding factors, the confounding bias can be significantly limited. As confounding factors are associated with the exposure, they can be identified based on previous association studies, and since they are a possible cause for the outcome, they can also be identified through their molecular pathology. At the design stage of the study, restricting the participation of volunteers who have a similar background with respect to confounding factors can effectively eliminate bias. For instance, in the above example, recruiting non-smokers to assess the effect of OSA on lung cancer would eliminate the confounding bias of smoking. Another method to minimize confounding effect is randomization. Random allocation of participants into two groups would also randomly distribute the known and unknown confounding factors thus leading to the confounding effect in both groups being similar. Mendelian randomization is a unique analytical technique that utilizes the randomly annotated genotype for an individual as an independent instrument variable to minimize confounding bias and accurately predicts the causal association between the exposure and outcome in a given population.

Mendelian Randomization

Mendelian randomization (MR) is a technique that overcomes confounding by leveraging independent instruments (such as genetic variations) to determine if an "exposure," such as sleep duration or insomnia, is responsible for a disease risk or "outcome." MR utilizes genetic variables that are randomly assigned to an individual at birth as independent instruments. Based on Mendel's second law of inheritance, "law of random assortment," MR assumes the inheritance of one trait is independent of the inheritance of other traits. When widely prevalent within a population, these genetic variations make a meaningful association between the exposure and the outcome in that population [10, 11]. MR assumes that the genetic variants are (1) not associated with any confounding factors; (2) significantly associated with exposure (genome wide significance $P < 5 \times 10^{-8}$); and (3) independently associate with the phenotype of interest (outcome) only through the exposure being investigated [12].

The concept of MR can be further clarified through a simple directed acyclic graph illustrated in Fig. 1. Assume exposure "*A*" (OSA in the above example) leads to the outcome "*B*" (lung cancer in the above example) (i), and we need to determine if this causal association is true. The association between A and B is heavily influenced by the confounding factors "*C*" (smoking in the above example) which can influence *A* (ii) or *B* (iii) or both. Assume the existence of a genetic variant, SNP "*S*", which robustly influences *A*, independent of any confounding factors and thus directly associates with *A* (iv). If *S* lacks any association with *C* (v), as well as *B* (vi), we can assume that *S* influences *B* only through A. Thus, SNP *S* can be considered a true independent instrument which can predict the true causal association between exposure *A* and outcome *B* (vii) without any bias.

How Are the Independent Instruments Identified?

To understand the effect of environment, individual genetic make-up and the impact of lifestyle choice on population health, several biobanks have been established. Biobanks collect biological material and associated information from consented

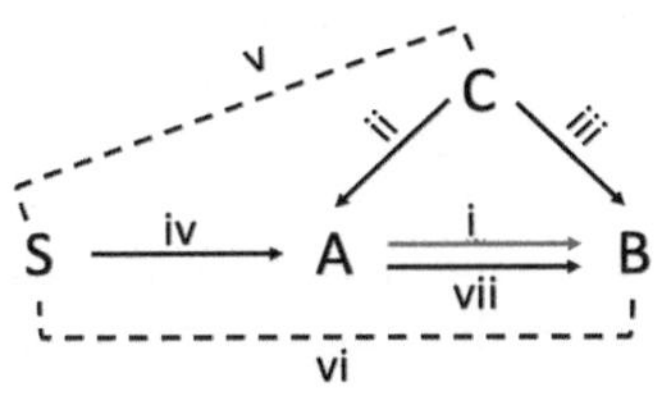

Fig. 1 Directed acyclic representation of MR. Red solid arrow represents assumed association; black solid arrows represent direct association; broken lines represent lack of association

participants for education and research purposes. These biobanks have progressively evolved over the years, with increasing participation of individuals of various ethnicities and they have become an invaluable resource for medical research. The genotypic and phenotypic data from these biobanks have been leveraged by several large-scale epidemiologic studies to assess a wide range of health-related outcomes.

The advent of these biobanks enabled genome wide association studies in large cohorts of individuals to identify associations between a disease phenotype and genotype. A genome wide association study (GWAS) involves scanning the entire genome of the participants in an attempt to identify genetic variants that may be responsible for the onset of a given disease [13] (Fig. 2). This has led to the uncovering of many informative associations between genetic variation (principally single nucleotide polymorphisms [SNPs]) and a given phenotype (disease of interest). The variant needs to pass the strict genome wide significance threshold (observed *P*-value is less than 5×10^{-8}) to be significantly associated with the phenotype [14].

GWAS datasets have been made available to the public to encourage research, and this accessibility has enabled and accelerated the identification of genetic loci associated with various diseases and traits, and evaluation of their putative causal relationships with various health outcomes. To date, more than 300,000 SNP-trait associations have been identified through GWAS [15]. The summary statistics utilized to identify genetic variants associated with sleep traits have been largely obtained from a mixed European population participating in the UK Biobank [16]. The UK Biobank utilizes self-assessment questionnaires to collect participant data.

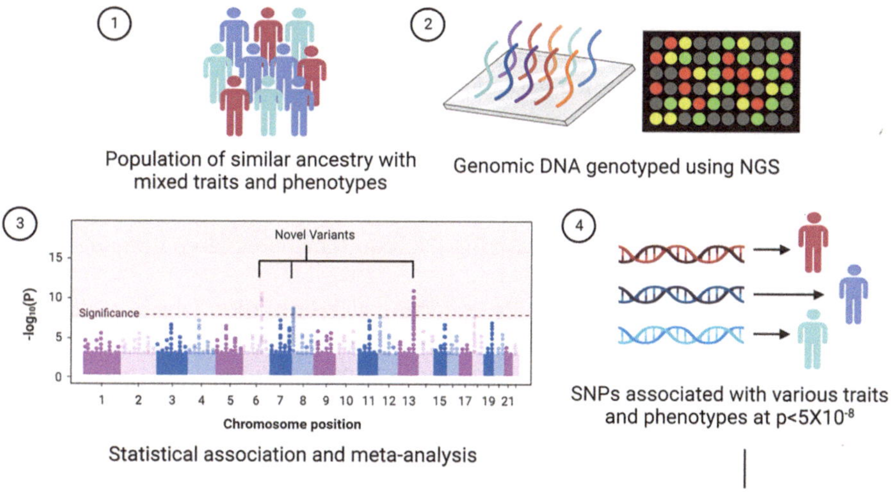

Fig. 2 Overview of identification of genetic variants using GWAS. (1) Participants of mixed phenotypes are considered for genotyping; (2) Genomic regions potentially associated with the phenotype of interest are identified through genotyping; (3) Statistical association testing and meta-analyses are leveraged to identify novel associated genetic variants (SNPs); (4) SNPs that pass the genome wide significance thresholds for each phenotype are considered significantly associated with that phenotype

Although the UK Biobank has a large sample size, it comprises of mostly individuals of European ancestry and lacks ethnic diversity. Thus, the validation of variants identified in this cohort may not necessarily replicate in other populations, warranting the requirement of more GWAS efforts in additional communities to identify disease associated variants in those ethnicities. Biobanks in other communities, such as Biobank Japan [17], China Kadoorie Biobank [18], have large sample sizes and have enabled identification of several risk loci in their respective cohorts. The associations identified by GWAS are important to subsequently elucidate causal relationships between various traits and disease pathogenesis.

Sleep GWAS

GWAS efforts in the field of sleep have revealed genetic variants associated with metrics related to sleep duration, sleep efficiency, and chronotype together with sleep disorders such as insomnia, narcolepsy, and restless leg syndrome (RLS). Some of the key GWAS efforts in the sleep field have been discussed in detail in a recent review [19] and are briefly summarized below.

Many early GWAS efforts aimed at identifying genetic variants likely associated with sleep disturbances in settings that included RLS and narcolepsy. RLS is a neurological disorder that is characterized by an urge to move one's legs in response to uncomfortable sensations, and as a result cause sleep disruption. Identification of a variant in the *BTBD9* gene associated with RLS in Icelandic and American populations marks the first successful GWAS milestone in the sleep field [20]. This association was replicated by another independent study in a cohort of European ancestry, which also identified two additional associations [21]. Subsequent meta-analyses performed in more recent GWAS efforts leveraging independent datasets confirmed the previously reported risk variants for RLS as well as identifying several novel signals [22]. Only 4 out of the 43 total variants associated with RLS reached genome wide significance when reassessed in a large dataset comprising of almost 20,000 participants, emphasizing the requirements of large sample sizes for deducing robust conclusions [23].

Narcolepsy is another commonly occurring sleep disorder, which is characterized by cataplexy, sleep paralysis, and excessive daytime sleepiness. Narcolepsy occurs across multiple ethnicities; however, it is most prevalent in Japan. The first narcolepsy GWAS was performed in a Japanese cohort and identified a significant association with a variant located between *CHKB* and *CPT1B*. This association was replicated in an independent Japanese cohort, as well as in Koreans, but not in Europeans or African Americans [24]. Reassessment of the Japanese variants identified in a large-scale dataset identified an additional variant near *TEAD4* that significantly associated with the age of onset of cataplexy associated with narcolepsy [25]. Following this study, several genome wide studies in independent cohorts identified multiple variants in *TCRA* [26], *CTSH*, *TNFSF4* [27], *P2RY11*, *TCRB*,

IL10RB-INFAR1, and *ZNF365* [28] significantly associating with narcolepsy bringing the total to >30 associated loci.

Several recent GWAS efforts have leveraged data from large scale biobanks to identify significant associations between different sleep traits and genetic variants. The GWAS by Jones et al. leveraged UK Biobank data from >100,000 participants and identified 3 variants significantly associated with sleep duration and 16 variants significantly associated with chronotype (the propensity of an individual to be a morning or evening person) [29]. Eleven of these variants have been replicated in an independent cohort of 89,283 participants from the 23andMe study by Hu et al. and retained their genome wide significance. In addition, this study identified nine additional variants significantly associated with chronotype [30]. In 2016, Lane et al leveraged the summary statistics from the UK Biobank (100,420 participants) and identified 12 additional significant variants associated with chronotype bringing the signal count to a total of 37 [31]. In 2017, Lane et al. then conducted a GWAS on 201,869 individuals participating in the UK Biobank and 23andMe studies and identified 9 significant variants associated with various sleep traits—one variant associated with sleep duration, five variants associated with insomnia, and three variants associated with excessive daytime sleepiness (one of the three variants, in *ROBO1,* was significant only after adjusting for depression; and another variant, in *TMEM132B* was significant only after adjusting for BMI) [32]. In 2019, Dashti et al. utilized the summary statistics from the UK Biobank (446,118 participants) and identified 78 risk variants for sleep duration. Thirteen additional variants significantly associated with short and lone sleep duration were identified from another GWAS in the same study [33]. Hammerschlag et al. conducted a large scale GWAS on 113,006 individuals (UK Biobank) and identified five novel genes associated with insomnia [34]. The largest GWAS in the sleep field to date was carried out by Janssen et al. on insomnia in 1,331,010 participants from the UK Biobank and 23andMe combined [35, 36], and identified 202 risk loci associated with the trait [37]. This study also conducted GWAS on six other sleep traits—morningness, sleep duration, ease of getting up in the morning, daytime napping, daytime dozing, and snoring and identified several loci overlapped with multiple sleep traits suggesting that there is a genotypic and phenotypic correlation between various sleep traits. The strongest overlap in loci was shared between sleep duration and insomnia.

While sleep disturbances such as insomnia, narcolepsy, and restless leg syndrome have been the focus of most GWAS efforts, recent studies have tried to expand the genome wide studies to other sleep disturbances including snoring and obstructive sleep apnea (OSA). Leveraging the data from the FinnGen database, five variants were identified for OSA [38]. BMI was a key confounding factor influencing the significance of these variants. Four out of these five variants were previously associated with BMI indicating an OSA association that is BMI-dependent for these signals [39]. The remaining locus stayed significantly associated with OSA before and after adjusting for BMI indicating a BMI-independent OSA association [38]. A more recent GWAS in a Chinese cohort (20,590 individuals) discovered two novel significant loci associated with OSA, one of which, *SLC53A3*, was identified as a potential therapeutic target [40]. A group from Children's Hospital of Philadelphia performed the first ever GWAS on pediatric OSA in children of European and African American descent and identified a strong association at a locus on chromosome 18p11.32. In addition, sex stratified analyses revealed separate loci associating with OSA in males and females. However, owing to the small sample size (1486 participants), further analysis using a larger sample size is required to validate these findings [41]. Forty-two genome-wide significant loci were identified for snoring in a recent GWAS conducted in 408,000 individuals participating in the UK Biobank.

GWAS efforts have thus successfully identified genetic variants associated with numerous sleep and circadian traits and sleep disorders. Although many GWAS consortia have made their results publicly available, these results are rarely consistent in format and output. Phenoscanner was developed as a centralized resource for accessing GWAS data to address these drawbacks. Phenoscanner allows the user to search for a genetic association across many phenotypes by including all available results [42]. In addition, web-based analytical platforms like MR-Base [15] and LD hub [43] make it possible to assess all possible pairwise relationships to yield targets that can be pursued in functional follow-up studies.

Types of MR

Initially, MR was carried out to assess the effect of an exposure in a sample cohort using the variants identified from the same cohort; however, it is not always possible to identify variants that independently associate with the exposure and outcome within a population [44]. The public availability of GWAS summary statistics enabled access to a pool of variants identified in independent sample cohorts, thus enabling MR (*two sample MR*) to estimate causal associations across different sample cohorts using variant-exposure and variant-outcome associations [45]. If an intermediate trait is involved in mediating the effect of an exposure on an outcome, *two-step MR analysis* can determine its involvement. The first step leverages SNPs associated with the exposure to determine the effect of exposure on the intermediate, and the second step utilizes SNPs associated with the intermediate to determine the effect of that intermediate on the outcome. Evidence of association in both steps points to the involvement of the intermediate in mediating the effect of an exposure on an outcome. Two-step MR analysis can thus estimate the direct effect (independent of the mediator) and the indirect effect (via the mediator) of the exposure on the outcome [46]. A *bidirectional MR analysis* can assist in estimating the direction of the causal effect, i.e., whether the exposure causes the outcome or if the outcome influences the exposure, by using SNPs robustly associated with exposure and outcome (from independent GWAS endeavors). Assuming an underlying mechanism exists between the exposure and the outcome, MR analysis is first performed in one direction (exposure > outcome), and then in the opposite direction (outcome > exposure) to elucidate a possible causal direction [47]. *Factorial MR analysis* is performed when there is a suspected interaction between the exposure and outcome. Factorial analysis is also carried out when investigating the interaction between a pharmacological agent and a disease risk (outcome). In this case, genetic variants are used as proxies for pharmacological agents, and using factorial MR can determine the causal effect between an exposure and outcome while considering a possible interaction between the exposure and outcome [48]. *Multivariable MR analysis* incorporates the variant's potential pleiotropic impact on other trait outcomes before estimating causality. Conversely, multivariable MR also considers the pleiotropic effect of other genes in addition to the exposure of interest that may affect a risk outcome when determining causality [49, 50].

Three methods—the inverse-variance weighted (IVW), weighted median (WM), and MR-Egger methods are generally used to estimate the odds ratio (OR), which determines whether the association is causal or not. OR is calculated from the beta coefficient (β), which is calculated through SNP-outcome and SNP-exposure association estimates [51]. The association between an exposure and outcome is causal when OR > 1. Estimates calculated through IVW are the most accurate; nevertheless, sensitivity analyses are often carried out to confirm the result. The weighted median (WM) method considers the validity of the instrument variables and generates an estimate if greater than 50% of the estimated weight is via valid variants. The employment of MR-Egger analysis is to detect and correct for any pleiotropy (discussed later). If all the three methods generate a similar estimate (if slopes are similar; and OR > 1 in all cases), the association between the two traits is most likely causal [52].

A recent study by Chen et al. illustrates how a typical two-sample MR analysis is performed (Fig. 3; adapted from a recent review [19]). This study sought out to determine the causality between various sleep traits and IBD. While seemingly unrelated, prior preclinical [53], clinical [54, 55], and observational studies [56] indicated an association between sleep disorders and IBD. However, the true causality between these two traits is unknown [57]. Instrument variables for sleep traits and IBD were obtained from the UK Biobank and 23andMe (~one million participants) and the IBD Genetics Consortium (n = 34,652) respectively. The genetic variants selected for MR passed the genome wide significance threshold ($P < 5 \times 10^{-8}$) and were not in linkage disequilibrium (LD) with other variants that may have significant association with the outcome. Use of all methods—IVW, weighted median, and the MR-Eggar analyses— showed that none of differential sleep traits had a causal effect on the pathogenesis of IBD (Fig. 3).

Mendelian Randomization of Sleep Traits and Common Complex Disease

Genetic correlation analysis of variants identified through GWAS efforts revealed associations with various neurodegenerative, cardiovascular, and metabolic traits. MR is a major resource to validate these associations and conclude whether they are causal. Determining causality can help tailor treatments to manage these conditions.

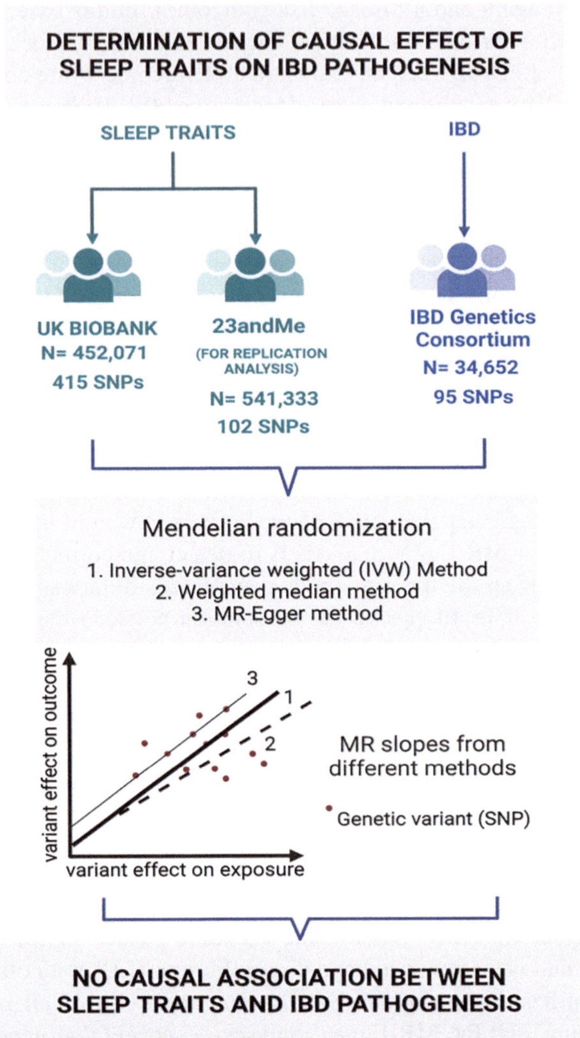

Fig. 3 This flowchart outlines a two-sample MR analysis executed by Chen et al. to elucidate the causality between various sleep traits and IBD pathogenesis. Three MR methods were leveraged to elucidate causality, as shown in the graph: (1) Inverse variance weighted method, (2) Weighted median method, (3) MR-Egger method. In the event of all three methods yielding a slope > 1.0, the association between the two traits is considered likely causal. (Adapted from a review by Sonti and Grant)

The principal MR findings in the field of sleep outlining the causal associations between various sleep phenotypes and disease consequences have been discussed in detail in a recent review [19]. They are summarized here along with new findings that have been published since the review.

MR of Sleep and Neurodegenerative Disease

Neurodegenerative diseases are characterized by progressive neuronal degeneration leading to the dysfunction of several sensory and motor systems. These diseases are generally triggered by environmental and genetic risk factors, or through a combination of both. Sleep disturbances in various forms manifest in up to 60% of the population impacted by any given neurodegenerative disease [58] and can often manifest much earlier than the actual disease symptoms are apparent [59]. Sleep disturbances lead to the disruption of circadian rhythm, which regulates gene expression in several brain regions, causing impaired neural function [60, 61]. Notable MR findings in the context of sleep and neurodegenerative diseases are listed in Table 1.

MR revealed that short sleep duration is causally associated with cognitive decline [62], and that chronic pain caused sleep disturbances including insomnia [63, 64]. Two-sample MR studies failed to show an association between sleep duration and Alzheimer's disease risk [62, 65, 66]; however, bidirectional MR concluded that Alzheimer's disease was causal for sleep disturbances, but not the other way around [67]. Moreover, increased sleep efficiency was significantly associated with decreased risk for Alzheimer's disease [58]. However, a recent bidirectional MR identified a positive association between insomnia and Alzheimer's disease [68]. The inconsistencies between these studies may be attributed to differences in definition and diagnostic criteria, forms of characterization of Alzheimer's disease, type of data used for analysis, type of MR analysis employed (linear or non-linear), and statistics employed to draw conclusions. OSA was neither associated with Alzheimer's disease nor Parkinson's disease [69]. Of the different sleep traits, only the morning chronotype was inversely associated with the age of onset of Parkinson's disorder [58]. Several independent MR studies confirmed that daytime sleepiness is associated with increased ALS risk [58, 70, 71]. Another recent MR has determined a causal association between morningness and age-related macular degeneration [72].

MR of Sleep and Psychiatric Disorders

Emerging evidence has associated dysregulation of the circadian system and sleep disturbances (insomnia, hypersomnia, nightmares, etc.) with psychiatric disorders such as depression, bipolar disorder, and schizophrenia [74, 75]. Sleep GWAS efforts identified several loci that were associated with both sleep and circadian rhythms as well as psychiatric disorders [33, 37, 76]. The overlap in loci suggests a strong underlying genetic connection between the two traits. MR can determine causality using this information which can in turn be leveraged for developing new therapeutic interventions for individuals presenting with sleep/psychiatric

Table 1 MR of sleep and neurodegenerative diseases

Disease outcome	Sleep exposure	MR model	Interpretation	Reference
Cognition	Sleep duration	One and two sample MR	Positive association between short sleep duration and cognitive decline	Henry et al. [62]
Alzheimer's disease	Sleep duration	Two sample MR	No causal association determined	Henry et al. [62]
	Multiple sleep traits	Two sample MR	No causal association determined	Anderson et al. [65]
	Sleep duration	Two sample MR	No causal association determined	Yuan et al. [66]
	Sleep disturbances	Bidirectional two-sample MR	AD had a causal effect on sleep disturbances, but sleep disturbances were not causal to AD	Huang et al. [67]
	Insomnia	Bidirectional two-sample MR	Insomnia was associated with a higher risk of AD	Chen et al. [68]
Parkinson's disease	Sleep traits	Two sample MR	No causal association determined	Noyce et al. [73]
Alzheimer's and Parkinson's disease	Obstructive sleep apnea (OSA)	Two sample MR	No causal association determined	Li et al. [69]
Neurodegenerative disease risk	Sleep/wake patterns	Two sample MR	Chronotype was inversely associated with Parkinson's age of onset Sleep efficiency was associated with decreased AD risk Daytime sleepiness was associated with increased ALS risk	Cullel et al. [58]
Amyotropic lateral sclerosis (ALS)	Sleep disturbances	Two sample MR	Daytime sleepiness was associated with increased ALS risk	Zhang et al. [70]
	Daytime sleepiness	Bidirectional two-sample MR	Daytime sleepiness was associated with increased ALS risk but ALS risk is not causal for daytime sleepiness	Di et al. [71]
Pain	Sleep disturbances	Bidirectional two-sample MR	Chronic pain had a causal effect on sleep disturbances, but sleep disturbances were not causal for pain	Sun et al. [63]

(continued)

Table 1 (continued)

Disease outcome	Sleep exposure	MR model	Interpretation	Reference
	Insomnia	Bidirectional two-sample MR	Causal association between insomnia and pain was determined in both directions	Broberg et al. [64]
Macular degeneration	Morningness	Two sample MR	Morningness was causally associated with risk of age-related macular degeneration	Grover et al. [72]

disturbances. Findings of notable MR analyses of sleep traits and psychiatric disorders are listed in Table 2 and summarized below.

MR analysis indicated that daytime napping and insomnia were positively associated with risk of depression [77–79], while longer sleep duration and morning preference were inversely associated with the risk of depression [80, 81]. In addition, depression significantly increased the risk of insomnia [82, 83] and sleep apnea [84], but the effect of sleep apnea on depression has not been explored. Despite the clinical and genetic evidence suggesting the involvement of sleep disturbances in psychiatric disorders, a causal relationship has not yet been established. Several recent MR analyses tried to assess causality between the two conditions. MR failed to identify a causal associations between schizophrenia and chronotype [31], and schizophrenia and insomnia in both directions [85]. However, a positive association of schizophrenia with long sleep duration and daytime napping was identified by two independent MRs [33, 86]. Bidirectional MR determined the positive association between insomnia and bipolar disorder as well as autism spectrum disorder [85]. A positive association was determined by a two sample MR between insomnia, daytime napping, short sleep duration, and ADHD [87].

MR of Sleep and Cardiometabolic Disease

Cardiovascular disease (CVD), obesity, and diabetes share common risk factors and are collectively known as cardiometabolic diseases. Cardiometabolic disorders share overlapping pathways with sleep, and thus any disruption in these pathways can lead to an elevated risk for cardiometabolic disease or sleep disturbances, or both [88]. Several epidemiological studies report contradicting observations and associations between various cardiometabolic traits and sleep disturbances. The true causal associations between sleep traits and cardiometabolic traits can be validated through MR. The findings are listed in Table 3 and summarized below.

Body mass index (BMI), when reported through accelerometer data, was significantly associated with long sleep duration [89, 90], and insomnia [91]. In contrast, in children the association between long sleep duration and BMI was reversed [92]. Snoring was also found to significantly associate with BMI [93]. A complex

Table 2 MR of sleep and psychiatric disorders

Disease outcome	Sleep exposure	MR model	Interpretation	Reference
Depression	Sleep duration	Two sample MR	Sleep duration was associated with decreased risk of depression	Choi et al. [77]
	Daytime napping	Two sample MR	Daytime napping was associated with increased risk of depression	Choi et al. [77]
	Chronotype	Two sample MR	Chronotype was associated with decreased risk of depression	Daghlas et al. [80]
	Chronotype	One sample MR	Chronotype was associated with decreased risk of depression	O'Loughlin et al. [81]
	Insomnia	Bidirectional multivariable MR	Causal association between insomnia and depression was determined in both directions	Cai et al. [78]
	Insomnia	Bidirectional MR	Causal association between insomnia and depression was determined in both directions	Zhou et al. [79]
	Insomnia	Two sample bidirectional MR	Depression was association with a higher risk of insomnia	Sun et al. [82]
	Insomnia	Bidirectional MR	Causal association between insomnia and depression was determined in both directions	Baranova et al. [83]
	Sleep apnea	Two sample MR	Depression was association with a higher risk of sleep apnea	Chen et al. [84]
Schizophrenia	Insomnia	Bidirectional MR	No causal association determined in either direction	Gao et al. [85]
	Chronotype	Two sample MR	No causal association determined	Lane et al. [31]
	Sleep duration	Bidirectional two-sample MR	Long sleep duration is associated with pathogenesis of schizophrenia	Dashti et al. [33]
	Sleep traits	Bidirectional two-sample MR	Long sleep duration and daytime napping were associated with a higher risk of schizophrenia	Wang et al. [86]

(continued)

Table 2 (continued)

Disease outcome	Sleep exposure	MR model	Interpretation	Reference
Bipolar disorder	Insomnia	Bidirectional MR	Insomnia had a causal effect on bipolar disorder, but bipolar disorder was not causal for insomnia	Gao et al. [85]
Autism spectrum disorder (ASD)	Insomnia	Bidirectional MR	Insomnia had a causal effect on ASD, but ASD was not causal for insomnia	Gao et al. [85]
Attention deficit hyperactive disorder (ADHD)	Sleep traits	Two sample MR	Causal association was seen between insomnia, daytime napping, short sleep duration and ADHD	Carpena et al. [87]

relationship between BMI, OSA, and atrial fibrillation had been identified through several MR studies. While some confirmed the positive association between OSA and atrial fibrillation, one study identified that a positive association between OSA and atrial fibrillation is independent of BMI [94–96]. In contrast, another study failed to establish the positive association between OSA and atrial fibrillation after adjusting for BMI, suggesting the key role of BMI in the relationship between the two traits [97]. Moreover, bidirectional MR detected no reverse causality, i.e., no causal effect of atrial fibrillation on OSA [95]. In addition, a causal association was identified between insomnia and adiposity [98] as well as between daytime napping and increased waist circumference [99].

Several risk factors were identified for type 2 diabetes, and multivariate MR identified a significant causal association between type 2 diabetes and insomnia, which was later confirmed by two other MR analyses [100–102]. However, no causal association was seen between type 2 diabetes and sleep duration [103] or morningness [76]. Sleep duration and insomnia are significantly associated with overall cardiovascular risk [104–106]. Short sleep duration was found to associate with myocardial infarction [107], heart failure [108], and hypertension [109]. Hypertension was also causally associated with insomnia [109] and daytime napping [99]. MR analysis conducted during a prospective study looking at the association between sleep duration and stroke identified a significant association with insomnia, but not sleep duration [110]. This association was confirmed by another independent study [111].

MR of Sleep and Cancer

The association between cancer and sleep disorders is becoming increasingly evident as more and more observational and epidemiological studies are identifying strong phenotypic correlations. MR has been an incredibly effective tool to determine true causal associations between sleep disturbances and various cancers. The

Table 3 MR of sleep and cardiometabolic disease

Disease outcome	Sleep exposure	MR model	Interpretation	Reference
BMI (adult)	Sleep duration	Two sample MR	Long sleep duration had a causal association with increased adult BMI	Doherty et al. [89]
	Sleep duration	Two sample MR	Long sleep duration had a causal association with increased adult BMI	Jones et al. [90]
	Insomnia	Two sample MR	Insomnia is positively associated with elevated BMI	Xiuyun et al. [91]
	Snoring	Multivariable MR	Causal association between high BMI and snoring was identified	Campos et al. [93]
	Obstructive sleep apnea (OSA)	Two sample MR	High BMI has a causal association with obstructive sleep apnea	Strausz et al. [38]
BMI (children)	Sleep duration	Two sample MR	Long sleep duration had a causal association with decreased BMI in children (2–10 years)	Wang et al. [92]
Atrial fibrillation	Obstructive sleep apnea (OSA)	Two sample MR	OSA had a causal association with increased risk of atrial fibrillation	Chen et al. [94]
	Obstructive sleep apnea (OSA)	Bidirectional MR	OSA had a causal effect on atrial fibrillation, but atrial fibrillation was not causal for OSA	Chen et al. [95]
	Obstructive sleep apnea (OSA)	Bidirectional MR/ multivariable MR	OSA had a causal effect on atrial fibrillation, independent of BMI	Li et al. [96]
	Obstructive sleep apnea (OSA)	Univariable MR/ multivariable MR	OSA had a causal effect on atrial fibrillation, which disappeared after adjusting for BMI	Ardissino et al. [97]
Adiposity	Insomnia	Two sample MR	Insomnia had a causal effect on increased measure of adiposity	Dashti et al. [98]
Increased waist circumference	Daytime napping	Two sample MR	Daytime napping had a causal association with increased waist circumference	Dashti et al. [99]
Type 2 diabetes (T2D)	Insomnia	Univariable MR/ Multivariable MR	Insomnia had a causal association with T2D	Yuan et al. [100]
	Insomnia	Two sample MR	Insomnia had a causal association with T2D	Gao et al. [101]

(continued)

Table 3 (continued)

Disease outcome	Sleep exposure	MR model	Interpretation	Reference
	Insomnia	Two sample MR	Insomnia had a causal association with T2D, independent of BMI	Jia et al. [102]
	Sleep duration	Two sample MR	No causal association determined	Wang et al. [103]
	Chronotype	Two sample MR	No causal association determined	Jones et al. [76]
Cardiovascular diseases (CVDs)	Sleep duration	Two sample MR	Short sleep duration was causally associated with increased risk of CVDs	Ai et al. [104]
	Insomnia	Two sample MR	Insomnia was causally associated with nine CVD traits	Liu et al. [105]
	Insomnia	Two sample MR	Insomnia was causally associated with overall CVD risk	Larsson et al. [106]
Myocardial infarction (MI)	Sleep duration	Two sample MR	Short sleep duration was causally associated with MI	Dashti et al. [107]
Heart failure	Sleep duration	Two sample MR	Longer sleep duration was causally associated with decreased risk for heart failure.	van Oort et al. [108]
Stroke	Insomnia	Two sample MR	Insomnia was causally associated with increased risk for Stroke	Titova et al. [110]
	Sleep duration	Two sample MR	No causal association determined	Titova et al. [110]
	Sleep duration	Two sample MR	No causal association determined	Lu et al. [111]
Hypertension	Insomnia	Two sample MR	Insomnia was causally associated with increased risk for hypertension	van Oort et al. [109]
	Sleep duration	Two sample MR	Longer sleep duration was causally associated with decreased risk for hypertension	van Oort et al. [109]
	Daytime napping	Two sample MR	Frequent daytime napping is causally associated with hypertension	Dashti et al. [99]

availability of consortia comprising information on cancer patients has been an invaluable resource of genetic variants that make MR analyses possible. Notable MR findings between sleep traits and risk of various cancers are summarized below and in Table 4.

Two independent MR studies identified that morning preference was associated with a decreased risk of breast [112], prostate [113], and digestive tract (stomach

Table 4 MR of sleep and cancer

Disease outcome	Sleep exposure	MR model	Interpretation	Reference
Breast cancer	Chronotype	One and two sample MR	Chronotype was causally associated with decreased risk for breast cancer	Richmond et al. [112]
	Obstructive sleep apnea (OSA)	Multivariable MR	OSA was causally associated with increased risk for breast cancer	Yap et al. [115]
	Obstructive sleep apnea (OSA)	Two sample MR	OSA was causally associated with increased risk for breast cancer	Gao et al. [116]
Digestive tract cancer	Chronotype	Multivariable MR	Chronotype was causally associated with decreased risk for digestive tract cancer	Yuan et al. [114]
Prostate cancer	Chronotype	Univariable/ Multivariable MR	Chronotype was causally associated with decreased risk for prostate cancer	Sun et al. [113]
Lung cancer	Insomnia	Two sample/ Multivariable MR	Insomnia was causally associated with increased risk for lung cancer	Wang et al. [117]
	Insomnia	Two sample MR	Insomnia was causally associated with increased risk for lung cancer	Shen et al. [118]
	Insomnia	Two sample MR	Insomnia was causally associated with increased risk for lung cancer	Huo et al. [119]

and colorectal) [114] cancers. While OSA was associated with increased risk of breast cancer [115], particularly in Chinese cohorts [116], insomnia was associated with increased risk of lung cancer [117–119].

MR of Sleep and Other Health Outcomes

We have discussed the potential of MR in identifying causal associations between sleep traits and major disease states in the above sections. MR analyses have been successful in also identifying meaningful associations between sleep traits and other health outcomes. These findings have been summarized in Table 5. Insomnia had been implicated as a risk factor for peptic ulcer disease [120]. A recent MR has demonstrated a causal association between insomnia and perinatal depression, thus stressing on the importance of sleep in pregnant women/new mothers [121]. Insomnia and short sleep duration have also been reported to causally associate with higher osteoarthritis risk [122]. Through its association with cardiometabolic conditions, short sleep duration has been hypothesized to impact renal function. Indeed, two-sample MR analysis showed that longer sleep duration was causally associated

Table 5 MR of sleep and other health outcomes

Disease outcome	Sleep exposure	MR model	Interpretation	Reference
Peptic ulcer	Insomnia	Bidirectional MR	Insomnia was causally associated with increased risk for peptic ulcer	Wu et al. [120]
Perinatal depression	Insomnia	Two sample MR	Insomnia was causally associated with increased risk of perinatal depression	Yang et al. [121]
Osteoarthritis	Insomnia	Univariable/ Multivariable MR	Insomnia was causally associated with increased risk for osteoarthritis	Ni et al. [122]
	Sleep duration	Univariable/ Multivariable MR	Short sleep duration was causally associated with increased risk for osteoarthritis	Ni et al. [122]
Chronic kidney disease	Sleep duration	Two sample MR	Long sleep duration was causally associated with increased risk for chronic kidney disease #	Mazidi et al. [123]

Table 6 MR of different sleep traits

Disease outcome	Sleep exposure	MR model	Interpretation	Reference
Restless leg syndrome (RLS)	Sleep duration	Bidirectional MR	Insomnia was causally associated with increased risk for peptic ulcer	Dashti et al. [124]
Restless leg syndrome (RLS)	Periodic limb movement syndrome (PLMS)	Two sample MR	Insomnia was causally associated with increased risk of perinatal depression	Edelson et al. [125]
Snoring	Obstructive sleep apnea (OSA)	Univariable/ multivariable MR	Insomnia was causally associated with increased risk for osteoarthritis	Campos et al. [93]

with an increased risk of presenting with chronic kidney disease in diabetic individuals [123]. MR analysis has also reported significant associations between different sleep traits (Table 6). Significant causal associations between long sleep duration and RLS [124]; RLS and periodic limb movement syndrome (PLMS) [125]; and snoring and sleep apnea [93] were identified by various MR studies.

Limitations of MR

MR has proven therefore to be an invaluable tool that can determine causality between two given traits; however, it has several limitations:

Pleiotropy

Pleiotropy, or the ability of one gene to express multiple phenotypes, is the major limitation of MR. The instrument variables selected for MR in general must satisfy the following three criteria—the instrument variable should significantly associate only with the exposure of interest; the instrument should *not* significantly associate with any other factor (confounding factors); and the instrument variable should be associated with the outcome only and only through the exposure of interest. Pleiotropy is the violation of the first assumption—the instrument variable should significantly associate only with the exposure of interest. Pleiotropy occurs through multiple mechanisms—modulation of different proteins, alternate splicing, random mutation, etc. and controls multiple traits [10].

Developmental Compensation

Developmental compensation, or canalization, is the compensation of the phenotypic end point through an alternate mechanism/gene expression when the original gene responsible for the phenotype malfunctions due to environmental context or an error during fetal segregation [126]. The alternate mechanism may be suppressed until the original mechanism malfunctions. This phenomenon is a limitation of MR because it violates the second assumption of MR—the instrument variable should not associate with confounding factors. The presence of developmental compensation introduces new variables that can interfere with analysis and output. Therefore, consideration must be given to alternate compensatory mechanisms and gene expression when trying to determine true causality of associations identified in observational studies.

Nature of Association

The nature of association between any two given traits is an important consideration when designing a MR analysis. It is often assumed that the association between the exposure and outcome under study is linear, and this is an oversimplification. There may be several underlying mechanisms supporting or against the relationship being explored. The possibility of non-linear association and the existence of a different shape of association should be considered to avoid a false positive or negative output. The only MR study to take both linear and non-linear associations into consideration was conducted by Henry et al. to determine causality between sleep duration and cognitive function/dementia [62].

Sample Diversity

MR utilizes genetic variants to determine causality, and these variants are identified through GWAS by leveraging summary statistics from large scale prospective studies. Diversity is an important consideration when considering a population sample for GWAS study. Diversity in all forms—age, sex, race—all play an important role in identifying truly significant loci which are instrumental for MR. Early GWAS efforts were performed in small cohorts with individuals of the same ethnic background. This was a disadvantage as the variants identified in one group may not be representative of another ethnic group. For example, leveraging the statistics from a European ancestry cohort one GWAS identified the *PAX8* and *VRK2* loci significantly associated with sleep duration. Another GWAS conducted in a Japanese cohort failed to identify the association of these loci with sleep duration clearly emphasizing the importance of diversity in a given population sample. Similarly, most studies use data from healthy adults as they are the most frequent participants in biobanks; however, not all associations are evident in adults and some may be seen only in childhood or some associations may change with age, making it important to evaluate and replicate these associations in other age groups as well.

Statistical Power

Statistical power is one of the many factors that govern the rigor of MR analysis. Statistical power determines the magnitude of effect of a variant on the health outcome. Identifying a single rare pathogenic variant in a small cohort is not possible if the cohort is not a proper reflection of the full population. In fact, it is now widely believed that the disease risk conferred by a multiple variants carrying low pathogenic score is much higher than the risk conferred by a single variant carrying high pathogenic score [127]. Including multiple variants, albeit their small pathogenic score, increases the statistical power and increases the rigor of analysis.

Conclusions and Future Perspectives

The advent of biobanks and GWAS revelations has greatly aided the discovery of putative genetic targets associated with sleep and circadian traits. This has enabled functional association studies that assist in understanding sleep phenotypes and pathogenesis of disease. While much progress has been made in identifying physiological and pathological associations of these variants, key questions remain to be answered yet. MR is as an effective tool that can determine causality between various sleep traits and other disease outcomes by utilizing genetic signals identified by GWAS. Although limitations occur, these can be adjusted and corrected for and true causal association can be determined. Determining true causal relationships can characterize underlying molecular mechanisms and aid in the development of therapeutic interventions.

References

1. Circadian Rhythms. October 2020; 1–3. Available from: https://nigms.nih.gov/education/fact-sheets/Documents/fact-sheet-circadian-rhythms.pdf.
2. Nath RD, et al. The Jellyfish Cassiopea exhibits a sleep-like state. Curr Biol. 2017;27(19):2984–2990 e3.
3. Kamangar F. Confounding variables in epidemiologic studies: basics and beyond. Arch Iran Med. 2012;15(8):508–16.
4. Seijo LM, et al. Obstructive sleep apnea and nocturnal hypoxemia are associated with an increased risk of lung cancer. Sleep Med. 2019;63:41–5.
5. Cheng L, et al. Obstructive sleep apnea and incidence of malignant tumors: a meta-analysis. Sleep Med. 2021;84:195–204.
6. Esen AD, Akpinar M. Relevance of obstructive sleep apnea and smoking: obstructive sleep apnea and smoking. Fam Pract. 2021;38(2):181–6.
7. Lin YN, Li QY, Zhang XJ. Interaction between smoking and obstructive sleep apnea: not just participants. Chin Med J. 2012;125(17):3150–6.
8. Aredo JV, et al. Tobacco smoking and risk of second primary lung cancer. J Thorac Oncol. 2021;16(6):968–79.
9. Atherton H. In: Health Do, editor. Confounding in epidemiological studies. Epidemiology for specialists. London: Imperial College; 2017. PHAST.
10. Smith GD, Ebrahim S. 'Mendelian randomization': can genetic epidemiology contribute to understanding environmental determinants of disease? Int J Epidemiol. 2003;32(1):1–22.
11. Zheng J, et al. Recent developments in Mendelian randomization studies. Curr Epidemiol Rep. 2017;4(4):330–45.
12. Emdin CA, Khera AV, Kathiresan S. Mendelian randomization. JAMA. 2017;318(19):1925–6.
13. Uffelmann E, Huang QQ, Munung NS. Genome-wide association studies. Nat Rev Methods Primers. 2021;1:59.
14. Garfield V. Sleep duration: a review of genome-wide association studies (GWAS) in adults from 2007 to 2020. Sleep Med Rev. 2021;56:101413.
15. Hemani G, et al. The MR-Base platform supports systematic causal inference across the human phenome. elife. 2018;7:e34408.
16. Sudlow C, et al. UK biobank: an open access resource for identifying the causes of a wide range of complex diseases of middle and old age. PLoS Med. 2015;12(3):e1001779.
17. Triendl R. Japan launches controversial Biobank project. Nat Med. 2003;9(8):982.
18. Chen Z, et al. China Kadoorie Biobank of 0.5 million people: survey methods, baseline characteristics and long-term follow-up. Int J Epidemiol. 2011;40(6):1652–66.
19. Sonti S, Grant SFA. Leveraging genetic discoveries for sleep to determine causal relationships with common complex traits. Sleep. 2022 Oct 10;45(10):zsac180.
20. Stefansson H, et al. A genetic risk factor for periodic limb movements in sleep. N Engl J Med. 2007;357(7):639–47.
21. Winkelmann J, et al. Genome-wide association study of restless legs syndrome identifies common variants in three genomic regions. Nat Genet. 2007;39(8):1000–6.
22. Didriksen M, et al. Large genome-wide association study identifies three novel risk variants for restless legs syndrome. Commun Biol. 2020;3(1):703.
23. Schormair B, et al. Reassessment of candidate gene studies for idiopathic restless legs syndrome in a large GWAS dataset of European ancestry. Sleep. 2022 Aug 11;45(8):zsac098.

24. Miyagawa T, et al. Variant between CPT1B and CHKB associated with susceptibility to narcolepsy. Nat Genet. 2008;40(11):1324–8.
25. Luca G, et al. Clinical, polysomnographic and genome-wide association analyses of narcolepsy with cataplexy: a European Narcolepsy Network study. J Sleep Res. 2013;22(5):482–95.
26. Hallmayer J, et al. Narcolepsy is strongly associated with the T-cell receptor alpha locus. Nat Genet. 2009;41(6):708–11.
27. Faraco J, et al. ImmunoChip study implicates antigen presentation to T cells in narcolepsy. PLoS Genet. 2013;9(2):e1003270.
28. Han F, et al. Genome wide analysis of narcolepsy in China implicates novel immune loci and reveals changes in association prior to versus after the 2009 H1N1 influenza pandemic. PLoS Genet. 2013;9(10):e1003880.
29. Jones SE, et al. Genome-wide association analyses in 128,266 individuals identifies new morningness and sleep duration loci. PLoS Genet. 2016;12(8):e1006125.
30. Hu Y, et al. GWAS of 89,283 individuals identifies genetic variants associated with self-reporting of being a morning person. Nat Commun. 2016;7:10448.
31. Lane JM, et al. Genome-wide association analysis identifies novel loci for chronotype in 100,420 individuals from the UK Biobank. Nat Commun. 2016;7:10889.
32. Lane JM, et al. Genome-wide association analyses of sleep disturbance traits identify new loci and highlight shared genetics with neuropsychiatric and metabolic traits. Nat Genet. 2017;49(2):274–81.
33. Dashti HS, et al. Genome-wide association study identifies genetic loci for self-reported habitual sleep duration supported by accelerometer-derived estimates. Nat Commun. 2019;10(1):1100.
34. Hammerschlag AR, et al. Genome-wide association analysis of insomnia complaints identifies risk genes and genetic overlap with psychiatric and metabolic traits. Nat Genet. 2017;49(11):1584–92.
35. Eriksson N, et al. Web-based, participant-driven studies yield novel genetic associations for common traits. PLoS Genet. 2010;6(6):e1000993.
36. Tung JY, et al. Efficient replication of over 180 genetic associations with self-reported medical data. PLoS One. 2011;6(8):e23473.
37. Jansen PR, et al. Genome-wide analysis of insomnia in 1,331,010 individuals identifies new risk loci and functional pathways. Nat Genet. 2019;51(3):394–403.
38. Strausz S, et al. Genetic analysis of obstructive sleep apnoea discovers a strong association with cardiometabolic health. Eur Respir J. 2021;57(5):2003091.
39. Hoffmann TJ, et al. A large multiethnic genome-wide association study of adult body mass index identifies novel loci. Genetics. 2018;210(2):499–515.
40. Xu H, et al. Genome-wide association study of obstructive sleep apnea and objective sleep-related traits identifies novel risk loci in Han Chinese individuals. Am J Respir Crit Care Med. 2022;206(12):1534–45.
41. Quinlan CM, et al. Identification of novel loci in obstructive sleep apnea in European American and African American children. Sleep. 2024 Mar 11;47(3):zsac182.
42. Staley JR, et al. PhenoScanner: a database of human genotype-phenotype associations. Bioinformatics. 2016;32(20):3207–9.
43. Zheng J, et al. LD Hub: a centralized database and web interface to perform LD score regression that maximizes the potential of summary level GWAS data for SNP heritability and genetic correlation analysis. Bioinformatics. 2017;33(2):272–9.
44. Burgess S, et al. Guidelines for performing Mendelian randomization investigations. Wellcome Open Res. 2019;4:186.
45. Hartwig FP, et al. Two-sample Mendelian randomization: avoiding the downsides of a powerful, widely applicable but potentially fallible technique. Int J Epidemiol. 2016;45(6):1717–26.
46. Burgess S, et al. Network Mendelian randomization: using genetic variants as instrumental variables to investigate mediation in causal pathways. Int J Epidemiol. 2015;44(2):484–95.

47. Haycock PC, et al. Best (but oft-forgotten) practices: the design, analysis, and interpretation of Mendelian randomization studies. Am J Clin Nutr. 2016;103(4):965–78.
48. Rees JMB, Foley CN, Burgess S. Factorial Mendelian randomization: using genetic variants to assess interactions. Int J Epidemiol. 2020;49(4):1147–58.
49. Montgomery AA, Peters TJ, Little P. Design, analysis and presentation of factorial randomised controlled trials. BMC Med Res Methodol. 2003;3:26.
50. Burgess S, Thompson SG. Multivariable Mendelian randomization: the use of pleiotropic genetic variants to estimate causal effects. Am J Epidemiol. 2015;181(4):251–60.
51. Burgess S, Butterworth A, Thompson SG. Mendelian randomization analysis with multiple genetic variants using summarized data. Genet Epidemiol. 2013;37(7):658–65.
52. Bowden J, Davey Smith G, Burgess S. Mendelian randomization with invalid instruments: effect estimation and bias detection through Egger regression. Int J Epidemiol. 2015;44(2):512–25.
53. Tang Y, et al. Sleep deprivation worsens inflammation and delays recovery in a mouse model of colitis. Sleep Med. 2009;10(6):597–603.
54. Jarasvaraparn C, et al. The relationship between sleep disturbance and disease activity in pediatric patients with inflammatory bowel disease. J Pediatr Gastroenterol Nutr. 2019;68(2):237–43.
55. Marinelli C, et al. Sleep disturbance in inflammatory bowel disease: prevalence and risk factors - a cross-sectional study. Sci Rep. 2020;10(1):507.
56. Hao G, et al. Sleep quality and disease activity in patients with inflammatory bowel disease: a systematic review and meta-analysis. Sleep Med. 2020;75:301–8.
57. Chen M, et al. Differential sleep traits have no causal effect on inflammatory bowel diseases: a Mendelian randomization study. Front Pharmacol. 2021;12:763649.
58. Cullell N, et al. Sleep/wake cycle alterations as a cause of neurodegenerative diseases: a Mendelian randomization study. Neurobiol Aging. 2021;106:320 e1–320 e12.
59. Iranzo A. Sleep in neurodegenerative diseases. Sleep Med Clin. 2016;11(1):1–18.
60. Schmidt C, Peigneux P, Cajochen C. Age-related changes in sleep and circadian rhythms: impact on cognitive performance and underlying neuroanatomical networks. Front Neurol. 2012;3:118.
61. Abbott SM, Videnovic A. Chronic sleep disturbance and neural injury: links to neurodegenerative disease. Nat Sci Sleep. 2016;8:55–61.
62. Henry A, et al. The relationship between sleep duration, cognition and dementia: a Mendelian randomization study. Int J Epidemiol. 2019;48(3):849–60.
63. Sun J, et al. Polygenic evidence and overlapped brain functional connectivities for the association between chronic pain and sleep disturbance. Transl Psychiatry. 2020;10(1):252.
64. Broberg M, et al. Mendelian randomization highlights insomnia as a risk factor for pain diagnoses. Sleep. 2021;44(7):zsab025.
65. Anderson EL, et al. Is disrupted sleep a risk factor for Alzheimer's disease? Evidence from a two-sample Mendelian randomization analysis. Int J Epidemiol. 2021;50(3):817–28.
66. Yuan S, et al. Sleep duration, genetic susceptibility, and Alzheimer's disease: a longitudinal UK Biobank-based study. BMC Geriatr. 2022;22(1):638.
67. Huang J, et al. Sleep, major depressive disorder, and Alzheimer disease: a Mendelian randomization study. Neurology. 2020;95(14):e1963–70.
68. Chen D, et al. Sleep and late-onset Alzheimer's disease: shared genetic risk factors, drug targets, molecular mechanisms, and causal effects. Front Genet. 2022;13:794202.
69. Li J, et al. Obstructive sleep apnea and the risk of Alzheimer's disease and Parkinson disease: a Mendelian randomization study OSA, Alzheimer's disease and Parkinson disease. Sleep Med. 2022;97:55–63.
70. Zhang G, et al. Daytime sleepiness might increase the risk of ALS: a 2-sample Mendelian randomization study. J Neurol. 2021;268(11):4332–9.

71. Di H, et al. Bidirectional Mendelian randomization to explore the causal relationships between Sleep traits, Parkinson's disease and Amyotrophic lateral sclerosis. Sleep Med. 2022;96:42–9.
72. Grover S, Sharma M. International age-related macular degeneration genomics, sleep, pain, and neurodegeneration: a Mendelian randomization study. Front Neurol. 2022;13:765321.
73. Noyce AJ, et al. The Parkinson's disease Mendelian randomization research portal. Mov Disord. 2019;34(12):1864–72.
74. Harvey AG. A transdiagnostic approach to treating sleep disturbance in psychiatric disorders. Cogn Behav Ther. 2009;38(Suppl 1):35–42.
75. Walker WH 2nd, et al. Circadian rhythm disruption and mental health. Transl Psychiatry. 2020;10(1):28.
76. Jones SE, et al. Genome-wide association analyses of chronotype in 697,828 individuals provides insights into circadian rhythms. Nat Commun. 2019;10(1):343.
77. Choi KW, et al. An exposure-wide and Mendelian randomization approach to identifying modifiable factors for the prevention of depression. Am J Psychiatry. 2020;177(10):944–54.
78. Cai L, et al. Causal links between major depressive disorder and insomnia: a Mendelian randomisation study. Gene. 2021;768:145271.
79. Zhou F, et al. Assessing the causal associations of insomnia with depressive symptoms and subjective well-being: a bidirectional Mendelian randomization study. Sleep Med. 2021;87:85–91.
80. Daghlas I, et al. Genetically proxied diurnal preference, sleep timing, and risk of major depressive disorder. JAMA Psychiatry. 2021;78(8):903–10.
81. O'Loughlin J, et al. Using Mendelian randomisation methods to understand whether diurnal preference is causally related to mental health. Mol Psychiatry. 2021;26(11):6305–16.
82. Sun X, et al. Sleep disturbance and psychiatric disorders: a bidirectional Mendelian randomisation study. Epidemiol Psychiatr Sci. 2022;31:e26.
83. Baranova A, Cao H, Zhang F. Shared genetic liability and causal effects between major depressive disorder and insomnia. Hum Mol Genet. 2022;31(8):1336–45.
84. Chen G, et al. Association between depression and sleep apnoea: a Mendelian randomisation study. ERJ Open Res. 2022;8(1):00394-2021.
85. Gao X, et al. The bidirectional causal relationships of insomnia with five major psychiatric disorders: a Mendelian randomization study. Eur Psychiatry. 2019;60:79–85.
86. Wang Z, et al. The causal relationship between sleep traits and the risk of schizophrenia: a two-sample bidirectional Mendelian randomization study. BMC Psychiatry. 2022;22(1):399.
87. Carpena MX, et al. Sleep-related traits and attention-deficit/hyperactivity disorder comorbidity: shared genetic risk factors, molecular mechanisms, and causal effects. World J Biol Psychiatry. 2021;22(10):778–91.
88. Rangaraj VR, Knutson KL. Association between sleep deficiency and cardiometabolic disease: implications for health disparities. Sleep Med. 2016;18:19–35.
89. Doherty A, et al. GWAS identifies 14 loci for device-measured physical activity and sleep duration. Nat Commun. 2018;9(1):5257.
90. Jones SE, et al. Genetic studies of accelerometer-based sleep measures yield new insights into human sleep behaviour. Nat Commun. 2019;10(1):1585.
91. Xiuyun W, et al. Network Mendelian randomization study: exploring the causal pathway from insomnia to type 2 diabetes. BMJ Open Diabetes Res Care. 2022;10(1):e002510.
92. Wang J, et al. Sleep duration and adiposity in children and adults: observational and Mendelian randomization studies. Obesity (Silver Spring). 2019;27(6):1013–22.
93. Campos AI, et al. Insights into the aetiology of snoring from observational and genetic investigations in the UK Biobank. Nat Commun. 2020;11(1):817.
94. Chen W, et al. Causal effect of obstructive sleep apnea on atrial fibrillation: a Mendelian randomization study. J Am Heart Assoc. 2021;10(23):e022560.
95. Chen L, et al. Obstructive sleep apnea and atrial fibrillation: insights from a bidirectional Mendelian randomization study. BMC Med Genet. 2022;15(1):28.

96. Li Y, et al. Assessment of the causal effects of obstructive sleep apnea on atrial fibrillation: a Mendelian randomization study. Front Cardiovasc Med. 2022;9:843681.
97. Ardissino M, et al. Sleep disordered breathing, obesity and atrial fibrillation: a Mendelian randomisation study. Genes (Basel). 2022;13(1):104.
98. Dashti HS, Ordovas JM. Genetics of sleep and insights into its relationship with obesity. Annu Rev Nutr. 2021;41:223–52.
99. Dashti HS, et al. Genetic determinants of daytime napping and effects on cardiometabolic health. Nat Commun. 2021;12(1):900.
100. Yuan S, Larsson SC. An atlas on risk factors for type 2 diabetes: a wide-angled Mendelian randomisation study. Diabetologia. 2020;63(11):2359–71.
101. Gao X, et al. Investigating causal relations between sleep-related traits and risk of type 2 diabetes mellitus: a Mendelian randomization study. Front Genet. 2020;11:607865.
102. Jia Y, et al. Diet, lifestyle behaviour and other risk factors associated with type 2 diabetes beyond body mass index: a Mendelian randomization study. Can J Diabetes. 2022;46:822–8.
103. Wang J, et al. Sleep duration and risk of diabetes: observational and Mendelian randomization studies. Prev Med. 2019;119:24–30.
104. Ai S, et al. Causal associations of short and long sleep durations with 12 cardiovascular diseases: linear and nonlinear Mendelian randomization analyses in UK Biobank. Eur Heart J. 2021;42(34):3349–57.
105. Liu X, et al. Genetically predicted insomnia in relation to 14 cardiovascular conditions and 17 cardiometabolic risk factors: a Mendelian randomization study. J Am Heart Assoc. 2021;10(15):e020187.
106. Larsson SC, Markus HS. Genetic liability to insomnia and cardiovascular disease risk. Circulation. 2019;140(9):796–8.
107. Daghlas I, et al. Sleep duration and myocardial infarction. J Am Coll Cardiol. 2019;74(10):1304–14.
108. van Oort S, et al. Modifiable lifestyle factors and heart failure: a Mendelian randomization study. Am Heart J. 2020;227:64–73.
109. van Oort S, et al. Association of cardiovascular risk factors and lifestyle behaviors with hypertension: a Mendelian randomization study. Hypertension. 2020;76(6):1971–9.
110. Titova OE, Michaelsson K, Larsson SC. Sleep duration and stroke: prospective cohort study and Mendelian randomization analysis. Stroke. 2020;51(11):3279–85.
111. Lu H, et al. Sleep duration and stroke: a Mendelian randomization study. Front Neurol. 2020;11:976.
112. Richmond RC, et al. Investigating causal relations between sleep traits and risk of breast cancer in women: Mendelian randomisation study. BMJ. 2019;365:l2327.
113. Sun X, et al. Genetically proxied morning chronotype was associated with a reduced risk of prostate cancer. Sleep. 2021;44(10):zsab104.
114. Yuan S, et al. Morning chronotype and digestive tract cancers: Mendelian randomization study. Int J Cancer. 2022;152:697–704.
115. Yap DWT, et al. The association of obstructive sleep apnea with breast cancer incidence and mortality: a systematic review and meta-analysis. J Breast Cancer. 2022;25(3):149–63.
116. Gao XL, et al. Obstructive sleep apnea syndrome and causal relationship with female breast cancer: a Mendelian randomization study. Aging (Albany NY). 2020;12(5):4082–92.
117. Wang J, et al. Association between sleep traits and lung cancer: a Mendelian randomization study. J Immunol Res. 2021;2021:1893882.
118. Shen J, et al. Genetic liability to insomnia and lung cancer risk: a Mendelian randomization analysis. Front Genet. 2021;12:756908.
119. Huo Z, et al. Genetically predicted insomnia and lung cancer risk: a Mendelian randomization study. Sleep Med. 2021;87:183–90.
120. Wu Y, et al. GWAS of peptic ulcer disease implicates Helicobacter pylori infection, other gastrointestinal disorders and depression. Nat Commun. 2021;12(1):1146.

121. Yang Q, et al. Associations between insomnia and pregnancy and perinatal outcomes: evidence from Mendelian randomization and multivariable regression analyses. PLoS Med. 2022;19(9):e1004090.
122. Ni J, et al. Evidence for causal effects of sleep disturbances on risk for osteoarthritis: a univariable and multivariable Mendelian randomization study. Osteoarthr Cartil. 2021;30:443–50.
123. Mazidi M, et al. Longer sleep duration may negatively affect renal function. Int Urol Nephrol. 2021;53(2):325–32.
124. Dashti HS, Redline S, Saxena R. Polygenic risk score identifies associations between sleep duration and diseases determined from an electronic medical record biobank. Sleep. 2019;42(3):zsy247.
125. Edelson JL, et al. The genetic etiology of periodic leg movement in sleep. Sleep. 2022;46(4):zsac121.
126. Wilkins AS. Canalization: a molecular genetic perspective. BioEssays. 1997;19(3):257–62.
127. Khera AV, et al. Genome-wide polygenic scores for common diseases identify individuals with risk equivalent to monogenic mutations. Nat Genet. 2018;50(9):1219–24.

Genetics of Insomnia

Rackeb Tesfaye and Philip Gehrman

Abstract Insomnia is the most prevalent sleep disorder, but surprisingly little attention has been paid to genetic risk factors until the past decade. Twin and family aggregation studies have found that 30–40% of variability in insomnia is due to genetic factors. Candidate gene studies found associations between insomnia and a range of genes involved in circadian rhythms and neurotransmitter function, but results have often failed to replicate. Rapid increases in available genome-wide association data have led to a burst of significant associations, with several replicating in multiple cohorts. The findings support substantial pleitropy with genetic risk factors for psychiatric disorder, suggesting that insomnia-specific genetic effects are unlikely to be found. There is now a need for these genetic variants to be explored in model systems to determine which are truly causal for insomnia.

Keywords Insomnia · Heritability · Candidate genes · Genome wide association study

Insomnia is the most common sleep disorder but, until recently, knowledge about the genetics of insomnia lagged behind that of other mental health diagnoses. In the past 10 years that has been a rapid uptick in this area of research and we are beginning to have some understanding of the contribution of genetic factors to the pathophysiology of insomnia. The goal of this chapter is to review our current understanding of the genetics underpinnings of insomnia.

R. Tesfaye
Department of Neurology and Neurosurgery, McGill University, Montreal, Canada
e-mail: rackeb.tesfaye@mail.mcgill.ca

P. Gehrman (✉)
Department of Psychiatry, Perelman School of Medicine of the University of Pennsylvania, Philadelphia, PA, United States
e-mail: philip.gehrman@pennmedicine.upenn.edu

P. Gehrman et al. (eds.), *Genetics of Sleep and Sleep Disorders*,
https://doi.org/10.1007/978-3-031-62723-1_10

Characterization of the Insomnia Phenotype

Clinical Definition

A formal diagnosis of an insomnia disorder is made by a clinical expert following diagnostic guidelines from the Diagnostic and Statistical Manual of Mental Disorders edition 5 (DSM-5; [1]) or the International Classification of Sleep Disorders Edition 3 (ICSD-3; [2]). Insomnia refers to at least one complaint of difficulty initiating sleep (i.e., sleep onset delay), difficulty maintaining sleep (i.e., nocturnal awakenings), or early-morning awakenings with an inability to return to sleep [1]. These difficulties must be present at least three nights per week for at least 3 months and have significant daytime functioning consequences. In children, insomnia typically presents as a behavioral disorder of two types (ICSD-3). Sleep-Onset Association Type refers to infants and children who cannot fall asleep without specific conditions (e.g., an item or parental interventions), or are unable to soothe themselves back to sleep resulting in prolonged night awakenings. Limit Setting Type characterizes children who exhibit bedtime resistance and often results from parents who struggle to enforce regular bedtimes or rules, which may delay sleep onset.

A diagnosis of insomnia disorder is based on subjective complaints, rather than objective measurements. Physiological recordings, including those captured by polysomnography (PSG), have demonstrated that individuals with insomnia are more likely than controls to have lower total sleep time, decreased sleep efficiency, and a lower portion of time spent in slow-wave and rapid eye movement sleep [3]. However, not all individuals with insomnia have atypical sleep architecture. Hence, physiological measures cannot be relied on in diagnostic decision-making for insomnia, but they may help rule out the contribution of other sleep disturbances (e.g., sleep apnea or restless legs syndrome).

Prevalence and Impact

Insomnia is the most prevalent sleep complaint in the general population. Accounts of insomnia symptoms have increased in the past decade [4, 5], with some reports suggesting a two-fold increase in 15 years from the early 1990s to the 2000s [6]. Current estimates suggest that ~30% of the adult population have occasional insomnia symptoms, while ~10% meet formal diagnostic criteria [7]. In clinical populations, estimates of insomnia symptoms rise to ~50% [8], with over one-third receiving a formal DSM diagnosis [9].

Insomnia is a concern throughout the lifespan, 20–25% of children and 25–35% of adolescents in the general population experience insomnia symptoms [10]. Longitudinal analyses reveal that insomnia symptoms at an early age often persist into adulthood, and the severity of insomnia symptoms increases with age [10].

Moreover, females are found to be at greater risk for insomnia, with one meta-analysis suggesting an Odds Ratio (OR) = 1.4 [95% confidence interval (CI): 1.28–1.55] for women compared to men [11].

Insomnia is a major public health concern and is associated with psychiatric and neurodevelopmental conditions [9, 12], poor health outcomes [13], lower quality of life, and mortality [14]. Hence, delineating factors contributing to insomnia is imperative.

Heterogeneity of Insomnia

The heterogeneity of the insomnia phenotype makes it complex to investigate. Previous versions of the DSM and ICD attempted to partition insomnia into diagnostic subtypes, yet this yielded low validity and reliability [15]. The phenotypic presentation of insomnia can differ over time and often accompanies multiple co-morbidities rather than in isolation [16]. Despite early attempts falling short, calls have been made to leverage large-scale datasets to inform meaningful subtypes of insomnia that consider both sleep and non-sleep characteristics [16, 17].

In the general population, recent multi-data-driven approaches have identified subtypes of insomnia based on behavior, environmental factors, and distinct sleep traits [18, 19]. Distinct insomnia subtypes have even been found to differ in developmental trajectories of sleep complaints, health and psychiatric risks, and response to sleep treatments [18]. These findings illustrate the dynamic and multifactorial nature of insomnia and its underlying causal factors. It will be important over time to validate subytpes and determine whether they have distinct genetic etiologies.

Insomnia Heritability: Evidence for Genetic Underpinnings

In 1990, the first twin study of insomnia was conducted in an Australian cohort of 1792 monozygotic (MZ) twins and 2101 dizygotic (DZ) twins aged 17–88 years [20]. This study provided the first evidence for insomnia being heritable, implicating genetics as a causal factor of insomnia (see Fig. 1. for a timeline of genetic discovery in insomnia). *Heritability* estimates the relative contribution of genetic variation versus environmental variation in a trait. However, it does not account for the interaction between genes and the environment. Twin studies are able to provide heritability estimates as they rely on the equal environments assumption, which presumes that MZ twins who share 100% of their genetic material and dizygotic DZ twins who share 50% of their genetic material are exposed to similar environments [21]. Therefore, differences in trait similarity between MZ and DZ twins can be attributed to genetic relatedness. The heritability coefficient (h^2) derived in twin studies represents the proportion of variance in a trait that can be attributed to

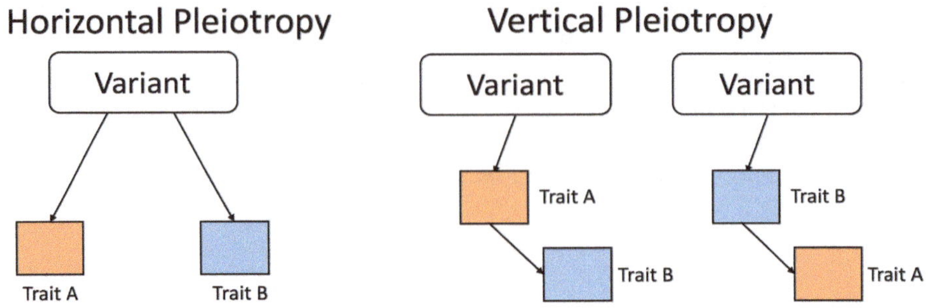

Fig. 1 Horizontal and vertical pleiotropic pathways

genetics. An h^2 of 0 (lower limit) represents no genetic contribution to a trait, while 1 (upper limit) represents a 100% genetic contribution to a trait.

Multiple twin studies since the 1990s have contributed to advancing the accuracy and reliability of heritability estimates of insomnia. The most recent meta-analysis of over 47,000 children and adults from twin studies suggests insomnia is approximately 40% heritable [22]. The authors found that heritability was higher for females (37%) compared to men (30%), suggesting females may carry a greater biological risk for insomnia. It's of note that while no age differences were found, most twin studies to date have been conducted in adult populations. Furthermore, the majority of these studies have used self-reports, as opposed to clinician respondents. Additional assessment of how heritability estimates may differ based on clinical confirmation of insomnia or with sleep traits captured by PSG is needed.

Family aggregation studies have also contributed to our understanding of the genetic contribution to insomnia, by testing the degree of family resemblance of a trait. For instance, given a family's shared genetic background, the genetic influence of insomnia would be higher if insomniac family members reported more insomnia symptoms compared to unrelated controls. This genetic influence would be stronger if symptom reporting was greater in individuals with a higher degree of relatedness (e.g., mom compared to cousin). However, a caveat is that due to the shared environment of family members, family aggregation could either be a result of genes or the environment, which cannot be parsed out. Regardless, concordant with twin studies, the few family aggregation studies to date have confirmed that insomnia likely "runs in the family" [23, 24]. One study suggests family aggregation may only occur for chronic cases of insomnia [25]. More recently these results have been confirmed in psychiatric populations, like autism and Attention Deficit Hyper Activity Disorder (ADHD). A recent familial co-aggregation (i.e., relatedness of two traits) study investigating family members of over 50,000 individuals with autism found relatives of autistic individuals compared to controls had an increased risk of sleep onset and maintenance difficulties [26]. The highest risk for sleep problems was found in MZ twins of individuals with ASD (OR = 6.6 [2.5–17.4]), followed by DZ twins (OR = 2.6 [1.5–4.5]) and full-siblings (OR = 2.5 [2.4–2.6]), with half-siblings and cousins least at risk. Based on twin analyses, the authors reported a moderate shared genetic etiology between ASD and sleep onset and maintenance difficulties, mirroring similar results in ADHD [27].

The Search for Insomnia Genes in Humans

Attempts to uncover the genetic underpinnings of insomnia in humans have broadly taken two approaches, candidate genes studies and Genome Wide Association Studies (GWAS) described below. At its crux, both approaches rely on being able to detect variations within the genome that can be linked to a trait—insomnia. The most common type of genetic variant studied is a single nucleotide polymorphism (SNP). SNPs represent a substitution in a single nucleotide base pair found in >1% of the population (representing ~1 in 1000 nucleotides).

Less studied variants include variable number tandem repeats (VNTRs), areas in the genome where short nucleotide sequences are repeated. The majority of the insomnia genetics literature has focused on common variants; however, a small subset has investigated rare variations found in <1% of the population. Although rare variations are more difficult to identify in populations due to their frequency, hence limiting study sample sizes, they are often linked to more deleterious (e.g., loss of gene function) and larger quantifiable effects.

Candidate Gene Studies and Insomnia

Early attempts to identify genes associated with insomnia primarily used candidate gene studies, an approach whereby genes are investigated due to a priori evidence of their influence on known sleep/wake mechanisms. This often means drawing from and validating mechanisms extensively studied in animal models. Likely candidates have included circadian pathway genes. The seminal Nobel Prize discovery of the circadian mechanism in *drosophila* and mice over 50 years ago led to our understanding that core genes expressed in the SCN regulate circadian rhythms in humans through an autoregulatory transcription-translation feedback loop (See [28] for a comprehensive review). Variants in core clock genes *CRY1* [29] and *PER3* [30, 31], along with the melatonin (a neuropeptide produced in the pineal gland that resets the circadian cycle) synthesis gene *AANAT* [32], have been implicated in delayed sleep phase disorder (DSPD) (see Chapter "Genetics of Chronotype and Circadian Rhythm Disorders"). DSPD is characterized by a biological delay in sleep onset and wake times that do not align with societal convention and often presents as insomnia. Conversely, advanced sleep phase disorder (ASPD), marked by a need to fall asleep and wake up earlier relative to societal norms, has also been linked to variants in multiple clock genes like *PER2* [33, 34], *PER3* [35], *CRY2* [36] and *CSNK1D* [37, 38]. Although preliminary evidence in small samples suggests clock genes like *PER2* [39], *PER3* [40], and *CLOCK* [41] are likely associated with insomnia traits, limited investigations have examined the impact of circadian-related genes and the insomnia phenotype. Moreover, despite supplemental melatonin being commonly used to ameliorate insomnia symptoms, no rigorous study examining the contributions of melatonin pathway genes to insomnia has been conducted. The limited investigations of genes involved in other mechanisms implicated with sleep, including serotonergic [42–44], GABAergic [45], and dopaminergic pathways [46, 47], have yielded mixed associations with insomnia and related sleep traits. As these genes are involved in mechanisms targeted by pharmacological treatments for insomnia, continuing to examine their contribution to insomnia is of significance.

Overall, due to small sample sizes and the reliance on known mechanisms, candidate gene approaches have not yet produced robust advances toward uncovering the genetic underpinnings of insomnia. Moving forward, complimentary animal model and human validation approaches provide an opportunity for pinpointing genes of interest, as illustrated in the first animal model study of insomnia [48]. After designing *drosophila* that exhibit human-like insomnia traits, the authors identified genes differentially expressed compared to *drosophila* without insomnia using whole-genome profiling [48]. The authors then sleep deprived healthy participants and found that two of the candidate genes identified in the insomnia *drosophila* model were upregulated after acute sleep deprivation. Experimentally manipulating sleep related to insomnia traits and subsequently examining genetic changes can provide insight into causal factors linked to insomnia; hence, similar approaches should continue to be applied at a larger scale.

Understanding how other factors increase the likelihood of a negative sleep experience in the presence of genomic variants will also need to be considered in future studies. Emerging literature suggests stress may be one such risk factor [49]. In one study, adults with high job-related stress and a polymorphism in the serotonin transport gene 5-HTTLPR resulting in an SS genotype were five times more likely to experience insomnia compared to individuals with low job-related stress and no SS phenotype [44]. Interestingly, low job stress was a protective factor for individuals with an SS genotype, as they were less likely to experience insomnia compared to individuals with low job-related stress and no SS genotype. A twin study with young adults found that sleep reactivity to stress (i.e., the likelihood of having disturbed sleep like insomnia in response to stressful situations) was moderately heritable (29–43%) and correlated with genetic variances of insomnia symptoms [50]. These findings suggest sleep reactivity to stress may be a genetic vulnerability that predisposes individuals to develop insomnia. Taken together, the interplay between insomnia, stress, and genetics can inform one potential pathway for how insomnia emerges and persists during the human lifespan.

Genome-Wide Association Studies (GWAS) and Insomnia

Over the past decade, the search for genetic variants underlying insomnia has largely relied on GWAS (see Table 1). A genome-wide association study (GWA) study is an approach used to identify genetic variants (i.e., SNPs) statistically associated with a trait or disease by scanning the genomes (millions of SNPs) of many individuals in a large population cohort. Unlike gene candidate studies, GWAS are unrestricted to biological mechanisms or functions, enabling the identification of any type of genetic variant overrepresented in individuals with the outcome of interest compared to case controls.

The Beginning of the GWAS Approach in Insomnia Research

The first (GWA) study of insomnia in 2011 [51] examined 10,038 Korean adults who self-reported insomnia symptoms. The authors found no significant SNPs associated with self-reported insomnia that passed the widely accepted P value threshold for GWAS, 5×10^{-8}. SNPs s11208305 in *ROR1* and rs718712 in *PLCB1* were the most significantly related to insomnia traits. Both ROR1 and PLCB1 genes have been linked to psychiatric conditions like bipolar disorder and schizophrenia; however, neither gene has been identified in subsequent GWAS of insomnia.

The following (GWA) study in 2013 was conducted on a smaller cohort of 2323 adult Australian twins and similarly found no SNPs were significantly associated with an insomnia factor score at the genome-wide level [52]. Further investigations of distinct self-reported sleep items that loaded onto the insomnia factor score revealed a set of SNPs within the *CACNA1C* gene were the most strongly associated with self-reported sleep latency. Like the previous GWAS, CACNA1C—a gene encoding a subunit of a voltage-dependent calcium channel—is also linked with psychiatric disorders, including bipolar disorder and depression.

Notably, CACNA1C is activated by orexin (hypocretin), a neuropeptide produces in the hypothalamus that regulates arousal [62], and *CACNA1C* variations have been related to atypical EEG oscillation during REM sleep in mice [63].

Although the authors were not able to replicate these *CACNA1C* findings in an independent cohort, subsequent investigations found SNP rs16929277 within *CACNA1C* was associated with self-reported sleep quality [64], while later GWAS of insomnia have since identified SNPs in other calcium channel genes that may contribute to insomnia traits [57, 58]. Genomic pathway analysis of SNPs linked to self-reported sleep duration in European adults also suggests calcium channel genes contribute to sleep duration variance [65], while mouse models have implicated the T-type calcium channel gene *CACNA1I* as disrupting electrical sleep spindle activity that occurs during slow wave sleep and is a suggested mediator of sleep spindle atypicality found in schizophrenia [66]. Interestingly, the first GWAS of insomnia was also the first to identify a calcium channel gene (*CACNA1A*) as one potential insomnia candidate gene, but their observations did not reach statistical significance

Table 1 Description of insomnia GWAS

Study	Country	*N*=	Sex (female/male%)	Age	Ethnicity or ancestry	Sleep measure	Insomnia traits	Replication group	Genes identified
[51]	Korea	Total = 10,038 Insomnia cases/controls = 1439/7280	N/A	Range = 40–69 years	East Asian	Self-report questionnaire (Unvalidated)	Insomnia factor score: "Difficulty falling asleep" "Wake up during the night and not be able to fall back asleep" "Wake up too early in the morning"	No	(1) No SNPs passed multiple comparison testing (2) Top gene hits[a]: *ROR1* & *PLCB1* (3) Weaker gene hits[a]: *CACNA1A, GNAS, NOS3,* & *ADCY8*
[52]	Australian	Total = 7616	62.7/37.3	Mean = 34.5 years (SD = 14.3)	N/A	Self-report questionnaire (Unvalidated)	*"On weekdays, how long in minutes do you think it usually takes you to fall asleep from when you first try to go to sleep?"* (Sleep Latency)	Four independent Cohorts (self-report insomnia items)	(1) No SNPs passed multiple comparison testing (2) Top gene hit[a]: *CACNA1C*, not replicated in independent cohorts
[53]	Netherlands	Total = 4242	60.6/39.5[c]	Range = 18–65 years	European	Self-report questionnaire (Munich Chronotype Questionnaire)	*"I need (x) minutes to fall asleep"* Sleep latency	Twelve independent cohorts (self-report insomnia items)	(1) No SNPs passed multiple comparison testing (2) Top gene hit[a]: *RBFOX3*. (3) Three SNPs in *RBFOX3* replicated in independent cohorts[a]
[54]	Germany	Total = 956	50.4/49.5	Range = 40–79 years	European	Actigraphy	Sleep Latency (minutes) Number of awakenings after sleep onset	Replication of SNPs in literature (previous GWA and candidate gene studies).	(1) SNP within *DMRT1* associated with sleep latency (2)SNP within *CSNK2A1* marginally associated with sleep latency, marginally replicates in the literature review
[55]	United States	Total = 59,128 Insomnia cases/controls = 32,155/26,973	51.5/48.5	Range = 40–69 years	European	Self-report questionnaire (Unvalidated)	"Do you have trouble falling asleep at night or do you wake up in the middle of the night?"	No	(1) Top gene hits[b]: *MEIS1, TMEM132E, CYCL1* and *TGFBI* (females) and *WDR27* (males)

[56]	Netherlands	Total = 113,006 Insomnia cases/ controls = 32,384/80,622	52.5/47.5	Range = 40–69 years	European	Self-report questionnaire (Unvalidated)	"Do you have trouble falling asleep at night or do you wake up in the middle of the night?"	One independent sample (clinically diagnosed with insomnia)	(1) Top gene hits[b]: *MEIS1*, *SCFD2*, *DCBLD1* and *MED27*. *HHEX and RHCG* for males and *IPO7* and *TSNARE* for females
[57]	United States	Total = 17,651 Insomnia cases/ controls = 3237/14,414	0/100	Mean = 21.0 years (SD = 3.3)	European American African American Latino American	Self-report questionnaire (Brief Insomnia Questionnaire)	"Did you ever in your life have a whole month or longer when you had insomnia at least three nights a week?" *Follow up (if yes):* "Think of a typical month when your insomnia was worst. During that month, how much did your insomnia interfere with your daytime functioning in the following ways": (1) daytime fatigue, sleepiness, or low motivation, (2) headaches, upset stomach, diarrhea or constipation, (3) moodiness (irritability, nerves, worry, or depression), (4) reduced performance at work or school, and (5) accident proneness."	GWA analyses for each ancestry group were run separately then pooled in meta- analysis	(1) Across ancestry groups 3 SNPs in chromosome 7 (not within a gene) were associated with insomnia[b] (2) African ancestry group top gene hits include[b]: NTF3 and CTD-2535I10.1 (LOC105371069; predicted non-coding gene) (3) European ancestry group had no genome-wide significant SNPs. Top gene hit[a]: *RFX3* (4) Latin American ancestry group had no genome-wide significant SNPs. Top gene hits include[a]: *WDR17*, *ENPP6* and *DEC1*

(continued)

Table 1 (continued)

Study	Country	*N*=	Sex (female/ male%)	Age	Ethnicity or ancestry	Sleep measure	Insomnia traits	Replication group	Genes identified
[58]	Netherlands	Total = 1,331,010 Insomnia cases/ controls = 398,004/933,006	53.3/46.7	Range = 40–69 years	European	Self-report questionnaire (Unvalidated)	*UkBiobank Cohort:* "Do you have trouble falling asleep at night or do you wake up in the middle of the night?" *23andMe cohort*: "Have you ever been diagnosed with, or treated for: Insomnia?" "Have you ever been 240 diagnosed with, or treated for, any of the following conditions: Insomnia but not Narcolepsy, 241 Sleep apnea or Restless leg syndrome" "Has a doctor ever told you that you have any of 242 these conditions: Insomnia (difficulty getting to sleep or staying asleep)?" "Do you routinely have trouble getting to sleep at night?"	GWA analyses for each cohort were run separately then pooled in a meta- analysis	(1) 956 genes associated with insomnia traits. Top gene hits include[b]: *MEIS1* and BTBD9

[59]	United States	Total = 453,379 Frequent insomnia symptoms/controls = 129,270/108,352 Any insomnia symptoms/controls = 345,022/108,352	61.2/38.8	Range = 40–69 years	European	Self-report questionnaire (Unvalidated)	"Do you have trouble falling asleep at night or do you wake up in the middle of the night?"	One independent cohort with self-reported insomnia symptoms One independent cohort physician diagnosed insomnia Replication with actigraphy derived sleep efficiency	(1) Up to 236 genes associated with insomnia trait[b]. (2) Top gene hit *MEIS1* was replicated in both independent cohorts and associated with lower sleep efficiency[b]
[60]	United States	Total = 18,055 Insomnia cases/controls = 3135/14,920	53.6/46.4	Mean = 59.7 (SD = 16.70)	European	ICD codes of insomnia (diagnostic records)	*307.4:* specific disorders of sleep of nonorganic origin *327.0:* organic disorders of initiating and maintaining sleep; *780.51:* insomnia with sleep apnea, unspecified *780.52:* insomnia, unspecified; *G47.0:* insomnia; *F51.0:* insomnia not due to a substance or known physiological condition.	No	(1) One SNP in a novel genetic locus on chromosome 8 (not within a gene) was associated with insomnia[b] (2) Top gene hits marginally associated with insomnia include[a]: *SORCS2* and *SLIT2*.

(continued)

Table 1 (continued)

Study	Country	*N*=	Sex (female/ male%)	Age	Ethnicity or ancestry	Sleep measure	Insomnia traits	Replication group	Genes identified
[61]	Netherlands	Total = 2,365,010 Insomnia cases/controls = 593,724/1,771,286	59.6/40.4	Range = 40–69 years	European	Self-report questionnaire (Unvalidated)	*UkBiobank Cohort:* "Do you have trouble falling asleep at night or do you wake up in the middle of the night?" *23andMe cohort*: "Have you ever been diagnosed with, or treated for: Insomnia?" "Have you ever been 240 diagnosed with, or treated for, any of the following conditions: Insomnia but not Narcolepsy, 241 Sleep apnea or Restless leg syndrome" "Has a doctor ever told you that you have any of 242 these conditions: Insomnia (difficulty getting to sleep or staying asleep)?" "Do you routinely have trouble getting to sleep at night?"	GWA analyses for each cohort were run separately then pooled in a meta- analysis	(1) Insomnia associated with 3893 genes or which 289 were classified as high confidence genes (i.e., potentially causal) (2) Top gene hits include[b]: *PTPRD and LSAMP, along with replicated MEIS1 and BTBD9 genes from previous GWAS*

[a]Marginally significant *p*-value (did not pass genome-wide significance $<5.0 \times 10^{-8}$)
[b]Passed genome-wide significance $<5.0 \times 10^{-8}$
[c]Sex ratio were calculated based on data from primary GWAS cohorts available

[51]. Taken together, there is accumulating evidence that calcium channel genes may contribute to various sleep and psychiatric traits.

It was evident in early genome-wide association studies that genetic variants related to self-reported insomnia traits were additionally related to other outcomes like psychiatric disorders. These findings also extended to physiologically derived insomnia traits. A 2016 GWAS investigating 11 actigraphic measures linked the *DMRT1* gene to sleep latency, while a common variation in the *UFL1* gene was associated with lower sleep efficiency, which is related to insomnia. Both genes are associated with schizophrenia [67, 68], in fact, a complete gain or loss of the *DMRT1* gene (known as a copy number variation; CNV) increases the risk of schizophrenia by an OR = 12 [CI: 1.6–98.8] [67].

Despite contributing to our understanding of the complex genetic landscape of insomnia, early genome-wide association studies were limited in sample size. This limitation made it difficult to identify multiple variants at the conservative genome-wide p-value estimate that accounts for multiple testing [69, 70]. Early attempts to circumvent this issue included GWAS meta-analyses, whereby data from multiple studies are combined, and subsequent replication in independent datasets, but this still yielded marginal results [53]. As sequencing technology advanced and became cheaper, genotyping larger cohorts became has become more feasible, leading to the inclusion of hundreds of thousands to millions of individuals in genome-wide association studies to date.

The Large-Scale Expansion of the GWAS Approach in Insomnia Research

The UK Biobank is an ongoing longitudinal study aimed to understand multidimensional factors related to health [71]. It currently represents the most expansive phenotyped and genotyped cohort in the world, with over 500,000 participants from the United Kingdom aged 37–74. In 2017, two separate genome-wide association studies of insomnia and other sleep traits included over 100,000 individuals from the UKBiobank [55, 56]. These studies ushered in the large-scale expansion of insomnia genetics research and the identification of novel loci.

Using subsets of UKBiobank participants, Hammerschlag et al. and Lane et al. published slightly different approaches in 2017. Lane et al. [55], categorized UkBiobank individuals who responded "usually" to having trouble falling asleep at night or waking up in the middle of the night as probable insomnia cases ($n = 32{,}155$), while those responding "never/rarely" were considered controls ($n = 26{,}973$). Hammerschlag et al. [56] included participants responding "sometimes" or "never/rarely" as controls ($n = 32{,}384$ cases/$n = 80{,}622$ controls) and further validated their insomnia phenotype using an independent European cohort containing patients clinically diagnosed with insomnia ($n = 1073$). Moreover, they expanded the comprehensiveness of their GWAS by including a genome-wide gene-association study (GWGAS) and meta-analysis replication with an independent cohort [56]. The GWGAS approach, now commonly used, allows for genes containing multiple

SNPs with mild effects (hence not likely to be picked up by GWAS individually) to be identified.

The GWAS performed by Lane et al. demonstrated significant associations between probable insomnia cases and SNPs near or within *MEIS1* (encoding a homeobox protein), TMEM132E (encoding a transmembrane protein), and *CYCL1* (encoding a cytoskeletal protein) across sexes, while *TGFB1* (encoding a secreted ligand) and *WDR27* (encoding a protein linked to cell signalling) were related to females and male cases respectively. *MEIS1* and *WDR27* (in males) were confirmed as insomnia candidate genes by the Hammerschlag et al. [56] GWAS, which also identified other novel loci, including *IPO7* (encoding a protein that plays a role in nuclear localization), and TSNARE1 (encoding a protein linked to synaptic signaling) in females, and HHEX (encoding a homeobox protein) and RHCG (encoding a protein that regulates ammonia transport) in males. Interestingly, the top gene hit in both GWAS—*MEIS1*—has been implicated with restless leg syndrome (RLS) by multiple studies [72–75]. Some researchers remain skeptical that MEIS1 contributes to both insomnia and RLS independently and argue that the effect is due to undiagnosed RLS, which is known to cause insomnia. A study re-evaluating these GWAS findings found genotypic distribution and allelic frequency differences of *MEIS1* SNPs between patients clinically diagnosed with chronic insomnia vs patients with chronic insomnia and RLS [76]. However, no variation differences were found between patients with chronic insomnia and controls, nor between patients with combined chronic insomnia and RLS and patients only diagnosed with RLS, leading authors to suggest *MEIS1* is likely an independent risk factor for RLS with limited contributions to insomnia. UKBiobank authors have acknowledged that insomnia traits are likely confounded by RLS in 12% of identified insomnia cases, yet suggest MEIS1 still remains implicated in the insomnia phenotype [56]. Subsequent larger GWAS also using the UKBiobank continue to identify *MEIS1* as a top insomnia candidate gene [58, 59, 77], which has also been implicated in an independent clinical cohort and associated with lower sleep efficiency derived from actigraphy [59]. Further research is needed to untangle the complex contributions of MEIS1 to insomnia, while more stringent clinical phenotyping of sleep in the UKBiobank may also be warranted.

GWAS of insomnia has now included millions of participants. In 2019, Jansen et al. pooled together data from the UKBiobank and the personalized genomics company 23andMe, yielding a large sample size of 1.3 million individuals. The GWAS uncovered 202 loci related to 965 genes that were associated with insomnia traits. While the majority of these genetic loci identified were novel, candidate genes previously identified, *MEIS1*, *MED27*, IPO7, were replicated. As of 2022, the current largest GWAS has doubled the previous sample size (n = 593,724 insomnia cases/n = 1,771,286 controls) by pooling updated information from the UKBiobank and 23andMe [77]. Watanabe and colleagues identified 554 insomnia risk loci (related to 3898 genes), of which 364 were novel. Top hit genes MEIS1 and BTBD9 from Jansen et al. were supported in the new GWAS. In mouse models, *BTBD9* has previously been associated with disrupted sleep, including RLS symptoms [78].

Key Takeaways from GWAS to Date

As GWAS of insomnia continue to grow and implicate an overwhelming number of risk loci, the current state of knowledge based on GWAS to date can be synthesized into the following key points:

1. *Risk for insomnia is polygenic, yet GWAS explain minimal variance*

   Multiple genetic variants contribute to the insomnia phenotype, highlighting the polygenicity of insomnia risk. However, these identified common genetic risk factors account for very little of the insomnia phenotype. Even with a sample of 1.3 million individuals, the phenotypic variation of insomnia based on GWAS results from [58] was estimated at 2.6%.

   SNP heritability (H^2), referring to the proportion of a trait's variation explained by inherited genetic variants, is estimated at 7% for insomnia [58] and has not substantially increased despite doubling GWAS sample size [77].

   The heritability for insomnia is ~40%, so if all genetic risk factors for insomnia were identified, their contributions would equal 40%. The discrepancy between heritability estimates and the proportion of variance attributed to GWAS for complex traits like insomnia is often referred to as "missing heritability" (196, 197). From genetic methodologies to limited sample sizes, many factors contribute to this "missingness" (196–198).

   Given common variants have minimal effects that are difficult to qualify on their own, one way to measure their cumulative contribution to a trait is using a polygenic risk score (PRS). PRS refers to an individual's genetic liability to a trait/disease based on an aggregated effect of common variants, which is calculated based on the weighted sum of SNPs detected by GWAS. Weights are based on the effect size of each SNP on a trait. PRS serves as a sanity check, to determine if the variance of insomnia traits explained by genetic risk in one population can be replicated in another. For instance, GWAS of soldiers in the Army Study To Assess Risk and Resilience in Service members (STARRS) confirmed PRS of insomnia risk in the UkBiobank was positively associated with insomnia disorders in the cohort [57], albeit with a small effect OR = 1.09 [95% CI 1.04–1.14].

2. *Genetic risk factors for insomnia are pleitropic and highly associated with complex disease*

   The most robust finding among GWAS [55–59] is that insomnia is highly genetically correlated with psychiatric symptoms, including depression ($r_g = 0.64$), anxiety ($r_g = 0.56$), and neuroticism ($r_g = 0.48$). Genetic correlations describe the shared genetic architecture between traits and diseases. Linkage disequilibrium is a common parameter used to understand the strength of correlations. LD occurs when SNPs are non-randomly correlated with SNPs in other regions. An LD score is estimated for each SNP by calculating the sum of correlations between a SNP and nearby SNPs.

This shared genetic architecture among insomnia and psychiatric disorders has also emerged in other studies. For instance, a PRS of various psychiatric disorders in youth has been significantly correlated with insomnia symptoms [79], highlighting the complex relationship between psychiatry and insomnia across development. Interestingly, except for sleep duration (rG=), insomnia is minimally associated with other sleep traits (e.g., chronotype, sleepiness, etc.). This suggests separate biological mechanisms may underlie distinct sleep traits.

In addition to psychiatric traits, insomnia is genetically correlated, albeit weakly, to metabolic traits like BMI ($r_g = 0.16$), risk of type 2 diabetes ($r_g = 0.20$) and risk of coronary artery disease ($r_g = 0.18$) [58].

Insomnia is genetically pleitropic (the same genetic variants that contribute to insomnia also contribute to other traits), which mirrors the heterogeneity of the insomnia phenotype itself. However, this makes it complex to untangle the directions of causation between insomnia and co-occurring traits, like psychiatric symptoms. In the case of *horizontal pleiotropy*, the same genetic variant can influence two traits independently (see Fig. 1), like insomnia and anxiety. This variant may then correspond to an anxiety/insomnia subtype. With *vertical pleiotropy,* a genetic variant can influence one trait (risk/exposure variable) which then exerts influence on another trait or "outcome" (see Fig. 1). Mendelian randomization is an epidemiological method that can identify pleiotropy by using genetic variants to determine if there is a causal association between a risk factor (e.g., insomnia) and an outcome (e.g., depression) (see Chapter "Approaches to Studying Human Genetics").

Mendelian randomizations in general population GWAS demonstrate a causal link between insomnia and multiple health outcomes it was genetically correlated with (e.g., depression, coronary artery disease, etc.) [58, 59]. However, the effects of vertical pleiotropy can be bi-directional (see Fig. 1). Although a moderate proportion of SNPs were found to be pleiotropic in current insomnia GWAS, this may increase in heterogeneous clinical groups. Other researchers have applied a top-bottom approach, by classifying subgroups of insomnia, then investigating neurophysiological biomarker differences between subgroups [18]. Hence, future genetic-driven investigations could follow a similar approach (Fig. 2).

3. *Insomnia candidate genes are largely expressed in the brain*

Together, the current largest GWAS suggest that candidate genes of insomnia are "brain genes" [58, 59, 77], likely exerting their effects via alterations in brain function. Insomnia candidate genes from GWAS are found to be enriched in multiple brain regions including cerebellum, frontal cortex, anterior cingulate cortex, insula, hypothalamus, basal ganglia, mesial temporal lobe, and many more. Genes associated with insomnia have also been found to be expressed by medium spiny neurons in the basal ganglia, and involved in synaptic signaling functions and neuronal differentiation.

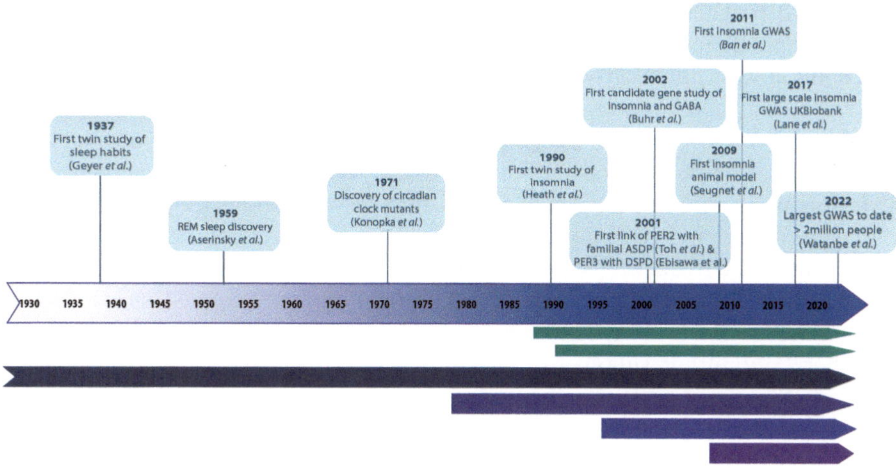

Fig. 2 Genetic timeline of insomnia and related sleep traits

4. *Casual genetic risk factors for insomnia are still lacking*

Genes extracted from GWAS can be based on methods that calculate SNPs proximity to genes, but SNPs associated with insomnia may not have any functional link to its gene neighbor (i.e., "insomnia gene"). Moreover, non-causal SNPs identified through linkage disequilibrium may show statistical significance with sleep traits as a result of being correlated with causal SNPs.

The *now* largest insomnia GWAS of over two million European adults used a novel cross-locus functional approach to identify high-confidence candidate genes [77]. This highlighted a selection preference for SNPs that had a structural or regulatory effect on genes, genes that were not close to other genes in their locus, and genes that were functionally related (protein-protein interactions) to other candidate genes. By using this strategy, the authors were able to reduce the 3898 genes from their initial GWAS to 289 genes. These newly identified high-confidence insomnia genes were found to largely be implicated in synaptic signaling and neuronal differentiation functions and expressed in cortex regions underlying cognitive and emotional regulation.

Future Directions

The identification of causal genetic mechanisms underlying insomnia and subsequent translation toward clinical applications is still in its infancy. The field will need to go through a period of refinement and diversification of approaches to reach this end goal. As such, many factors will need to be considered moving forward:

Measurement of insomnia The variability in measuring insomnia traits, as seen in Table 1, presents as a barrier for replicating and comparing studies. As we move toward larger datasets, community consensus is needed to harmonize measures, which should better align with diagnostic criteria. Moreover, insomnia traits should be captured with granularity, rather than clumped into one item. For instance, the separation between initiating sleep and night awakenings might be useful as current GWAS indicates that separate genetic mechanisms may underlie distinct sleep traits. Recent evidence from a genetics study with autistic youth supports this, revealing rare genetic variants map onto distinct insomnia traits [80]. The chronicity and the evolution of insomnia should also be taken into account. As Lane et al. [59] demonstrate, the frequency of insomnia traits may be dependent on different genetic variants, indicating that longitudinal capture of symptom progression would be a helpful addition to GWAS. The measurement of insomnia also needs to address the likelihood of multiple subtypes of insomnia, which may have differing genetic etiologies. While past efforts have attempted to identify subtypes of insomnia using patterns of clinical symptoms or self-reported psychological traits, it may be possible to leverage the increasingly large datasets to take a bottom-up approach to identify genetic subtypes which then can be compared in terms of their clinical presentation.

The search for causal factors Targeting specific molecular pathways or functions [81, 82] and establishing strategies to identify causal SNPs [77] are needed to meaningfully interpret GWAS and understand connections with other disorders or traits. Animal models and deep phenotyping of individuals with variations in high-confidence candidate genes will also provide a better understanding of risk factors underlying insomnia, or potential protective factors that have yet to be determined. The search for genetic risk factors will also require adopting other genetic approaches. For instance, combining both common and rare variants may be beneficial for refining predictions in sleep research, as seen in the neurodevelopmental literature [83, 84]. Structural genetic variations like CNVs (i.e., large deletions and duplications in the genome) and accompanying metrics annotating the severity of variants that have been linked to psychiatry and complex disease [61, 83, 85] remain another avenue to investigate in sleep research generally. It is also important to consider that the expansion of next-generation sequencing methods, like whole exome sequencing and whole genome sequencing, will contribute to the identification of more genetic variants, particularly those with low allele frequencies and within non-coding genes that contribute to 99% of the genes in the genome. As such, a new wave of exploration still remains.

Leveraging multidimensional and transdiagnostic lens The field needs to move beyond studying insomnia in isolation, or correlating it with a singular outcome (e.g., genetic variation → insomnia) without a multidimensional lens. As described in [86], dissecting the impacts of sleep in humans requires a whole systems approach, capturing socio-ecological factors that impact an individual. Sleep disturbances like insomnia are one of the most universal concerns found across general and clinical populations [9, 87], they should be leveraged as a transdiagnostic factor, hence pooling across clinical and general population cohorts can be mutually beneficial for delineating complex heterogenetic and multidimensional factors influencing sleep. This transdiagnostic and multidimensional approach can enable the application of known computational methods [18, 19, 88] to derive profiles of sleep-risk informed by biology, behavior, health, and environmental variables simultaneously and across developmental periods and populations. Moreover, it aligns with NIMH Research Domain Criteria (RDoC), which encourages the deconstruction of diagnostic categories to focus on core behavioral and neurobiological features that cuts across diagnostic categories [89].

Scaling up with open source and diverse partnerships Due to the extreme polygenicity of insomnia, is it estimated that at least 50 million people will be needed to detect SNPs at the genome-wide significance level to explain 90% of the genetic variance of insomnia [90]. Given advances in sequencing technology, this is feasible, yet will rely on open data collaborations with researchers across the research domains and globally. These open-source collaborations can lead us toward unity in harmonizing processes and measures to increase reproducibility [91]. However, this will require a truly diverse global effort, which has been neglected. The majority of research has been conducted on European individuals (see Table 1), and previous attempts to stratify GWAS based on "ancestral groups" have been underpowered [57]. This remains a major gap in genetic research generally, where 80% of GWAS involve individuals of European ancestry who make up only ~16% of the world population, while regions with the largest genetic diversity remain excluded [92]. Sleep is influenced by light/temperature (i.e., geographic location), culture (e.g., siesta), policies [93], racial/economic disparities [94], and insomnia datasets need to reflect this global diversity.

References

1. American Psychiatric Association. Diagnostic and statistical manual of mental disorders (DSM-5®). Washington, DC: American psychiatric association; 2013.
2. American Academy of Sleep Medicine. International classification of sleep disorders. In: Diagnostic and coding manual; 2014. p. 51–5.
3. Baglioni C, et al. Sleep changes in the disorder of insomnia: a meta-analysis of polysomnographic studies. Sleep Med Rev. 2014;18(3):195–213.
4. Ford ES, Cunningham TJ, Giles WH, Croft JB. Trends in insomnia and excessive daytime sleepiness among U.S. adults from 2002 to 2012. Sleep Med. 2015;16(3):372–8. https://doi.org/10.1016/j.sleep.2014.12.008.
5. Garland SN, Rowe H, Repa LM, Fowler K, Zhou ES, Grandner MA. A decade's difference: 10-year change in insomnia symptom prevalence in Canada depends on sociodemographics and health status. Sleep Health. 2018;4(2):160–5. https://doi.org/10.1016/j.sleh.2018.01.003.
6. Calem M, et al. Increased prevalence of insomnia and changes in hypnotics use in England over 15 years: analysis of the 1993, 2000, and 2007 National Psychiatric Morbidity Surveys. Sleep. 2012;35(3):377–84.

7. Morin CM, Drake CL, Harvey AG, Krystal AD, Manber R, Riemann D, Spiegelhalder K. Insomnia disorder. Nat Rev Dis Primers. 2015;1:15026. https://doi.org/10.1038/nrdp.2015.26.
8. Smith MT, Perlis ML, Park A, Smith MS, Pennington J, Giles DE, Buysse DJ. Comparative meta-analysis of pharmacotherapy and behavior therapy for persistent insomnia. Am J Psychiatry. 2002;159(1):5–11. https://doi.org/10.1176/appi.ajp.159.1.5.
9. Seow LSE, Verma SK, Mok YM, Kumar S, Chang S, Satghare P, Hombali A, Vaingankar J, Chong SA, Subramaniam M. Evaluating DSM-5 insomnia disorder and the treatment of sleep problems in a psychiatric population. J Clin Sleep Med. 2018;14(02):237–44. https://doi.org/10.5664/jcsm.6942.
10. Fernandez-Mendoza J, et al. Trajectories of insomnia symptoms from childhood through young adulthood. Pediatrics. 2022;149(3):e2021053616.
11. Zhang B, Wing Y-K. Sex differences in insomnia: a meta-analysis. Sleep. 2006;29(1):85–93. https://doi.org/10.1093/sleep/29.1.85.
12. Agar G, Brown C, Sutherland D, Coulborn S, Oliver C, Richards C. Sleep disorders in rare genetic syndromes: a meta-analysis of prevalence and profile. Molecular Autism. 2021;12(1). Article 1 https://doi.org/10.1186/s13229-021-00426-w.
13. Kyle SD, Morgan K, Espie CA. Insomnia and health-related quality of life. Sleep Med Rev. 2010;14(1):69–82. https://doi.org/10.1016/j.smrv.2009.07.004.
14. Kripke DF, Garfinkel L, Wingard DL, Klauber MR, Marler MR. Mortality associated with sleep duration and insomnia. Arch Gen Psychiatry. 2002;59(2):131–6. https://doi.org/10.1001/archpsyc.59.2.131.
15. Edinger JD, Wyatt JK, Stepanski EJ, Olsen MK, Stechuchak KM, Carney CE, Chiang A, Crisostomo MI, Lineberger MD, Means MK, Radtke RA, Wohlgemuth WK, Krystal AD. Testing the reliability and validity of DSM-IV-TR and ICSD-2 insomnia diagnoses: results of a multitrait-multimethod analysis. Arch Gen Psychiatry. 2011;68(10):992–1002. https://doi.org/10.1001/archgenpsychiatry.2011.64.
16. Riemann D, Benz F, Dressle RJ, Espie CA, Johann AF, Blanken TF, Leerssen J, Wassing R, Henry AL, Kyle SD, Spiegelhalder K, Van Someren EJW. Insomnia disorder: state of the science and challenges for the future. J Sleep Res. 2022;31(4):e13604. https://doi.org/10.1111/jsr.13604.
17. Benjamins JS, Migliorati F, Dekker K, Wassing R, Moens S, Blanken TF, Te Lindert BHW, Sjauw Mook J, Van Someren EJW. Insomnia heterogeneity: characteristics to consider for data-driven multivariate subtyping. Sleep Med Rev. 2017;36:71–81. https://doi.org/10.1016/j.smrv.2016.10.005.
18. Blanken TF, Benjamins JS, Borsboom D, Vermunt JK, Paquola C, Ramautar J, Dekker K, Stoffers D, Wassing R, Wei Y, Someren EJWV. Insomnia disorder subtypes derived from life history and traits of affect and personality. Lancet Psychiatry. 2019;6(2):151–63. https://doi.org/10.1016/S2215-0366(18)30464-4.
19. Katori M, Shi S, Ode KL, Tomita Y, Ueda HR. The 103,200-arm acceleration dataset in the UK Biobank revealed a landscape of human sleep phenotypes. Proc Natl Acad Sci. 2022;119(12):e2116729119. https://doi.org/10.1073/pnas.2116729119.
20. Heath AC, Kendler KS, Eaves LJ, Martin NG. Evidence for genetic influences on sleep disturbance and sleep pattern in twins. Sleep. 1990;13(4):318–35. https://doi.org/10.1093/sleep/13.4.318.
21. Boomsma D, Busjahn A, Peltonen L. Classical twin studies and beyond. Nat Rev Genet. 2002;3(11):872–82.
22. Barclay NL, Kocevska D, Bramer WM, Van Someren EJW, Gehrman P. The heritability of insomnia: a meta-analysis of twin studies. Genes Brain Behav. 2021;20(4):e12717. https://doi.org/10.1111/gbb.12717.
23. Dauvilliers Y, et al. Family studies in insomnia. J Psychosom Res. 2005;58(3)271–78.
24. Wing YK, et al. Familial aggregation and heritability of insomnia in a community-based study. Sleep Med. 2012;13(8):985–90.

25. Beaulieu-Bonneau S, et al. Family history of insomnia in a population-based sample. Sleep. 2007;30(12):1739–45.
26. Taylor MJ, et al. Etiological links between autism and difficulties in initiating and maintaining sleep: a familial co-aggregation and twin study. J Child Psychol Psychiatry. 2022;63(3):315–23.
27. Gregory AM, et al. ADHD and sleep quality: longitudinal analyses from childhood to early adulthood in a twin cohort. J Clin Child Adolesc Psychol. 2017;46(2):284–94.
28. Zhang EE, Kay SA. Clocks not winding down: unravelling circadian networks. Nat Rev Mol Cell Biol. 2010;11(11):764–76.
29. Patke A, Murphy PJ, Onat OE, Krieger AC, Özçelik T, Campbell SS, Young MW. Mutation of the human circadian clock gene CRY1 in familial delayed sleep phase disorder. Cell. 2017;169(2):203–215.e13. https://doi.org/10.1016/j.cell.2017.03.027.
30. Archer SN, Carpen JD, Gibson M, Lim GH, Johnston JD, Skene DJ, von Schantz M. Polymorphism in the PER3 promoter associates with diurnal preference and delayed sleep phase disorder. Sleep. 2010;33(5):695–701. https://doi.org/10.1093/sleep/33.5.695.
31. Ebisawa T, Uchiyama M, Kajimura N, Mishima K, Kamei Y, Katoh M, Watanabe T, Sekimoto M, Shibui K, Kim K, Kudo Y, Ozeki Y, Sugishita M, Toyoshima R, Inoue Y, Yamada N, Nagase T, Ozaki N, Ohara O, et al. Association of structural polymorphisms in the human period3 gene with delayed sleep phase syndrome. EMBO Rep. 2001;2(4):342–6. https://doi.org/10.1093/embo-reports/kve070.
32. Hohjoh H, Takasu M, Shishikura K, Takahashi Y, Honda Y, Tokunaga K. Significant association of the arylalkylamine N-acetyltransferase (AA-NAT) gene with delayed sleep phase syndrome. Neurogenetics. 2003;4(3):151–3. https://doi.org/10.1007/s10048-002-0141-9.
33. Toh KL, Jones CR, He Y, Eide EJ, Hinz WA, Virshup DM, Ptácek LJ, Fu YH. An hPer2 phosphorylation site mutation in familial advanced sleep phase syndrome. Science (New York, N.Y.). 2001;291(5506):1040–3. https://doi.org/10.1126/science.1057499.
34. Xu Y, Toh KL, Jones CR, Shin J-Y, Fu Y-H, Ptácek LJ. Modeling of a human circadian mutation yields insights into clock regulation by PER2. Cell. 2007;128(1):59–70. https://doi.org/10.1016/j.cell.2006.11.043.
35. Zhang L, Hirano A, Hsu P-K, Jones CR, Sakai N, Okuro M, McMahon T, Yamazaki M, Xu Y, Saigoh N, Saigoh K, Lin S-T, Kaasik K, Nishino S, Ptáček LJ, Fu Y-H. A PERIOD3 variant causes a circadian phenotype and is associated with a seasonal mood trait. Proc Natl Acad Sci USA. 2016;113(11):E1536–44. https://doi.org/10.1073/pnas.1600039113.
36. Hirano A, Shi G, Jones CR, Lipzen A, Pennacchio LA, Xu Y, Hallows WC, McMahon T, Yamazaki M, Ptáček LJ, Fu Y-H. A Cryptochrome 2 mutation yields advanced sleep phase in humans. elife. 2016;5:e16695. https://doi.org/10.7554/eLife.16695.
37. Brennan KC, Bates EA, Shapiro RE, Zyuzin J, Hallows WC, Huang Y, Lee H-Y, Jones CR, Fu Y-H, Charles AC, Ptáček LJ. Casein kinase iδ mutations in familial migraine and advanced sleep phase. Sci Transl Med. 2013;5(183):1–11. https://doi.org/10.1126/scitranslmed.3005784.
38. Xu Y, Padiath QS, Shapiro RE, Jones CR, Wu SC, Saigoh N, Saigoh K, Ptáček LJ, Fu Y-H. Functional consequences of a CKIδ mutation causing familial advanced sleep phase syndrome. Nature. 2005;434(7033). Article 7033 https://doi.org/10.1038/nature03453.
39. Li J, Huang C, Lan Y, Wang Y. A cross-sectional study on the relationships among the polymorphism of period2 gene, work stress, and insomnia. Sleep Breath. 2015;19(4):1399–406. https://doi.org/10.1007/s11325-015-1229-4.
40. Brower KJ, Wojnar M, Sliwerska E, Armitage R, Burmeister M. PER3 polymorphism and insomnia severity in alcohol dependence. Sleep. 2012;35(4):571–7. https://doi.org/10.5665/sleep.1748.
41. Ziv-Gal A, Flaws JA, Mahoney MM, Miller SR, Zacur HA, Gallicchio L. Genetic polymorphisms in the aryl hydrocarbon receptor-signaling pathway and sleep disturbances in middle-aged women. Sleep Med. 2013;14(9):883–7. https://doi.org/10.1016/j.sleep.2013.04.007.
42. Deuschle M, Schredl M, Schilling C, Wüst S, Frank J, Witt SH, Rietschel M, Buckert M, Meyer-Lindenberg A, Schulze TG. Association between a serotonin transporter length polymorphism and primary insomnia. Sleep. 2010;33(3):343–7. https://doi.org/10.1093/sleep/33.3.343.

43. Du L, Bakish D, Ravindran A, Hrdina PD. MAO-A gene polymorphisms are associated with major depression and sleep disturbance in males. Neuroreport. 2004;15(13):2097–101. https://doi.org/10.1097/00001756-200409150-00020.
44. Huang C, Li J, Lu L, Ren X, Li Y, Huang Q, Lan Y, Wang Y. Interaction between serotonin transporter gene-linked polymorphic region (5-HTTLPR) and job-related stress in insomnia: a cross-sectional study in Sichuan, China. Sleep Med. 2014;15(10):1269–75. https://doi.org/10.1016/j.sleep.2014.01.023.
45. Buhr A, Bianchi MT, Baur R, Courtet P, Pignay V, Boulenger JP, Gallati S, Hinkle DJ, Macdonald RL, Sigel E. Functional characterization of the new human GABA(A) receptor mutation beta3(R192H). Hum Genet. 2002;111(2):154–60. https://doi.org/10.1007/s00439-002-0766-7.
46. Cade BE, Gottlieb DJ, Lauderdale DS, Bennett DA, Buchman AS, Buxbaum SG, De Jager PL, Evans DS, Fülöp T, Gharib SA, Johnson WC, Kim H, Larkin EK, Lee SK, Lim AS, Punjabi NM, Shin C, Stone KL, Tranah GJ, et al. Common variants in DRD2 are associated with sleep duration: the CARe consortium. Hum Mol Genet. 2016;25(1):167–79. https://doi.org/10.1093/hmg/ddv434.
47. Jawinski P, Tegelkamp S, Sander C, Häntzsch M, Huang J, Mauche N, Scholz M, Spada J, Ulke C, Burkhardt R, Reif A, Hegerl U, Hensch T. Time to wake up: no impact of COMT Val158Met gene variation on circadian preferences, arousal regulation and sleep. Chronobiol Int. 2016;33(7):893–905. https://doi.org/10.1080/07420528.2016.1178275.
48. Seugnet L, Suzuki Y, Thimgan M, Donlea J, Gimbel SI, Gottschalk L, Duntley SP, Shaw PJ. Identifying sleep regulatory genes using a drosophila model of insomnia. J Neurosci. 2009;29(22):7148–57. https://doi.org/10.1523/JNEUROSCI.5629-08.2009.
49. Harvey C-J, Gehrman P, Espie CA. Who is predisposed to insomnia: a review of familial aggregation, stress-reactivity, personality and coping style. Sleep Med Rev. 2014;18(3):237–47. https://doi.org/10.1016/j.smrv.2013.11.004.
50. Drake CL, Friedman NP, Wright KP, Roth T. Sleep reactivity and insomnia: genetic and environmental influences. Sleep. 2011;34(9):1179–88. https://doi.org/10.5665/SLEEP.1234.
51. Ban H-J, Kim SC, Seo J, Kang H-B, Choi JK. Genetic and metabolic characterization of insomnia. PLoS One. 2011;6(4):e18455. https://doi.org/10.1371/journal.pone.0018455.
52. Byrne EM, Gehrman PR, Medland SE, Nyholt DR, Heath AC, Madden PA, Hickie IB, Van Duijn CM, Henders AK, Montgomery GW, Martin NG, Wray NR. A genome-wide association study of sleep habits and insomnia. Am J Med Genet B Neuropsychiatr Genet. 2013:439–51. https://doi.org/10.1002/ajmg.b.32168.
53. Amin N, Allebrandt KV, van der Spek A, Müller-Myhsok B, Hek K, Teder-Laving M, Hayward C, Esko T, van Mill JG, Mbarek H, Watson NF, Melville SA, Del Greco FM, Byrne EM, Oole E, Kolcic I, Chen T, Evans DS, Coresh J, et al. Genetic variants in RBFOX3 are associated with sleep latency. Eur J Hum Genet. 2016;24(10):1488–95. https://doi.org/10.1038/ejhg.2016.31.
54. Spada J, et al. Genome-wide association analysis of actigraphic sleep phenotypes in the LIFE Adult Study. J Sleep Res. 2016;25(6):690–701.
55. Lane JM, Liang J, Vlasac I, Anderson SG, Bechtold DA, Bowden J, Emsley R, Gill S, Little MA, Luik AI, Loudon A, Scheer FAJL, Purcell SM, Kyle SD, Lawlor DA, Zhu X, Redline S, Ray DW, Rutter MK, Saxena R. Genome-wide association analyses of sleep disturbance traits identify new loci and highlight shared genetics with neuropsychiatric and metabolic traits. Nat Genet. 2017;49(2):274–81. https://doi.org/10.1038/ng.3749.
56. Hammerschlag AR, Stringer S, de Leeuw CA, Sniekers S, Taskesen E, Watanabe K, Blanken TF, Dekker K, te Lindert BH, Wassing R, Jonsdottir I, Thorleifsson G, Stefansson H, Gislason T, Berger K, Schormair B, Wellmann J, Winkelmann J, Stefansson K, et al. Genome-wide association analysis of insomnia complaints identifies risk genes and genetic overlap with psychiatric and metabolic traits. Nat Genet. 2017;49(11):1584–92. https://doi.org/10.1038/ng.3888.
57. Stein MB, McCarthy MJ, Chen C-Y, Jain S, Gelernter J, He F, Heeringa SG, Kessler RC, Nock MK, Ripke S, Sun X, Wynn GH, Smoller JW, Ursano RJ. Genome-wide analysis of insomnia disorder. Mol Psychiatry. 2018;23(11):2238–50. https://doi.org/10.1038/s41380-018-0033-5.

58. Jansen PR, Watanabe K, Stringer S, Skene N, Bryois J, Hammerschlag AR, de Leeuw CA, Benjamins JS, Muñoz-Manchado AB, Nagel M, Savage JE, Tiemeier H, White T, Tung JY, Hinds DA, Vacic V, Wang X, Sullivan PF, van der Sluis S, et al. Genome-wide analysis of insomnia in 1,331,010 individuals identifies new risk loci and functional pathways. Nat Genet. 2019;51(3):394. https://doi.org/10.1038/s41588-018-0333-3.
59. Lane JM, Jones SE, Dashti HS, Wood AR, Aragam KG, van Hees VT, Strand LB, Winsvold BS, Wang H, Bowden J, Song Y, Patel K, Anderson SG, Beaumont RN, Bechtold DA, Cade BE, Haas M, Kathiresan S, Little MA, et al. Biological and clinical insights from genetics of insomnia symptoms. Nat Genet. 2019;51(3):387. https://doi.org/10.1038/s41588-019-0361-7.
60. Song W, et al. Genome-wide association analysis of insomnia using data from Partners Biobank. Sci Rep. 2020;10(1):6928.
61. Wainberg M, Merico D, Huguet G, Zarrei M, Jacquemont S, Scherer SW, Tripathy SJ. Deletion of loss-of-function–intolerant genes and risk of 5 psychiatric disorders. JAMA Psychiatry. 2022;79(1):78–81. https://doi.org/10.1001/jamapsychiatry.2021.3211.
62. Tsujino N, Sakurai T. Orexin/hypocretin: a neuropeptide at the interface of sleep, energy homeostasis, and reward system. Pharmacol Rev. 2009;61(2):162–76. https://doi.org/10.1124/pr.109.001321.
63. Kumar D, Dedic N, Flachskamm C, Voulé S, Deussing JM, Kimura M. Cacna1c (Cav1.2) modulates electroencephalographic rhythm and rapid eye movement sleep recovery. Sleep. 2015;38(9):1371–80. https://doi.org/10.5665/sleep.4972.
64. Parsons MJ, Lester KJ, Barclay NL, Nolan PM, Eley TC, Gregory AM. Replication of Genome-Wide association studies (GWAS) loci for sleep in the British G1219 cohort. Am J Med Genet B Neuropsychiatr Genet. 2013;162(5):431–8. https://doi.org/10.1002/ajmg.b.32106.
65. Ollila HM, Kettunen J, Pietiläinen O, Aho V, Silander K, Kronholm E, Perola M, Lahti J, Räikkönen K, Widen E, Palotie A, Eriksson JG, Partonen T, Kaprio J, Salomaa V, Raitakari O, Lehtimäki T, Sallinen M, Härmä M, et al. Genome-wide association study of sleep duration in the finnish population. J Sleep Res. 2014;23(6):609–18. https://doi.org/10.1111/jsr.12175.
66. Manoach DS, Pan JQ, Purcell SM, Stickgold R. Reduced sleep spindles in schizophrenia: a treatable endophenotype that links risk genes to impaired cognition? Biol Psychiatry. 2016;80(8):599–608. https://doi.org/10.1016/j.biopsych.2015.10.003.
67. Marshall CR, Howrigan DP, Merico D, Thiruvahindrapuram B, Wu W, Greer DS, Antaki D, Shetty A, Holmans PA, Pinto D, Gujral M, Brandler WM, Malhotra D, Wang Z, Fajarado KVF, Maile MS, Ripke S, Agartz I, Albus M, et al. Contribution of copy number variants to schizophrenia from a genome-wide study of 41,321 subjects. Nat Genet. 2017;49(1):27–35. https://doi.org/10.1038/ng.3725.
68. Rubio MD, Wood K, Haroutunian V, Meador-Woodruff JH. Dysfunction of the ubiquitin proteasome and ubiquitin-like systems in Schizophrenia. Neuropsychopharmacology. 2013;38(10):1910–20. https://doi.org/10.1038/npp.2013.84.
69. International HapMap Consortium. A haplotype map of the human genome. Nature. 2005;437(7063):1299–320. https://doi.org/10.1038/nature04226.
70. Pe'er I, Yelensky R, Altshuler D, Daly MJ. Estimation of the multiple testing burden for genome-wide association studies of nearly all common variants. Genet Epidemiol. 2008;32(4):381–5. https://doi.org/10.1002/gepi.20303.
71. Sudlow C, Gallacher J, Allen N, Beral V, Burton P, Danesh J, Downey P, Elliott P, Green J, Landray M, Liu B, Matthews P, Ong G, Pell J, Silman A, Young A, Sprosen T, Peakman T, Collins R. UK Biobank: an open access resource for identifying the causes of a wide range of complex diseases of middle and old age. PLoS Med. 2015;12(3):e1001779. https://doi.org/10.1371/journal.pmed.1001779.
72. Winkelmann J, Czamara D, Schormair B, Knauf F, Schulte EC, Trenkwalder C, Dauvilliers Y, Polo O, Högl B, Berger K, Fuhs A, Gross N, Stiasny-Kolster K, Oertel W, Bachmann CG, Paulus W, Xiong L, Montplaisir J, Rouleau GA, et al. Genome-wide association study identifies novel restless legs syndrome susceptibility loci on 2p14 and 16q12.1. PLOS Genetics. 2011;7(7):e1002171. https://doi.org/10.1371/journal.pgen.1002171.

73. Winkelmann J, Schormair B, Xiong L, Dion PA, Rye DB, Rouleau GA. Genetics of restless legs syndrome. Sleep Med. 2017;31:18–22. https://doi.org/10.1016/j.sleep.2016.10.012.
74. Xiong L, Catoire H, Dion P, Gaspar C, Lafrenière RG, Girard SL, Levchenko A, Rivière J-B, Fiori L, St-Onge J, Bachand I, Thibodeau P, Allen R, Earley C, Turecki G, Montplaisir J, Rouleau GA. MEIS1 intronic risk haplotype associated with restless legs syndrome affects its mRNA and protein expression levels. Hum Mol Genet. 2009;18(6):1065–74. https://doi.org/10.1093/hmg/ddn443.
75. Yang Q, Li L, Chen Q, Foldvary-Schaefer N, Ondo WG, Wang QK. Association studies of variants in MEIS1, BTBD9 and MAP2K5/SKOR1 with restless legs syndrome in a U.S. population. Sleep Med. 2011;12(8):800–4. https://doi.org/10.1016/j.sleep.2011.06.006.
76. El Gewely M, Welman M, Xiong L, Yin S, Catoire H, Rouleau G, Montplaisir JY, Desautels A, Warby SC. Reassessing GWAS findings for the shared genetic basis of insomnia and restless legs syndrome. Sleep. 2018;41(11):zsy164. https://doi.org/10.1093/sleep/zsy164.
77. Watanabe K, Jansen PR, Savage JE, Nandakumar P, Wang X, Hinds DA, Gelernter J, Levey DF, Polimanti R, Stein MB, Van Someren EJW, Smit AB, Posthuma D. Genome-wide meta-analysis of insomnia prioritizes genes associated with metabolic and psychiatric pathways. Nat Genet. 2022;54(8). Article 8 https://doi.org/10.1038/s41588-022-01124-w.
78. DeAndrade MP, Johnson RL, Unger EL, Zhang L, van Groen T, Gamble KL, Li Y. Motor restlessness, sleep disturbances, thermal sensory alterations and elevated serum iron levels in Btbd9 mutant mice. Hum Mol Genet. 2012;21(18):3984–92. https://doi.org/10.1093/hmg/dds221.
79. Ohi K, Ochi R, Noda Y, Wada M, Sugiyama S, Nishi A, Shioiri T, Mimura M, Nakajima S. Polygenic risk scores for major psychiatric and neurodevelopmental disorders contribute to sleep disturbance in childhood: Adolescent Brain Cognitive Development (ABCD) Study. Transl Psychiatry. 2021;11(1):1–11. https://doi.org/10.1038/s41398-021-01308-8.
80. Earl RK, Ward T, Gerdts J, Eichler EE, Bernier RA, Hudac CM. Sleep problems in children with ASD and gene disrupting mutations. J Genet Psychol. 2021;182(5):317–34. https://doi.org/10.1080/00221325.2021.1922869.
81. Doldur-Balli F, Imamura T, Veatch OJ, Gong NN, Lim DC, Hart MP, Abel T, Kayser MS, Brodkin ES, Pack AI. Synaptic dysfunction connects autism spectrum disorder and sleep disturbances: a perspective from studies in model organisms. Sleep Med Rev. 2022;62:101595. https://doi.org/10.1016/j.smrv.2022.101595.
82. Hari Dass SA, McCracken K, Pokhvisneva I, Chen LM, Garg E, Nguyen TTT, Wang Z, Barth B, Yaqubi M, McEwen LM, MacIsaac JL, Diorio J, Kobor MS, O'Donnell KJ, Meaney MJ, Silveira PP. A biologically-informed polygenic score identifies endophenotypes and clinical conditions associated with the insulin receptor function on specific brain regions. EBioMedicine. 2019;42:188–202. https://doi.org/10.1016/j.ebiom.2019.03.051.
83. Alexander-Bloch A, Huguet G, Schultz LM, Huffnagle N, Jacquemont S, Seidlitz J, Saci Z, Moore TM, Bethlehem RAI, Mollon J, Knowles EK, Raznahan A, Merikangas A, Chaiyachati BH, Raman H, Schmitt JE, Barzilay R, Calkins ME, Shinohara RT, et al. Copy number variant risk scores associated with cognition, psychopathology, and brain structure in youths in the Philadelphia Neurodevelopmental Cohort. JAMA Psychiatry. 2022; https://doi.org/10.1001/jamapsychiatry.2022.1017.
84. Jacquemont S, Huguet G, Klein M, Chawner SJRA, Donald KA, van den Bree MBM, Sebat J, Ledbetter DH, Constantino JN, Earl RK, McDonald-McGinn DM, van Amelsvoort T, Swillen A, O'Donnell-Luria AH, Glahn DC, Almasy L, Eichler EE, Scherer SW, Robinson E, et al. Genes To Mental Health (G2MH): a framework to map the combined effects of rare and common variants on dimensions of cognition and psychopathology. Am J Psychiatry. 2022;179(3):189–203. https://doi.org/10.1176/appi.ajp.2021.21040432.
85. Douard E, Zeribi A, Schramm C, Tamer P, Loum MA, Nowak S, Saci Z, Lord M-P, Rodríguez-Herreros B, Jean-Louis M, Moreau C, Loth E, Schumann G, Pausova Z, Elsabbagh M, Almasy L, Glahn DC, Bourgeron T, Labbe A, et al. Effect sizes of deletions and duplications on autism risk across the genome. Am J Psychiatry. 2021;178(1):87–98. https://doi.org/10.1176/appi.ajp.2020.19080834.

86. Meltzer LJ, Williamson AA, Mindell JA. Pediatric sleep health: It matters, and so does how we define it. Sleep Med Rev. 2021;57:101425. https://doi.org/10.1016/10.1016/j.smrv.2021.101425.
87. Ohayon MM. Epidemiological overview of sleep disorders in the general population. Sleep Med Res. 2011;2(1):1–9. https://doi.org/10.17241/smr.2011.2.1.1.
88. Zeighami Y, Fereshtehnejad S-M, Dadar M, Collins DL, Postuma RB, Mišić B, Dagher A. A clinical-anatomical signature of Parkinson's disease identified with partial least squares and magnetic resonance imaging. NeuroImage. 2019;190:69–78. https://doi.org/10.1016/j.neuroimage.2017.12.050.
89. Cuthbert BN, Insel TR. Toward the future of psychiatric diagnosis: the seven pillars of RDoC. BMC Med. 2013;11:126. https://doi.org/10.1186/1741-7015-11-126.
90. Watanabe K, Stringer S, Frei O, Umićević Mirkov M, de Leeuw C, Polderman TJC, van der Sluis S, Andreassen OA, Neale BM, Posthuma D. A global overview of pleiotropy and genetic architecture in complex traits. Nat Genet. 2019;51(9):1339–48. https://doi.org/10.1038/s41588-019-0481-0.
91. Mazzotti DR, Haendel MA, McMurry JA, Smith CJ, Buysse DJ, Roenneberg T, Penzel T, Purcell S, Redline S, Zhang Y, Merikangas KR, Menetski JP, Mullington J, Boudreau E, on behalf of the Sleep Research Network Task Force. Sleep and circadian informatics data harmonization: a workshop report from the Sleep Research Society and Sleep Research Network. Sleep. 2022;45(6):zsac002. https://doi.org/10.1093/sleep/zsac002.
92. Peterson RE, Kuchenbaecker K, Walters RK, Chen C-Y, Popejoy AB, Periyasamy S, Lam M, Iyegbe C, Strawbridge RJ, Brick L, Carey CE, Martin AR, Meyers JL, Su J, Chen J, Edwards AC, Kalungi A, Koen N, Majara L, et al. Genome-wide association studies in ancestrally diverse populations: opportunities, methods, pitfalls, and recommendations. Cell. 2019;179(3):589–603. https://doi.org/10.1016/j.cell.2019.08.051.
93. Dunster GP, de la Iglesia L, Ben-Hamo M, Nave C, Fleischer JG, Panda S, de la Iglesia HO. Sleepmore in Seattle: later school start times are associated with more sleep and better performance in high school students. Sci Adv. 2018;4(12):eaau6200. https://doi.org/10.1126/sciadv.aau6200.
94. Grandner MA, Williams NJ, Knutson KL, Roberts D, Jean-Louis G. Sleep disparity, race/ethnicity, and socioeconomic position. Sleep Med. 2016;18:7–18. https://doi.org/10.1016/j.sleep.2015.01.020.

Genetics of Chronotype and Circadian Rhythm Disorders

Lovemore Kunorozva and Jacqueline Lane

Abstract Circadian rhythms are important biological processes that are essential to human health and well-being. These rhythms are generated by the internal "body clock" and regulate the sleep-wake cycle and numerous physiological processes. Mutations (i.e., changes in circadian genes) from one generation to the next including underlying genetic issues which may result in one's inability to receive or process environmental cues can affect the clock's timing. However, circadian rhythm disruption may also be related to external factors and an individual's behavior as a result of their sleep timing (chronotype) or due to activities such as trans-meridian travel and shift-work that puts sleep schedules out of sync with daylight exposure. Dysregulation of the body's internal circadian timekeeping mechanism is an established risk factor for circadian rhythm sleep-wake disorders (CRSWD). This dysregulation is common in modern society and is associated with a number of highly prevalent population diseases including CRSWD. CRSWDs involve a disruption in the timing of physiological processes, including sleep and wake timing. There is evidence of genetic variation playing a role in chronotype, with circadian, behavioral and photo-transduction pathways implicated using large scale GWAS, providing opportunities to better understand the causes and consequences of circadian rhythm disturbances on human physiology. As well, both Advanced and Delayed Sleep Wake Phase Disorders have been linked to rare genetic variation in circadian clock genes in families. In this chapter, we will focus on the genetics of chronotype and CRSWDs in humans, additionally looking at the role of chronomedicine in

L. Kunorozva
Brigham and Women's Hospital, Division of Sleep and Circadian Disorders, Boston, MA, USA

Broad Institute, Molecular and Population Genetics Program, Cambridge, MA, USA

J. Lane (✉)
Brigham and Women's Hospital, Division of Sleep and Circadian Disorders, Boston, MA, USA

Broad Institute, Molecular and Population Genetics Program, Cambridge, MA, USA

Massachusetts General Hospital, Center for Genomic Medicine, Boston, MA, USA
e-mail: jlane@broadinstitute.org

P. Gehrman et al. (eds.), *Genetics of Sleep and Sleep Disorders*,
https://doi.org/10.1007/978-3-031-62723-1_11

providing precision medicine for the treatment of CRSWDs. Thus, this chapter seeks to expand our knowledge of the genetic basis of chronotype and CRSWDs.

Keywords Chronotype · Circadian rhythm · Circadian rhythm sleep wake phase disorder · Human rare gene variants · Chronomedicine

Overview

Chronotype refers to entrainment and/or an individual's proclivity for earlier or later sleep timing, which is influenced by circadian rhythms in physiology, cognition, behavior, and the need for sleep driven by the sleep homeostasis in a given 24 h cycle [1–4]. Thus, chronotype is closely related to the circadian rhythm (i.e., body clock), which regulates cyclical changes in cellular, molecular, and biological processes that repeat once every 24 h. Chronotype varies from person to person and is typically measured on a continuous scale [5, 6]; however, individuals are often grouped into three categories: early chronotypes (or morning larks) who prefer going to bed and waking earlier, late chronotypes (night owls) who prefer a later bedtime and later rising, and neither-types (intermediates) who lie between the two extremes [3, 7, 8]. This variation arises from differences in the body clock's timing in a number of biological (i.e., core body temperature regulation, hormonal secretion, sleep wake cycles etc.) and behavioral processes including sleep and wake timing [8, 9].

While chronotype variation can be influenced by an individual's genetics [10–14], other factors including age [15–17], environmental [18–20], and societal [20, 21] factors may play a role. Several studies have shown timing or phase differences in an individual's "body clock" resulting from differences in physiological and behavioral factors, as well as, sleep timing [1, 7, 22]. This notion is supported by findings which showed different phase angles of entrainment between early and late chronotypes [22, 23]. Additionally, variation in the circadian clock genes has been shown to be associated with markers of sleep homeostasis following sleep deprivation and with behavioral consequences following this deprivation [24–27], which is consistent with studies in mice and *Drosophila* [28]. For example, a recent screen for circadian gene mutations in families with habitual short sleep duration identified a mutation in a critical basic-helix-loop-helix domain of DEC (a repressor of CLOCK/BMAL1 activity) [29], which causes increased daily wakefulness in transgenic mice, as well as, dominant effects in transgenic flies.

Growing evidence points to a link between chronotype, differences in the sleep homeostat and circadian rhythm sleep-wake disorder (CRSWD) [16, 30], with chronotype variation observed at a population level and CRSWD such as advanced sleep wake phase disorder (ASWPD) and delayed sleep wake phase disorder (DSWPD) observed in extreme chronotype cases. Rare genetic variations that shorten circadian period have been reported for familial ASWPD, a CRSWD with habitual sleep times earlier than the societal norm [31–34]. In contrast, limited comparable evidence exists for familial DSWPD and the association of proposed genetic polymorphisms with later chronotype [35–37]. For example, some classical twin studies' reported a strong hereditary component to chronotype preference in the range of 40–60%, suggesting an important role for genetic predisposition to DSWPD etiology [35, 36].

CRSWDs are complications that occur when the "body clock" is out of sync with the external environment [12, 30, 38]. This disruption can be temporary and caused by changes in one's sleep habits, shift work, or travel across multiple time zones. Alternatively, it can be long-term and caused by one's genes [24, 25, 27], medical conditions [39], or aging [40, 41]. There are different types of CRSWDs namely, ASWPD, DSWPD, Non-24-h sleep-wake phase disorder (N24-SWPD), irregular sleep-wake rhythm disorder (ISWRD), and shift work sleep disorder (SWD). Of these CRSWDs, ASWPD and DSWPD are thought to be heritable from one generation to the next. While there is vast evidence of reported associations between disrupted circadian rhythms and CRSWDs, there is limited evidence for a causal role of chronotype on CRSWDs. Additionally, very few studies have explored the use of chronomedicine in providing precision treatment of CRSWDs. Thus, this chapter will focus on recent discoveries of the genetics of chronotype, circadian and sleep physiology, and CRSWD.

Mechanism of the Molecular Circadian Clock

The self-sustaining nature of the ~24 h "body clock" is primarily attributed to circadian molecular clocks and in humans each cell has its own circadian clock [37, 42]. The cellular circadian rhythm is generated via a molecular network of essential transcriptional-translational feedback loops, with a single cycle taking nearly 24 h to complete [43, 44]. The Circadian Locomotor Output Cycles Kaput (CLOCK) and Brain Muscle ARyl Hydrocarbon Receptor Nuclear Translocator-Like 1 (BMAL1, encoded by the ARNTL gene) genes encode a heterodimeric transcription factor known as CLOCK/BMAL1 [37, 43, 45]. This CLOCK/BMAL1-initiated loop is considered the core loop of the mammalian clock (Fig. 1) [37]. These genes belong to the basic-helix-loop-helix-PER-ARNT-SIM (bHLH-PAS) transcription factor family. CLOCK and BMAL1 heterodimerize and activate E-box promoter containing genes (by binding to E-box enhancer elements 5'-CACGTG-3' in the promoter region) including the *PERIOD* (*PER1*, *PER2,* and *PER3*) and *CRYPTOCHROME* (*CRY1* and *CR2*) core clock genes and a variety of clock controlled genes (CCG) that couple the body clock to physiological and biological processes [37, 45, 46]. Accumulated PER and CRY proteins enter the nucleus and form repressive complexes that suppress E-box-mediated transcription, including their own by binding to CLOCK/BMAL1 heterodimers (Fig. 1), whereas PER and CRY degradation terminates this repression and allows the transcription cycle to commence [37, 45]. The oscillating transcriptome directly influences most biological pathways,

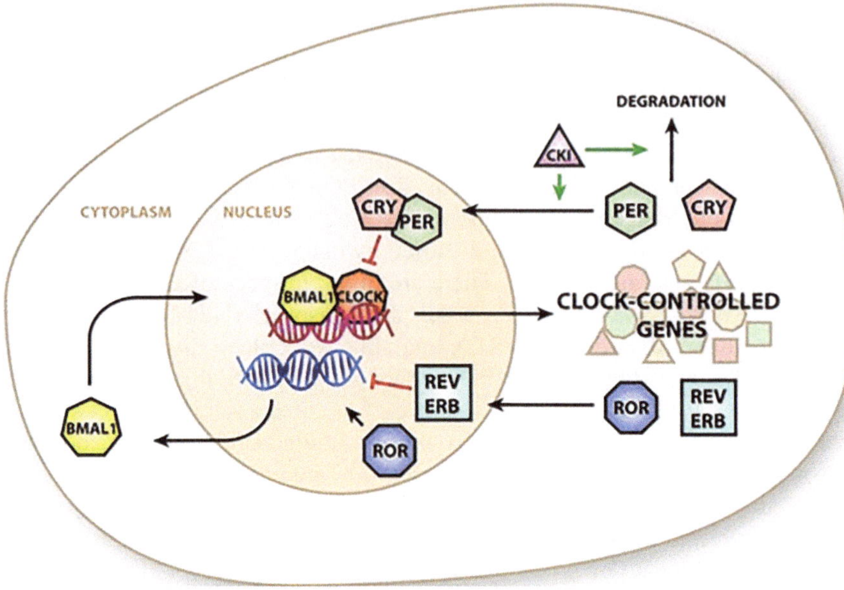

Fig. 1 Diagram showing the mammalian circadian molecular clock. The mammalian clock is composed of two interlocking transcriptional/translational feedback loops, the core, and stabilizing loops, respectively. The main components of the core loop are CLOCK and BMAL1, which form heterodimers and induce E-box-mediated transcription of their negative regulators *PERIOD* (*PERs*) and *CRYPTOCHROMES* (*CRYs*). Accumulated PER and CRY proteins repress E-box mediated transcription until they are sufficiently cleared by proteosome-mediated degradation. CLOCK and BMAL1 also control expression of circadian nuclear receptors such as retinoic acid receptor related orphan nuclear receptors (*RORs)* and *REV-ERBs*, which regulate *BMAL1* levels by competitive actions on the RORs/REV-ERBs-responsive elements (RREs) in the *BMAL1* promoter. Cycling of clock components determines the periodic mRNA expression levels of numerous clock-controlled genes (CCGs) through E-box, RRE, and/or other cis-elements recognized by secondary circadian transcription factors, thus generating rhythmic physiological outputs [55]

ultimately shaping measurable indices including body temperature [47], brainwave activity [48], immunity [49], and hormonal secretion [50].

The stability of PER and CRY proteins is linked to their post translational modifications and is important for proper circadian period length. In addition to CLOCK, Neuronal PAS 2 (NPAS2) is another bHLH-PAS protein that can also form heterodimers with BMAL1 to control the E-box element dependent gene transcription [37]. Through negative feedback of BMAL1/CLOCK, PER (PER1, 2 and 3) and CRY (CRY1 and CRY2) inhibit their own expression over ~24 h cycles leading to rhythms in gene expression [37, 51]. A secondary stabilizing loop consisting of the circadian nuclear receptors REV-ERBs (REV-ERBα and β) and retinoic acid receptor related orphan nuclear receptors (RORs: RORα-γ) adjusts the amounts of bHLH-PAS proteins under the transcriptional control of CLOCK/BMAL1 heterodimers [43, 52, 53]. RORs and REV-ERBs compete to occupy the promoter regions of their target genes with RORs activating REV-ERBs-responsive elements (RREs) mediated transcription, while REV-ERBs suppress transcription [33]. REV-ERBs control circadian outputs by cooperating with cell type specific transcriptional regulators [54].

Mice with mutations in clock genes were shown to have less sleeping time than normal mice in a sleep restriction study [56]. In other studies mutations in *CRY1*, *CRY2, PER1, PER3* and *CK1ε* were shown to result in altered period length, with mutations in *CLOCK* and *PER2* reported to result in shorter period length while a mutation in BMAL1 led to complete loss of circadian rhythmicity [57]. A study by Ralph and Menaker [58] showed a shortened free running period in locomotor activities of Tau-mutant hamsters bearing a mutation in the CK1 gene. PER proteins are phosphorylated by CK1 prior to their proteosomal degradation, contributing to regulation of circadian period length [52]. Likewise, CRY protein phosphorylation by adenosine monophosphate-activated protein kinase leads to degradation mediated by paralogous F-box proteins, FBXL3 and FBXL21 [31]. Mutations in the fbxl3 gene have been shown to result in long period phenotypes in mice, while fblx21-mutant mice have been shown to display short period phenotypes [31, 59].

Genetics of Circadian Rhythms

Genetic variation may regulate how fast or slow one's "body clock" runs or entrains and, as a result, how closely the clock and bodily functions align with the 24 h day [53, 60]. This variation can be a result of mutations or polymorphisms in genes involved in the core clock, input/out pathways, and has the potential to affect the quality, timing, and amount of sleep, which may result in daytime impairment and/or CRSWDs [40, 41, 61]. Genetic investigations to date have investigated the relationship between genetic variation and population level traits, such as chronotype or sleep mid-point, and clinician diagnosed circadian disorders. These studies have paved the way toward understanding extreme cases of chronotype and CRSWDs often observed at an individual level. Individual level differences may result from genetic mutations and/or polymorphisms in circadian clock genes subsequently influencing chronotype and CRSWDs, the current genetic variants' circadian pathways have been elucidated using genome-wide association studies (GWAS) [11, 62]. This genetic variants information is fundamental in determining causal consequential or coincidental pathways in the disease process and important in the clinical diagnosis, treatment, and surveillance of CRSWDs.

Natural Variation in Circadian Physiology

Chronotype is most often evident in timing of bed and wake times. In the population, chronotype is normally distributed and impacted by age and sex [15]. Chronotype varies across the lifespan and is typically most evening type in teens and early adolescents and tends towards morningness as one ages [15]. It is also heritable, with heritability estimates ranging from 12% to 21% [11, 13, 62]. To date, several GWAS studies have been conducted for chronotype, investigating the association of common genetic variation (largely single-nucleotide polymorphisms and small insertions/deletions) with self-reported chronotype and objectively measured sleep timing [11, 13, 62–64]. Thus, the majority of studies discussed in this section are mostly rare variant studies. Rare variants alter gene function and occur at low frequencies in a population and have increasingly been studied using GWAS [11, 60, 65]. These rare variants are kept at low frequencies in populations by purifying selection and play an important role in both complex and Mendelian conditions [66, 67].

A consistent theme across these studies is the involvement of genetic variation in the molecular clock, including variants in *RGS16*, *PER1*, *PER2*, *PER3*, *PIGK/AK5*, *CRY1*, *ARNTL*, *INADL*, *FBXL3*, *HCRTR2,* and *HTR6* [13, 62, 63]. Within the *PER3* regions, two missense variants previously reported to be associated with familial ASWPD are associated with morningness [54]—although with a much lower effect size than reported in familial cases [68]. Furthermore, the genetic regions associated with chronotype are associated with sleep timing, but typically not quality or duration of sleep using both self-reported and actigraphy measures. In the *RGS16* gene, a low frequency missense variant in the minor allele previously associated with chronotype [13] was shown to be strongly associated with morningness [11]. As well, missense variants in the *INADL, HCRTR2, PLCL1,* and *CLN5* genes were shown to be strongly associated with morningness [11, 62, 63].

The *RGS16* and *INADL* variants have been reported to play an important role in the entrainment of circadian rhythms by external stimuli by ensuring correct formation and functioning of retinal ganglion cells [11]. RGS16 deficiency leads to delayed *PER1* expression resulting in circadian period lengthening which has been shown in mice [69]. Thus, some natural variation in chronotype may be explained by differences in how well the retina detects external light signals and transmits these signals to the SCN [11]. Genetic variation in the circadian core clock genes (*PER1, PER2, PER3, CRY1, FBXL3*) has been attributed to subtle differences in the biochemical mechanism of the circadian clock [11]. These core clock variants may precisely affect modification and regulation of the circadian clock and/or the sleep-wake homeostasis contributing to population variation of chronotype. For example, *FBXL3* located in the ubiquitin-proteasome pathway was shown to be involved in the degradation of several core clock genes via protein (de)ubiquitination [70, 71], ultimately influencing the robustness and speed of core clock proteins [72]. The *FBXL3* acts in the nucleus where it mediates ubiquitination and subsequent degradation of CRY1 and CRY2 [73].

HCRTR2 regulates the expression of the G-protein coupled hypocretin receptor type 2, with deficient hypocretin systems thought to be associated with a reduction in wakefulness and circadian alertness signals in narcolepsy [74]. Another variant *HTR6* has been reported to function in sleep regulation in rats with 5-HTR6 agonists resulting in sleep reduction [75]. APH1A plays an important role in cleaving the amyloid- precursor protein, which is higher during wakefulness compared to when asleep [13, 63, 76] and is affected by manipulations of the hypocretin system [77]. A recent study reported an association between *CRY1* and sleep mid-point timing (~40 mins) in an independent cohort of Turkish ancestry participants [78]. A missense in the *PER2* gene was shown to alter circadian period and is associated with familial ASWPD [32, 79]. These genetic studies are shedding light on the variation driving population level differences in circadian rhythm timing, and complement genetic findings of families with the more extreme phenotypes seen in CRSWD.

Circadian Rhythm Sleep Wake Disorders

A circadian rhythm sleep wake disorder (CRSWD) is an illness which occurs when the internal "body clock" is out of sync (desynchronized) with the external environmental time [16, 31]. It occurs as a result of the body's inability to entrain to the 24 h light-dark cycle or misalignment between one's internal time and that of the external environment (Fig. 2). CRSWDs often involve difficulty falling asleep, waking up during sleep, waking up too early or being unable to fall back to sleep once awake, and sleepiness during waking hours [9, 30]. There are six main types of CRSWDs: advanced sleep-wake phase disorder, delayed sleep-wake phase disorder, irregular sleep-wake type disorder, non-24-h sleep-wake disorder, which is most often associated with blindness, jet lag type, and shift work disorder [9, 44, 45, 80, 81]. Each CRSWD is defined by specific patterns of circadian and sleep disruption that result in insomnia or excessive daytime sleepiness. These conditions can have a negative effect on mental and physical health, as well as, daily functioning. While genetics may be a factor, the etiology of CRSWD is complex and often involves both genetics and environmental factors [38, 80]. Of the CRSWDs, ASWPD and DSWPD have demonstrated genetic susceptibility [9, 24, 25, 32]. Individuals who suffer from CRSWDs have shifted physiology and sleep or wake timing. ASWPD and DSWPD are the most common types of CRSWD with DSWPD accounting for the majority of cases (3.3–4.6%) followed by free running type and ASWPD, with ISWT being relatively rare (1%) [80].

Mendelian forms of ASWPD and DSWPD have been identified, and linkage or sequencing studies have implicated rare variants in the genes *PER2*, *PER3*, *CRY1*, *CRY2*, *CSNK1D*, and *TIMELESS*. These studies confirm the involvement of circadian clock genes that were previously identified in model organism studies [24–26, 33, 82]. For example, sleep timing that is advanced about 3–4 h earlier has been reported in familial ASWPD for variants in *PER2* [32], *PER3* [54], *CRY2* [83], *CSNK1D* [33], and *TIMELESS* [82]. In contrast, DSWPD has been reported in

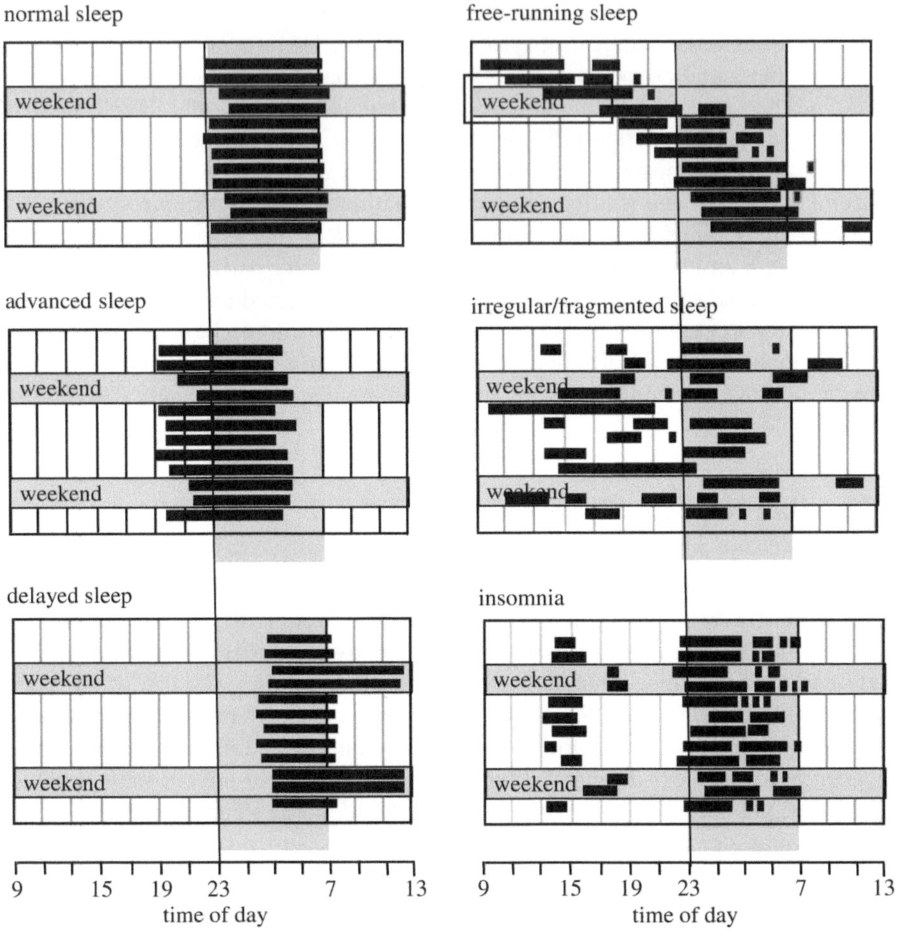

Fig. 2 Graphs showing societal normal and abnormal sleep patterns. Black horizontal bars represent periods of sleep on consecutive work days and on the weekends. In abnormal sleep patterns, insomnia leads to fragmented sleep as a result of difficulty falling asleep or staying asleep, even when one has access to many sleep opportunities. In contrast, individuals who have circadian timing system defects and normally have difficulty sleeping are often referred to as pathological extremes with those preferring to retire to bed earlier than usual (advanced sleep-wake phase disorder), later than usual (delayed sleep-wake phase disorder), and those with progressive delays due to external rhythm desynchronization (non-24-h sleep-wake phase disorder). Lastly those who lack regular 24 h sleep-wake cycles (irregular sleep type) as observed in individuals who lack a circadian clock. The severity of these disorders may vary as one's circadian system drifts in and out of phase with societal norms [88]

CRY1 caused by a gain of function, with affected individuals sleeping nearly an hour later than unaffected individuals [84]. Numerous rare variants have been functionally linked to physiological changes in circadian rhythm cycle length (period), as well as, molecular changes in phosphorylation in mice [85]. However, to date only a limited number of families with CRSWDs have been studied and for several rare variants, evidence for causality needs strengthening [86].

Given that the pathogenicity of rare genetic variants may be overestimated in families ascertained for CRSWD, it will also be important to assess the impact of these rare functional variants in population-scale databases [87]. The threshold of minor allele frequency (MAF) has not yet been clearly defined with this value varying between studies (0.1–5%). Notàbly, some circadian forms of ASWPD and DSWPD may represent extremes at the end of the chronotype distribution, e.g., rare variants (MAF = 0.4%) in *PER3* contribute to both diurnal preference and associate with clinically defined ASWPD and DSWPD [11, 25, 54].

Advanced Sleep Wake Phase Disorder

Individuals with ASWPD have advanced circadian rhythms that lead to a tendency to go to sleep and wake up early [89–91]. This disorder is relatively rare affecting nearly 1% of individuals in middle to older age and occurs more frequently in older adults [90]. In some cases, this syndrome may be caused by an inherited genetic disorder (i.e., it has a family tendency, FASWPD) [9]. The first case of familial ASWPD was a missense mutation (S662G) in the *PER2* gene [32]. In this study, transgenic mice carrying FASWPD *hPER2* S662G had a shorter period, while those carrying S662D mutation (i.e., mimics the process of phosphorylation) exhibited a longer period. FASWPD has also been linked with a missense mutation (T44A) in the human CK1δ gene, which reduces kinase activity in vitro and results in a shorter circadian period in mice [33]. The casein kinases 1δ and 1ε (CKIδ/ε) regulate the levels of PER2 through phosphorylation mediated degradation and cellular localisation [45]. Another study reported a missense mutation in the human *CRY2* (A260T) gene that is associated with FASWPD [31]. In humans, FASWPD was demonstrated when researchers tracked patients from three families' sleep logs and measured their core body temperature and melatonin secretion, they concluded that ASWPD has a strong relation with genes [45]. Specifically, affected individuals have an amino acid change from serine to glycine within the CK1ε binding region of hPER2 [32].

Delayed Sleep Wake Phase Disorder

Patients with DSWPD tend to fall asleep very late and have delayed wake times compared with conventional times because of delayed circadian rhythms, which may result in chronic sleep insufficiency and daytime impairment due to differences between the "timing" of the world and the physiology [44, 92]. Clinical diagnosis of DSWPD requires that the affected individual (1) show a significant delay in relation to desired sleep and wake times as evidenced by a chronic or recurrent complaint, (2) symptoms are present for at least 3 months, and (3) data obtained from sleep logs and actigraphy for at least 7 days wherever possible (preferably 14 days) demonstrate a delay in the timing of the habitual sleep period. DSWPD is often associated with night owls who stay up late at night and sleep in the early hours of the morning [81]. It is very rare among the general population affecting 1–2 people out of every 1000 individuals, and it's primarily experienced by teens and young adults (3.3–10%) [92, 93]. Intrinsic circadian periods (tau) longer than 24 h require greater adjustment to entrain to the 24 h day-night cycle [22], and there is some evidence that adolescence is associated with a physiologic lengthening of tau [94, 95]. Such lengthening may partially explain the propensity for DSWPD. This is consistent with findings from a study which found a longer median tau in patients with DSWPD compared with healthy controls [96]. However, comparison of individual studies is complicated by the use of inconsistent diagnostic criteria and the use of evening chronotype as a proxy for DSWPD. The cause of DSWPD is unknown, but may be related to genetics [24–26, 32, 91, 97, 98], underlying physical conditions and a person's behavior [99].

Although the prevalence of familial DSWPD has not been established, a family history is often present among afflicted individuals [100]. Familial cases of DSWPD have been described, suggesting that Mendelian inheritance of DSWPD may exist with polymorphisms in the CLOCK or *PER3* genes [101]. Two studies of unrelated families showed strong heritability of DSWPD and described a gain of function mutation in the *CRY1* clock gene, which creates a transcriptional inhibitor with enhanced affinity for circadian activator proteins CLOCK and BMAL1 [84, 100], and lengthens the period of circadian rhythms as a result. In one study, *GSK3B* was associated with DSWPD in the sleep center sample. In another study, *MTNR1A*, *RORC*, *CRY2*, and *NFIL3* were associated with DSWPD using the older sleep phenotype form [102]. NFIL3 inhibits transcription factor binding at D sites, which promote transcription of genes on both negative and positive sides of the circadian clock feedback loop. RORC binding at RRE elements may promote the positive arm of the CLOCK-ARNTL transcription; however, SNPs in both *NFIL3* and *RORC* were reported to be associated with relative phase advance. Analysis of sleep electroencephalography suggests that slow wave activity (a proxy of the sleep homeostatic mechanism) decreases in magnitude in the first sleep cycle among evening chronotypes compared with those with normal or intermediate chronotypes, despite similar baseline sleep among groups [99]. Management of DSWPD may include eliminating caffeine, nicotine and alcohol, avoiding day time naps, and refraining from stimulating activities before desired bed time. For individuals who fail to respond to behavioral modifications and sleep wake scheduling, timed melatonin administration in the evenings may be a better alternative. Timed morning bright light exposure combined with behavioral education is another viable strategy for some patients.

Non-24 Hour Sleep Wake Rhythm Disorder

This occurs mostly in totally blind individuals (up to 70%) who are not able to receive light-based cues for their circadian rhythm [39, 103]. This disorder has also been reported in sighted individuals, very rarely and often caused by traumatic brain injury or damage to cells in the retina. These individuals experience a circadian rhythm that is shorter or, more often, slightly longer than 24 h [104]. N-24-SWRD may also be due to lesions of the circadian systems, which result in reduced activation of intact neuronal circuits by circadian synchronizers. Individuals with N-24-SWRD still follow a 24 h cycle; however, their sleeping hours constantly shift backwards or forward by minutes or hours at a time [39].

This results in circadian processes being desynchronized due to the absence of light input into the body clock [105]. Additionally, this leads to an abnormal phase angle between the body clock and sleep wake behavior altering the pattern of alertness, mood, and performance across the day, which interferes with social, academic, and professional life. Several studies have reported sleep insomnia among blind people, with the frequency of CRSWD ranging between 60% and 80% [39, 106, 107]. This is consistent with a study which reported that the majority (76%, $n = 50$) of blind individuals experienced CRSWD, with 40% of these individuals reporting cyclical symptoms [108]. No Mendelian inheritance for N-24-SWRD has been demonstrated; however, one study reported minor allele changes in *BHLHE40* in four patients with N-24-SWRD [102]. BHLHE40 inhibits the positive limb of the circadian clock transcription translation feedback loop by competitively inhibiting ANRT-CLOCK transcription factor binding to E-box promoter sites. A separate study consisting of 67 N-24-SWRD patients showed an association of N-24-SWRD with *PER3* gene, with aspects related to homeostatic sleep drive and response to light shown to be abnormal in N-24-SWRD.

Irregular Sleep Wake Rhythm Disorder

This is a rare disorder and individuals who suffer from it have no consistent pattern to their sleep and may take many naps or have short sleeping periods throughout a 24 h day [89, 109]; thus, sleep is fragmented and frequently insufficient. This is often associated with conditions that affect the brain, such as dementia or traumatic brain injury, that limit the proper functioning of the "master clock" in the hypothalamus [89]. Documentation (sleep diaries and/or actigraphy) of multiple non-circadian sleep wake bouts for a period of at least 7 days is required for diagnosis.

Shift Work Sleep Disorder

This arises from work obligations which cause major disruptions in a person's circadian rhythm and is characterized by sleepiness or insomnia accompanied by a total reduction in sleep time [61, 110]. Specifically, shift work, which requires one to work through the night and sleep during the day, puts an individual's sleep schedule at odds with the local day-night cycle [110, 111]. An individual's chronotype has been shown to influence one's tolerance to shift work [80], with late chronotypes reported to show better adjustment to night work [24, 111–114]. The evening type chronotype has been shown to be related to variations in clock genes including SNPs in *PER2* [115] and *VNTR* in *PER3* [24]. In one study, nurses with *PER3* and *NPAS2* polymorphisms were reported to have reduced dozing during work timings and the *PER2* polymorphism was associated with long sleep duration during a daytime shift [116]. This is consistent with findings which showed an association between *PER3* VNTR polymorphism and disrupted sleep architecture, as well as, compromised cognitive performance following sleep disruption [27].

Individuals carrying the longer version of the *PER3* 5-repeat allele may experience elevated levels of sleepiness during night shift and correspondingly be at greater risk for car and occupational accidents. In general, the dysregulation of the sleep homeostasis and circadian rhythms can lead to psychiatric disorders, metabolic and cardiomyopathies [117]. Other studies showed associations in *CRY1* and *CRY2* with shift work and psychiatric disorders that may be affected by the circadian clock, although no relation to CRSWD was shown [118, 119]. One of these studies reported different ratios of *CRY* gene polymorphism between day and night shift nurses and midwives [119]. Finally, a recent study demonstrated an association between the *PER1* (C/C) intronic polymorphism with susceptibility to SWD in factory workers [120].

Jet Lag Disorder

This is experienced frequently by individuals of all ages and can occur when one undergoes rapid transition across time zones, resulting in asynchrony between local and internal "circadian clock" time [121–123]. The body often has difficulty adjusting to new time zones because the internal signals for wakefulness and sleep are disrupted due to the mismatch with the local day and night cycle [123]. The more time zones crossed particularly in an eastward direction the more pronounced the sleep disturbance and other jet lag symptoms such as sleep and/or excessive daytime sleepiness, gastrointestinal discomfort, general malaise, and somatic complaints [123, 124]. The clinical diagnosis of JLD includes the identification of excessive daytime sleepiness or disruption of the sleep wake cycle along with decreased total sleep time that coincides with rapid transition across time zones. Individuals differ in their tolerance to circadian phase misalignment and this tolerance decreases with age [125]. Synchronization to the new environment may require timed bright light exposure in the evening following westward travel and in the morning following eastward travel [126]. Simulated jet lag studies in mice showed that the Lhx1 gene was suppressed in response to light following an 8 h phase shift of the sleep-wake cycle, and the mice with little or no Lhx1 readjusted faster to the phase shift than normal mice [127].

Circadian Rhythm Disorder Clinical Diagnosis

To properly diagnose CRSWDs, a physician will need information on an individual's underlying physiology, and sleep habits. An individual's underlying physiology, for example, dim-light melatonin onset (DLMO) is more reliable than a recorded history of sleep habits and can be used as an objective biomarker to assess CRSWD [128, 129]. Using DLMO to establish a patient's circadian timing is also relevant for distinguishing CRSWDs from other non-circadian related sleep disorders [128]. This is consistent with findings from a study, which showed that a diminished initial surge of melatonin secretion following DLMO contributed to the etiology of DSWPD [129]. Also, DLMO is useful in determining if an individual is entrained to a 24 h day-night cycle and assessing phase advances or delayed rhythms of entrained individuals. DLMO is also important for identifying optimal application times for therapies including timed bright light exposure and/or exogenous melatonin administration. Best practices will make use of both underlying physiological data and an individual's sleep-wake history collected for 7–14 days.

An individual's sleep history may be obtained through a daily sleep log, sleep questionnaire [130], sleep study, actigraphy [131], medical history, and conducting a physical exam, all meant to help demonstrate sleep irregularities and/or shifted timing. However, it is important to note that each disorder has unique symptoms, and its diagnosis may be constrained by social and occupational obligations [44]. As a result, some people may exhibit symptoms identical to CRSWD due to shifted timing, but are not clinically disordered. Shifted timing may lead to a range of medical conditions including obesity, metabolic syndrome, type II diabetes, gastrointestinal dysfunction, cardiovascular disease, excessive sleepiness, compromised immune function, and social disorders [89, 91, 98, 112]. The majority of people with CRSWDs exhibit one or more of the following symptoms: difficulty falling or staying asleep, feeling tired after sleep, feeling less alert, and experiencing issues with memory [44, 80, 107], each of which can present as other sleep disorders, leading individuals to often get misdiagnosed. There are many causes of CRSWDs with conditions that influence sleep patterns being internal or external [107]. External conditions include light exposure [61], time zone travel [121, 123], non-traditional work hours, late night social activities [61, 125], levels of physical activity [132], medical conditions [39], and medications [133].

Additionally, individuals who suffer from blindness [108], depression [134, 135], brain injury [136], neurodegenerative diseases (e.g., dementia) [137, 138], pregnancy [139–141], and other genetic conditions (e.g., Huntington's disease) are at greater risk for a CRSWD. Treatment for CRSWDs will depend on the type and cause of the disorder, with the goal of fitting the individual's sleep pattern into a schedule that matches their lifestyle. Treatment may include bright light therapy, sleep hygiene, chronotherapy, lifestyle changes, and medications such as melatonin. A combination of treatments to promote better sleep may help one to find relief. For example, one can adjust exposure to daylight, making changes in the timing of their daily routines and strategically scheduling naps. Alternatively, they can reset their circadian rhythm to a desired pattern by exposing the eyes to safe levels of intense, bright light for short durations at precise times of the day [142]. Home remedies including avoiding naps during the day, following a set sleep schedule, avoiding bright lights at night, sleeping in a cool quiet room, making one's bed more comfortable may reduce sleep deprivation for individuals with CRSWDs. Obtaining adequate sleep is important for health as having a CRSWD places one at a greater risk of depression, a weaker immune system, diabetes, cardiovascular diseases, and cognitive impairment.

Circadian/Sleep Genetic Link to Disease

Several studies have linked circadian or sleep traits and disorders with characteristics of human health and disease including psychiatric disorders and cardio metabolic diseases [45, 119, 143–147]. However, ASWPD [31, 45, 60] and DSWPS [25, 27, 148, 149] have been reported to have an established genetic basis. At present, it is not clear if these disorders are markers, consequences, or causal in disease pathology. Thus, the directionality of association could be resolved with the use of genetics [150]. Genetic relationships to health and disease and causal effect have been established using circadian and sleep GWAS [60, 146]. For example, genetic associations between insomnia and psychiatric disorders including anxiety and depression are reported to be higher (rq = 0.45–0.59) than between insomnia and most other sleep traits. This suggests that the etiology of insomnia may share more biological processes with psychiatric disorders than it does with sleep wake regulatory processes [13, 151]. Furthermore, genetic sharing with different sleep phenotypes differs between CRSWD, suggesting that diseases may be distinguishable by their distinct patterns of sleep disruption.

Observational studies on circadian rhythms have reported associations between chronotype and psychiatric disorders; however, research in this area has been hampered by confounding variables. Mendelian randomization studies [11, 13, 62, 152, 153] confirm a causal effect of the lifelong intervention of genetic variation for monitoring chronotype as protective for psychiatric disorders, especially major depressive disorder (MDD). One study reported that morningness decreased the liability of depression, with increasing morningness from moderate- to definite morning-type associated with an odds ratio of 0.65 (0.44–095; p = 0.003) for MDD. Several interventions including sleep and meal timing, napping, caffeine intake, light exposure, and exercise behavior have been shown to provide viable evidence for the feasibility of interventions that produce clinically relevant shifts in circadian rhythms [154]. In other short sleep studies, a causal link to coronary artery disease and myocardial infarction was demonstrated, further supporting several observational studies linking CRSWD with cardiac outcome [155–158]. This connection emphasizes the need for RCTs in validating novel treatment methods and preventative therapies for CRSWD and the importance of advocacy to reduce the health disparities around sleep and circadian disruption. The above-mentioned studies provide strong evidence that supports the use of human genetics to understand the causal pathways of circadian rhythms, sleep disruption in health and disease. This is consistent with genetic studies which support the bidirectional causal links between CRSWDs and disease, and congruent with the models which suggest that sleep disturbances contribute to worsening of disease [159].

Chronomedicine

Chronomedicine is the application of chronobiology (Fig. 3) in order to understand the pattern of disease (i.e., prevention, causation, diagnosis, and treatment), which can be related to disturbances of circadian rhythms [7, 160, 161]. It dates back to the 1970s, when researchers noticed how mice with cancer responded better to treatment given in decreasing doses over a 24 h cycle [162]. Chronomedicine has gained momentum in recent years and focuses on the role of time (Greek: chronos) in our physiology, endocrinology, metabolism, and behavior at many organizational levels [161, 163–166]. Additionally, it explores the interaction between biological rhythms, medicine, and drugs at an individual level, thus enabling personalized and precise treatment [7], which also includes preventive chronomedicine (population-based)

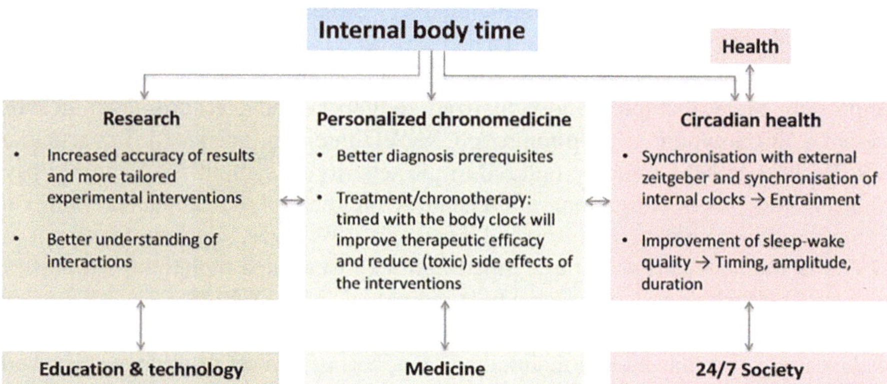

Fig. 3 Determining internal body time followed by circadian precision medicine has beneficial effects on research, therapeutic, and health outcomes [7]

[161, 165, 166]. Thus, it recognizes that there are optimal times and less optimal or even hazardous times for certain medical interventions [7, 166]. Chronomedicine has not gained much traction in the pharmaceutical industry for various reasons. Firstly, the translation of research findings into practice is often complicated by individual variations in the optimum dosing times [160]. Secondly, differences in gender, and age groups and the fact that pharmaceutical companies likely need to test many products to determine optimal times for all potential variations hinder the transition to individual-based precision medicine.

Clinicians need to become more familiar with normal circadian rhythm patterns in key physiological functions (e.g., cortisol, melatonin, body temperature). They also need to develop pattern recognition strategies that can identify critical circadian disruption, similar to the way electrocardiograms are used to identify a rhythmic disturbance [167]. A complex approach may require quantitative analysis of rhythms, perhaps using a software that is able to process real-time physiological data to identify the timing of peaks (or troughs). This information would enable the analysis of phase shifts, amplitude variation from peak to trough, and the consistency of the circadian rhythm over time [160, 166, 168]. It is important to monitor circadian rhythms and consider the use of chronotherapeutics, which involves timing an intervention or administration of a drug at the time of day at which it is likely to have the optimal effect.

The majority of chronomedicine research has been performed using animal models that help us to better understand human health and disease [169–171]. For example, the administration of certain chemotherapies is organized around timelines that recognize the potential interaction of the medication with liver enzymes under circadian control [172, 173]. Optimizing the timing of medicine with a narrow therapeutic index and significant circadian fluctuation, such as antibiotics, steroids, and anticoagulants, could lead to significant clinical benefit from improvements in efficacy to minimization of toxicity. For example, benzodiazepines phase shift the clock according to the time of day at which they are administered [174], thus the circadian disruption effect could be minimized by administering medicine at the correct time of day relative to an individual's circadian phase.

Chronomedicine, however, is not a panacea, as it is not applicable for all medications or ailments. Even so, it may still offer an important avenue for honing and improving many therapies. Individualized zeitgeber timing according to an individual's chronotype may help improve CRSWD interventions [160]. In particular, CRSWD treatment should be tailored to the severity of symptoms, comorbid psychopathology, work obligations, ability, and willingness of the patient and family to comply with treatment and societal pressures [89, 175]. Current treatment of DSWPD includes chronotherapy, timed light exposure, and melatonin administration [175]. A behavioral approach where individuals with DSWPD's circadian clock are reset by progressively delaying (i.e., 3 h every 2 days) until the desired sleep and wake times are achieved and maintained by adhering to a set of sleep-wake schedules and good sleep hygiene schedules is preferred. In contrast, ASWPD may be treated by progressively advancing the sleep time by 2 h every day until the desired sleep time is achieved [175].

In all DSWPD patients, adherence to a regular sleep wake schedule, avoidance of bright light in the evening, and sleep hygiene education are important elements in the approach to clinical management. Timed bright light exposure has the potential to stabilize the circadian rhythm with light exposure in the morning (i.e., advance portion of the phase response curve) and exposure in the evening resulting in phase delays. This is consistent with findings from studies which reported that individuals who had earlier sleep times had better subjective alertness in the morning after 2 weeks of daily morning (7–9 a.m.) bright light [176, 177]. Chronotherapy is often impractical in clinical settings given patients' poor compliance to light exposure; this has resulted in an increase in studies investigating melatonin [89, 175] as an alternative treatment. Melatonin administration in the evening advances the timing of circadian rhythms and the opposite is true when consumed in the morning [103]. One study in patients with DSWPD showed that the largest phase advance in the timing of circadian rhythms occurred when melatonin 0.3 or 3 mg was administered nearly 6–7 h before sleep time [178]. Behavioral adjustments such as 2 days adjustment of evening light exposure were reported to improve sleep and daytime performance in older adults with early morning insomnia [179].

For individuals with ASWPD, exposure to light in the evening has the potential to improve sleep quality, increase total sleep time, and daytime performance in older adults [89, 180]. In contrast, the effectiveness of melatonin can be improved by using a combination of chronotherapy, morning phototherapy, and late afternoon or early evening melatonin for individuals with DSWPD [181, 182]. Treatment of another CRSWD, the N-24-SWRD may require synchronizing the circadian clock using behavioral approaches, such as maintaining a regular sleep wake schedule and social/work activities, as well as, melatonin (10 mg before bed time or 0.5 mg melatonin nightly) administration [175]. In one case report, normal sleep wake time was restored by administering melatonin in the evening to a sighted adolescent with N-24-SWRD [183]. Limiting caffeine consumption in the evenings, exercising during the day, avoiding long naps later in the day, sleep hygiene education, and maintaining a regular schedule of activities is essential in the treatment of CRSWDs. Additionally, use of improved diagnostics and therapeutic devices such as light therapy glasses and smart lighting could play an important role in the treatment of CRSWD.

Summary and Conclusion

It is evident that while chronotype is influenced by numerous factors, genetics can play a role. Among the genetic variants which have been reported to be associated with chronotype are known genes including *RGS16*, *PER3*, *ANRTL*, *FBXL3*, *HCRTR2*, *PER1*, and *CRY1* and polymorphisms/mutations in these genetic variants have been shown to influence chronotype and CRSWDs. However, the evidence for chronotype's association with CRSWD is limited and affected by numerous confounders, warranting further research in this area. Given the negative impact the shift in circadian timing and physiology has on an individual's health and well-being, CRSWD should be considered in the differential diagnosis of every patient presenting with symptoms of excessive sleepiness and/or insomnia. A misalignment between the internal body clock and the external environment time constitutes the basis of most CRSWD. The management and treatment of these disorders typically focus on realigning of the circadian clock to the desired or socially acceptable sleep and work times. Thus, social and behavioral factors when at odds with the norm play an important role in perpetuating or exacerbating the risk of CRSWD. In summary, the successful diagnoses, surveillance, and treatment of individuals rely on a multimodal approach.

This approach requires a deep understanding of CRSWD and an individualized rather than a population-based approach using a combination of environmental factors, behavioral (e.g., sleep timing, time of exposure to light) and physiological (i.e., use of melatonin) strategies to treat affected individuals. For example, when treating DSWPD use of morning bright light exposure, avoidance of evening light and/or appropriately timed melatonin administration may be a viable combination. In general, treatment of CRSWD remains a challenge in part because there is limited evidence from RCT studies utilizing chronotherapy and/or pharmacological products to treat CRSWD. Given the recent advances in GWAS, exome sequencing, and systematic mutation screening of the circadian clock genes in families and/or patients with extreme sleep phenotypes, we may be closer to finding a population-based approach treatment for CRSWDs. This understanding of the neurobiology and genetics of chronotype, circadian rhythm, and sleep regulation has the potential to lead to improved diagnostic tools and treatment of CRSWD. Additionally, it will help establish evidence based and practical practice guidelines for the diagnosis, surveillance, and treatment of CRSWDs. More research investigating the genetics of chronotype and its link to CRSWDs is warranted and may be useful in determining the causality, direction, and association between the genetics of chronotype and the course of circadian rhythm sleep disorders.

References

1. Bauducco S, Richardson C, Gradisar M. Chronotype, circadian rhythms and mood. Curr Opin Psychol. 2020;34:77–83. https://doi.org/10.1016/j.copsyc.2019.09.002.
2. Khan S, Duan P, Yao L, Hou H. Shiftwork-mediated disruptions of circadian rhythms and sleep homeostasis cause serious health problems. Int J Genomics. 2018;2018:1–11. https://doi.org/10.1155/2018/8576890.
3. Lack L. Chronotype differences in circadian rhythms of temperature, melatonin, and sleepiness as measured in a modified constant routine protocol. Nat Sci Sleep. 2009:1. https://doi.org/10.2147/NSS.S6234.
4. Skeldon AC, Derks G, Dijk DJ. Modelling changes in sleep timing and duration across the lifespan: changes in circadian rhythmicity or sleep homeostasis? Sleep Med Rev. 2016;28:96–107. https://doi.org/10.1016/j.smrv.2015.05.011.
5. Putilov AA, Marcoen N, Neu D, Pattyn N, Mairesse O. There is more to chronotypes than evening and morning types: results of a large-scale community survey provide evidence for

high prevalence of two further types. Personal Individ Differ. 2019;148:77–84. https://doi.org/10.1016/j.paid.2019.05.017.
6. Roenneberg T, Pilz LK, Zerbini G, Winnebeck EC. Chronotype and social jetlag: a (self-) critical review. Biology. 2019;8(3):54. https://doi.org/10.3390/biology8030054.
7. Münch M, Kramer A. Timing matters: new tools for personalized chronomedicine and circadian health. Acta Physiol. 2019;227(2) https://doi.org/10.1111/apha.13300.
8. Vitale JA, Roveda E, Montaruli A, et al. Chronotype influences activity circadian rhythm and sleep: differences in sleep quality between weekdays and weekend. Chronobiol Int. 2015;32(3):405–15. https://doi.org/10.3109/07420528.2014.986273.
9. Rijo-Ferreira F, Takahashi JS. Genomics of circadian rhythms in health and disease. Genome Med. 2019;11(1):82. https://doi.org/10.1186/s13073-019-0704-0.
10. Dmitrzak-Węglarz M, Pawlak J, Wiłkość M, et al. Chronotype and sleep quality as a subphenotype in association studies of clock genes in mood disorders. Acta Neurobiol Exp (Warsz). 2016;76(1):32–42. https://doi.org/10.21307/ane-2017-003.
11. Jones SE, Lane JM, Wood AR, et al. Genome-wide association analyses of chronotype in 697,828 individuals provides insights into circadian rhythms. Nat Commun. 2019;10(1):343. https://doi.org/10.1038/s41467-018-08259-7.
12. Kalmbach DA, Schneider LD, Cheung J, et al. Genetic basis of chronotype in humans: insights from three landmark GWAS. Sleep. 2017;40(2) https://doi.org/10.1093/sleep/zsw048.
13. Lane JM, Vlasac I, Anderson SG, et al. Genome-wide association analysis identifies novel loci for chronotype in 100,420 individuals from the UK Biobank. Nat Commun. 2016;7(1):10889. https://doi.org/10.1038/ncomms10889.
14. Maukonen M, Havulinna AS, Männistö S, Kanerva N, Salomaa V, Partonen T. Genetic associations of chronotype in the finnish general population. J Biol Rhythm. 2020;35(5):501–11. https://doi.org/10.1177/0748730420935328.
15. Fischer D, Lombardi DA, Marucci-Wellman H, Roenneberg T. Chronotypes in the US—Influence of age and sex. Tosini G (ed) Plos One. 2017;12(6):e0178782. https://doi.org/10.1371/journal.pone.0178782.
16. Kivelä L, Papadopoulos MR, Antypa N. Chronotype and psychiatric disorders. Curr Sleep Med Rep. 2018;4(2):94–103. https://doi.org/10.1007/s40675-018-0113-8.
17. Randler C, Engelke J. Gender differences in chronotype diminish with age: a meta-analysis based on morningness/chronotype questionnaires. Chronobiol Int. 2019;36(7):888–905. https://doi.org/10.1080/07420528.2019.1585867.
18. Brockmann PE, Gozal D, Villarroel L, Damiani F, Nuñez F, Cajochen C. Geographic latitude and sleep duration: a population-based survey from the Tropic of Capricorn to the Antarctic Circle. Chronobiol Int. 2017;34(3):373–81. https://doi.org/10.1080/07420528.2016.1277735.
19. Randler C, Rahafar A. Latitude affects morningness-eveningness: evidence for the environment hypothesis based on a systematic review. Sci Rep. 2017;7(1):39976. https://doi.org/10.1038/srep39976.
20. Ujma PP, Kirkegaard EOW. The overlapping geography of cognitive ability and chronotype. PsyCh J. 2021;10(5):834–46. https://doi.org/10.1002/pchj.477.
21. Allebrandt KV, Teder-Laving M, Kantermann T, et al. Chronotype and sleep duration: the influence of season of assessment. Chronobiol Int. 2014;31(5):731–40. https://doi.org/10.3109/07420528.2014.901347.
22. Papatsimpa C, Schlangen LJM, Smolders KCHJ, Linnartz JPMG, de Kort YAW. The inter-individual variability of sleep timing and circadian phase in humans is influenced by daytime and evening light conditions. Sci Rep. 2021;11(1):13709. https://doi.org/10.1038/s41598-021-92863-z.
23. Lang C, Richardson C, Micic G, Gradisar M. Understanding sleep-wake behavior in late chronotype adolescents: the role of circadian phase, sleep timing, and evening vigilance. PsyArXiv. 2021; https://doi.org/10.31234/osf.io/npkeu.

24. Archer SN, Robilliard DL, Skene DJ, et al. A length polymorphism in the circadian clock gene Per3 is linked to delayed sleep phase syndrome and extreme diurnal preference. Sleep. 2003;26(4):413–5. https://doi.org/10.1093/sleep/26.4.413.
25. Ebisawa T, Uchiyama M, Kajimura N, et al. Association of structural polymorphisms in the human *period3* gene with delayed sleep phase syndrome. EMBO Rep. 2001;2(4):342–6. https://doi.org/10.1093/embo-reports/kve070.
26. Pereira SD, Tufik S, Louzada FM. Association of the length polymorphism in the human Per3 gene with the delayed sleep-phase syndrome: does latitude have an influence upon it? Sleep. 2005; https://doi.org/10.1093/sleep/28.1.29.
27. Viola AU, Archer SN, James LM, et al. PER3 polymorphism predicts sleep structure and waking performance. Curr Biol. 2007;17(7):613–8. https://doi.org/10.1016/j.cub.2007.01.073.
28. Mignot E, Takahashi JS. A circadian sleep disorder reveals complexity in the clock. Published online 2013:4.
29. He Y, Jones CR, Fujiki N, et al. The transcriptional repressor DEC2 regulates sleep length in mammals. Science. 2009;325(5942):866–70. https://doi.org/10.1126/science.1174443.
30. Durmuş FB, Arman AR, Ayaz AB. Chronotype and its relationship with sleep disorders in children with attention deficit hyperactivity disorder. Chronobiol Int. 2017;34(7):886–94. https://doi.org/10.1080/07420528.2017.1329207.
31. Hirano A, Shi G, Jones CR, et al. A cryptochrome 2 mutation yields advanced sleep phase in humans. elife. 2016;5:e16695. https://doi.org/10.7554/eLife.16695.
32. Toh KL, Jones CR, He Y, et al. An hPer2 phosphorylation site mutation in familial advanced sleep phase syndrome. Science. 2001;291(5506):1040–3. https://doi.org/10.1126/science.1057499.
33. Xu Y, Padiath QS, Shapiro RE, et al. Functional consequences of a CKId mutation causing familial advanced sleep phase syndrome. Nature. 2005;434:5.
34. Xu Y, Toh KL, Jones CR, Shin JY, Fu YH, Ptáček LJ. Modeling of a human circadian mutation yields insights into clock regulation by PER2. Cell. 2007;128(1):59–70. https://doi.org/10.1016/j.cell.2006.11.043.
35. Barclay NL, Eley TC, Buysse DJ, Archer SN, Gregory AM. Diurnal preference and sleep quality: same genes? a study of young adult twins. Chronobiol Int. 2010;27(2):278–96. https://doi.org/10.3109/07420521003663801.
36. Koskenvuo M, Hublin C, Partinen M, Heikkilä K, Kaprio J. Heritability of diurnal type: a nationwide study of 8753 adult twin pairs. J Sleep Res. 2007;16(2):156–62. https://doi.org/10.1111/j.1365-2869.2007.00580.x.
37. Patke A, Young MW, Axelrod S. Molecular mechanisms and physiological importance of circadian rhythms. Nat Rev Mol Cell Biol. 2020;21(2):67–84. https://doi.org/10.1038/s41580-019-0179-2.
38. Faulkner SM, Bee PE, Meyer N, Dijk DJ, Drake RJ. Light therapies to improve sleep in intrinsic circadian rhythm sleep disorders and neuro-psychiatric illness: a systematic review and meta-analysis. Sleep Med Rev. 2019;46:108–23. https://doi.org/10.1016/j.smrv.2019.04.012.
39. Quera Salva MA, Hartley S, Léger D, Dauvilliers YA. Non-24-hour sleep–wake rhythm disorder in the totally blind: diagnosis and management. Front Neurol. 2017;8:686. https://doi.org/10.3389/fneur.2017.00686.
40. Gulia KK, Kumar VM. Sleep disorders in the elderly: a growing challenge: sleep in elderly. Psychogeriatrics. 2018;18(3):155–65. https://doi.org/10.1111/psyg.12319.
41. Suzuki K, Miyamoto M, Hirata K. Sleep disorders in the elderly: diagnosis and management. J Gen Fam Med. 2017;18(2):61–71. https://doi.org/10.1002/jgf2.27.
42. Albrecht U. Invited review: regulation of mammalian circadian clock genes. J Appl Physiol. 2002;92(3):1348–55. https://doi.org/10.1152/japplphysiol.00759.2001.
43. Gekakis N, Staknis D, Nguyen HB, et al. Role of the CLOCK protein in the mammalian circadian mechanism. Science. 1998;280(5369):1564–9. https://doi.org/10.1126/science.280.5369.1564.

44. Walker WH, Walton JC, DeVries AC, Nelson RJ. Circadian rhythm disruption and mental health. Transl Psychiatry. 2020;10(1):28. https://doi.org/10.1038/s41398-020-0694-0.
45. Liu W. Mini-review: genetics of common types of sleep disorders. AIP Conf Proc. 2019;2058:020015. https://doi.org/10.1063/1.5085528.
46. Fagiani F, Di Marino D, Romagnoli A, et al. Molecular regulations of circadian rhythm and implications for physiology and diseases. Signal Transduct Target Ther. 2022;7(1):41. https://doi.org/10.1038/s41392-022-00899-y.
47. Goh GH, Mark PJ, Maloney SK. Altered energy intake and the amplitude of the body temperature rhythm are associated with changes in phase, but not amplitude, of clock gene expression in the rat suprachiasmatic nucleus in vivo. Chronobiol Int. 2016;33(1):85–97. https://doi.org/10.3109/07420528.2015.1112395.
48. Blume C, Garbazza C, Spitschan M. Effects of light on human circadian rhythms, sleep and mood. Somnologie. 2019;23(3):147–56. https://doi.org/10.1007/s11818-019-00215-x.
49. Haspel JA, Anafi R, Brown MK, et al. Perfect timing: circadian rhythms, sleep, and immunity — an NIH workshop summary. JCI Insight. 2020;5(1):e131487. https://doi.org/10.1172/jci.insight.131487.
50. Angelousi A, Kassi E, Nasiri-Ansari N, Weickert MO, Randeva H, Kaltsas G. Clock genes alterations and endocrine disorders. Eur J Clin Investig. 2018;48(6):e12927. https://doi.org/10.1111/eci.12927.
51. Xie Y, Tang Q, Chen G, et al. New insights into the circadian rhythm and its related diseases. Front Physiol. 2019;10:682. https://doi.org/10.3389/fphys.2019.00682.
52. Cha HK, Chung S, Lim HY, Jung JW, Son GH. Small molecule modulators of the circadian molecular clock with implications for neuropsychiatric diseases. Front Mol Neurosci. 2019;11:496. https://doi.org/10.3389/fnmol.2018.00496.
53. Narasimamurthy R, Virshup DM. The phosphorylation switch that regulates ticking of the circadian clock. Mol Cell. 2021;81(6):1133–46. https://doi.org/10.1016/j.molcel.2021.01.006.
54. Zhang L, Hirano A, Hsu PK, et al. A *PERIOD3* variant causes a circadian phenotype and is associated with a seasonal mood trait. Proc Natl Acad Sci. 2016;113(11) https://doi.org/10.1073/pnas.1600039113.
55. Hoyle NP, O'Neill JS. Oxidation–reduction cycles of peroxiredoxin proteins and non-transcriptional aspects of timekeeping. Biochemistry. 2015;54(2):184–93. https://doi.org/10.1021/bi5008386.
56. Saré RM, Song A, Levine M, et al. Chronic sleep restriction in developing male mice results in long lasting behavior impairments. Front Behav Neurosci. 2019;13:90. https://doi.org/10.3389/fnbeh.2019.00090.
57. Aryal RP, Kwak PB, Tamayo AG, et al. Macromolecular assemblies of the mammalian circadian clock. Mol Cell. 2017;67(5):770–782.e6. https://doi.org/10.1016/j.molcel.2017.07.017.
58. Ralph MR, Menaker M. A mutation of the circadian system in golden hamsters. Science. 1988;241(4870):1225–7. https://doi.org/10.1126/science.3413487.
59. Yoo SH, Mohawk JA, Siepka SM, et al. Competing E3 ubiquitin ligases govern circadian periodicity by degradation of CRY in nucleus and cytoplasm. Cell. 2013;152(5):1091–105. https://doi.org/10.1016/j.cell.2013.01.055.
60. Jones CR, Huang AL, Ptáček LJ, Fu YH. Genetic basis of human circadian rhythm disorders. Exp Neurol. 2013;243:28–33. https://doi.org/10.1016/j.expneurol.2012.07.012.
61. Wickwire EM, Geiger-Brown J, Scharf SM, Drake CL. Shift work and shift work sleep disorder. Chest. 2017;151(5):1156–72. https://doi.org/10.1016/j.chest.2016.12.007.
62. Jones SE, Tyrrell J, Wood AR, et al. Genome-wide association analyses in 128,266 individuals identifies new morningness and sleep duration loci. Shi J (ed) PLOS Genet. 2016;12(8):e1006125. https://doi.org/10.1371/journal.pgen.1006125.
63. Hu Y, Shmygelska A, Tran D, Eriksson N, Tung JY, Hinds DA. GWAS of 89,283 individuals identifies genetic variants associated with self-reporting of being a morning person. Nat Commun. 2016;7(1):10448. https://doi.org/10.1038/ncomms10448.

64. Leocadio-Miguel MA, Ruiz FS, Ahmed SS, et al. Compared heritability of chronotype instruments in a single population sample. J Biol Rhythm. 2021;36(5):483–90. https://doi.org/10.1177/07487304211030420.
65. Cano-Gamez E, Trynka G. From GWAS to function: using functional genomics to identify the mechanisms underlying complex diseases. Front Genet. 2020;11:424. https://doi.org/10.3389/fgene.2020.00424.
66. Farhan S. Gene discovery in Mendelian and complex diseases. Electron Diss. 2016;4077:313.
67. Jordan DM, Do R. Using full genomic information to predict disease: breaking down the barriers between complex and Mendelian diseases. Annu Rev Genomics Hum Genet. 2018;19(1):289–301. https://doi.org/10.1146/annurev-genom-083117-021136.
68. The Electronic Medical Records and Genomics (eMERGE) Consortium, The MIGen Consortium, The PAGE Consortium, et al. Defining the role of common variation in the genomic and biological architecture of adult human height. Nat Genet. 2014;46(11):1173–86. https://doi.org/10.1038/ng.3097.
69. Doi M. Circadian regulation of intracellular G-protein signalling mediates intercellular synchrony and rhythmicity in the suprachiasmatic nucleus. Nat Commun. 2011:9.
70. Tong X, Zhang D, Guha A, et al. CUL4-DDB1-CDT2 E3 ligase regulates the molecular clock activity by promoting ubiquitination-dependent degradation of the mammalian CRY1. Brodsky JL (ed) Plos one. 2015;10(10):e0139725. https://doi.org/10.1371/journal.pone.0139725.
71. Yang Y, Duguay D, Bédard N, et al. Regulation of behavioral circadian rhythms and clock protein PER1 by the deubiquitinating enzyme USP2. Biol Open. 2012;1(8):789–801. https://doi.org/10.1242/bio.20121990.
72. Stojkovic K, Wing SS, Cermakian N. A central role for ubiquitination within a circadian clock protein modification code. Front Mol Neurosci. 2014;7 https://doi.org/10.3389/fnmol.2014.00069.
73. Scott DC, Rhee DY, Duda DM, et al. Two distinct types of E3 ligases work in unison to regulate substrate ubiquitylation. Cell. 2016;166(5):1198–1214.e24. https://doi.org/10.1016/j.cell.2016.07.027.
74. Tyree SM, Borniger JC, de Lecea L. Hypocretin as a hub for arousal and motivation. Front Neurol. 2018;9:413. https://doi.org/10.3389/fneur.2018.00413.
75. Ly S, Pishdari B, Lok LL, Hajos M, Kocsis B. Activation of 5-HT6 receptors modulates sleep–wake activity and hippocampal theta oscillation. ACS Chem Neurosci. 2013;4(1):191–9. https://doi.org/10.1021/cn300184t.
76. Francis R, McGrath G, Zhang J, et al. aph-1 and pen-2 are required for Notch pathway signaling, g-secretase cleavage of bAPP, and presenilin protein accumulation. Dev Cell. 3:85–97.
77. Kang JE, Lim MM, Bateman RJ, et al. Amyloid-β dynamics are regulated by orexin and the sleep-wake cycle. Science. 2010;11
78. Smieszek SP, Brzezynski JL, Kaden AR, et al. An observational study investigating the CRY1Δ11 variant associated with delayed sleep–wake patterns and circadian metabolic output. Sci Rep. 2021;11(1):20103. https://doi.org/10.1038/s41598-021-99418-2.
79. Weedon MN, Jones SE, Lane Jacqueline M, et al. The impact of Mendelian sleep and circadian genetic variants in a population setting. PLoS Genet. 2022; https://doi.org/10.1101/2022.01.04.21268199.
80. Sack RL, Auckley D, Auger RR, et al. Circadian rhythm sleep disorders: part I, basic principles, shift work and jet lag disorders. Sleep. 2007;30(11):1460–83. https://doi.org/10.1093/sleep/30.11.1460.
81. Zee PC, Attarian H, Videnovic A. Circadian rhythm abnormalities. Continuum (Minneap Minn). 2013;16:132–47.
82. Kurien P, Hsu PK, Leon J, et al. TIMELESS mutation alters phase responsiveness and causes advanced sleep phase. Proc Natl Acad Sci. 2019;116(24):12045–53. https://doi.org/10.1073/pnas.1819110116.

83. Bae K, Jin X, Maywood ES, Hastings MH, Reppert SM, Weaver DR. Differential functions of mPer1, mPer2, and mPer3 in the SCN circadian clock. Neuron. 2001;30(2):525–36. https://doi.org/10.1016/S0896-6273(01)00302-6.
84. Patke A, Murphy PJ, Onat OE, et al. Mutation of the human circadian clock gene CRY1 in familial delayed sleep phase disorder. Cell. 2017;169(2):203–215.e13. https://doi.org/10.1016/j.cell.2017.03.027.
85. Brenna A, Albrecht U. Phosphorylation and circadian molecular timing. Front Physiol. 2020;11:612510. https://doi.org/10.3389/fphys.2020.612510.
86. MacArthur DG, Manolio TA, Dimmock DP, et al. Guidelines for investigating causality of sequence variants in human disease. Nature. 2014;508(7497):469–76. https://doi.org/10.1038/nature13127.
87. Wright CF, West B, Tuke M, et al. Assessing the pathogenicity, penetrance, and expressivity of putative disease-causing variants in a population setting. Am J Hum Genet. 2019;104(2):275–86. https://doi.org/10.1016/j.ajhg.2018.12.015.
88. Foster RG. Sleep, circadian rhythms and health. Interface Focus. 2020;10(3):20190098. https://doi.org/10.1098/rsfs.2019.0098.
89. Auger RR, Burgess HJ, Emens JS, Deriy LV, Thomas SM, Sharkey KM. Clinical practice guideline for the treatment of intrinsic circadian rhythm sleep-wake disorders: advanced sleep-wake phase disorder (ASWPD), delayed sleep-wake phase disorder (DSWPD), non-24-hour sleep-wake rhythm disorder (N24SWD), and irregular sleep-wake rhythm disorder (ISWRD). An update for 2015: an American academy of sleep medicine clinical practice guideline. J Clin Sleep Med. 2015;11(10):1199–236. https://doi.org/10.5664/jcsm.5100.
90. Curtis BJ, Ashbrook LH, Young T, et al. Extreme morning chronotypes are often familial and not exceedingly rare: the estimated prevalence of advanced sleep phase, familial advanced sleep phase, and advanced sleep–wake phase disorder in a sleep clinic population. Sleep. 2019;42(10):zsz148. https://doi.org/10.1093/sleep/zsz148.
91. Hudec M, Dankova P, Solc R, Bettazova N, Cerna M. Epigenetic regulation of circadian rhythm and its possible role in diabetes mellitus. Int J Mol Sci. 2020;21(8):3005. https://doi.org/10.3390/ijms21083005.
92. Sivertsen B, Harvey AG, Gradisar M, Pallesen S, Hysing M. Delayed sleep–wake phase disorder in young adults: prevalence and correlates from a national survey of Norwegian university students. Sleep Med. 2021;77:184–91. https://doi.org/10.1016/j.sleep.2020.09.028.
93. Saxvig IW, Pallesen S, Wilhelmsen-Langeland A, Molde H, Bjorvatn B. Prevalence and correlates of delayed sleep phase in high school students. Sleep Med. 2012;13(2):193–9. https://doi.org/10.1016/j.sleep.2011.10.024.
94. Hummer DL, Jechura TJ, Mahoney MM, Lee TM. Gonadal hormone effects on entrained and free-running circadian activity rhythms in the developing diurnal rodent *Octodon degus*. Am J Physiol-Regul Integr Comp Physiol. 2007;292(1):R586–97. https://doi.org/10.1152/ajpregu.00043.2006.
95. McGinnis MY, Lumia AR, Tetel MJ, Molenda-Figueira HA, Possidente B. Effects of anabolic androgenic steroids on the development and expression of running wheel activity and circadian rhythms in male rats. Physiol Behav. 2007;92(5):1010–8. https://doi.org/10.1016/j.physbeh.2007.07.010.
96. Micic G, de Bruyn A, Lovato N, et al. The endogenous circadian temperature period length (tau) in delayed sleep phase disorder compared to good sleepers. J Sleep Res. 2013;22(6):617–24. https://doi.org/10.1111/jsr.12072.
97. Ancoli-Israel S, Schnierow B, Kelsoe J, Fink R. A pedigree of one family with delayed sleep phase syndrome. Chronobiol Int. 2001;18(5):831–40. https://doi.org/10.1081/CBI-100107518.
98. Sack RL, Auckley D, Auger RR, et al. Circadian rhythm sleep disorders: part II, advanced sleep phase disorder, delayed sleep phase disorder, free-running disorder, and irregular sleep-wake rhythm. Sleep. 2007;30(11):1484–501. https://doi.org/10.1093/sleep/30.11.1484.

99. Crowley SJ, Acebo C, Carskadon MA. Sleep, circadian rhythms, and delayed phase in adolescence. Sleep Med. 2007;8(6):602–12. https://doi.org/10.1016/j.sleep.2006.12.002.
100. Nesbitt AD. Delayed sleep-wake phase disorder. J Thorac Dis. 2018;10(S1):S103–11. https://doi.org/10.21037/jtd.2018.01.11.
101. Jagannath A, Taylor L, Wakaf Z, Vasudevan SR, Foster RG. The genetics of circadian rhythms, sleep and health. Hum Mol Genet. 2017;26(R2):R128–38. https://doi.org/10.1093/hmg/ddx240.
102. Kripke DF, Klimecki WT, Nievergelt CM, et al. Circadian polymorphisms in night owls, in bipolars, and in non-24-hour sleep cycles. Psychiatry Investig. 2014;11(4):345. https://doi.org/10.4306/pi.2014.11.4.345.
103. Lewy AJ, Ahmed S, Jackson JML, Sack RL. Melatonin shifts human orcadian rhythms according to a phase-response curve. Chronobiol Int. 1992;9(5):380–92. https://doi.org/10.3109/07420529209064550.
104. Abbott SM, Choi J, Wilson J, Zee PC. Melanopsin-dependent phototransduction is impaired in delayed sleep–wake phase disorder and sighted non–24-hour sleep–wake rhythm disorder. Sleep. 2021;44(2):zsaa184. https://doi.org/10.1093/sleep/zsaa184.
105. Emens JS, Eastman CI. Diagnosis and treatment of non-24-h sleep–wake disorder in the blind. Drugs. 2017;77(6):637–50. https://doi.org/10.1007/s40265-017-0707-3.
106. Flynn-Evans EE, Lockley SW. A pre-screening questionnaire to predict non-24-hour sleep-wake rhythm disorder (N24HSWD) among the blind. J Clin Sleep Med. 2016;12(05):703–10. https://doi.org/10.5664/jcsm.5800.
107. Tamura N, Sasai-Sakuma T, Morita Y, Okawa M, Inoue S, Inoue Y. Prevalence and associated factors of circadian rhythm sleep-wake disorders and insomnia among visually impaired Japanese individuals. BMC Public Health. 2021;21(1):31. https://doi.org/10.1186/s12889-020-09993-8.
108. Miles LEM, Raynal DM, Wilson MA. Blind man living in normal society has circadian rhythms of 24.9 hours. Science. 1977;198(4315):421–3. https://doi.org/10.1126/science.910139.
109. Pokrzywinski R, Abel C, Lenderking WR, Yardley J, Cheng J, Moline M. Irregular sleep wake rhythm disorder (ISWRD) signs and symptoms reported directly from patients with dementia and caregivers. Alzheimers Dement. 2021;17(S6) https://doi.org/10.1002/alz.049917.
110. Silva A, Silva A, Duarte J, Torres da Costa J. Shift-work: a review of the health consequences. Int J Occup Environ Saf. 2020;4(2):48–79. https://doi.org/10.24840/2184-0954_004.002_0005.
111. Drake CL, Belcher R, Howard R, Roth T, Levin AM, Gumenyuk V. Length polymorphism in the period 3 gene is associated with sleepiness and maladaptive circadian phase in night-shift workers. J Sleep Res. 2015;24(3):254–61. https://doi.org/10.1111/jsr.12264.
112. Gumenyuk V, Howard R, Roth T, Korzyukov O, Drake CL. Sleep loss, circadian mismatch, and abnormalities in reorienting of attention in night workers with shift work disorder. Sleep. 2014;37(3):545–56. https://doi.org/10.5665/sleep.3494.
113. Juda M, Vetter C, Roenneberg T. Chronotype modulates sleep duration, sleep quality, and social jet lag in shift-workers. J Biol Rhythm. 2013;28(2):141–51. https://doi.org/10.1177/0748730412475042.
114. Lo JC, Groeger JA, Santhi N, et al. Effects of partial and acute total sleep deprivation on performance across cognitive domains, individuals and circadian phase. Yamazaki S (ed) PLoS One. 2012;7(9):e45987. https://doi.org/10.1371/journal.pone.0045987.
115. Lee HJ, Kim L, Kang SG, et al. PER2 variation is associated with diurnal preference in a korean young population. Behav Genet. 2011;41(2):273–7. https://doi.org/10.1007/s10519-010-9396-3.
116. Gamble KL, Motsinger-Reif AA, Hida A, et al. Shift work in nurses: contribution of phenotypes and genotypes to adaptation. Goel N (ed) PLoS One. 2011;6(4):e18395. https://doi.org/10.1371/journal.pone.0018395.

117. Allebrandt KV, Amin N, Müller-Myhsok B, et al. A KATP channel gene effect on sleep duration: from genome-wide association studies to function in Drosophila. Mol Psychiatry. 2013;18(1):122–32. https://doi.org/10.1038/mp.2011.142.
118. Partonen T. Clock gene variants in mood and anxiety disorders. J Neural Transm. 2012;119(10):1133–45. https://doi.org/10.1007/s00702-012-0810-2.
119. Reszka E, Peplonska B, Wieczorek E, et al. Rotating night shift work and polymorphism of genes important for the regulation of circadian rhythm. Scand J Work Environ Health. 2013;39(2):178–86. https://doi.org/10.5271/sjweh.3299.
120. Taniyama Y, Yamauchi T, Takeuchi S, Kuroda Y. *PER1* polymorphism associated with shift work disorder: *PER1* polymorphism in shift work disorder. Sleep Biol Rhythms. 2015;13(4):342–7. https://doi.org/10.1111/sbr.12123.
121. Ambesh P, Shetty V, Ambesh S, Gupta S, Kamholz S, Wolf L. Jet lag: heuristics and therapeutics. J Fam Med Prim Care. 2018;7(3):507. https://doi.org/10.4103/jfmpc.jfmpc_220_17.
122. Choy M, Salbu RL. Jet lag: current and potential therapies. P & T: A Peer-Reviewed Journal for Formulary Management. 2011;36(4):221–31.
123. Drake CL, Wright KP. Shift work, shift-work disorder, and jet lag. In: Principles and practice of sleep medicine. Elsevier; 2011. p. 784–98. https://doi.org/10.1016/B978-1-4160-6645-3.00071-2.
124. Thornton HR, Miller J, Taylor L, Sargent C, Lastella M, Fowler PM. Impact of short- compared to long-haul international travel on the sleep and wellbeing of national wheelchair basketball athletes. J Sports Sci. 2018;36(13):1476–84. https://doi.org/10.1080/02640414.2017.1398883.
125. Li P, Morris CJ, Patxot M, et al. Reduced tolerance to night shift in chronic shift workers: insight from fractal regulation. Sleep. 2017;40(7) https://doi.org/10.1093/sleep/zsx092.
126. Khalsa SBS, Jewett ME, Cajochen C, Czeisler CA. A phase response curve to single bright light pulses in human subjects. J Physiol. 2003;549(3):945–52. https://doi.org/10.1113/jphysiol.2003.040477.
127. Hatori M, Gill S, Mure LS, Goulding M, O'Leary DDM, Panda S. Lhx1 maintains synchrony among circadian oscillator neurons of the SCN. elife. 2014;3:e03357. https://doi.org/10.7554/eLife.03357.
128. Keijzer H, Spruyt K, Smits MG, de Geest A, Curfs LMG. Can dim light melatonin onset be predicted by the timing of sleep in patients with possible circadian sleep-wake rhythm disorders? Biol Rhythm Res. 2017;48(4):557–66. https://doi.org/10.1080/09291016.2017.1287817.
129. Micic G, Lovato N, Gradisar M, et al. Nocturnal melatonin profiles in patients with delayed sleep-wake phase disorder and control sleepers. 2015;30:12.
130. van Geijlswijk IM, Mol RH, Egberts TCG, Smits MG. Evaluation of sleep, puberty and mental health in children with long-term melatonin treatment for chronic idiopathic childhood sleep onset insomnia. Psychopharmacol. 2011;216:10.
131. Vasko MY, Tkachenko IM, Pavlenkova OV, Pysarenko OA. Interrelation of biological rhythms and circadian hormones producement and their impact on medicine usage. Wiad Lek. 2021;74(8):1970–4. https://doi.org/10.36740/WLek202108133.
132. Difrancesco S, Lamers F, Riese H, et al. Sleep, circadian rhythm, and physical activity patterns in depressive and anxiety disorders: a 2-week ambulatory assessment study. Depress Anxiety. 2019;36(10):975–86. https://doi.org/10.1002/da.22949.
133. Xie Z, Chen F, Li WA, et al. A review of sleep disorders and melatonin. Neurol Res. 2017;39(6):559–65. https://doi.org/10.1080/01616412.2017.1315864.
134. Baiardi S, Cirignotta F, Cicolin A, et al. Chronobiology, sleep-related risk factors and light therapy in perinatal depression: the "Life-ON" project. BMC Psychiatry. 2016;16(1):374. https://doi.org/10.1186/s12888-016-1086-0.
135. Franzen PL, Buysse DJ. Sleep disturbances and depression: risk relationships for subsequent depression and therapeutic implications. Dialogues Clin Neurosci. 2008;10(4):473–81. https://doi.org/10.31887/DCNS.2008.10.4/plfranzen.

136. Grima N, Ponsford J, Rajaratnam SM, Mansfield D, Pase MP. Sleep disturbances in traumatic brain injury: a meta-analysis. J Clin Sleep Med. 2016;12(03):419–28. https://doi.org/10.5664/jcsm.5598.
137. Mayer G, Happe S, Evers S, et al. Insomnia in neurological diseases. Neurol Res Pract. 2021;3(1):15. https://doi.org/10.1186/s42466-021-00106-3.
138. Zuzuárregui JRP, Bickart K, Kutscher SJ. A review of sleep disturbances following traumatic brain injury. Sleep Sci Pract. 2018;2(1):2. https://doi.org/10.1186/s41606-018-0020-4.
139. Smyka M, Kosińska-Kaczyńska K, Sochacki-Wójcicka N, Zgliczyńska M, Wielgoś M. Sleep problems in pregnancy—a cross-sectional study in over 7000 pregnant women in Poland. Int J Environ Res Public Health. 2020;17(15):5306. https://doi.org/10.3390/ijerph17155306.
140. Wang WJ, Hou CL, Jiang YP, et al. Prevalence and associated risk factors of insomnia among pregnant women in China. Compr Psychiatry. 2020;98:152168. https://doi.org/10.1016/j.comppsych.2020.152168.
141. Wong PF, D'Cruz R, Hare A. Sleep disorders in pregnancy. Breathe. 2022;18(2):220004. https://doi.org/10.1183/20734735.0004-2022.
142. Richardson C, Cain N, Bartel K, Micic G, Maddock B, Gradisar M. A randomised controlled trial of bright light therapy and morning activity for adolescents and young adults with Delayed Sleep-Wake Phase Disorder. Sleep Med. 2018;45:114–23. https://doi.org/10.1016/j.sleep.2018.02.001.
143. Agar G, Brown C, Sutherland D, Coulborn S, Oliver C, Richards C. Sleep disorders in rare genetic syndromes: a meta-analysis of prevalence and profile. Mol Autism. 2021;12(1):18. https://doi.org/10.1186/s13229-021-00426-w.
144. Bidaki R, Zarei M, Khorram Toosi A, Hakim Shooshtari M. A review on genetics of sleep disorders. Iran J Psychiatry Behav Sci. 2012;6(1):12–19.
145. Dashti HS, Ordovás JM. Genetics of sleep and insights into its relationship with obesity. Annu Rev Nutr. 2021;41(1):223–52. https://doi.org/10.1146/annurev-nutr-082018-124258.
146. Doherty A, Smith-Bryne K, Ferreira T, et al. GWAS identifies 10 Loci for objectively-measured physical activity and sleep with causal roles in cardiometabolic disease. Genetics. 2018; https://doi.org/10.1101/261719.
147. Parsons M. On the genetics of sleep disorders: genome-wide association studies and beyond. Adv Genomics Genet. 2015:293. https://doi.org/10.2147/AGG.S57139.
148. Ebisawa T. Circadian rhythms in the CNS and peripheral clock disorders: human sleep disorders and clock genes. J Pharmacol Sci. 2007;103(2):150–4. https://doi.org/10.1254/jphs.FMJ06003X5.
149. Pedrazzoli M, Louzada FM, Pereira DS, et al. Clock polymorphisms and circadian rhythms phenotypes in a sample of the Brazilian population. Chronobiol Int. 2007;24(1):1–8. https://doi.org/10.1080/07420520601139789.
150. Pingault JB, O'Reilly PF, Schoeler T, Ploubidis GB, Rijsdijk F, Dudbridge F. Using genetic data to strengthen causal inference in observational research. Nat Rev Genet. 2018;19(9):566–80. https://doi.org/10.1038/s41576-018-0020-3.
151. Byrne EM. The relationship between insomnia and complex diseases—insights from genetic data. Genome Med. 2019;11(1):57. https://doi.org/10.1186/s13073-019-0668-0.
152. Daghlas I, Lane JM, Saxena R, Vetter C. Genetically proxied diurnal preference, sleep timing, and risk of major depressive disorder. JAMA Psychiatry. 2021;78(8):903. https://doi.org/10.1001/jamapsychiatry.2021.0959.
153. O'Loughlin J, Casanova F, Jones SE, et al. Using Mendelian Randomisation methods to understand whether diurnal preference is causally related to mental health. Mol Psychiatry. 2021;26(11):6305–16. https://doi.org/10.1038/s41380-021-01157-3.
154. Facer-Childs ER, Middleton B, Skene DJ, Bagshaw AP. Resetting the late timing of 'night owls' has a positive impact on mental health and performance. Sleep Med. 2019;60:236–47. https://doi.org/10.1016/j.sleep.2019.05.001.
155. Javaheri S, Redline S. Insomnia and risk of cardiovascular disease. Chest. 2017;152(2):435–44. https://doi.org/10.1016/j.chest.2017.01.026.

156. Lane JM, Liang J, Vlasac I, et al. Genome-wide association analyses of sleep disturbance traits identify new loci and highlight shared genetics with neuropsychiatric and metabolic traits. Nat Genet. 2017;49(2):274–81. https://doi.org/10.1038/ng.3749.
157. The 23andMe Research Team, Jansen PR, Watanabe K, et al. Genome-wide analysis of insomnia in 1,331,010 individuals identifies new risk loci and functional pathways. Nat Genet. 2019;51(3):394–403. https://doi.org/10.1038/s41588-018-0333-3.
158. Zheng B, Yu C, Lv J, et al. Insomnia symptoms and risk of cardiovascular diseases among 0.5 million adults: a 10-year cohort. Neurology. 2019;93(23):e2110–20. https://doi.org/10.1212/WNL.0000000000008581.
159. Zhang F, Cao H, Baranova A. Shared genetic liability and causal associations between major depressive disorder and cardiovascular diseases. Front Cardiovasc Med. 2021;8:735136. https://doi.org/10.3389/fcvm.2021.735136.
160. Duffy JF, Abbott SM, Burgess HJ, et al. Workshop report. Circadian rhythm sleep–wake disorders: gaps and opportunities. Sleep. 2021;44(5):zsaa281. https://doi.org/10.1093/sleep/zsaa281.
161. Simko F, Baka T. Chronotherapy of cardiovascular pathologies: a hopeful strategy. Ther Adv Chronic Dis. 2022;13:204062232210923. https://doi.org/10.1177/20406223221092372.
162. Haus E, Halberg F, Scheving LE, et al. Increased tolerance of leukemic mice to arabinosyl cytosine with schedule adjusted to circadian system. Science. 1972;177(4043):80–2. https://doi.org/10.1126/science.177.4043.80.
163. Dong D, Yang D, Lin L, Wang S, Wu B. Circadian rhythm in pharmacokinetics and its relevance to chronotherapy. Biochem Pharmacol. 2020;178:114045. https://doi.org/10.1016/j.bcp.2020.114045.
164. Kaur M, Bala R. Chronotherapy: a review. Int J Pharma Sci Res. 2013;4:13.
165. Kenig A, Ilan Y. A personalized signature and chronotherapy-based platform for improving the efficacy of sepsis treatment. Front Physiol. 2019;10:1542. https://doi.org/10.3389/fphys.2019.01542.
166. Selfridge JM, Gotoh T, Schiffhauer S, et al. Chronotherapy: intuitive, sound, founded…but not broadly applied. Drugs. 2016;76(16):1507–21. https://doi.org/10.1007/s40265-016-0646-4.
167. Finocchiaro G, Merlo M, Sheikh N, et al. The electrocardiogram in the diagnosis and management of patients with dilated cardiomyopathy. Eur J Heart Fail. 2020;22(7):1097–107. https://doi.org/10.1002/ejhf.1815.
168. McKenna H, van der Horst GTJ, Reiss I, Martin D. Clinical chronobiology: a timely consideration in critical care medicine. Crit Care. 2018;22(1):124. https://doi.org/10.1186/s13054-018-2041-x.
169. Zhang SL, Lahens NF, Yue Z, et al. A circadian clock regulates efflux by the blood-brain barrier in mice and human cells. Nat Commun. 2021;12(1):617. https://doi.org/10.1038/s41467-020-20795-9.
170. Forrestel AC, Miedlich SU, Yurcheshen M, Wittlin SD, Sellix MT. Chronomedicine and type 2 diabetes: shining some light on melatonin. Diabetologia. 2017;60(5):808–22. https://doi.org/10.1007/s00125-016-4175-1.
171. Honma A, Yamada Y, Nakamaru Y, Fukuda S, Honma K, Honma S. Glucocorticoids reset the nasal circadian clock in mice. Endocrinology. 2015;156(11):4302–11. https://doi.org/10.1210/en.2015-1490.
172. Kuo TT, Ladurner AG. Exploiting the circadian clock for improved cancer therapy: perspective from a cell biologist. Front Genet. 2019;10:1210. https://doi.org/10.3389/fgene.2019.01210.
173. Sulli G, Lam MTY, Panda S. Interplay between circadian clock and cancer: new frontiers for cancer treatment. Trends Cancer. 2019;5(8):475–94. https://doi.org/10.1016/j.trecan.2019.07.002.
174. Iii CEG, Kaye AM, Kaye AD. Benzodiazepine pharmacology and central nervous system–mediated effects. Ochsner J. 2013;13(2):10.
175. Barion A, Zee PC. A clinical approach to circadian rhythm sleep disorders. Sleep Med. 2007;8(6):566–77. https://doi.org/10.1016/j.sleep.2006.11.017.

176. Chesson AL, Littner M, Davila D, et al. Practice parameters for the use of light therapy in the treatment of sleep disorders. Sleep. 1999;22(5):641–60. https://doi.org/10.1093/sleep/22.5.641.
177. Rosenthal NE. Phase-shifting effects of bright morning light as treatment for delayed sleep phase syndrome. Sleep. Published online July 1, 1990. doi:https://doi.org/10.1093/sleep/13.4.354.
178. Mundey K, Benloucif S, Harsanyi K, Dubocovich ML, Zee PC. Phase-dependent treatment of delayed sleep phase syndrome with melatonin. Sleep. 2005;28(10):1271–8. https://doi.org/10.1093/sleep/28.10.1271.
179. Lack L, Wright H, Kemp K, Gibbon S. The treatment of early-morning awakening insomnia with 2 evenings of bright light. Sleep. 2005;28(5):8.
180. Rubiño JA, Gamundí A, Akaarir M, Canellas F, Rial R, Nicolau MC. Bright light therapy and circadian cycles in institutionalized elders. Front Neurosci. 2020;14:359. https://doi.org/10.3389/fnins.2020.00359.
181. Burgess HJ. Partial sleep deprivation reduces phase advances to light in humans. J Biol Rhythm. 2010;25(6):460–8. https://doi.org/10.1177/0748730410385544.
182. Crowley SJ, Eastman CI. Phase advancing human circadian rhythms with morning bright light, afternoon melatonin, and gradually shifted sleep: can we reduce morning bright-light duration? Sleep Med. 2015;16(2):288–97. https://doi.org/10.1016/j.sleep.2014.12.004.
183. Dagan Y, Ayalon L. Case study: psychiatric misdiagnosis of non-24-hours sleep-wake schedule disorder resolved by melatonin. J Am Acad Child Adolesc Psychiatry. 2005;44(12):1271–5. https://doi.org/10.1097/01.chi.0000181040.83465.48.

Genetics of Central Disorders of Hypersomnolence

Chenyang Li, Fulong Xiao, and Fang Han

Abstract Central disorders of hypersomnolence (CDH) are characterized by a primary complaint of excessive daytime sleepiness (EDS) not attributable to another sleep disorder disturbing nocturnal sleep (e.g., obstructive sleep apnea), insufficient sleep, and/or circadian dysrhythmias. EDS represents the common background feature of these disorders which have different pathophysiology and clinical features. Due to dysfunction of the central nervous system, CDH impairs daily functions. Familial cases of EDS have been widely seen indicating genetic influence on these sleep disorders with hypersomnia. Great effort has been put to find biomarkers of CDH and it is well recognized the genetic variations affect the timing of sleep, duration of sleep, and EEG patterns. In this review, we specifically focused on a group of central disorders such as narcolepsy, idiopathic hypersomnia, and Kleine-Levin syndrome. This section will provide a detailed background on the genetics of these disorders.

Keywords Narcolepsy · Idiopathic hypersomnia · Kleine-Levin syndrome · Genetics

The central disorders of hypersomnolence (CDH) are a group of disorders marked by pathologic daytime sleepiness, most likely due to dysfunction of the central nervous system. CDH impair performance in school and work, impact social interactions and family dynamics, and present real risks with driving. The International Classification of Sleep Disorders third edition (ICSD-3) has included narcolepsy type 1 (NT1) and type 2 (NT2), idiopathic hypersomnia (IH), and Kleine-Levin syndrome (KLS) under the umbrella of CDH, where excessive daytime sleepiness (EDS) is a cardinal and common feature to all of them [1]. These disorders have

C. Li
Division of Sleep Medicine, Peking University People's Hospital, Beijing, China

School of Nursing, Capital Medical University, Beijing, China

F. Xiao · F. Han (✉)
Division of Sleep Medicine, Peking University People's Hospital, Beijing, China

P. Gehrman et al. (eds.), *Genetics of Sleep and Sleep Disorders*,
https://doi.org/10.1007/978-3-031-62723-1_12

distinct clinical features and diagnostic criteria. Patients with NT1 typically sleep about 7.5 to 8 h per 24 h, but they have great difficulty sustaining wakefulness and complain of severe daytime sleepiness. IH is characterized by longer sleep durations, higher nocturnal sleep efficiency, and quicker time to fall asleep in the daytime. NT2 seems to be a more heterogeneous condition that lies in the borderland of NT1 and IH. Patients with KLS have recurrent periods of hypersomnia with about 18 h of sleep per 24 h during an episode; insomnia with short sleep time may occur at the end of an episode. An overnight polysomnogram (PSG) followed by Multiple Sleep Latency Test (MSLT) is an excellent method to objectively assess EDS and sleep-onset rapid eye movement (REM) periods, and it has high diagnostic validity and reliability in people with NT1. In NT2, the severity of EDS and REM sleep dysregulation may be more variable, contributing to low reliability of the PSG and MSLT. The utility in the diagnosis of IH and KLS of this technique is of limited value, as routine PSG and MSLT do not measure the high amount of sleep per 24 h. Great effort has been made to find biomarkers of CDH [2], especially the finding of the link between orexin/hypocretin and NT1. It is well recognized the genetic variations affect the timing of sleep, duration of sleep, and EEG patterns. Specific genes may affect circadian hemostasis, neurotransmission, and immunological reaction in wake-sleep cycle, whereas gene expression can be influenced by sleep [3, 4]. In addition, familial cases of excessive sleepiness, including narcolepsy, IH, or KLS, have been widely seen indicating genetic influence on these sleep disorders with hypersomnia [5]. Over the past 30 years, a remarkable research finding was the strong association between narcolepsy and specific HLA allele, DQB1*0602. In this review, we specifically focused on the genetics of hypersomnolence, including the three major sleep disorders from ICSD-3, i.e., narcolepsy, IH, and KLS, especially narcolepsy, which is the most studied. In contrast to NT1, the genetic backgrounds of other central disorders of hypersomnolence have not been well elucidated, most likely due to difficulty with their diagnoses, their heterogeneity, and their low prevalence compared with those of NT1.

Narcolepsy

Clinical Features

Narcolepsy is a disabling neurodegenerative condition with the core symptoms of excessive daytime sleepiness (EDS), cataplexy, sleep paralysis, hallucinations, and disrupted nocturnal sleep (DNS). ICSD-3 categorizes narcolepsy as narcolepsy type 1 (NT1) and narcolepsy type 2 (NT2). NT1 accounts for 70% of the cases of narcolepsy, and 10% of patients with initially diagnosed NT2 may be later diagnosed as NT1 due to the symptoms of cataplexy development. NT1 is biologically marked by low cerebrospinal fluid hypocretin-1 levels (CSF hcrt-1), caused by the selective loss of hypothalamic hypocretin producing neurons, while NT2 is characterized by normal levels of CSF hcrt-1 [1].

Genetic Associations with Predisposing Effect

Familial Narcolepsy

The concordance of NT1 is approximately 1–2% in first-degree relatives of affected individuals; however, it represents 10–40 folds higher risk than the general population. Less than 2% of individuals with narcolepsy have more than 1 affected family member, and multiplex families are rare. The concordance rate in monozygotic twins of NT1 is approximately 20–35%. All these implicate the genetic predisposition in narcolepsy [5].

Mapping genes in narcolepsy families for genome-wide linkage analysis with sufficient power is hampered by the small size of the pedigrees. One study linked the disease to a locus on chromosome 4p13-q21 in eight small Japanese families [6]. Another linkage study in a large French family identified a 5·15 Mb candidate region on chromosome 21q [7]. So far, mutation analysis of genes identified in those regions has failed to detect any causative gene or pathogenic mutation. Linkage analysis followed by exome sequencing in a large family with 12 affected members enabled identification of a causal mutation in myelin oligodendrocyte glycoprotein [8]. DNA (cytosine-5)- methyltransferase 1 mutations were also found in families with rare disorder of autosomal dominant cerebellar ataxia, deafness, and narcolepsy. It was found that orexin, MOG, and DNMT1 mutations seem to be exceptional in orexin-deficient DQB1*06:02-negative cases [8, 9].

HLA Related Genes

Genetic studies in narcolepsy have focused mainly on NT1. Since 1984, it has been known that narcolepsy is associated with specific class II HLA antigens, 100% were HLADR2 (HLA DRB1*15:01 by high resolution analysis) positive in Japanese narcolepsy patients [10, 11]. This was replicated in Caucasian narcoleptics, but substantially less so in African Americans [12]. Instead, HLA Class II allele DQB1*06:02 (a subtype of DQ1) is expressed in 86–98% of patients with NT1 across all ethnic groups, which is higher than population frequency in northern European populations (25%) and African Americans (38%). The effect is most likely explained primarily by DQB1*06:02 as the haplotypes with DQB1*06:02 alone increase risk the most in different populations, including African Americans [12–14]. HLA DRB1*15:01 and DQB1*06:02 alleles are in strong linkage disequilibrium in both Japanese and Caucasians, but much less in African Americans [14, 15]. The results are also in line with a previously proposed "HLA-DQ allelic competition model" that involves competition between non-DQA1*01:02, non-DQB1*06:02 "competent" (able to dimerize together), DQ1 alleles and the major DQA*01:02/DQB*06:02 narcolepsy heterodimer to reduce susceptibility [16]. Further study in Chinese NT1 confirmed the presence of DQB1*06:02 and not

DRB1*15:01 was associated with narcolepsy [17]. In particular, haplotypes such as the DRB1*15:01-DQA1*01:02-DQB1*06:01 and DRB1*15:01-DQA1*01:02-DQB1*05 were not associated with narcolepsy from Southern China [17]. These results illustrate the extraordinary conservation of HLA class II effects in narcolepsy across populations and show that DRB1*15:01 has no effect on narcolepsy susceptibility in the absence of DQB1*06:02. In Japanese, Chinese, and Europeans, the predisposing DQB1*06:02 allele occurs together with DQA1*01:02 on a haplotype with DRB1*15:01. In African Americans, however, these DQB1 and DQA1 alleles are found with distinct DRB1 haplotypes, particularly DRB1*15:03, DRB1*15:01, DRB1*11:01, and DRB1*08:06 [14, 18, 19]. DQA1*01:02 occurs in conjunction with a variety of non-DQB1*06:01 haplotypes, and these do not increase risk for narcolepsy. Therefore, DQA1*01:02 by itself does not predispose to the disease. The DQB1 and DQA1 loci are located adjacent to each other within a 20-kb segment of the HLA class II region of human chromosome 6 and encode DQB and DQA, respectively. The translated product forms a functional cell surface αβ heterodimer, called HLA DQ antigen, which binds to foreign or self-antigens and presents these to the T-cell receptor to effect either immune response or tolerance.

In the general population, the frequency of HLA-DQB1*06:02 allele varies from 5% to 38% and only 1 in 1000 carriers of this allele will develop narcolepsy. The frequency is high in northern European populations (25%) and African Americans (38%), but a low frequency is found in Jewish and Sardinian populations (<5%), which partially correspond with higher prevalence of narcolepsy in African Americans and a low prevalence of narcolepsy in Jewish and Sardinian populations.

It is also worth noting that the association of DQB1*06:02 is particularly strong only in typical cataplexy cases, while DQB1*06:02 was in turn associated with low or undetectable levels of cerebrospinal fluid (CSF) hypocretin-1. In fact, less than 1% of cases with low hypocretin levels are DQB1*06:02 negative [20]. The HLA association is markedly decreased or absent in cases without cataplexy, suggesting that disease heterogeneity is present among DQB1*06:02 negative subjects and those without cataplexy. DQB1*06:02 allele conferring an increased risk of 200 folds. In those who carriers of two HLA-DQB1*06:02 alleles (homozygotes) have an additional two to fourfold risk for narcolepsy compared to that of heterozygotes, and may exhibit more severe symptoms, thus the effect is not merely additive [21].

DQB1*06:02-DRB1*15:01, the same HLA haplotypes that predispose people to NT1, also affects the response to influenza-A infections. The differences have been identified in HLA haplotype frequencies among cases with onset following the 2009 H1N1 influenza pandemic as compared to before the outbreak, with fewer HLA DQB1*06:02 homozygotes [22].

DQB1*06:02 is now known to be the most impactful risk factor, but not only, associated allele Other HLA class II alleles are also associated with narcolepsy, albeit to a lesser extent. Previous studies in European ancestry, Japanese, Korean, and Chinese populations have identified DQB1*03:01 as a second susceptibility allele, HLA-DQB1*03:01 allele increases the risk for NT1 by 1.5-fold and also lowers the disease onset age (nearly 2 years) in Chinese individuals [22, 23]. Differences in HLA-DQ frequencies across populations may thus contribute to differences in age at onset.

These results suggest that complex HLA interactions, including both HLA class I and II molecules, contribute to the development of NT1, although DQB1*06:02 is an essential condition. Logistic regression analyses controlling for the effect of the HLA-DQB1 alleles were also conducted to identify other independent HLA associations. As a result, the susceptibility effect of DPB1*05:01 was identified across multiple ethnic groups [23, 24]. Subsequent analyses that considered alleles of HLA class II genes, such as HLADQB1 and –DPB1, demonstrated an independent association for HLA class I gene alleles [24, 25]. Indeed, a recent study using an animal model reported that cytotoxic CD8+ T cells but not Th1 CD4+ cells were able to induce a selective loss of orexin neurons [26]. Therefore, cytotoxic responses may also play a central role in NT1 pathogenesis because orexin neurons can express HLA class I molecules. Weak associations with narcolepsy have also been reported for HLA class I genes, which encode molecules that present antigenic peptides to CD8+ T cells. HLA-A*11:01, B*14:02, B*35:01, B*35:03, B*51:01, C*01:02, C*04:01, and C*07:01 were associated with susceptibility to narcolepsy [24, 27–30]. This observation suggests that patients with narcolepsy might have an increased genetic susceptibility to infections and that this disease involves cytotoxic immune mechanisms.

As HLA class II alleles are recognized by helper CD4+ T cells and class I alleles by cytotoxic CD8+ T cells, these findings suggest the involvement of both helper and cytotoxic CD8+ T cells in the development of NT1.

Non-HLA Related Genes

NT1 is 86–98% associated with HLA DQA1*01:02~ DQB1*06:02, alleles encoding the DQ heterodimer DQ0602 [14, 24]. In our dataset with around 3000 narcolepsy patients with typical cataplexy and positive for MSLT testing, only 1% were DQB1*06:02 negative, fewer of them also has low CSF hypocretin level (Han, unpublished data). During studies of multiplex families, it was found that more than one third of patients are negative for HLA DQB1*06:02. These findings indicate that genetic predisposition to narcolepsy cannot not be fully explained by HLA allele association, and that additional non-HLA gene components contribute to the development of the disease in at least some cases.

The link between orexin dysfunction and narcolepsy is consistent in both humans and animals. Unlike canine narcolepsy, the patients did not have disease causative variants in the prepro-orexin and orexin receptor genes except for one early-onset case with an onset age < 6 months and negative HLA DQ0602, who has Hcrt gene mutation, impairing peptide trafficking and processing [31]. Sequencing of a large number of early-onset narcolepsy subjects in China revealed three novel nonsynonymous substitutions within the pre-prohypocretin protein, two of which were only found in patients with early-onset narcolepsy but are not likely to be functionally significant, especially in heterozygote subjects. These are all inherited from their parent(s) without any symptoms of narcolepsy [32].

The significant progress of genome-wide association studies (GWAS) and bioinformatics have extended the field of genetics in hypersomnolence. Early GWAS studies suggested that CPT1B or CHKB (rs5770917C) are common genetic susceptibility factors to narcolepsy in Koreans and Japanese populations, indicating that fatty acid β-oxidation may be altered in NT1 [33]. In contrast, the association was not significant in European descent and African Americans, a finding possibly partially explained by the lower frequency of rs5770917[C] in these populations [34]. However, it is worth noting that Chinese population has similar rs5770917[C] frequency as Koreans and Japanese, but Chinese narcoleptics does not associate with CPT1B/CHKB polymorphisms [35].

Other predisposing loci include T cell receptor (TCR) loci *TRA* and *TRB*, interferon alpha and beta receptor subunit 1 (IFNAR1) as well as other autoimmune associated genes including Cathepsin H (*CTSH),* purinergic receptor P2Y11 (P2RY11), zinc-finger protein 365 (ZNF365), interleukin 10 receptor, beta (IL10RB) and tumor necrosis factor superfamily member 4 (TNFSF4) [22, 34, 36–38]. Most of the genes mentioned above are involved in the immune system. Purinergic receptor P2Y11 is correlated with the P2RY11 expression levels in CD8+ T cells and natural killer cells [38]. Another GWAS on two cohorts of European and Chinese ancestry found a novel SNP, rs3826784 (c.596-260A4G), in the EIF3G gene (associated with NT1), which increases the mRNA expression of EIF3G and in turn correlates with the expression of PPAN and P2RY11 [39]. CTSH is a cathepsin protease thought to be involved in antigen processing. TNFSF4 is mainly expressed by antigen-presenting cells, such as dendritic cells, B cells, and macrophages, thereby promoting expansion, differentiation, and survival of T cells. Recently, in a multiethnic sample of 6073 cases and 84,856 controls, fine-mapped GWAS signals within HLA (DQ0602, DQB1*03:01, and DPB1*04:02), seven novel associations (CD207, NAB1, IKZF4-177 ERBB3, CTSC, DENND1B, SIRPG, PRF1) were detected. Significant signals at TRA and DQB1*06:02 loci were found in vaccination-related cases, who also shared polygenic risk. T cell receptor associations in NT1 modulated TRAJ*24, TRAJ*28, and TRBV*4-2 chain-usage. Partitioned heritability and immune cell enrichment analyses found genetic signals to be driven by dendritic and helper T cells [40]. Variants in the HLA-DQ region varied significantly among cases with onset after the 2009 H1N1 influenza pandemic compared to previous years [22]. HLA class I and II molecules interact with the TCR to present peptides to T cells. When the strong association between HLA-DQB1*06:02 and NT1 is taken into consideration, together with the H1N1 virus infection and/or vaccination as triggers, a logical assumption is that DQ0602 molecules interact with a specific TCR associated with the risk genotypes of rs1154155 and rs9648789. Indeed, mechanisms studies demonstrated that a unique immune-related pathway may be responsible for the pathological mechanism of NT1 [41–43].

NT1 genetic variants shape autoimmunity and response to environmental triggers, including influenza A infection and immunization with Pandemrix®. One gene, however, does not fit this pattern.

Genetic Associations with Protective Effect

Protective alleles for both HLA class I and class II genes have also been identified, which might play a part in the delay (or even absence) of evolution of NT2 to NT1 seen in some patients. While closely related alleles to DQB1*06:02 such as DQB1*05:01, DQB1*06:01, DQB1*06:03, DQB1*06:09, DQB1*02 were also identified to be protective alleles. HLA DQB1*06:03 confers a 50-fold increase in protection against narcolepsy [25]. It is noteworthy that DQB1*06:01 is protective but is structurally very similar to HLA DQB1*06:02, which indicates that very minor changes in the peptide-binding pocket may determine disease risk. However, several differences were found between European countries, e.g., DQB1*02 is significantly protective in Dutch but not in others [25, 44]. The protective association is also seen with a SNP located in the HLA-DQA2 region (rs2858884) in European individuals with narcolepsy independent of DQB1*06:02 [44]. In addition, DQA1*01:01 and DQA1*01:03 alleles are protective and several other independent alleles from both HLA class I and II regions affect NT1 risk. These include HLA-DPB1*05:01 HLA-DPB1*04:02 [24, 25, 27]. Two of the haplotypes with the strongest protective effects carry DQA1*01 alleles that are non-DQA1*01:02, namely DQA1*01:03 (with DQB1*06:01) and DQA1*01:01 (with DQB1*05:01). Structurally similar DQA and DQB chains (e.g., those from the same family, such as DQ1) can form stable trans heterodimers; thus, DQB1*05 and DQB1*06 can form heterodimers with DQA1*01 subtypes, but not other DQA1 alleles (all these alleles belong to the large group DQ1). These haplotypes may therefore exert a protective effect by reducing the incidence of the pathogenic.

DQB1*06:02 homozygotes or DQB1*06:02/03:01 heterozygotes are at higher risk for narcolepsy compared to DQB1*06:02 heterozygotes in general, whereas DQB1*06:02/06:01, DQB1*06:02/05:01 and DQB1*06:02/06:03 are at decreased risk. Heterodimerization of DQB1*06:02 with other DQA1 and DQB1 alleles may explain these protective effects by reducing abundance of the disease susceptibility DQA1*01:02/DQB1*06:02 heterodimer. There is a complex influence of HLA-DQ or other loci in high linkage disequilibrium with HLA-DQ on narcolepsy.

Epigenetics of NT1

Mutations affecting the DNA methyl-transferase type 1 (DNMT1) gene, which encodes a DNA methyl-transferase involved in maintenance of DNA methylation patterns, have been associated with two distinct autosomal dominant diseases:

hereditary sensory and autonomic neuropathy with dementia and hearing loss type IE (HSAN IE) and autosomal dominant cerebellar ataxia, deafness and narcolepsy (ADCA-DN). Narcolepsy is a common phenotype in HSAN IE and ADCA-DN, GWAS study in patients with sporadic narcolepsy-cataplexy identified polymorphisms in a region including the DNMT1 gene, and its expression was found to be lower as compared to control subjects [38]. This, together with the finding of narcolepsy in DNMT1-related disease, suggests that impaired DNMT1 activity could influence hypocretin expression, which is regulated by epigenetic factors influenced by metabolic intermediates in mice [45].

Integrative analysis of GWAS and epigenome-wide association study (EWAS) data identified that the CCR3 region, as both a single methylation site and multiple SNPs, were found to be associated with the disease as a candidate region responsible for narcolepsy [29, 30]. The CCR3 region encodes a G protein-coupled receptor for C-C-type chemokines, and play a central role in inflammatory recruitment of leukocytes and other cells in the central nervous system. Interestingly, the number of orexin neurons in the lateral hypothalamus was slightly but significantly decreased and fragmented sleep patterns were observed in the Ccr3 knockout mice during the light phase. Therefore, Ccr3 may be involved in sleep/wake regulation through the orexin system. Further pathway analysis indicated that the CpG site, located in the first intron of IL22RA2, was significantly hypomethylated in patients with narcolepsy, especially upstream of the adenosine A1 receptor (ADORA1) gene [30]. In a recent study with 246 NT1 patients, mitochondrial DNA (mtDNA) haplogroups J and K were demonstrated to modulate metabolic characteristics in NT1, linking mtDNA variation to the dysmetabolic phenotype [46]. Research has been conducted on miRNA alterations in NT1 suggesting an involvement of various miRNA expression in the pathophysiology of narcolepsy [47–49].

Genetics and Epigenetics of NT2

In contrast to the genetics of NT1, little is known about the genetics of NT2. NT2 is also associated with DQB1*06:02. 30–50% of patients with NT2 are DQB1*06:02-positive, and the frequency is significantly higher than that in the general population but is lower than that in NT1. Limited studies were conducted for the evaluation of the role of miRNA in NT2, all those findings are still lacking replication by other groups, and further validation in tissue and brain remains to be done. A recent study on the evolutionarily conserved sleep–wake regulatory role of miRNA-137(miR-137) indicated lowering of miR-137 levels increased wakefulness in mice. The miR-137: Hcrt interaction is conserved across mice and humans, and miR-137 locus is genetically associated with sleep duration in humans. With viral vector-mediated overexpression of the miRs, up to a 20% decrease in Hcrt-1 levels have been seen. The miR-137 are unlikely to play a major role in narcolepsy, as they most likely cannot induce such a massive lowering of HCRT-1 levels as is normally seen in NT1. However, it was speculated miR-137 might be involved in the rare instances of narcolepsy with intermediate HCRT-1 levels [50].

Idiopathic Hypersomnia

Clinical Features

Idiopathic hypersomnia (IH) is a rare central disorder of hypersomnolence characterized by disabling EDS, in the absence of another identified cause. Although EDS may sometimes be the only manifestation, it is often accompanied by long sleep duration at night, long naps, a sense of unrefreshing sleep, and great difficulty in awakening, known as sleep drunkenness or pronounced sleep inertia. Fatigue, cognitive symptoms, and autonomic symptoms are common and contribute to the burden of disease [1].

Family History

A familial factor of IH has been proposed, as a family history of EDS is common in patients with IH, family history of excessive sleepiness, IH, or another central disorder of hypersomnolence is present in 34–38% of IH patients [51–53]. 12.5% of IH patients have at least one parent who routinely sleeps more than 9.5 h per night, which suggests the parent-child transmission [54]. Even in patients with narcolepsy, approximately one third of probands or parents have a positive family history of EDS meeting the diagnostic criteria of idiopathic hypersomnia. The familial patterns suggest a genetic contribution to IH.

Genetic Studies

In contrast to narcolepsy, the underlying genetic contribution to IH is poorly understood, and results are inconsistent, despite the tendency for familial aggregation. Meanwhile, no significant associations between IH and HLA have been observed, and no other genetic variants associated with IH have been clearly identified.

The prevalence of the HLA-DQB1*06:02 allele was reported to be increased in IH when compared to healthy subjects. However, no consistent data are available [55, 56]. The HLA-DQB1*02 frequency was higher in the Portuguese population with IH when compared with the control population (55% vs. 34%; $p = 0.004$) [57]. However, HLA analyses did not reveal any significant associations between IH and specific HLA allele in Japanese [58].

Most of the genetic studies on IH were conducted in the Japanese population, with a definition different with the diagnostic criteria by ICSD-3 in some degree [1, 59]. Although no single-nucleotide polymorphisms (SNPs) reached the genome-wide significance level, a GWAS study including a total of 414 Japanese patients with IH and 6587 healthy Japanese individuals identified several candidate SNPs for IH, such as a common genetic variant (rs2250870) within a PDE9A intron, rs2250870 was significantly associated with expression levels of *PDE9A* in not only whole blood but also brain tissues. PDE9A is a potential target in the treatment of several brain diseases, such as depression, schizophrenia, and Alzheimer's disease [60]. Polygenic risks analysis in Japanese patients with hypersomnia indicated that the pathogenesis of narcolepsy and hypersomnia might be partially shared. Hypersomnia patients with HLA-DQB1*06:02 were estimated to have higher shared genetic background to narcoleptic patients than hypersomnia patients without HLA-DQB1*06:02 even when the effects of HLA-DQB1*06:02 were excluded [58]. A rare missense variant (g.42184347T>C; p.Lys68Arg; rs537376938) in the cleavage site of prepro-orexin that was identified to be associated with IH, indicating that a subgroup of IH is associated with decreased orexin signaling [61]. Furthermore, LPGDS as a sleep promotion substance increase in both CSF/serum in type1, type 2 narcolepsy, and IH, indicating this may serve as a new common biomarker for EDS of central origin and imply a common pathogenetic association, but would complement rather than replaces orexin markers [62].

Clinical features and melatonin profiles suggest abnormalities of the circadian system may be a contributor to the pathogenesis of IH. IH patients tend to have both a late chronotype and prominent sleep inertia or sleep drunkenness. Melatonin and cortisol profiles in people with IH confirm this tendency toward phase delay. Recently, limited studies suggested that genetic influence may exist in the development of IH. First, a highly significant damping in the cyclic expression of BMAL1, PER1, and PER2 was found in patients with IH, with a significant reduction in the expression of BMAL1 [63–65]. Second, an interesting case of parthenogenetic mosaicism indicated the homozygous PER2 variant in parthenogenetic cells as a probable causal factor in IH [66]. In this case, due to the p.Arg468Gln-PER2 variant, relative accumulation of the heterodimer PER2-CRY in the nucleus is likely and, consequently, prolonged repression by this on the CLOCK-BMAL1 transactivation function, which might slow the production of PER2 and CRY, leading to hypersomnia [66]. Epigenetic studies indicate only quantitative differences in miRNAs between IH and controls; miR-224 and miR-643 were different in both NT2 and IH; furthermore, the results suggested that NT2 was a heterogeneous entity, including NT1 and IH [47]. The same team then extended the search to CSF, finding no variations in miRNAs [48].

Kleine-Levin Syndrome

Clinical Features

Kleine-Levin syndrome (KLS) is a rare sleep disorder characterized by relapsing–remitting episodes of hypersomnia, and is associated with cognitive disturbances and behavioral abnormalities. It is characterized by sudden recurrent episodes, including major hypersomnia as well as cognitive and behavioral issues alternating with asymptomatic periods. Eighty-one percent of patients with KLS develop symptoms during the second decade. Adults and younger children may also be affected. It is rare for patients over the age of 30 years to present with their first episode, and, although though cases have been reported with ages ranging from 4 to 82 [1].

Family History

Although most cases are sporadic, family members account for about 5% of reported cases including two sets of monozygotic twins, suggesting a genetic predisposition for KLS in some patients resulting in an estimated 800-to-4000-fold increased risk for first-degree relatives for those with the disease [67–70]. There is no increased history of mood disorders in family members of patients [71].

Up to date, there are 19 multiplex families with 2–6 affected members and 4 pairs of monozygotic twins several complex families (those with 2–6 patients) with KLS have been published in the form of case reports. One study reviewed retrospectively 260 patients with confirmed KLS from the USA and France for familial occurrence and yielded the following results [71]. Out of a total of 249 families, 10 (4%) were found to contain at least two patients, providing that there was a first-or second-degree familial relation. Twenty-one out of the 260 patients were familial cases (8%). In all families there was no consanguinity between the parents. The frequency of familial cases was similar for the French and American groups.

Genetic Studies

No specific genes have been implicated as causative for KLS; there is some evidence to suggest a genetic predisposition. An association of KLS with HLA DQB1*02 suggestive of autoimmune origin has been reported in the European series [72], which is consistent with the finding that upper airway infection as trigger for KLS episode [73]. However, this was not confirmed in two controlled series from the USA [74], France [72], and Taiwan [73]. There was similar HLA DQB1 distribution in familial versus sporadic cases and no abnormal karyotypes.

Non-HLA genes related to KLS have been reported. In a large family with seven affected members, linkage analysis mapped the disease locus to chromosome 3 and exome analysis identified a heterozygous missense variant in LMOD3 in the linkage interval [75]. In a more recent case of atypical KLS with positive familiarity, a mutation in the LMOD3 gene was also identified [76]. Animal study suggests LMOD3 gene is related to the brain structures involved with the control of alertness, indicating that KLS may be a neurodevelopmental structural brain disease. Further GWAS study on 673 KLS cases from different ethnics shows that variants in the TRANK1 gene region may predispose to KLS when patients have had a difficult birth, suggesting that TRANK1 gene region modulates newborns' response to brain injury, with consequences for mental and neurological health in adulthood [77]. TRANK1 might be a genetic predisposition links KLS, circadian regulation, and bipolar disorder.

Summary

This review summarized the genetic studies that have emerged over the past decade on narcolepsy, IH, and KLS, with most of which focus on NT1. The data allows us to better understand the etiopathogenesis and the possible biological pathways specifically involved in hypersomnia. One success example is the identification of genetics in narcolepsy leads the findings of the possible role for immune mechanisms in disease development. In consideration of the heterogeneity of patients with hypersomnia, search for genetic markers therefore assumes considerable importance for diagnostics, and provides further clarifications and targeted treatment on hypersomnia disorders, in the context of a precision medicine approach. Finally, findings from the study of the genetics and epigenetics of hypersomnia might also expand to the field of chronotherapy. However, future research work needs to be done in the light of promising prospects offered by the emerging knowledge and techniques of genetics study.

Bibliography

1. American Academy of Sleep Medicine. International classification of sleep disorders. 3rd ed. Darien: American Academy of Sleep Medicine; 2014.
2. Trotti LM. Are we there yet? getting closer to certainty in idiopathic hypersomnia diagnosis. J Clin Sleep Med: JCSM. 2019;15(4):539–40. https://doi.org/10.5664/jcsm.7706.
3. Goel N. Genetics of sleep timing, duration and homeostasis in humans. Sleep Med Clin. 2011;6(2):171–82. https://doi.org/10.1016/j.jsmc.2011.03.004.
4. Lane JM, Qian J, Mignot E, Redline S, Scheer F, Saxena R. Genetics of circadian rhythms and sleep in human health and disease. Nat Rev Genetics. 2023;24(1):4–20. https://doi.org/10.1038/s41576-022-00519-z.
5. Mogavero MP, DelRosso LM, Bruni O, Salemi M, Salsone M, Novellino F, et al. Genetics and epigenetics of rare hypersomnia. Trends Genet. 2023;39(5):415–29. https://doi.org/10.1016/j.tig.2023.02.003.
6. Nakayama J, Miura M, Honda M, Miki T, Honda Y, Arinami T. Linkage of human narcolepsy with HLA association to chromosome 4p13-q21. Genomics. 2000;65(1):84–6. https://doi.org/10.1006/geno.2000.6143.
7. Dauvilliers Y, Blouin JL, Neidhart E, Carlander B, Eliaou JF, Antonarakis SE, et al. A narcolepsy susceptibility locus maps to a 5 Mb region of chromosome 21q. Ann Neurol. 2004;56(3):382–8. https://doi.org/10.1002/ana.20208.
8. Winkelmann J, Lin L, Schormair B, Kornum BR, Faraco J, Plazzi G, et al. Mutations in DNMT1 cause autosomal dominant cerebellar ataxia, deafness and narcolepsy. Hum Mol Genet. 2012;21(10):2205–10. https://doi.org/10.1093/hmg/dds035.

9. Hor H, Bartesaghi L, Kutalik Z, Vicário JL, de Andrés C, Pfister C, et al. A missense mutation in myelin oligodendrocyte glycoprotein as a cause of familial narcolepsy with cataplexy. Am J Hum Genet. 2011;89(3):474–9. https://doi.org/10.1016/j.ajhg.2011.08.007.
10. Juji T, Satake M, Honda Y, Doi Y. HLA antigens in Japanese patients with narcolepsy. All the patients were DR2 positive. Tissue Antigens. 1984;24(5):316–9. https://doi.org/10.1111/j.1399-0039.1984.tb02144.x.
11. Planelles D, Puig N, Beneto A, Gomez E, Rubio P, Mirabet V, et al. HLA-DQA, -DQB and -DRB allele contribution to narcolepsy susceptibility. Eur J Immunogenet. 1997;24(6):409–21. https://doi.org/10.1046/j.1365-2370.1997.d01-115.x.
12. Matsuki K, Grumet FC, Lin X, Gelb M, Guilleminault C, Dement WC, et al. DQ (rather than DR) gene marks susceptibility to narcolepsy. Lancet (London, England). 1992;339(8800):1052. https://doi.org/10.1016/0140-6736(92)90571-j.
13. Mignot E, Hayduk R, Black J, Grumet FC, Guilleminault C. HLA DQB1*0602 is associated with cataplexy in 509 narcoleptic patients. Sleep. 1997;20(11):1012–20.
14. Mignot E, Lin L, Rogers W, Honda Y, Qiu X, Lin X, et al. Complex HLA-DR and -DQ interactions confer risk of narcolepsy-cataplexy in three ethnic groups. Am J Hum Genet. 2001;68(3):686–99. https://doi.org/10.1086/318799.
15. Sasai T, Inoue Y, Komada Y, Sugiura T, Matsushima E. Comparison of clinical characteristics among narcolepsy with and without cataplexy and idiopathic hypersomnia without long sleep time, focusing on HLA-DRB1(*)1501/DQB1(*)0602 finding. Sleep Med. 2009;10(9):961–6. https://doi.org/10.1016/j.sleep.2008.12.007.
16. Kwok WW, Kovats S, Thurtle P, Nepom GT. HLA-DQ allelic polymorphisms constrain patterns of class II heterodimer formation. J Immunol (Baltimore, Md : 1950). 1993;150(6):2263–72.
17. Han F, Lin L, Li J, Dong SX, An P, Zhao L, et al. HLA-DQ association and allele competition in Chinese narcolepsy. Tissue Antigens. 2012;80(4):328–35. https://doi.org/10.1111/j.1399-0039.2012.01948.x.
18. Chabas D, Taheri S, Renier C, Mignot E. The genetics of narcolepsy. Annu Rev Genomics Hum Genet. 2003;4:459–83. https://doi.org/10.1146/annurev.genom.4.070802.110432.
19. Mignot E, Kimura A, Lattermann A, Lin X, Yasunaga S, Mueller-Eckhardt G, et al. Extensive HLA class II studies in 58 non-DRB1*15 (DR2) narcoleptic patients with cataplexy. Tissue Antigens. 1997;49(4):329–41. https://doi.org/10.1111/j.1399-0039.1997.tb02761.x.
20. Han F, Lin L, Schormair B, Pizza F, Plazzi G, Ollila HM, et al. HLA DQB1*06:02 negative narcolepsy with hypocretin/orexin deficiency. Sleep. 2014;37(10):1601–8. https://doi.org/10.5665/sleep.4066.
21. Weiner Lachmi K, Lin L, Kornum BR, Rico T, Lo B, Aran A, et al. DQB1*06:02 allele-specific expression varies by allelic dosage, not narcolepsy status. Hum Immunol. 2012;73(4):405–10. https://doi.org/10.1016/j.humimm.2012.01.004.
22. Han F, Faraco J, Dong XS, Ollila HM, Lin L, Li J, et al. Genome wide analysis of narcolepsy in China implicates novel immune loci and reveals changes in association prior to versus after the 2009 H1N1 influenza pandemic. PLoS Genet. 2013;9(10):e1003880. https://doi.org/10.1371/journal.pgen.1003880.
23. Han F, Lin L, Li J, Aran A, Dong SX, An P, et al. Presentations of primary hypersomnia in Chinese children. Sleep. 2011;34(5):627–32. https://doi.org/10.1093/sleep/34.5.627.
24. Ollila HM, Ravel JM, Han F, Faraco J, Lin L, Zheng X, et al. HLA-DPB1 and HLA class I confer risk of and protection from narcolepsy. Am J Hum Genet. 2015;96(1):136–46. https://doi.org/10.1016/j.ajhg.2014.12.010.
25. Tafti M, Hor H, Dauvilliers Y, Lammers GJ, Overeem S, Mayer G, et al. DQB1 locus alone explains most of the risk and protection in narcolepsy with cataplexy in Europe. Sleep. 2014;37(1):19–25. https://doi.org/10.5665/sleep.3300.
26. Bernard-Valnet R, Yshii L, Quériault C, Nguyen XH, Arthaud S, Rodrigues M, et al. CD8 T cell-mediated killing of orexinergic neurons induces a narcolepsy-like phenotype in mice. PNAS. 2016;113(39):10956–61. https://doi.org/10.1073/pnas.1603325113.

27. Tafti M, Lammers GJ, Dauvilliers Y, Overeem S, Mayer G, Nowak J, et al. Narcolepsy-associated HLA class I alleles implicate cell-mediated cytotoxicity. Sleep. 2016;39(3):581–7. https://doi.org/10.5665/sleep.5532.
28. Juvodden HT, Viken MK, Nordstrand SEH, Viste R, Westlye LT, Thorsby PM, et al. HLA and sleep parameter associations in post-H1N1 narcolepsy type 1 patients and first-degree relatives. Sleep. 2020;43(3) https://doi.org/10.1093/sleep/zsz239.
29. Ouyang H, Han F, Zhou ZC, Zhang J. Differences in clinical and genetic characteristics between early- and late-onset narcolepsy in a Han Chinese cohort. Neural Regen Res. 2020;15(10):1887–93. https://doi.org/10.4103/1673-5374.280322.
30. Shimada M, Miyagawa T, Toyoda H, Tokunaga K, Honda M. Epigenome-wide association study of DNA methylation in narcolepsy: an integrated genetic and epigenetic approach. Sleep. 2018;41(4) https://doi.org/10.1093/sleep/zsy019.
31. Peyron C, Faraco J, Rogers W, Ripley B, Overeem S, Charnay Y, et al. A mutation in a case of early onset narcolepsy and a generalized absence of hypocretin peptides in human narcoleptic brains. Nat Med. 2000;6(9):991–7. https://doi.org/10.1038/79690.
32. Dong XS, Ma SF, Cao CW, Li J, An P, Zhao L, Liu NY, Yan H, Hu QT, Mignot E, Strohl KP, Gao ZC, Zeng C, Han F. Hypocretin (orexin) neuropeptide precursor gene, HCRT, polymorphisms in early-onset narcolepsy with cataplexy. Sleep Med. 2013;14(6):482–7. https://doi.org/10.1016/j.sleep.2013.01.016. Epub 2013 May 3.
33. Miyagawa T, Kawashima M, Nishida N, Ohashi J, Kimura R, Fujimoto A, et al. Variant between CPT1B and CHKB associated with susceptibility to narcolepsy. Nat Genet. 2008;40(11):1324–8. https://doi.org/10.1038/ng.231.
34. Hallmayer J, Faraco J, Lin L, Hesselson S, Winkelmann J, Kawashima M, et al. Narcolepsy is strongly associated with the T-cell receptor alpha locus. Nat Genet. 2009;41(6):708–11. https://doi.org/10.1038/ng.372.
35. Han F, Lin L, Li J, Aran A, Dong SX, An P, et al. TCRA, P2RY11, and CPT1B/CHKB associations in Chinese narcolepsy. Sleep Med. 2012;13(3):269–72. https://doi.org/10.1016/j.sleep.2011.06.020.
36. Faraco J, Lin L, Kornum BR, Kenny EE, Trynka G, Einen M, et al. ImmunoChip study implicates antigen presentation to T cells in narcolepsy. PLoS Genet. 2013;9(2):e1003270. https://doi.org/10.1371/journal.pgen.1003270.
37. Mahlios J, De la Herrán-Arita AK, Mignot E. The autoimmune basis of narcolepsy. Curr Opin Neurobiol. 2013;23(5):767–73. https://doi.org/10.1016/j.conb.2013.04.013.
38. Kornum BR, Kawashima M, Faraco J, Lin L, Rico TJ, Hesselson S, et al. Common variants in P2RY11 are associated with narcolepsy. Nat Genet. 2011;43(1):66–71. https://doi.org/10.1038/ng.734.
39. Holm A, Lin L, Faraco J, Mostafavi S, Battle A, Zhu X, et al. EIF3G is associated with narcolepsy across ethnicities. Eur J Hum Genet: EJHG. 2015;23(11):1573–80. https://doi.org/10.1038/ejhg.2015.4.
40. Ollila HM, et al. Narcolepsy risk loci outline role of T cell autoimmunity and infectious triggers in narcolepsy. Nat Commun. 2023;14(1):2709.
41. Luo G, Zhang J, Lin L, Mignot EJ. Characterization of T cell receptors reactive to HCRT(NH2), pHA(273-287), and NP(17-31) in control and narcolepsy patients. Proc Natl Acad Sci USA. 2022;119(32):e2205797119. https://doi.org/10.1073/pnas.2205797119.
42. Latorre D, Kallweit U, Armentani E, Foglierini M, Mele F, Cassotta A, et al. T cells in patients with narcolepsy target self-antigens of hypocretin neurons. Nature. 2018;562(7725):63–8. https://doi.org/10.1038/s41586-018-0540-1.
43. Luo G, Ambati A, Lin L, Bonvalet M, Partinen M, Ji X, et al. Autoimmunity to hypocretin and molecular mimicry to flu in type 1 narcolepsy. Proc Natl Acad Sci USA. 2018;115(52):E12323–e12332. https://doi.org/10.1073/pnas.1818150116.
44. Hor H, Kutalik Z, Dauvilliers Y, Valsesia A, Lammers GJ, Donjacour CE, et al. Genome-wide association study identifies new HLA class II haplotypes strongly protective against narcolepsy. Nat Genet. 2010;42(9):786–9. https://doi.org/10.1038/ng.647.

45. Takasugi M, Hayakawa K, Arai D, Shiota K. Age- and sex-dependent DNA hypomethylation controlled by growth hormone in mouse liver. Mech Ageing Dev. 2013;134(7–8):331–7. https://doi.org/10.1016/j.mad.2013.05.003.
46. Caporali L, Moresco M, Pizza F, La Morgia C, Fiorini C, Strobbe D, et al. The role of mtDNA haplogroups on metabolic features in narcolepsy type 1. Mitochondrion. 2022;63:37–42. https://doi.org/10.1016/j.mito.2022.01.005.
47. Holm A, Bang-Berthelsen CH, Knudsen S, Kornum BR, Modvig S, Jennum P, et al. miRNA profiles in plasma from patients with sleep disorders reveal dysregulation of miRNAs in narcolepsy and other central hypersomnias. Sleep. 2014;37(9):1525–33. https://doi.org/10.5665/sleep.4004.
48. Holm A, Bang-Berthelsen CH, Knudsen S, Modvig S, Kornum BR, Gammeltoft S, et al. MiRNA profiles in cerebrospinal fluid from patients with central hypersomnias. J Neurol Sci. 2014;347(1–2):199–204. https://doi.org/10.1016/j.jns.2014.09.047.
49. Liu Z, Yang L, Zhao Y, Tang M, Wang F, Wang X, et al. Reproducibility of quantitative real-time PCR assay in microRNA expression profiling and comparison with microarray analysis in narcolepsy. Springerplus. 2015;4:812. https://doi.org/10.1186/s40064-015-1613-3.
50. Holm A, Possovre ML, Bandarabadi M, Moseholm KF, Justinussen JL, Bozic I, et al. The evolutionarily conserved miRNA-137 targets the neuropeptide hypocretin/orexin and modulates the wake to sleep ratio. Proc Natl Acad Sci USA. 2022;119(17):e2112225119. https://doi.org/10.1073/pnas.2112225119.
51. Billiard M, Merle C, Carlander B, Ondze B, Alvarez D, Besset A. Idiopathic hypersomnia. Psychiatry Clin Neurosci. 1998;52(2):125–9. https://doi.org/10.1111/j.1440-1819.1998.tb00987.x.
52. Takei Y, Komada Y, Namba K, Sasai T, Nakamura M, Sugiura T, et al. Differences in findings of nocturnal polysomnography and multiple sleep latency test between narcolepsy and idiopathic hypersomnia. Clin Neurophysiol. 2012;123(1):137–41. https://doi.org/10.1016/j.clinph.2011.05.024.
53. Billiard M, Dauvilliers Y. Idiopathic hypersomnia. Sleep Med Rev. 2001;5(5):349–58. https://doi.org/10.1053/smrv.2001.0168.
54. Vernet C, Leu-Semenescu S, Buzare MA, Arnulf I. Subjective symptoms in idiopathic hypersomnia: beyond excessive sleepiness. J Sleep Res. 2010;19(4):525–34. https://doi.org/10.1111/j.1365-2869.2010.00824.x.
55. Coelho FM, Pradella-Hallinan M, Predazzoli Neto M, Bittencourt LR, Tufik S. Prevalence of the HLA-DQB1*0602 allele in narcolepsy and idiopathic hypersomnia patients seen at a sleep disorders outpatient unit in São Paulo. Revista brasileira de psiquiatria (Sao Paulo, Brazil : 1999). 2009;31(1):10–4. https://doi.org/10.1590/s1516-44462009000100004.
56. Vrana M, Siffnerova V, Pecherkova P, Ratajova E, Sonka K. Distribution of HLA-DQB1 in Czech patients with central hypersomnias. Arch Immunol Ther Exp. 2016;64(Suppl 1):89–98. https://doi.org/10.1007/s00005-016-0435-5.
57. Martins-da-Silva A, Lopes J, Ramalheira J, Carvalho C, Cunha D, Costa PP, et al. Usefulness of genetic characterization of narcolepsy and hypersomnia on phenotype definition: a study in Portuguese patients. Rev Neurol. 2014;58(2):49–54.
58. Yamasaki M, Miyagawa T, Toyoda H, Khor SS, Liu X, Kuwabara H, et al. Evaluation of polygenic risks for narcolepsy and essential hypersomnia. J Hum Genet. 2016;61(10):873–8. https://doi.org/10.1038/jhg.2016.65.
59. Honda Y, Honda M. Idiopathic hypersomnia--review of literatures and clinical experiences on 16 Japanese cases. Nihon Rinsho Jpn J Clin Med. 1998;56(2):371–5.
60. Tanida K, Shimada M, Khor S-S, Toyoda H, Kato K, Kotorii N, et al. Genome-wide association study of idiopathic hypersomnia in a Japanese population. Sleep Biol Rhythms. 2022;20(1):137–48. https://doi.org/10.1007/s41105-021-00349-2.
61. Miyagawa T, Tanaka S, Shimada M, Sakai N, Tanida K, Kotorii N, et al. A rare genetic variant in the cleavage site of prepro-orexin is associated with idiopathic hypersomnia. NPJ Genom Med. 2022;7(1):29. https://doi.org/10.1038/s41525-022-00298-w.

62. Wang P, Li Q, Dong X, An H, Li J, Zhao L, et al. Lipocalin-type prostaglandin D synthase levels increase in patients with narcolepsy and idiopathic hypersomnia. Sleep. 2021;44(4) https://doi.org/10.1093/sleep/zsaa234.
63. Landzberg D, Trotti LM. Is idiopathic hypersomnia a circadian rhythm disorder? Curr Sleep Med Rep. 2019;5(4):201–6. https://doi.org/10.1007/s40675-019-00154-x.
64. Lippert J, Halfter H, Heidbreder A, Röhr D, Gess B, Boentert M, et al. Altered dynamics in the circadian oscillation of clock genes in dermal fibroblasts of patients suffering from idiopathic hypersomnia. PLoS One. 2014;9(1):e85255. https://doi.org/10.1371/journal.pone.0085255.
65. Materna L, Halfter H, Heidbreder A, Boentert M, Lippert J, Koch R, et al. Idiopathic hypersomnia patients revealed longer circadian period length in peripheral skin fibroblasts. Front Neurol. 2018;9:424. https://doi.org/10.3389/fneur.2018.00424.
66. Masunaga Y, Kagami M, Kato F, Usui T, Yonemoto T, Mishima K, et al. Parthenogenetic mosaicism: generation via second polar body retention and unmasking of a likely causative PER2 variant for hypersomnia. Clin Epigenetics. 2021;13(1):73. https://doi.org/10.1186/s13148-021-01062-0.
67. Ueno T, Fukuhara A, Ikegami A, Ohishi F, Kume K. Monozygotic twins concordant for Kleine-Levin syndrome. BMC Neurol. 2012;12:31. https://doi.org/10.1186/1471-2377-12-31.
68. Peraita-Adrados R, Vicario JL, Tafti M, García de León M, Billiard M. Monozygotic twins affected with Kleine-Levin syndrome. Sleep. 2012;35(5):595–6. https://doi.org/10.5665/sleep.1808.
69. Arnulf I, Rico TJ, Mignot E. Diagnosis, disease course, and management of patients with Kleine-Levin syndrome. Lancet Neurol. 2012;11(10):918–28. https://doi.org/10.1016/s1474-4422(12)70187-4.
70. Lavault S, Golmard JL, Groos E, Brion A, Dauvilliers Y, Lecendreux M, et al. Kleine-Levin syndrome in 120 patients: differential diagnosis and long episodes. Ann Neurol. 2015;77(3):529–40. https://doi.org/10.1002/ana.24350.
71. Nguyen QT, Groos E, Leclair-Visonneau L, Monaca-Charley C, Rico T, Farber N, et al. Familial Kleine-Levin syndrome: a specific entity? Sleep. 2016;39(8):1535–42. https://doi.org/10.5665/sleep.6014.
72. Dauvilliers Y, Mayer G, Lecendreux M, Neidhart E, Peraita-Adrados R, Sonka K, et al. Kleine-Levin syndrome: an autoimmune hypothesis based on clinical and genetic analyses. Neurology. 2002;59(11):1739–45. https://doi.org/10.1212/01.wnl.0000036605.89977.d0.
73. Huang YS, Guilleminault C, Lin KL, Hwang FM, Liu FY, Kung YP. Relationship between Kleine-Levin syndrome and upper respiratory infection in Taiwan. Sleep. 2012;35(1):123–9. https://doi.org/10.5665/sleep.1600.
74. Arnulf I, Lin L, Gadoth N, File J, Lecendreux M, Franco P, et al. Kleine-Levin syndrome: a systematic study of 108 patients. Ann Neurol. 2008;63(4):482–93. https://doi.org/10.1002/ana.21333.
75. Al Shareef SM, Basit S, Li S, Pfister C, Pradervand S, Lecendreux M, et al. Kleine-Levin syndrome is associated with LMOD3 variants. J Sleep Res. 2019;28(3):e12718. https://doi.org/10.1111/jsr.12718.
76. Wenz E, Tafti M, Bassetti CLA. LMOD3 gene variant in familial periodic hypersomnolence. Sleep Med. 2022;91:105–8. https://doi.org/10.1016/j.sleep.2022.02.019.
77. Ambati A, Hillary R, Leu-Semenescu S, Ollila HM, Lin L, During EH, et al. Kleine-Levin syndrome is associated with birth difficulties and genetic variants in the TRANK1 gene loci. Proc Natl Acad Sci USA. 2021;118(12) https://doi.org/10.1073/pnas.2005753118.

Restless Leg Syndrome Through the Magnifying Glass of Genetics

L. Ferini-Strambi, Combi Romina, and Salsone Maria

Abstract Restless leg syndrome (RLS) is a sensorimotor disorder clinically characterized by unpleasant sensations associated to urge to move the legs, predominantly manifested at resting or in the evening. RLS is considered a complex genetic disease, and a strong familial component has been proposed also in the primary RLS. While a role of genetics in this form is well established, the aetiopathogenesis of the disease remains somewhat uncertain, and genetic studies are still an important pillar of RLS research. This chapter discusses the contribution of genetics in the pathogenesis of primary RLS, with a focus on how genetics is able to drive the clinical and imaging phenotype. Genome-wide association studies have been decisive in the identification of genes (MEIS1, BTBD9, and MAP 2K5/SKOR1) strongly related with RLS in the Caucasian and Asian populations. These disease-associated genes encode for proteins involved in the ferritin expression and iron metabolism and dopaminergic system activity. Finally, using a neuroimaging approach, we also provide to identify potential vulnerable areas for the development of RLS. Structural/functional white and gray matter changes and changes due to the iron storage alterations have been questioned, in order to decode the possible association to RLS neuropathology.

Keywords RLS · Genetics · MEIS1 · BTBD9 · MAP 2K5

L. Ferini-Strambi (✉)
Sleep Disorders Center, Division of Neuroscience, San Raffaele Scientific Institute, Milan, Italy

Vita-Salute San Raffaele University, Milan, Italy
e-mail: ferinistrambi.luigi@hsr.it

C. Romina
Department of Medicine and Surgery, University of Milano-Bicocca, Milan, Italy

S. Maria
Sleep Disorders Center, Division of Neuroscience, San Raffaele Scientific Institute, Milan, Italy

Department of Medicine and Surgery, University of Milano-Bicocca, Milan, Italy

P. Gehrman et al. (eds.), *Genetics of Sleep and Sleep Disorders*,
https://doi.org/10.1007/978-3-031-62723-1_13

Introduction

Restless leg syndrome (RLS), also known as Willis-Ekbom disease, is a common neurological disorder with a prevalence of 5–10% in the White population, characterized by unpleasant sensations in the lower extremities resulting in an irresistible urge to move the legs [1]. It is clinically manifested particularly at night-time and during quiet wakefulness causing nocturnal discomfort and sleep fragmentation with consequent cognitive and mood symptoms that negatively impact the quality of life [1]. RLS is most frequently reported in the female gender at a young age (<40 years old), and a strong familial component has been proposed in these cases. Although the exact pathophysiology remains unknown unti now, complex environmental and genetic interactions play a crucial role in the development of this peculiar movement disorder. Comorbidities such as cardiovascular disease, diabetes, arterial hypertension, iron deficiency anaemia, renal insufficiency, and pregnancy are considered environmental factors complicating diagnosis and treatment [1, 2]. Genetic research has widely documented that RLS is a complex genetic disease, characterized by genetic heterogeneity, decreased penetrance, and variable expression [1, 2]. The most successful genetic study has been the genome-wide association approach, not only for the opportunity to perform screening of common variants across the genome but also to identify novel risk loci as potential therapeutic targets [1, 2].

According to the ethology, RLS could be classified into two forms [3]: primary RLS represents the common idiopathic form of RLS accounting for 70–80% of cases [4] frequently showing a family history and an earlier onset of symptoms (before 45 age) and secondary RLS that occurs secondary to various medical conditions. It is also interesting to note that a positive family history of RLS has been reported in 20–60% of RLS patients, depending on the study and varying due to differences in sample size, ethnicity, and study design. The diagnosis of RLS remains a clinical diagnosis, although the presence of periodic limb movements (PLMs) during sleep detected on polysomnography and actigraphy is considered a potential biomarker for RLS [5]. This said, RLS remains to date a neurological disorder clinically unrecognized and underestimated. The time between the clinical diagnosis and the initiation of the treatment increases over time. Dopaminergic treatment is formerly the first-line pharmacological therapy, although prolonged use can result in two serious complications, impulse control disorders such as pathological gambling/shopping and hypersexuality and the worsening of RLS clinical symptoms known as augmentation [1, 6].

Despite all the efforts made in the last years, there are several issues needed to be addressed: what are the genetic risk factors for RLS? How does genetics impact the clinical and neuroimaging phenotype? What is the role of epigenetics? With this in mind, in this chapter we have the ambition to answer all these questions with a focus on the contribution of genetics in the pathogenesis of this disabling disorder.

Genetic Risk Factors for RLS

The higher frequency of positive familiar history (up to 60%) was reported for primary RLS patients and up to 80% of monozygotic twins [7–9]. This heritability estimates indicate a significant influence of genetic risk factors on RLS susceptibility [10, 11].

Familial cases of RLS showed symptoms with an earlier onset in life than in sporadic RLS, with a mean onset before 30–40 years of age [12]. The majority of reported families with RLS are consistent with a typical autosomal dominant inheritance pattern, with different manifestations [13, 14]. However, recessive inheritance patterns have also been reported in some families, and 2.8% of families show also bilinear inheritance [15].

Although monogenic forms may exist in individual families, it is well established that the majority of idiopathic RLS cases are probably due to a multifactorial form for which common small-effect size variants determine the genetic risk. RLS is considered a complex condition in which genetic background, environmental factors, and gene–environment interactions predispose people to disease and affect expression of the phenotype [16]. Several lines of evidence point to a complex mode of inheritance. One of these is the occurrence by chance of phenotypic duplication which was reported to be relatively frequent in RLS patients and RLS families where affected individuals with similar manifestations are due to different genetic and/or environmental risks.

The approaches commonly used to study genetics of RLS depend on the available sample: in the case of multigenerational families supposed to follow a monogenic inheritance, linkage studies and next-generation sequencing of exomes or whole genomes are considered the best approaches to find rare variants with strong effects [11, 17, 18]. Otherwise, case-control association studies are considered the method of choice in identifying common risk variants with small effects for the multifactorial or complex form of RLS [16]. These studies can be broadly performed using two approaches: hypothesis-driven candidate gene studies and hypothesis-free genome-wide association studies (GWAS). The latter is actually considered the gold standard being now available large phenotype-specific consortia and comparably cheap and reliable microarrays for genotyping [19].

In the monogenic form of RLS, linkage studies allowed the identification of several linked regions but no causal genes [11], and exome sequencing or whole-genome sequencing studies, which were limited to a small number of families, did not identify unequivocal causal genetic variants, and no confirmation was then obtained by independent studies [17, 18, 20, 21]. At least eight susceptibility loci have been reported in the OMIM database. RLS1 locus (OMIM 102300) has been mapped to chromosome 12q [22]. Other susceptibility loci for RLS include RLS2 (OMIM 608831) on chromosome 14q13-q31 [23], RLS3 (OMIM 610438) on chromosome 9p24-p22 [24], RLS4 (OMIM 610439) on chromosome 2q33 [25], RLS5 (OMIM 611242) on chromosome 20p13 [26], RLS6 (OMIM 611185) on chromosome 6p21 [27], RLS7 (612853) on chromosome 2p14 [27], and RLS8 (615197) on chromosome 5q31 [16].

The first large-scale GWAS in RLS was reported in 2007 by Winkelmann et al. [27] and identified three associated loci including MEIS1, BTBD9, and MAP 2 K5/SKOR1 genes that were subsequently confirmed in several follow-up studies. In a recent study combining GWAS and exome sequencing approaches, Ergun et al. [28] found six new candidate genes (SEMA3A, PPP1R9A, PUS7, CDHR3, HBP1, and COG5) having high LOD scores in linkage and significant *P*-values in association analysis.

Among the most recent GWAS metanalysis reports, the first [29], besides yielding 20 genome-wide significant risk variants, estimated also a heritability contribution due to common variants of 19.6% thus providing a measure of their importance in RLS susceptibility. The second GWAS meta-analysis confirmed the known loci and identified three novel risk loci revealing also significant cis-eQTL effects which suggested candidate causal genes in four risk loci [30].

To date several GWAS studies and meta-analyses uncovered and replicated 23 common genetic risk variants in 22 genomic loci and nominated genes for functional follow-up [27, 29–32] as summarized in Table 1. The identified genes are linked to several neuronal processes: neurogenesis (e.g. MDGA1, MYT1, NTNG1, and SEMA6D genes), synaptogenesis, changes in neuronal circuit formation, cell-junction organization (e.g. PKP4 and SMAD3 genes), and axonal guidance (e.g. NTNG1 and SEMA6D genes) [29]. Moreover, locomotor behaviour (BTBD9, CLN6, HOXB8, and MEIS1 genes) and pathways related to DNA repair and maintenance (APLF, ASTE1, DIS3, PRMT6, and RNF8 genes) were also involved [29].

Hypothesis-based candidate gene studies have been also performed for RLS. In general, these studies included limited samples (with a size below 1000 individuals per study) and analysed a few selected single nucleotide polymorphisms (SNPs) and a variable number of tandem repeats (VNTRs) located in individual candidate genes. In particular, a transcriptome-wide association study (TWAS, which integrated RLS GWAS summary statistics and publicly available genome-wide gene expression data) reported 5 candidate genes in known loci as well as 6 novel genes [33], while another study identified 14 candidate causal genes by testing for a significant burden of rare coding variants using a targeted resequencing approach of 84 genes in known RLS risk loci [34]. The latter study was aimed at verifying the role of rare and low-frequency variants in RLS using the ExomeChip design, but it demonstrated that only common variants representing the known GWAS loci reached genome-wide significance and suggested that the used technology might be avoided in studying RLS genetics [34].

The majority of studies that suggested a role of both common and rare variants in RLS susceptibility have been obtained by studying a cohort population of European ancestry. Some studies indicate a higher susceptibility to RLS of North American and White populations as compared to the Asian and African populations [35]. However, only a few genetic studies are available for Asian individuals. These are mainly studies of small sample size validating single SNP associations in known RLS loci, for example, for BTBD9 and PTPRD [36, 37]. One of them was

Table 1 Association results for lead single-nucleotide polymorphisms reaching genome-wide significance in GWAS meta-analyses [29, 30]

SNP	Chromosome	Candidate genes	*P*-values[a] (combined analysis)	Odds ratio[a] (95% CI) (combined analysis)
rs12046503	1	PRMT6, NTNG1	3.25×10^{-63}	0.88 (0.86–0.89)
rs10208712	2	DCDC2C	1.41×10^{-34}	0.91 (0.90–0.92)
rs113851554	2	MEIS1	2.00×10^{-280}	1.92 (1.85–1.99)
rs1820989	2	MEIS1, C1D, APLF	1.39×10^{-58}	0.88 (0.87–0.90)
rs80319144	2	CCDC148, PKP4, TANC1	2.55×10^{-26}	0.90 (0.88–0.92)
rs1848460	3	CNTN4, CRBN, LRRN1	2.01×10^{-13}	1.07 (1.05–1.10)
rs35987657	3	ATP2C1, ASTE1	3.96×10^{-38}	0.90 (0.89–0.92)
rs10068599	5	RANBP17	6.9×10^{-10}	1.09 (1.06–1.12)
rs17636328	6	RNF8, CCDC167, MDGA1	2.55×10^{-26}	0.90 (0.88–0.92)
rs61192259	6	BTBD9, GLO1	3.58×10^{-202}	1.26 (1.25–1.28)
rs10952927	7	ADAM2, STEAP4, ZNF804B	1.73×10^{-34}	1.13 (1.11–1.15)
rs112716420	7	MICALL2/UNCX	1.5×10^{-18}	1.25 (1.19–1.31)
rs1836229	9	PTPRD	7.36×10^{-42}	0.90 (0.89–0.91)
rs62535767	9	PTPRD	3.23×10^{-9}	0.94 (0.93–0.96)
rs10769894	11	LMO1	9.4×10^{-14}	0.90 (0.88–0.93)
rs340561	13	DACH1, DIS3	3.23×10^{-9}	1.06 (1.04–1.08)
rs996064	15	DPH6, MEIS2	3.39×10^{-27}	1.22 (1.17–1.26)
rs111652004	15	SEMA6D	2.69×10^{-16}	0.86 (0.83–0.89)
rs868036	15	SMAD3, MAP2K5, SKOR1, CLN6	5.48×10^{-69}	0.84 (0.83–0.86)
rs45544231	16	TOX3	7.27×10^{-133}	0.83 (0.81–0.84)
rs12450895	17	HOXBcluster, PRAC1	4.27×10^{-14}	1.08 (1.06–1.10)

(continued)

Table 1 (continued)

SNP	Chromosome	Candidate genes	*P*-values[a] (combined analysis)	Odds ratio[a] (95% CI) (combined analysis)
rs12962305	18	SETBP1	1.11×10^{-7}	1.05 (1.03–1.07)
rs365032	20	MYT1	1.73×10^{-34}	1.13 (1.11–1.15)

[a]The *P*-values and OR are relevant to the combined results of the first metanalysis where the variant was reported

conducted on a Chinese population sample composed by 116 Chinese individuals suffering from RLS and 200 control subjects and pointed to a SNP variant of MAP 2 K5/SKOR1 as a prime cause of RLS in the Chinese population [37]. Another study on a Chinese population reported single-variant and haplotype associations at the BST1 gene coding for the bone marrow stromal cell antigen 1 [38]. A similar hypothesis-driven approach was used in a study on a South Korean sample and identified a positive association of RLS with the circadian rhythm regulator CLOCK gene [39]. Generally, the way the associated genes impact RLS in different racial groups is still unknown, and more data are needed.

To date, the results of targeted studies have not been validated in independent cohorts. Recently, Schormair et al. [40] performed a GWAS-based re-evaluation of published candidate gene studies for idiopathic RLS in populations of European ancestry and reported that none of the published significant associations in candidate gene studies were replicated in their dataset. The authors suggested that the non-replication of the variants could be most likely explained by false-positive associations reported in the candidate gene studies. They also reported the limitations of their study. One of these limits was represented by the lack of large-scale genetic data for populations of non-European ancestry. Therefore, they could not assess the signals reported in two candidate gene studies in Asian populations. A further limitation is the fact that the VNTR variants were not directly measured in their GWAS dataset.

In a recent editorial paper, Akçimen and colleagues [41] highlighted that the currently known variants account for only 20% of the RLS genetic risk, thus remaining uncovered a large fraction of the heritability. The authors suggested three possible causes of this lack of knowledge: the first is that one source of unknown heritability may be intermediate frequency variants with moderate effect sizes; the second may be due to the fact that some variants (e.g. repeat sequences and insertion/deletion) might be difficult to find using technology used to date; the third could be represented by the fact that up to now identified variants through GWAS studies could be likely not the biologically causative variants responsible for the association signal detected [41]. A possible way to overcome these limits would be the massive use of whole-genome sequencing analysis, the inclusion of diverse populations, and biological studies aimed at unravelling the genetic mechanisms leading to RLS. The study of RLS genetics is further complicated by the fact that, being RLS also possibly secondary to a variety of medical conditions (i.e. uraemia, iron deficiency, pregnancy, neuropathy), any genetic or environmental factor that promotes any of these conditions can also be associated with RLS.

Genes Associated to RLS

All genetic findings have led to the development of several animal and cellular models of RLS. In-depth functional studies have mainly focused on the three genes (MEIS1, BTBD9, and PTPRD) with strongest association data in Caucasian populations and on MAP 2K5 that was prevalently correlated with RLS in the Asian population.

MEIS1 Gene

In 2011, Catoire and colleagues reported an association between MEIS1 and RLS. The study was performed using *Caenorhabditis elegans* as model organism and a correlated a decrease in MEIS1 amount with a deficiency of iron [42]. The association of RLS with MEIS1 was subsequently replicated also in humans.

Among GWAS, one of the first studies reporting a direct role of the MEIS1 role as a prime cause of RLS was published in 2018. The study analysed 646 patients affected by insomnia disorder with or without RLS and revealed a strong association of three MEIS1 variants with RLS in all patients [43].

MEIS1, which is now considered the strongest identified genetic risk factor for the disease [44, 45], encodes a homeodomain-containing transcription factor of the three amino acid loop extension (TALE) superclass, which is an important regulator of cellular proliferation and differentiation during development playing a role in the ganglionic eminences of the developing forebrain [46–48]. MEIS proteins are broadly expressed in embryonic development and regulate diverse developmental processes, including eye and neural development, limb and vascular patterning, and hematopoiesis and cardiogenesis [49].

The strongest association with RLS has been reported for an intronic regulatory element within MEIS1. Interestingly, the MEIS1 locus was demonstrated to harbour an array of highly evolutionarily conserved noncoding elements with presumed regulatory function, and several elements in the disease-associated intronic region are active in forebrain neural development [50]. In their recent paper, by dissecting the MEIS1 regulatory region to go deep inside to the involved mechanisms, Lam et al. identified the precise regulatory module, mostly within intron 8 but extending to neighbouring introns, prominently associated with sleep/wake phenotypes [50]. They also reported a higher DNA methylation rate across the MEIS1 locus in neural cells expressing MEIS1.

It was demonstrated that RLS-associated variants of MEIS1 affect gene activity during embryonic development, specifically in the abovementioned ganglionic eminences, from which a diverse population of neurons derives [51]. The hypothesized mechanism underlying RLS is that the change in MEIS1 activity may change the timing of differentiation of basal ganglia cells that could in turn result in differences in the type, number, and localization of cells in the central nervous system, possibly affecting also the sensorimotor circuits thought to be involved in the disease [52]. Since MEIS1 regulation is organized in a modular pattern (i.e. disease-associated intronic regulatory elements control MEIS1 expression with cell type and maturation stage specificity, particularly in the inhibitory neuron lineage), the spatiotemporal activity of these elements may be relevant to RLS pathogenesis.

Several studies have been reported on murine models aimed at highlighting the possible mechanism which correlates RLS with MEIS1. In particular, a knockout (KO) model where both alleles of Meis1 (the mouse orthologue of MEIS1) showed lethality in the embryos [53]. This was not the case of the heterozygous KO model which was then successfully used to study MEIS1-related phenotypes. To date three heterozygous lines of mice have been created and characterized, and all show RLS-like phenotypes.

The first Meis1+/− model, created by knocking in the modified estrogen receptor hormone-binding (ERT2) domain in-frame with the MEIS1a isoform coding region, while disrupting the MEIS1b coding region and the C-terminal of MEIS1 protein [53], showed hyperactivity, impaired prepulse inhibition of startle response, and a trend of reduced delta power in the electroencephalogram [51, 54, 55]. However, this line revealed the expression of a novel Meis1aERT2 fusion protein with a possible toxic gain-of-function effect. A second line was then derived from Meis1 loxP mice [56]. This line resulted to be hyperactive and with increased probability of waking during the light phase but not during the dark phase [57]. Interestingly, these mice did not have changes in serum levels of iron and transferrin while exhibiting increased ferritin levels and an increased dopamine turnover and decreased tyrosine hydroxylase levels in the striatum [58]. A third line of Meis1+/− mice was reported also showing spontaneous motor hyperactivity [59, 60].

In 2017, Thireau et al. [61] demonstrated that one RLS-associated MEIS1 SNP (the rs2300478) in the homozygous state (G/G) modifies the sleep stage-dependent sympathovagal balance in RLS patients. In particular, they showed in RLS patients that the sympathovagal balance tended to shift in favour of a predominant sympathetic activation, associated with reduced vagal influences, being instead the pattern of activity of the parasympathetic nervous system similar between RLS and control subjects. The authors suggested that RLS patients should be thus considered differently depending on their MEIS1 genotype opening new perspectives for a personalized medical approach. The MEIS1 variant might in fact identify people with RLS at risk for cardiovascular disorders [61].

PTPRD Gene

The receptor-type protein tyrosine phosphatase PTPRD, which is also located at an RLS risk locus, encodes a cell adhesion molecule likely involved in neurodevelopment [62]. PTPRD variants have been deeply studied as being attractive candidates to play roles in sleep, locomotor control, and addiction. RLS-associated variants in PTPRD have been demonstrated to modify its expression level in an analysis performed on mRNA extracted from 119 post-mortem brain samples, and a knockout mouse model showed behavioural changes such as increased locomotion and reduced sleep during the sleep onset period [63]. The Ptprd KO mice showed growth retardation, semi-lethality, enhanced hippocampal long-term potentiation, impaired learning and memory, enhanced locomotion, and reduced sleep [63]. The authors suggested that changes in PTPRD expression levels may favour differences in connectivity in dopaminergic systems and consequently contribute to the pathophysiology of RLS [63].

On the other hand, no significant differences in PTPRD expression were detected in the striatum and spinal cord of spontaneously hypertensive rats [64], which have been proposed as a possible model of sleep-related movement disorders [65].

BTBD9 Gene

BTBD9 is a gene ubiquitously expressed [66] encoding a protein that contains a BTB/POZ domain, which is responsible for protein–protein interaction [67]. The mouse orthologue of BTBD9 was the first among the RLS risk genes to be genetically inactivated in animal models. The obtained Btbd9 complete KO mice showed altered hippocampal synaptic transmission and plasticity, with enhanced memory and long-term potentiation [68]. Moreover, these mice showed increased motor activity, decreased dopamine levels, motor restlessness, thermal hypersensitivity, and a disruption in sleep structure disrupted sleep [69–71]. Changes in locomotor behaviour and sleep have been observed also *Drosophila melanogaster* knockout models of BTBD9 [72]. The observed phenotypes in both these models were partially restored by treating with a D2-like receptor agonist [72, 73]. Interestingly, the D2R protein and mRNA levels were reported to be altered in the Btbd9 KO mice in a circadian-dependent manner [74]. Through the use of manganese-enhanced magnetic resonance imaging, an increased neural activity in the cerebral cortex and striatum of the Btbd9 KO mice was demonstrated, indicating an elevated cortical input to the striatum [73]. In the same manuscript, a striatal KO (sKO) murine model was also obtained and characterized showing a hyperactivity consistent with RLS phenotype as well as motor restlessness and a similar circadian predominance to patients [66]. Additional studies on the same sKO model found also a higher sensitivity to the heat stimuli in the tail-flick test, indicating altered thermal sensation similar to complete Btbd9 KO mice [74].

Animal models suggested the existence of a correlation between BTBD9 functions and iron metabolism [75, 76]. These data are supported also by studies on cellular models: in HEK cells a BTBD9 overexpression demonstrated that this protein regulates levels of iron-responsive element-binding protein 2 levels in the cell together with Cullin-3, which control ferritin expression and iron metabolism [70].

Several region-specific KO murine models were generated to study the role of Btbd9. The cerebral cortex-specific Btbd9 KO (cKO) showed a rest phase-specific motor restlessness, a thinner primary somatosensory and primary motor cortex, and decreased thermal sensation, which is opposite to the findings in complete Btbd9 KO mice, while the cholinergic neuron-specific Btbd9 KO (ChKO) mice did not show any sleep and sensory deficits [72, 77]. Dopamine neuron- and Purkinje cell (PC)-specific conditional Btbd9 KO mice have also been generated although they didn't show sensory deficits [73, 78]. The knockout PCs mice showed hyperactivity and a very recent study correlated this phenotype to large conductance Ca2+- activated K+ (BK) channels and small conductance Ca2+-activated K+ (SK) channels [79]. Moreover, the authors highlighted that PC-specific Btbd9 knockout mice exhibited deficits in motor coordination and balance similar to Btbd9 knockout mice, suggesting that the motor effect of BTBD9 in PCs is cell-autonomous [79].

MAP 2K5 Gene

Murine MAP 2K5 KO models were also generated and characterized. The only viable mice resulted to be the heterozygous Map 2k5+/−, being instead lethal the homozygous KO [74]. Viable mice showed decreased levels of Map 2k5 mRNA, of dopamine cell survival, and of tyrosine hydroxylase levels, highlighting the role of MAP 2K5 in dopamine neuron function [80]. A gender-related difference in phenotype was reported for these mice. In particular, males showed impairments in open-field exploration and prepulse inhibition of acoustic startle response, while females showed motor deficits at the rotarod performance test and decreased locomotion during the rest period, thus an opposite change than those observed in most genetic RLS models [80].

Genetic Overlap of RLS and Comorbid Diseases

To date several papers reporting on GWAS studies of different RLS comorbid disease retrieved an overlap of associated loci with RLS.

In the case of insomnia, a strong association with variants in MEIS1 and a significant positive correlation of the overall association signals for RLS and insomnia symptoms [43, 44, 52, 81]. Results obtained in further studies on the association between MEIS1 and insomnia symptoms were consistent with the hypothesis that this association may be due to the presence of a subgroup of individuals in the sample showing comorbid RLS. This hypothesis was also supported by a negative GWAS of the same variants in a population of patients clinically diagnosed with insomnia [37]. In the case of other risk loci shared between RLS and insomnia symptoms, possible pleiotropic effects of the variants and shared pathologic mechanisms were suggested [52, 81]. Additional studies are needed to verify these hypotheses in large and carefully phenotyped cohorts.

Migraine is a highly prevalent and disabling neurological disorder, which is comorbid with a variety of neuropsychiatric disorders, including RLS [82]. Migraine is known to have high heritability, and GWAS have made substantial progress in identifying susceptibility genes for the disease [83–87]. The genetic constituents contributing to RLS in migraineurs remain to be fully explored. A candidate gene approach identified a nucleotide variant at MEIS1 (the rs2300478) increasing the risk of RLS in migraine subjects [88]. More recently, a small-scaled GWAS suggested additional genes which may contribute to RLS in migraineurs [89]. By using a two-stage GWAS, two novel susceptibility genes, VSTM2L and CCDC141, linked to an increased risk of RLS in patients with migraine were also identified [84]. The authors suggested that susceptibility genes for RLS in migraineurs might not be completely the same as those for RLS in general population [90]. Restless leg syndrome showed some positive correlations with neuropsychiatric diseases, whereas correlations were mainly negative with neurodegenerative disorders [29].

Some reports trying to verify the existence of a genetic overlap with RLS are available for attention deficit hyperactivity disorder (ADHD) in children. In particular, in 2008 an association between neuronal nitric oxide synthase (nNOS1/NOS1) genes that generate nitric oxide (NO) and both diseases was reported [91]. Further studies were then performed, testing the association of nNOS and RLS. In the first one, Paulus and colleagues [92] provide evidence for an association of variants in the NOS1 gene and RLS and suggest the involvement of the NO/arginine pathway in the pathogenesis of RLS. This was then supported by the observation that a deficiency or mutation in the NOS1 gene results in RLS and by a case-control study of 918 Caucasian patients with RLS [93, 94]. In 2009, Schimmelmann et al. [95] starting from the hypothesis that RLS and ADHD share common risk genes found no statistically significant associations between three RLS risk genes (MEIS1, BTBD9, and MAP 2K5) and ADHD. Commenting on this paper, Reif and colleagues concluded that a NOS1 variant (the ex1fVNTR) might be the molecular connection between at least some forms of RLS and ADHD and suggested that the feature of (motor) impulsivity might be the underlying neurobiological endophenotype moderated by NOS1 [96]. More recently, no significant association between NOS1 variants and RLS was found in a study on Spanish population [97] and in recent reanalyses on the EU-RLS-GENE GWAS dataset therefore suggesting that most of the reported association in target studies were likely false-positive results [40]. Another potential candidate gene explaining the possible association between having a child with ADHD and risk of RLS is BTBD9. Interestingly, the RLS risk allele of BTBD9 because of the iron deficiency reported for variants in that gene has been suggested as a risk factor of ADHD [98, 99]. Other potential overlapping candidate genes include PTRPD9 [100].

An association has been frequently reported between RLS and PD even though the true nature of it remains unclear. In fact, although RLS and PD co-occur at a rate higher than expected and share several traits such as dopaminergic treatment response, multiple lines of evidence have shown differences between RLS and PD from a pathophysiological perspective. In a recent study using Mendelian randomization and linkage disequilibrium (LD) score regression, a lack of a causal relationship between RLS and PD and a lack of a genetic correlation were demonstrated [101]. This result was consistent with previous genetic studies reporting no association between known RLS-associated variants and PD [102, 103].

Epigenetics and RLS

Moreover, in almost all RLS families with proven or suggested linkage in whom haplotypes were analysed, apparent phenocopies were reported, thus impairing the relevance of some of the found loci [104]. In particular, a phenocopy rate as high as 14% was suggested to occur in RLS families [105]. The high phenocopy rate might be due to a higher incidence of affected individuals among some RLS families. In fact, some papers hypothesized an apparent non-Mendelian pattern of inheritance for RLS because the proportion of those affected (>50%) is higher than would be expected from typical autosomal dominant inheritance. Winkelmann et al. [106] reported an increased frequency of affected relatives for patients with RLS who have a younger age of onset and that evaluating a large subset of RLS families in some generations an affection status of up to 100% could be observed [106, 107]. This apparently non-Mendelian ratio may indicate a more complex genetic architecture underlying familial RLS and raise the suspicion that the disease in some families is transmitted in a non-Mendelian fashion. Zimprich proposed the possible role of epigenetic mechanisms as candidate factors explaining this complexity and a non-Mendelian inheritance pattern of inheritance [104].

Recently, Roy et al. reported a study aimed at developing a genetic diagnostic marker for RLS by means of a whole-genome DNA methylation analysis in four independent cohorts [108]. The authors outlined the importance of epigenetic changes in understanding RLS, wherein both genetic and environmental factors contribute to its development. Results from t-tests showed 19,666 probes (from 13,927 genes) to be differentially methylated in subjects with RLS versus controls, and the biomarker model was able to distinguish RLS cases from controls with accuracies of 90% using lymphocytes and 70% when using neural tissues suggesting a possible diagnostic use of it. Although not yet replicated in younger sample without a previous diagnosis of RLS, this marker appears to be promising. The model is composed by several probes located in genes that are biologically relevant to RLS; it includes also a gene involved in the synthesis of tetrahydrobiopterin, a cofactor for tyrosine hydroxylase in the synthesis of dopamine whose metabolism is dysregulated in RLS [108]. Interestingly, none of the genes previously associated with RLS risk were detected in this study and this could suggest a polygenic nature of RLS.

From Genetics to Neuroimaging of RLS

One of the most fascinating aspects of genetics is how these genes could drive the phenotype. What are the effects on the brain of the genetic mutations associated to RLS? With this in mind, at least three main aspects should be discussed. The genetic RLS profile could result in (i) white matter changes, (ii) gray matter changes, and (iii) white/gray matter changes due to the alterations in the brain iron storage, considering that RLS neuropathology is strictly linked to iron metabolism. It is also interesting to note that, while for the primary RLS there are in literature exhaustive answers, little is known on the neuroimaging of the familial cases. To describe the neuroimaging pattern of primary RLS, however, is relevant for some reasons: (i) de novo genetic mutations may occur in the sporadic forms and contribute to the pathogenesis of the disease; (ii) in the most cases, a positive family history has been recognized. Being said, there is strong evidence about the crucial role of the advanced neuroimaging tools as structural/functional MRI and molecular tracers imaging, in understanding of the pathophysiological mechanisms underlying RLS pathology in the brain.

White Matter Changes in RLS

Although the involvement of white matter (WM) in RLS has been widely debated in the last years, there is evidence of alterations. Several structural MRI studies revealed the presence of widespread WM changes in the sensorimotor cortices and key somatosensory circuits of the patients with RLS (Fig. 1).

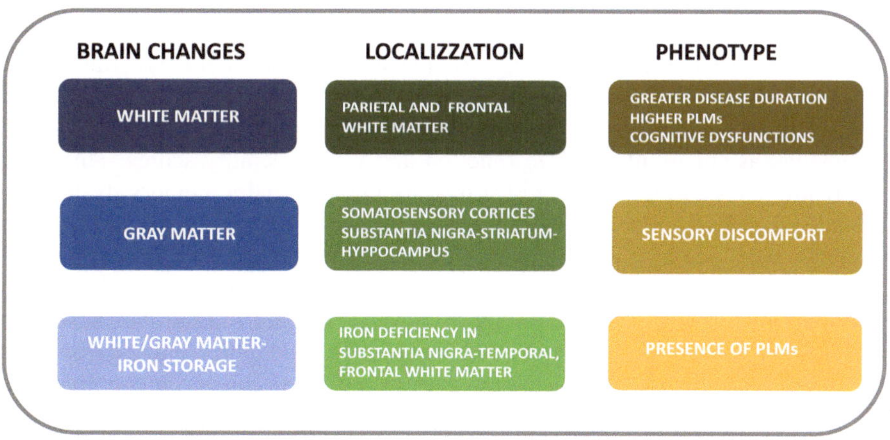

Fig. 1 From brain changes to clinical phenotype of RLS

In a diffusion tensor imaging (DTI) study, a neuroimaging tool used to investigate the WM tract integrity, some authors [109] found significant DTI parameters decrease (especially, fractional anisotropy, FA) in widespread parietal and frontal WM of both hemispheres. These alterations were also associated with clinical variables as the disease duration. Frontal WM microstructural changes, likely due to the loss of axonal density and myelin, have been subsequently detected in RLS patients in comparison with controls [109]. It is also remarkable that the areas reported by both DTI studies [109, 110] are functionally and anatomically connected to motor/premotor and somatosensory cortical areas and for this reason considered part of the motor and the somatosensory network [109]. The findings of Unrath A et al. [109] are solid, since the authors performed a whole-brain approach to investigate the integrity of WM, thus not needing an a priori hypothesis as occurs for ROI-based approach. Overall, their results support the concept that RLS may be considered a sensorimotor disorder of the central nervous system with a focus on the sensorimotor system [109]. In line with this, a recent multimodal imaging study including DTI investigation also revealed that the volume of the WM compartment adjacent to the post- and precentral cortex and DTI-parameters (FA) of the fronto-pontine tract were both significantly reduced in RLS patients as compared to controls [111]. Thus the WM decline of key somatosensory circuits could underlie the perception of sensory leg discomfort that typically occurs in RLS patients [111]. Finally, a combined DTI-VBM study [112] demonstrated that RLS patients may have altered diffusion properties in the right internal capsule, which include the corticospinal tract to the lower limbs, without associated gray matter changes.

To better clarify the pathophysiological mechanisms of RLS, other authors have used the tract-specific statistical analysis (TSSA), a more advanced tool compared to DTI, able to capture local deficits more precisely in anatomical-specific WM tracts [113]. Their protocol also included the assessment of associations between neuroimaging parameters and clinical, polysomnographic, and neuropsychological parameters. TSSA analysis revealed that RLS patients showed altered FA values in multiple major WM tracts, including the left cortico-spinaltract (CST) and cingulum and in the right anterior thalamic radiation (ATR) and inferior fronto-occipital fasciculus (IFO). It is also interesting to note that, CST is a major neural tract known to convey sensorimotor information, as reported by the same authors [113]. In addition, FA scores in the right ATR were positively correlated with neuropsychological parameters as attention/executive function and visual memory scores. This is not surprising, considering that cognitive impairment, especially memory dysfunction has been previously reported in RLS patients [114, 115]. Finally, FA scores in the left CST were negatively correlated with the number of periodic leg movements (PLMs) and movement arousal index [113]. It is possible to speculate that altered integrity apparent in WM tracts might result in abnormal neuronal transmission of sensorimotor output and consequently contribute to main motor and cognitive RLS manifestations [113]. The clinical relevance of this study consists in the identification, for the first time, of the neuroanatomical correlates that can underlie a specific RLS clinical phenotype.

Taken together, these MRI findings suggest that a significant decrease of the diffusion parameters may occur in several WM regions of RLS patients, especially in regions that are a part of sensorimotor networks. A distinct clinical phenotype characterized by greater disease duration, higher number of PLMs, and cognitive dysfunction could be the result of the microstructural WM alterations detected in these patients.

Gray Matter Changes in RLS

Similarly to WM changes, the presence of gray matter (GM) alterations in patients with RLS has been long debated with conflicting evidence. In a recent coordinate-based meta-analysis, some authors have quantitatively evaluated whether there were consistent findings on GM involvement in RLS [116]. According to their main results, obtained by the analysis of voxel-based morphometry (VBM) studies, they were not consistent GM changes in RLS. These findings, however, should be taken with caution, since probably due to several factors differentiating the studies: sample size, demographic characteristics genetic variability, clinical heterogeneity, imaging investigation, and methodological strategies [116].

Recent structural and functional neuroimaging studies put under the spotlight the cortical and subcortical changes in RLS, especially in the motor and somatosensory cortices, striatum, substantia nigra (SN), and thalamus, according to the pathophysiology of the disease (Fig. 1). An analysis of high-resolution three-dimensional MRI has been performed in 63 patients with RLS using an optimized VBM to evaluate potential alterations in the cortical volume. Compared to the controls, RLS patients showed significant decreased GM volume in the primary somatosensory cortices of both hemispheres, extending to the left-sided motor cortex [117]. Although the role of cortical areas in RLS pathophysiology has been widely debated, there is evidence for the involvement of various cortical areas, particularly those that are strictly connected to thalamic nuclei and the sensory cortices. It is a generally accepted view that sensory input is gated by the thalamus and processed in the sensory cortex [117]. This is not surprising, considering the clinical manifestation of RLS with the central symptom of sensory discomfort [117]. Another DTI study also revealed significant regional decreases of GM volume both in cortical and subcortical regions in RLS patients. More in detail, structural alterations that have been detected in certain parts of the brain in RLS patients are relevant to RLS, including the left hippocampal gyrus, both the parietal lobes, the medial frontal areas, and the cerebellum [118]. Finally, in a multimodal imaging study, some authors performed VBM and surface-based morphometry (SBM) to explore the alterations in cortical and subcortical GM volume and resting-state functional connectivity (RSFC) to detect the functional brain alterations in the regions characterized by the microstructural changes [118]. Interestingly, the authors found significant structural and functional changes in the subcortical rather than cortical regions of RLS patients, especially in the bilateral putamen thus suggesting this nucleus as a potential biomarker for the early diagnosis of RLS [119].

An increased scientific interest has been recently focused on the subcortical structures involved in RLS, in particular the striatum and thalamus. Molecular imaging techniques with various binding tracers have been widely applied to the investigation of the functionality of the striatum, basing on the concept that RLS could be caused by a central nervous system dopaminergic (DA) dysregulation. With this in mind, some authors assessed both striatal pre- and postsynaptic DA systems in patients with RLS, by using [123I] beta-CIT and [123I] IBZM SPECT, respectively [120]. Compared to the controls, no difference has been seen in DA transporter in RLS patients [120]. By contrast, a significantly lower binding has been detected in these patients probably due to an increased level of synaptic DA with attendant downregulation of D2 receptors [120]. In disagreement with the first result, a SPECT imaging study using 99mTc-TRODAT-1 SPECT revealed a reduced uptake of the tracer in the striatum profile of RLS patients thus supporting the hypothesis that RLS symptoms could result from a dopaminergic system dysregulation [121]. In agreement with the second result, separate morning and evening PET scans reported changes in striatal dopamine 2 receptor (D2R) binding indicative of an increase in synaptic DA especially in the putamen of RLS patients [122]. Finally, an involvement of medial thalamic nuclei, of a functional nature, has been demonstrated by proton magnetic resonance spectroscopic imaging, thus suggesting that a dysfunction of the limbic system may play a role in the pathophysiology of RLS [123].

In summary, neuroimaging investigation demonstrated that structural and functional changes in cortical and subcortical GM regions may occur in RLS patients. Cortical regions mainly involved were primary somatosensory cortices, whereas the subcortical nuclei were striatum, in particular putamen and thalamus. A dysregulation of central dopaminergic neurotransmission may underlie the pathogenesis of RLS.

White/Gray Changes Due to Brain Iron Storage Alterations

There is evidence for a multiregional brain iron deficiency in patients with RLS. This is not surprising, since iron insufficiency plays a central role in the RLS pathophysiology, especially in the contest of the conditions strictly correlated to altered iron metabolism such as iron deficiency anemia and pregnancy. Thus, several iron-sensitive MRI techniques have been developed over time in order to assess the possible association between brain iron content and RLS neuropathology.

Altered brain iron homeostasis with regional iron deficiency especially in the SN has been confirmed by pathological findings and subsequently detected by imaging study (Fig. 1). On this basis, SN echogenicity has been proposed as a potential morphological biomarker for brain iron status in the early diagnosis of RLS. A transcranial sonography (TCS) study reported that SN echogenicity was significantly decreased in RLS patients as compared to the controls, more commonly in the idiopathic rather than secondary RLS form [124]. In addition, some authors have compared SN echogenicity at baseline and after intravenous (IV) iron therapy [125]. Interestingly, IV iron therapy caused an increased SN echogenicity that was greater in RLS patients with more severe hypoechogenic at baseline [125]. Finally, MRI phase analysis revealed a significantly higher phase, suggestive of reduced iron, in several regions of the brain including not only the SN but also the thalamus, putamen, and pallidum [126]. Although not significant, multiregional iron deficiency has been found in temporal and frontal WM [127]. Little is known about the potential correlations between altered iron storage and different clinical phenotypes in RLS patients. In a recent study, some authors investigated the relation between brain iron concentrations and clinical variables such as disease severity and periodic limb movements during sleep, by using quantitative magnetic susceptibility [128]. Of interest, a significant correlation was found between magnetic susceptibility in SN and the PLMs, indicative of a potential link with this specific RLS phenotype [129]. Finally, it is worth pointing out that there are recent conflicting findings. In a quantitative 3T-MRI study, some authors found that patients with RLS showed increased signals of R2* and quantitative susceptibility in subcortical areas, as compared to the controls [128]. These MRI parameters are highly sensitive to magnetic field variations caused by iron particles. These results did not support the hypothesis of altered brain iron mobilization, and homeostasis may contribute to pathogenesis of RLS [129].

In conclusion, the iron-sensitive MRI techniques revealed that a regional iron deficiency in SN that significantly improved after iron treatment may occur in patients with RLS. A specific RLS clinical phenotype characterized by the presence of PLMs appears to be correlated to the SN iron alterations. SN echogenicity could be considered a morphological biomarker for an early detection of RLS patients.

Conclusions and Future Directions

The most successful of the genetics of the last decades has been the identification of genetic risk factors, such as disease-associated loci, each of them potentially involved in the pathogenesis of RLS. Three genes (MEIS1, BTBD9, and PTPRD) with strongest association data in Caucasian populations and one (MAP 2K5) in Asian population have been identified. In details, MEIS1 gene encodes a transcription factor that plays a role in the ganglionic eminences of the developing forebrain. An interesting hypothesis is that in RLS patients, MEIS1 activity changes the timing of differentiation of basal ganglia cells, thus affecting also the sensorimotor circuits involved in the disease. BTBD9 gene encodes a protein that regulates the levels of iron-responsive element-binding protein 2, thus controlling ferritin expression and iron metabolism. PTPRD gene encodes a cell adhesion molecule likely involved in neurodevelopment, prompting differences in connectivity in dopaminergic systems, and consequently may contribute to the pathophysiology of RLS. MAP

2K5 gene encodes for a protein involved in dopamine neuron function. A gender-related difference in clinical phenotype was reported in animal models. Finally, a possible role of epigenetic mechanisms as candidate factors explaining a non-Mendelian inheritance pattern in RLS has been proposed.

A specific neuroimaging pattern involving WM and GM alterations, often correlated to brain iron storage, has been recognized in patients with primary RLS, associated to specific clinical phenotype. A decrease of the structural parameters in WM regions that are a part of sensorimotor networks, structural, and functional GM changes in primary somatosensory cortices and subcortical nuclei as putamen, thalamus, and a regional iron deficiency in SN may occur in patients with primary RLS. A greater disease duration, higher number of PLMs, and cognitive dysfunction are the salient clinical features in association with these MRI changes.

The lack of further knowledge about the genetic forms of RLS represents, however, an urgent call for geneticists and neurologists in the future vision of a personalized therapeutical strategy both including effective treatments and preventive approaches.

References

1. Scholz SW. Restless legs syndrome: is it all in the genes? Lancet Neurol. 2017;16(11):859–60.
2. Trenkwalder C, Allen R, Högl B, Clemens S, Patton S, Schormair B, Winkelmann J. Comorbidities, treatment, and pathophysiology in restless legs syndrome. Lancet Neurol. 2018;17:994–1005.
3. Guo S, Huang J, Jiang H, Han C, Li J, Xu X, et al. Restless legs syndrome: from pathophysiology to clinical diagnosis and management. Front Aging Neurosci. 2017;9:171.
4. Bassetti CL, Mauerhofer D, Gugger M, Mathis J, Hess CW. Restless legs syndrome: a clinical study of 55 patients. Eur Neurol. 2001;45(2):67–74.
5. Sethi KD. Restless legs syndrome sees the light of day. Lancet Neurol. 2008;7(7):564–5.
6. Deerenberg C, Gould J, Andrews R. A call for better training in restless legs syndrome. Lancet Neurol. 2021;20(9):700–1.
7. Ondo WG, Vuong KD, Wang Q. Restless legs syndrome in monozygotic twins: clinical correlates. Neurology. 2000;55(9):1404–6.
8. Allen RP, La Buda MC, Becker P, Earley CJ. Family history study of the restless legs syndrome. Sleep Med. 2002;3(Suppl):S3–7.
9. Ohayon MM, O'Hara R, Vitiello MV. Epidemiology of restless legs syndrome: a synthesis of the literature. Sleep Med Rev. 2012;16(4):283–95.
10. Allen RP, Picchietti DL, Garcia-Borreguero D, Ondo WG, Walters AS, Winkelman JW, et al. Restless legs syndrome/Willis-Ekbom disease diagnostic criteria: updated International Restless Legs Syndrome Study Group (IRLSSG) consensus criteria – history, rationale, description, and significance. Sleep Med. 2014;15(8):860–73.
11. Jiménez-Jiménez FJ, Alonso-Navarro H, García-Martín E, Agúndez JAG. Genetics of restless legs syndrome: an update. Sleep Med Rev. 2018;39:108–21.
12. Whittom S, Dauvilliers Y, Pennestri MH, Vercauteren F, Molinari N, Petit D, et al. Age-at-onset in restless legs syndrome: a clinical and polysomnographic study. Sleep Med. 2007;9(1):54–9.
13. Winkelmann J. The genetics of restless legs syndrome. Sleep Med. 2002;3(Suppl):S9–12.

14. Wang B, Duanyang L, Jie W, et al. Meta-analysis of single nucleotide polymorphisms in restless legs syndrome risk genes. Chin J Behav Med Brain Sci. 2022;31:187–92.
15. Wang B, Wei W. Research progress on single nucleotide polymorphisms of restless legs syndrome risk genes. Chin J Neurol. 2021;54:1187–93.
16. Trenkwalder C, Allen R, Högl B, Paulus W, Winkelmann J. Restless legs syndrome associated with major diseases: A systematic review and new concept. Neurology. 2016;86(14):1336–43.
17. Akçimen F, Spiegelman D, Dionne-Laporte A, Gan-Or Z, Dion PA, Rouleau GA. Screening of novel restless legs syndrome-associated genes in French-Canadian families. Neurol Genet. 2018;4(6):e296.
18. Weissbach A, Siegesmund K, Brüggemann N, Schmidt A, Kasten M, Pichler I, et al. Exome sequencing in a family with restless legs syndrome. Mov Disord. 2012;27(13):1686–9.
19. Duncan LE, Ostacher M, Ballon J. How genome-wide association studies (GWAS) made traditional candidate gene studies obsolete. Neuropsychopharmacology. 2019;44(9):1518–23.
20. Aridon P, De Fusco M, Winkelmann JW, Zucconi M, Arnao V, Ferini-Strambi L, et al. A TRAPPC6B splicing variant associates to restless legs syndrome. Parkinsonism Relat Disord. 2016;31:135–8.
21. Gan-Or Z, Zhou S, Ambalavanan A, Leblond CS, Xie P, Johnson A, et al. Analysis of functional GLO1 variants in the BTBD9 locus and restless legs syndrome. Sleep Med. 2015;16(9):1151–5.
22. Winkelmann J, Lichtner P, Pütz B, Trenkwalder C, Hauk S, Meitinger T, et al. Evidence for further genetic locus heterogeneity and confirmation of RLS-1 in restless legs syndrome. Mov Disord. 2006;21(1):28–33.
23. Bonati MT, Ferini-Strambi L, Aridon P, Oldani A, Zucconi M, Casari G. Autosomal dominant restless legs syndrome maps on chromosome 14q. Brain. 2003;126(Pt 6):1485–92.
24. Chen S, Ondo WG, Rao S, Li L, Chen Q, Wang Q. Genomewide linkage scan identifies a novel susceptibility locus for restless legs syndrome on chromosome 9p. Am J Hum Genet. 2004;74(5):876–85.
25. Pichler I, Marroni F, Volpato CB, Gusella JF, Klein C, Casari G, et al. Linkage analysis identifies a novel locus for restless legs syndrome on chromosome 2q in a south Tyrolean population isolate. Am J Hum Genet. 2006;79(4):716–23.
26. Levchenko A, Provost S, Montplaisir JY, Xiong L, St-Onge J, Thibodeau P, et al. A novel autosomal dominant restless legs syndrome locus maps to chromosome 20p13. Neurology. 2006;67(5):900–1.
27. Winkelmann J, Schormair B, Lichtner P, Ripke S, Xiong L, Jalilzadeh S, et al. Genome-wide association study of restless legs syndrome identifies common variants in three genomic regions. Nat Genet. 2007;39(8):1000–6.
28. Ergun U, Say B, Ergun SG, Percin FE, Inan L, Kaygisiz S, et al. Genome-wide association and whole exome sequencing studies reveal a novel candidate locus for restless legs syndrome. Eur J Med Genet. 2021;64(4):104186.
29. Schormair B, Zhao C, Bell S, Tilch E, Salminen AV, Pütz B, et al. Identification of novel risk loci for restless legs syndrome in genome-wide association studies in individuals of European ancestry: a meta-analysis. Lancet Neurol. 2017;16(11):898–907.
30. Didriksen M, Nawaz MS, Dowsett J, Bell S, Erikstrup C, Pedersen OB, et al. Large genome-wide association study identifies three novel risk variants for restless legs syndrome. Commun Biol. 2020;3(1):703.
31. Stefansson H, Rye DB, Hicks A, Petursson H, Ingason A, Thorgeirsson TE, et al. A genetic risk factor for periodic limb movements in sleep. N Engl J Med. 2007;357(7):639–47.
32. Winkelmann J, Czamara D, Schormair B, Knauf F, Schulte EC, Trenkwalder C, et al. Genome-wide association study identifies novel restless legs syndrome susceptibility loci on 2p14 and 16q12.1. PLoS Genet. 2011;7(7):e1002171.
33. Akçimen F, Sarayloo F, Liao C, Ross JP, Oliveira RB, Dion PA, et al. Transcriptome-wide association study for restless legs syndrome identifies new susceptibility genes. Commun Biol. 2020;3(1):373.

34. Tilch E, Schormair B, Zhao C, Salminen AV, Antic Nikolic A, Holzknecht E, et al. Identification of Restless Legs Syndrome Genes by Mutational Load Analysis. Ann Neurol. 2020;87(2):184–93.
35. Kutner NG, Zhang R, Huang Y, Bliwise DL. Racial differences in restless legs symptoms and serum ferritin in an incident dialysis patient cohort. Int Urol Nephrol. 2012;44(6):1825–31.
36. Kim TJ, Yoon JE, Park JA, Lee SK, Chu MK, Yang KI, et al. Prevalence and characteristics of restless legs syndrome in Korean adults: A study in two independent samples of the general population. Neuroepidemiology. 2019;52(3–4):193–204.
37. Li G, Tang H, Wang C, Qi X, Chen J, Chen S, et al. Association of BTBD9 and MAP2K5/SKOR1 with restless legs syndrome in Chinese population. Sleep. 2017;40(4) https://doi.org/10.1093/sleep/zsx028.
38. Huang Y, Wang P, Luo Q, Ma J. Association of BST1 polymorphism with idiopathic restless legs syndrome in Chinese population. Sleep Breath. 2021;25(4):1987–93.
39. Seo JE, Yeom JW, Jeon S, Cho CH, Jeong S, Lee HJ. Association between CLOCK gene variants and restless legs syndrome in Koreans. Psychiatry Investig. 2021;18(11):1125–30.
40. Schormair B, Zhao C, Salminen AV, Oexle K, Winkelmann J, Consortium IE-R-G. Reassessment of candidate gene studies for idiopathic restless legs syndrome in a large genome-wide association study dataset of European ancestry. Sleep. 2022;45(8):zsac098.
41. Akçimen F, Dion PA, Rouleau GA. Progress in the genetics of restless legs syndrome: the path ahead in the era of whole-genome sequencing. Sleep. 2022;45(8):zsac136.
42. Catoire H, Dion PA, Xiong L, Amari M, Gaudet R, Girard SL, et al. Restless legs syndrome-associated MEIS1 risk variant influences iron homeostasis. Ann Neurol. 2011;70(1):170–5.
43. El Gewely M, Welman M, Xiong L, Yin S, Catoire H, Rouleau G, et al. Reassessing GWAS findings for the shared genetic basis of insomnia and restless legs syndrome. Sleep. 2018;41(11):zsy164.
44. Sarayloo F, Dion PA, Rouleau GA. Restless legs syndrome: a comprehensive review. Front Neurol. 2019;10:935.
45. Salminen AV, Lam DD, Winkelmann J. Role of MEIS1 in restless legs syndrome: from GWAS to functional studies in mice. Adv Pharmacol. 2019;84:175–84.
46. Bouilloux F, Thireau J, Ventéo S, Farah C, Karam S, Dauvilliers Y, et al. Loss of the transcription factor Meis1 prevents sympathetic neurons target-field innervation and increases susceptibility to sudden cardiac death. elife. 2016;5:e11627.
47. Owa T, Taya S, Miyashita S, Yamashita M, Adachi T, Yamada K, et al. Meis1 coordinates cerebellar granule cell development by regulating Pax6 transcription, BMP signaling and Atoh1 degradation. J Neurosci. 2018;38(5):1277–94.
48. Rataj-Baniowska M, Niewiadomska-Cimicka A, Paschaki M, Szyszka-Niagolov M, Carramolino L, Torres M, et al. Retinoic acid receptor β controls development of Striatonigral projection neurons through FGF-dependent and Meis1-dependent mechanisms. J Neurosci. 2015;35(43):14467–675.
49. Schulte D, Geerts D. MEIS transcription factors in development and disease. Development. 2019;146(16):dev174706.
50. Lam DD, Antic Nikolic A, Zhao C, Mirza-Schreiber N, Krężel W, Oexle K, et al. Intronic elements associated with insomnia and restless legs syndrome exhibit cell-type-specific epigenetic features contributing to MEIS1 regulation. Hum Mol Genet. 2022;1(11):1733–46.
51. Spieler D, Kaffe M, Knauf F, Bessa J, Tena JJ, Giesert F, et al. Restless legs syndrome-associated intronic common variant in Meis1 alters enhancer function in the developing telencephalon. Genome Res. 2014;24(4):592–603.
52. Hammerschlag AR, Stringer S, de Leeuw CA, Sniekers S, Taskesen E, Watanabe K, et al. Genome-wide association analysis of insomnia complaints identifies risk genes and genetic overlap with psychiatric and metabolic traits. Nat Genet. 2017;49(11):1584–92.
53. Azcoitia V, Aracil M, Martínez-A C, Torres M. The homeodomain protein Meis1 is essential for definitive hematopoiesis and vascular patterning in the mouse embryo. Dev Biol. 2005;280(2):307–20.

54. Salminen AV, Schormair B, Flachskamm C, Torres M, Müller-Myhsok B, Kimura M, et al. Sleep disturbance by pramipexole is modified by Meis1 in mice. J Sleep Res. 2018;27(4):e12557.
55. Salminen AV, Garrett L, Schormair B, Rozman J, Giesert F, Niedermeier KM, et al. Effects on motor phenotypes and the sensorimotor system in mice. Dis Model Mech. 2017;10(8):981–91.
56. Kocabas F, Zheng J, Thet S, Copeland NG, Jenkins NA, DeBerardinis RJ, et al. Meis1 regulates the metabolic phenotype and oxidant defense of hematopoietic stem cells. Blood. 2012;120(25):4963–72.
57. Meneely S, Dinkins ML, Kassai M, Lyu S, Liu Y, Lin CT, et al. Differential dopamine D1 and D3 receptor modulation and expression in the spinal cord of two mouse models of restless legs syndrome. Front Behav Neurosci. 2018;12:199.
58. Lyu S, Xing H, Liu Y, Girdhar P, Zhang K, Yokoi F, et al. Deficiency of Meis1, a transcriptional regulator, in mice and worms: neurochemical and behavioral characterizations with implications in the restless legs syndrome. J Neurochem. 2020;155(5):522–37.
59. González-Lázaro M, Roselló-Díez A, Delgado I, Carramolino L, Sanguino MA, Giovinazzo G, et al. Two new targeted alleles for the comprehensive analysis of Meis1 functions in the mouse. Genesis. 2014;52(12):967–75.
60. Cathiard L, Fraulob V, Lam DD, Torres M, Winkelmann J, Krężel W. Investigation of dopaminergic signalling in Meis homeobox 1 (Meis1) deficient mice as an animal model of restless legs syndrome. J Sleep Res. 2021;30(5):e13311.
61. Thireau J, Farah C, Molinari N, Bouilloux F, Torreilles L, Winkelmann J, et al. MEIS1 variant as a determinant of autonomic imbalance in restless legs syndrome. Sci Rep. 2017;7:46620.
62. Uhl GR, Martinez MJ. PTPRD: neurobiology, genetics, and initial pharmacology of a pleiotropic contributor to brain phenotypes. Ann N Y Acad Sci. 2019;1451(1):112–29.
63. Drgonova J, Walther D, Wang KJ, Hartstein GL, Lochte B, Troncoso J, et al. Mouse model for protein tyrosine phosphatase D. Mol Med. 2015;21(1):717–25.
64. Morais MA, Franco B, Holanda ASS, Simino LAP, Manconi M, Torsoni A, et al. PTPRD as a candidate druggable target for therapies for restless legs syndrome? J Sleep Res. 2021;30(4):e13216.
65. Esteves AM, Lopes C, Frussa-Filho R, Frank MK, Cavagnolli D, Arida RM, et al. Spontaneously hypertensive rats: possible animal model of sleep-related movement disorders. J Mot Behav. 2013;45(6):487–93.
66. Fagerberg L, Hallström BM, Oksvold P, Kampf C, Djureinovic D, Odeberg J, et al. Analysis of the human tissue-specific expression by genome-wide integration of transcriptomics and antibody-based proteomics. Mol Cell Proteomics. 2014;13(2):397–406.
67. Stogios PJ, Downs GS, Jauhal JJ, Nandra SK, Privé GG. Sequence and structural analysis of BTB domain proteins. Genome Biol. 2005;6(10):R82.
68. DeAndrade MP, Zhang L, Doroodchi A, Yokoi F, Cheetham CC, Chen HX, et al. Enhanced hippocampal long-term potentiation and fear memory in Btbd9 mutant mice. PLoS One. 2012;7(4):e35518.
69. Freeman A, Pranski E, Miller RD, Radmard S, Bernhard D, Jinnah HA, et al. Sleep fragmentation and motor restlessness in a drosophila model of restless legs syndrome. Curr Biol. 2012;22(12):1142–8.
70. DeAndrade MP, Johnson RL, Unger EL, Zhang L, van Groen T, Gamble KL, et al. Motor restlessness, sleep disturbances, thermal sensory alterations and elevated serum iron levels in Btbd9 mutant mice. Hum Mol Genet. 2012;21(18):3984–92.
71. Gao Z, Wang A, Zhao Y, Zhang X, Yuan X, Li N, et al. Integrative proteome and ubiquitinome analyses reveal the substrates of BTBD9 and its underlying mechanism in sleep regulation. ACS Omega. 2022;7(14):11839–52.
72. Lyu S, Xing H, DeAndrade MP, Perez PD, Zhang K, Liu Y, et al. The role of BTBD9 in the cerebral cortex and the pathogenesis of restless legs syndrome. Exp Neurol. 2020;323:113111.

73. Lyu S, Doroodchi A, Xing H, Sheng Y, DeAndrade MP, Yang Y, et al. BTBD9 and dopaminergic dysfunction in the pathogenesis of restless legs syndrome. Brain Struct Funct. 2020;225(6):1743–60.
74. Garcia Borreguero D, Winkelmann J, Allen RP. Introduction: towards a better understanding of the science of RLS/WED. Sleep Med. 2017;31:1–2.
75. Jellen LC, Beard JL, Jones BC. Systems genetics analysis of iron regulation in the brain. Biochimie. 2009;91(10):1255–9.
76. Muramatsu K, Chikahisa S, Shimizu N, Séi H, Inoue Y. Rotigotine suppresses sleep-related muscle activity augmented by injection of dialysis patients' sera in a mouse model of restless legs syndrome. Sci Rep. 2019;9(1):16344.
77. Silvani A, Ghorayeb I, Manconi M, Li Y, Clemens S. Putative animal models of restless legs syndrome: A systematic review and evaluation of their face and construct validity. Neurotherapeutics. 2022;20:154. https://doi.org/10.1007/s13311-022-01334-4.
78. Lyu S, Xing H, DeAndrade MP, Perez PD, Yokoi F, Febo M, et al. The role of BTBD9 in the cerebellum, sleep-like behaviors and the restless legs syndrome. Neuroscience. 2020;440:85–96.
79. Lyu S, Xing H, Liu Y, Girdhar P, Yokoi F, Li Y. Further studies on the role of BTBD9 in the cerebellum, sleep-like behaviors and the restless legs syndrome. Neuroscience. 2022;505:78–90.
80. Huang Y, Wang P, Morales R, Luo Q, Ma J. Map2k5-deficient mice manifest phenotypes and pathological changes of dopamine deficiency in the central nervous system. Front Aging Neurosci. 2021;13:651638.
81. Jansen PR, Watanabe K, Stringer S, Skene N, Bryois J, Hammerschlag AR, et al. Genome-wide analysis of insomnia in 1,331,010 individuals identifies new risk loci and functional pathways. Nat Genet. 2019;51(3):394–403.
82. Chen PK, Fuh JL, Chen SP, Wang SJ. Association between restless legs syndrome and migraine. J Neurol Neurosurg Psychiatry. 2010;81(5):524–8.
83. Chasman DI, Schürks M, Anttila V, de Vries B, Schminke U, Launer LJ, et al. Genome-wide association study reveals three susceptibility loci for common migraine in the general population. Nat Genet. 2011;43(7):695–8.
84. Freilinger T, Anttila V, de Vries B, Malik R, Kallela M, Terwindt GM, et al. Genome-wide association analysis identifies susceptibility loci for migraine without aura. Nat Genet. 2012;44(7):777–82.
85. Anttila V, Winsvold BS, Gormley P, Kurth T, Bettella F, McMahon G, et al. Genome-wide meta-analysis identifies new susceptibility loci for migraine. Nat Genet. 2013;45(8):912–7.
86. Gormley P, Anttila V, Winsvold BS, Palta P, Esko T, Pers TH, et al. Meta-analysis of 375,000 individuals identifies 38 susceptibility loci for migraine. Nat Genet. 2016;48(8):856–66.
87. Chen SP, Fuh JL, Chung MY, Lin YC, Liao YC, Wang YF, et al. Genome-wide association study identifies novel susceptibility loci for migraine in Han Chinese resided in Taiwan. Cephalalgia. 2018;38(3):466–75.
88. Fuh JL, Chung MY, Yao SC, Chen PK, Liao YC, Hsu CL, et al. Susceptible genes of restless legs syndrome in migraine. Cephalalgia. 2016;36(11):1028–37.
89. Lin GY, Lin YK, Liang CS, Lee JT, Tsai CL, Hung KS, et al. Association of genetic variants in migraineurs with and without restless legs syndrome. Ann Clin Transl Neurol. 2020;7(10):1942–50.
90. Jiang YJ, Fann CS, Fuh JL, Chung MY, Huang HY, Chu KC, et al. Genome-wide analysis identified novel susceptible genes of restless legs syndrome in migraineurs. J Headache Pain. 2022;23(1):39.
91. Gao Y, Heldt SA. Lack of neuronal nitric oxide synthase results in attention deficit hyperactivity disorder-like behaviors in mice. Behav Neurosci. 2015;129(1):50–61.
92. Paulus W, Dowling P, Rijsman R, Stiasny-Kolster K, Trenkwalder C. Update of the pathophysiology of the restless-legs-syndrome. Mov Disord. 2007;22(Suppl 18):S431–9.
93. Winkelmann J. Genetics of restless legs syndrome. Curr Neurol Neurosci Rep. 2008;8(3):211–6.

94. Winkelmann J, Lichtner P, Schormair B, Uhr M, Hauk S, Stiasny-Kolster K, et al. Variants in the neuronal nitric oxide synthase (nNOS, NOS1) gene are associated with restless legs syndrome. Mov Disord. 2008;23(3):350–8.
95. Schimmelmann BG, Friedel S, Nguyen TT, Sauer S, Ganz Vogel CI, Konrad K, et al. Exploring the genetic link between RLS and ADHD. J Psychiatr Res. 2009;43(10):941–5.
96. Reif A. Is NOS1 a genetic link between RLS and ADHD? J Psychiatr Res. 2010;44(1):60–1.
97. Jiménez-Jiménez FJ, Alonso-Navarro H, Martínez C, Zurdo M, Turpín-Fenoll L, Millán-Pascual J, et al. Neuronal nitric oxide synthase (nNOS, NOS1) rs693534 and rs7977109 variants and risk for restless legs syndrome. J Neural Transm (Vienna). 2015;122(6):819–23.
98. Oner P, Dirik EB, Taner Y, Caykoylu A, Anlar OA. Sociation between low serum ferritin and restless legs syndrome in patients with attention deficit hyperactivity disorder. Tohoku J Exp Med. 2007;213(3):269–76.
99. Konofal E, Cortese S, Marchand M, Mouren MC, Arnulf I, Lecendreux M. Impact of restless legs syndrome and iron deficiency on attention-deficit/hyperactivity disorder in children. Sleep Med. 2007;8(7–8):711–5.
100. Elia J, Gai X, Xie HM, Perin JC, Geiger E, Glessner JT, et al. Rare structural variants found in attention-deficit hyperactivity disorder are preferentially associated with neurodevelopmental genes. Mol Psychiatry. 2010;15(6):637–46.
101. Estiar MA, Senkevich K, Yu E, Varghaei P, Krohn L, Bandres-Ciga S, et al. Lack of causal effects or genetic correlation between restless legs syndrome and Parkinson's disease. Mov Disord. 2021;36(8):1967–72.
102. Vilariño-Güell C, Soto AI, Young JE, Lin SC, Uitti RJ, Wszolek ZK, et al. Susceptibility genes for restless legs syndrome are not associated with Parkinson disease. Neurology. 2008;71(3):222–3.
103. Gan-Or Z, Alcalay RN, Bar-Shira A, Leblond CS, Postuma RB, Ben-Shachar S, et al. Genetic markers of restless legs syndrome in Parkinson disease. Parkinsonism Relat Disord. 2015;21(6):582–5.
104. Zimprich A. Phenocopies in families with essential tremor and restless legs syndrome challenge Mendelian laws. Epigenetics might provide answers. Parkinsonism Relat Disord. 2012;18(6):711–6.
105. Mathias RA, Hening W, Washburn M, Allen RP, Lesage S, Wilson AF, et al. Segregation analysis of restless legs syndrome: possible evidence for a major gene in a family study using blinded diagnoses. Hum Hered. 2006;62(3):157–64.
106. Winkelmann J, Müller-Myhsok B. Genetics of restless legs syndrome: a burning urge to move. Neurology. 2008;70(9):664–5.
107. Winkelmann J, Polo O, Provini F, Nevsimalova S, Kemlink D, Sonka K, et al. Genetics of restless legs syndrome (RLS): state-of-the-art and future directions. Mov Disord. 2007;22(Suppl 18):S449–58.
108. Roy A, Earley CJ, Allen RP, Kaminsky ZA. Developing a biomarker for restless leg syndrome using genome wide DNA methylation data. Sleep Med. 2021;78:120–7.
109. Unrath A, Müller H-P, Ludolph AC, Riecker A, Kassubek J. Cerebral white matter alterations in idiopathic restless legs syndrome, as measured by diffusion tensor imaging. Mov Disord. 2008;23(9):1250–5.
110. Chang Y, Paik JS, Lee HJ, Chang HW, Moon HJ, Allen RP, Earley CJ, Cho YW. Altered white matter integrity in primary restless legs syndrome patients: diffusion tensor imaging study. Neurol Res. 2014;36(8):769–74.
111. Stefani A, Mitterling T, Heidbreder A, Steiger R, Kremser C, Frauscher B, et al. Multimodal magnetic resonance imaging reveals alterations of sensorimotor circuits in restless legs syndrome. Sleep. 2019;42(12):zsz171.
112. Belke M, Johannes T, Heverhagen JT, Keil B, Rosenow F, Oertel WH, Karin Stiasny-Kolster K, et al. DTI and VBM reveal white matter changes without associated gray matter changes in patients with idiopathic restless legs syndrome. Brain Behav. 2015;5(9):e00327.

113. Park HR, Kim HR, Seunghwan OS, Seong JK, Joo EY. White matter tract-specific alterations in patients with primary restless legs syndrome. Sci Rep. 2021;11(1):16116.
114. Galbiati A, Marelli S, Giora E, Zucconi M, Oldani A, Ferini-Strambi L. Neurocognitive function in patients with idiopathic restless legs syndrome before and after treatment with dopamine-agonist. Int J Psychophysiol. 2015;95:304–9.
115. Kim SM, Choi JW, Lee C, Lee BU, Koo YS, Kim KH, Jung KY. Working memory deficit in patients with restless legs syndrome: an event-related potential study. Sleep Med. 2014;15:808–15.
116. Sheng L, Zhao P, Ma H, Qi L, Yi Z, Shi Y, Zhong J, et al. Grey matter alterations in restless legs syndrome: A coordinate-based meta-analysis. J Sleep Res. 2021;30(5):e13298.
117. Unrath A, Juengling FD, Schork M, Kassubek J. Cortical grey matter alterations in idiopathic restless legs syndrome: an optimized voxel-based morphometry study. Mov Disord. 2007;22(12):1751–6.
118. Chang Y, Won Chang H, Song H, Ku J, Earley CJ, Allen RP, et al. Gray matter alteration in patients with restless legs syndrome: a voxel-based morphometry study. Clin Imaging. 2015;39(1):20–5.
119. Li T, Liu C, Lyu H, Xu Z, Hu Q, Xu B, et al. Alterations of sub-cortical gray matter volume and their associations with disease duration in patients with restless legs syndrome. Front Neurol. 2018;9:1098.
120. Michaud M, Soucy JP, Chabli A, Lavigne G, Montplaisir J. SPECT imaging of striatal pre- and postsynaptic dopaminergic status in restless legs syndrome with periodic leg movements in sleep. J Neurol. 2002;249(2):164–70.
121. Lin CC, Fan YM, Lin G-Y, Yang FC, Cheng CA, Lu KC, et al. 99mTc-TRODAT-1 SPECT as a potential neuroimaging biomarker in patients with restless legs syndrome. Clin Nucl Med. 2016;41(1):e14–7.
122. Earley JC, Kuwabara H, Wong DF, Gamaldo C, Salas ER, Brašić JR, et al. Increased synaptic dopamine in the putamen in restless legs syndrome. Sleep. 2013;36(1):51–7.
123. Rizzo G, Tonon C, Testa C, Manners D, Vetrugno R, Pizza F, et al. Abnormal medial thalamic metabolism in patients with idiopathic restless legs syndrome. Brain. 2012;135:3712–20.
124. Godau J, Schweitzer KJ, Liepelt I, Gerloff C, Berg D. Substantia nigra hypoechogenicity: definition and findings in restless legs syndrome. Mov Disord. 2007;22(2):187–92.
125. Garcia-Malo C, Wanner V, Miranda C, Peralta SR, Agudelo L, Cano-Pumarega I, et al. Quantitative transcranial sonography of the substantia nigra as a predictor of therapeutic response to intravenous iron therapy in restless legs syndrome. Sleep Med. 2020;66:123–9.
126. Rizzo G, Manners D, Testa C, Tonon C, Vetrugno R, Marconi S, et al. Low brain iron content in idiopathic restless legs syndrome patients detected by phase imaging. Mov Disord. 2013;2013(28):1886–90.
127. Godau J, Klose U, Di Santo A, Schweitzer K, Berg D. Multiregional brain iron deficiency in restless legs syndrome. Mov Disord. 2008;23(8):1184–7.
128. Li X, Allen RP, Earley CJ, Liu H, Cruz TE, Edden RAE, et al. Brain iron deficiency in idiopathic restless legs syndrome measured by quantitative magnetic susceptibility at 7 tesla. Sleep Med. 2016;22:75–82.
129. Beliveau V, Stefani A, Birkl C, Kremser C, Gizewski ER, Högl B, et al. Revisiting brain iron deficiency in restless legs syndrome using magnetic resonance imaging. Neuroimage Clin. 2022;34:103024.

Part III
Other Topics

Circadian Regulation of Sleep: From Genes to Circuits

Naureen A. Hameed and Annika F. Barber

Abstract The circadian clock is a genetically controlled transcriptional timing system that drives approximately 24-h rhythms in gene expression and consequently in physiology and behavior. At the molecular level, the circadian clock consists of a cell-autonomous transcription-translation feedback loop that maintains a 24-h cycle of gene expression. A master clock in the brain serves as a pacemaker for molecular clocks throughout the body to coordinate circadian rhythms with respect to one another and the external environment. Sleep is a circadian behavior, though sleep is coordinately regulated by both circadian and homeostatic processes. The interplay between brain clocks, clock output signals such as core body temperature and melatonin signaling, and sleep regulation are complex, and while research reveals some mechanistic underpinnings, other aspects require further research. Additional environmental factors regulating the clock and consequently sleep, include aging, circadian desynchrony, and immune activation. Finally, the circadian nature of so many physiological properties offers unique opportunities for chronotherapy for sleep and circadian disorders.

Keywords Chronobiology · Circadian rhythm · Clock genes · Suprachiasmatic nucleus · Entrainment

N. A. Hameed
Molecular Biosciences Graduate Program, Rutgers, the State University of New Jersey, New Brunswick, NJ, USA

A. F. Barber (✉)
Waksman Institute, Department of Molecular Biology and Biochemistry, Rutgers, the State University of New Jersey, New Brunswick, NJ, USA
e-mail: annika.barber@waksman.rutgers.edu

P. Gehrman et al. (eds.), *Genetics of Sleep and Sleep Disorders*,
https://doi.org/10.1007/978-3-031-62723-1_14

Definition of a Genetic Clock

The Three Rules of Clocks

The circadian clock is a genetically controlled transcriptional timing system that drives approximately 24-h rhythms in gene expression and consequently in physiology and behavior. These rhythms are termed circadian rhythms, from the Latin *circa diem*, for about a day. Life on earth evolved in a world in which many environmental factors vary with the 24-h day/night cycle. The circadian clock confers substantial fitness advantages by allowing organisms to anticipate these regularly occurring environmental changes and by allowing temporal coordination of internal physiological processes. The fitness advantage of circadian rhythms is reflected in the conservation of circadian clocks across phylogeny from yeast and bacteria to plants, insects, and mammals.

The circadian clock has three essential features: first, the clock keeps running with approximately 24-h periodicity even when environmental conditions are constant. Second, the circadian clock can align its phase, that is, the stage of the molecular clock cycle, to environmental cues, which are termed zeitgebers. Zeitgebers can both advance and delay the phase of the clock, keeping it in alignment with the solar day throughout seasonal changes. In natural conditions, this allows the circadian clock to appropriately time physiological processes and behaviors to the optimal circadian time. However, in modern society many people experience mistimed exposure to zeitgebers such as light and food, resulting in detrimental effects on health. Finally, the circadian clock can maintain 24-h rhythms across the full range of temperatures an organism may experience.

A Brief History of Chronobiology

Circadian rhythms are so fundamental to the physiology and behavior of many organisms that their research has a long history. As early as 1729, French scientist Jean-Jacques d'Ortous de Mairan noted that the daily opening and nightly closing of mimosa plant leaves persisted in constant darkness and tentatively postulated that plants may have some internal timekeeping capabilities. "Modern" chronobiology took off in the 1950s as researchers conducted studies in plants, fruit flies, and humans that formalized the properties of biological clocks. The genetic basis of the clock was first uncovered via forward genetic screens in *Drosophila* that identified the first clock gene *period* (*per*) in 1971. Further work in *Drosophila* identified additional components of a 24-h transcriptional feedback loop, for which the 2017 Nobel Prize in Physiology and Medicine was awarded to Jeffrey Hall, Michael Rosbash, and Michael Young. The first mammalian clock gene, *Clock*, was identified in 1997. In the intervening years, many additional clock genes and mechanisms have been uncovered across model organisms. In this chapter we will describe the fundamental principles of core clock transcription-translation feedback loop in the brain and peripheral tissues, with a focus on the role of clocks in regulating behavior and physiology in health and disease.

Circadian Regulation of Physiology and Behavior

Because molecular circadian clocks are present in nearly every tissue and cell type in animals, a vast array of physiological processes are clock-regulated. This reflects the two major advantages conferred by circadian function: anticipation of environmental changes and coordination of internal processes. Many circadian processes interact with and reinforce one another, such as a drop in body temperature preceding sleep or a rise in cortisol prior to wake, allowing complementary processes to co-occur, while antagonistic processes are temporally separated. Though circadian physiological outputs are nearly ubiquitous, here we highlight a few outputs that intersect with sleep-regulated processes.

Sleep-Wake

The sleep-wake cycle is the most obvious circadian rhythm that comes to mind for most people. As discussed elsewhere in this volume, sleep is regulated by both the circadian and sleep homeostatic processes. The so-called two-process model of sleep regulation posits that sleep homeostatic pressure increases during wakefulness and declines during sleep, and the thresholds that drive the transitions between sleep and wake oscillate with a circadian periodicity defined by the circadian clock. In the original formulation of this model, sleep homeostatic (Process S) and circadian (Process C) processes were posited to be independent; however, further work has demonstrated complex and bidirectional regulatory interactions between sleep pressure and circadian rhythms [31, 32]. A notable demonstration of the clock's influence on sleep is that knockout of molecular clock genes in rodents and mutations in core clock genes in humans not only impair the clock itself but also compromise sleep homeostasis [64, 90]. In turn, sleep deprivation alters clock gene expression in the brain and can attenuate the response of the clock to light-induced phase shifts [39, 45, 195]. The genetic and neurological underpinnings of the circadian regulation of sleep-wake cycles remain elusive. Work in model organisms suggests roles for direct projections from the brain clock to sleep regulatory brain areas, as well as indirect regulation of sleep through circadian regulation of circulating hormones, such as melatonin and cortisol, discussed later.

Core Body Temperature

Rhythms in human body temperature were recorded by physicians in the 1840s, and circadian rhythms in body temperature are noted across homeothermic species [214]. Circadian fluctuation of core body temperature (CBT) occurs through circadian regulation of metabolic processes that generate heat and regulation of processes such as vasodilation that dissipate heat. While the range of temperature variation is small, it is an entrainable rhythm with a ~24-h free-running period. In healthy humans, CBT increases during the daytime and decreases at night. Temperature regulates sleep initiation, with sleep onset occurring around the time when body temperature is at its maximum rate of decline [42]. The vasodilation that occurs during the cooling phase of the CBT rhythm is also a good predictor of sleep initiation [162]. Sleep states also regulate body temperature, with warming during rapid eye movement (REM) sleep and cooling during non-rapid eye movement (NREM) sleep, with longer bouts of NREM sleep resulting in a greater degree of cooling [58, 119]. Sleep deprivation also results in dysregulation of core body temperature, possibly due to changing metabolic requirements. It seems likely that in physiological conditions, circadian and sleep-driven regulations of core body temperature are synergistic, with circadian fluctuation in CBT promoting vasodilation and sleep initiation at the optimal time and NREM sleep further reducing CBT. Environmental disruptions such as immune activation, sleep deprivation, and nutritional stress can disrupt this synergistic regulation of CBT, with consequent feedback effects on both sleep and CBT itself. See Table 1 for a list of selected abbreviations used throughout this chapter.

Table 1 List of selected abbreviations used

Other abbreviations	
bHLH	Basic helix-loop-helix
CBT	Core body temperature
CCG	Clock-controlled gene
CDK5	Cyclin dependent kinase 5
CREB	Cyclic AMP-responsive element (CRE)-binding protein
CRTC1	CREB-regulated transcription coactivator 1
EEG	Electroencephalogram
FASPS	Familial advanced sleep phase syndrome
FDSPS	Familial delayed sleep phase syndrome
HDAC	Histone deacetylase
NREM	Non-rapid eye movement sleep
PAS	PER-ARNT-SIM domain
PTM	Posttranslational modification
REM	Rapid eye movement sleep
SIK1	Salt-inducible kinase 1
TTFL	Transcription-translation feedback loop

Immune Function

Lymphoid organs and immune cells all express the molecular clock [140, 262], and many aspects of immune function are under circadian control, including cytokine and chemokine production, immune cell trafficking, activation of both innate and adaptive immunity, and host-pathogen interactions (reviewed in [251]). Experiments as early as the 1960s showed that mouse susceptibility to endotoxin administration depends on the time of day that the toxin is administered [104]. More recent work has demonstrated specific roles for multiple molecular clock components in circadian control of immune function. Given the widespread circadian control of immunity, it is not surprising that chronic circadian misalignment may have detrimental effects on immune function; retrospective human studies have demonstrated that night-shift work is associated with elevated levels of inflammatory markers [12, 230]. In turn, infection can result in loss of circadian control of immune function, resulting in a shift toward a pro-inflammatory state [78, 109].

It seems likely that sleep, clocks, and the immune system are tightly interlinked, and indeed many of the same aspects of immune function are regulated by both the clock and by sleep.

Sleep and immunity are bidirectionally linked, with sleep-regulating immunity and immune activation altering sleep architecture. Sleep deprivation is linked to increased susceptibility to infection, and chronic short sleep is associated with reduced lifespan, inflammatory cytokine production, and increased vulnerability to viral infection [43, 84, 85, 203, 227]. Prolonged waking is associated with increases in circulating immune cells, which decrease with recovery sleep, suggesting a role for sleep in regulation of immune cell trafficking [29, 127]. In turn, immune substances are involved in regulation of sleep, and consequently immune activation can result in changes in sleep architecture. Many pathogens increase the amount and intensity of NREM sleep while decreasing REM sleep, though the mediators of these effects are complex and pathogen-specific [29, 183]. Future work examining the convergence of sleep and circadian regulation of immunity may identify new avenues for improving the health of at-risk populations such as shift workers, as well as identify new approaches to disease treatment that consider circadian time as a factor.

Feeding and Metabolism

Though circadian regulation is only one factor regulating meal timing and metabolism, circadian rhythms across tissues optimize energy homeostasis by temporally coordinating opposing and cooperative metabolic processes. Circadian rhythmicity of feeding has been noted for many species from lab model organisms to humans [218, 250]. The circadian clock regulates multiple metabolic processes, which results in diurnal rhythms in energy expenditure, appetite, insulin sensitivity, and lipid and glucose metabolism [44, 161, 211, 226, 245, 266, 290].

In the face of metabolic challenges, animals must reorganize sleep patterns to optimize food intake, likely via changes in both homeostatic and circadian regulations of sleep. Circadian rhythms, metabolism, and sleep are tightly interconnected, though the underlying biology is not yet fully understood. Genetic or environmental disruption of circadian rhythms and chronic sleep deficiency are associated with metabolic disorders including obesity and diabetes [66]. Because circadian disruption often results in reduced sleep time or quality, it can be difficult to distinguish effects of sleep loss vs. circadian misalignment. However, some direct effects of circadian misalignment include insulin resistance, impaired glucose tolerance, and increased inflammatory markers, which may compound with reduced sleep to drive metabolic disease [184, 185, 286].

Conservation of Clock Architecture Across Phylogeny

Input-Oscillator-Output Model

Although the genetic mechanisms vary, across organisms and tissues, circadian clocks follow a three-part structure consisting of environmental input pathways, a central oscillator, and outputs from the central oscillator (Fig. 1a). Environmental input pathways allow the clock to synchronize circadian phase and period to environmental cues, the central oscillator generates the endogenous circadian cycle, and the output pathways couple rhythmic gene expression to biological processes.

In single-celled organisms such as cyanobacteria, this scheme is very simple, with direct entrainment of the clock in each cell driving transcriptional oscillations. In multicellular organisms such as fungi and plants, oscillators in each cell or tissue remain relatively independent, albeit with some coupling between clocks within a tissue. In animals, the central oscillator is in the brain, and it coordinates the timing of molecular clocks in other tissues. This “central clock” in the brain has input pathways from the eyes and other sensory organs that allow entrainment to the external environment and output pathways that communicate time to the rest of the organism (Fig. 1b). These outputs include neuronal efferent pathways from the brain clock, secreted factors that regulate distal tissue physiology, and cues that entrain peripheral molecular clocks located in nearly every cell type. While brain clocks across organisms are primarily entrained by light, additional input pathways to clocks in peripheral tissues permit entrainment by additional zeitgebers including food, exercise, stress, and more. The final circadian outputs include the rhythmic physiology and behavior that we can quantify, such as sleep-wake cycles, daily locomotor activity, food intake, body temperature, and more.

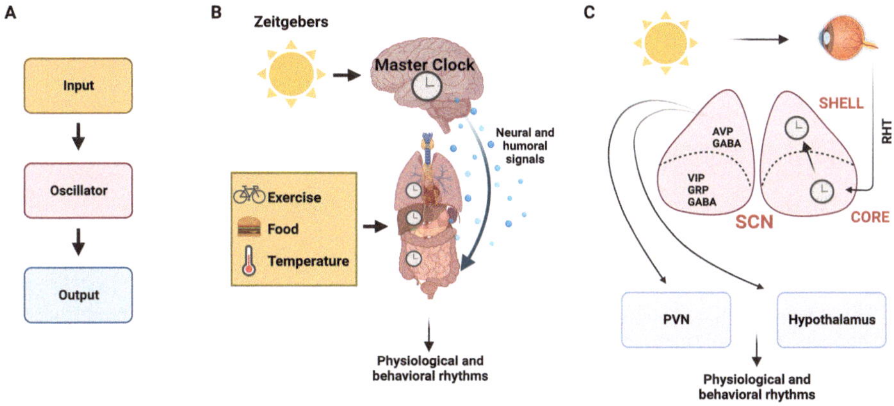

Fig. 1 The structure of circadian systems. (**a**) The input-oscillator-output structure of circadian timekeeping is conserved from the molecular clock to the brain clock to entrainment of clocks throughout the body. Inputs (orange), oscillators (pink), and outputs (blue) are shaded accordingly in panels b and c. (**b**) Environmental light is the primary input to the master clock in the brain, which provides outputs in the form of neuronal and humoral signals. Additional inputs to peripheral clocks, such as exercise, food, and temperature, contribute to peripheral clock timing. Together, the activity of central and peripheral clocks coordinates physiological and behavioral rhythms. (**c**) Light provides input to the SCN via the retinohypothalamic tract. Light inputs enter the core and are subsequently relayed to the shell and then to additional brain areas including the PVN and hypothalamus. Major neurotransmitters of the SCN core and shell are labeled

Transcription-Translation Feedback Loop Structure

The core molecular clock in all organisms consists of two interlinked transcription-translation feedback loops (TTFL) that take approximately 24 h to complete. At the center of the two loops are transcriptional activators, which are regulated at the protein and transcript levels. In the core clock negative feedback loop, transcriptional activators drive expression of protein products that reenter the nucleus to inhibit the transcriptional activators (Fig. 2a). In the secondary feedback loop, the transcriptional activators drive expression of additional gene regulatory proteins that regulate transcription of the activators. Additional cytoplasmic regulatory factors control stability and nuclear entry of molecular clock components to maintain the period at ~24 h. While the essential architecture of the TTFL is conserved across species, the genes and proteins involved are very different when comparing clock genes between domains of life. For comparison, a simplified diagram of the molecular clock loops in flies (B) and mammals (C) is depicted in Fig. 2. We will further discuss molecular clock components in the next section.

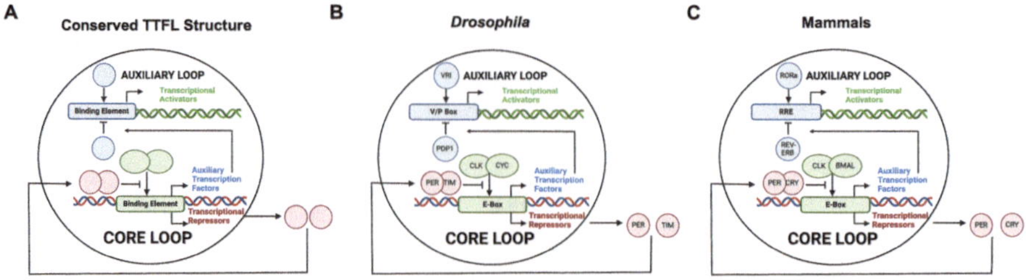

Fig. 2 Conservation of the clock transcription-translation feedback loop structure. (**a**) The clock TTFL consists of a core loop and an auxiliary loop whose structure is highly conserved from flies to humans. The core repressive loop of the clock consists of negative regulators (red circles) whose transcription is initiated by positive regulators (green ovals) that bind to a conserved upstream binding element. These negative regulators are translated, undergo cytosolic posttranslational modification and dimerization (outside the black circle), and reenter the nucleus to repress their own transcription. The auxiliary loop consists of both positive and negative transcriptional regulators (blue circles) of the transcriptional activators. (**b**) In flies, the negative regulators of the core loop are PERIOD and TIMELESS, while the positive regulators are CLOCK and CYCLE which bind to an E-box to initiate repressor transcription. The auxiliary regulators are PDP1 and VRILLE. (**c**) In mammals, the negative regulators of the core loop are PERIOD and CRYPTOCHROME, while the positive regulators are CLOCK and BMAL which bind to an E-box to initiate repressor transcription. The auxiliary regulators are RORa/b and REV/ERB. Colors in (**b**) and (**c**) match those shown in (**a**) to highlight homologous elements

The Core Molecular Clock

Discovery of Repressive Elements of the Molecular Clock in Flies

In 1971, Konopka and Benzer identified the first single-gene mutant that altered a complex behavior, ushering in the era of behavioral genetics [273]. Konopka and Benzer used a forward genetics approach to screen mutagenized flies for altered daily rhythms in emergence from their pupal case, ultimately finding three mutants with no periodicity, long period, and short periods of daily emergence [154]. These same mutants also showed arrhythmicity, long, and short periods in circadian locomotor behavior in constant conditions. Complementation assays determined that all mutations were in the same gene, which was named *period.* The field of molecular genetics took some time to catch up, and the period gene was not cloned until 1984 [21, 22, 232, 320].

After the identification of period, further screens from the Young, Hall, and Rosbash labs identified additional core molecular clock components in *Drosophila* [237, 253]. Molecular biology assays identified rhythmic gene transcription and nuclear localization of the repressors PERIOD and TIMELESS, leading to the hypothesis of a negative transcription-translation feedback loop (TTFL) driving circadian rhythms. These circadian Drosophilists were awarded the Nobel Prize for

Physiology or Medicine in 2017 for discovery of core molecular clock components and function. Additional components of the fly molecular clock allowed further elucidation of circadian timekeeping mechanisms in the fly throughout the 1990s. See Table 2 for a list of acronyms for key core clock components.

Discovery of the Core Molecular Clock Feedback Loop

The 1988 discovery of a spontaneous short-period mutant in hamsters, named *Tau*, facilitated experiments demonstrating the location of the molecular clock in the mammalian brain [231]. However, the lack of molecular markers in hamsters delayed the cloning and molecular of the Tau gene until 2000 [179]. In the interim, the Takahashi lab leaned on the forward genetic approach that had successfully identified clock genes in flies to screen mutagenized mice. This screen yielded the first molecular identification of a mammalian clock gene, and the first positive regulator of the TTFL, *Clock* (circadian locomotor output cycles kaput) [15, 146, 294]. With the Clock gene in hand, a yeast two-hybrid screen for interaction partners identified brain and muscle ARNT-Like 1 (Bmal1), which forms heterodimers with CLOCK at E-box sites and drives *per* transcriptional oscillation [95]. The *Clock* homolog in *Drosophila* was identified shortly thereafter, and the discovery that PER and TIM inhibit CLOCK activity closed the circadian feedback loop [61]. Synthesizing discoveries in flies and mammals, researchers assembled the core TTFL model of the molecular clock in which a transcriptional activator dimer (CLK and CYC in *Drosophila,* CLOCK and BMAL1 in mammals) drives expression of their own negative regulators (PER and TIM in *Drosophila* and PER1–3 and CRYPTOCHROME (CRY) 1–2 in mammals). This core TTFL structure is highly conserved from flies to mammals, albeit with more complexity due to the existence of paralogs for several core clock genes [10, 61].

Table 2 List of acronyms for key core clock components

BMAL	Brain and muscle ARNT-like
CLK	CLOCK
CRY	CRYPTOCHROME
CYC	CYCLE
NFIL3	Nuclear factor, interleukin 3 regulated
PER	PERIOD
ROR	Retinoic acid-related orphan receptor
TIM	TIMELESS
DBP	D-box protein

The Mammalian Core Clock Transcription-Translation Feedback Loop

In mammals, the forward arm of the TTFL consists of the transcription factors CLOCK and BMAL1 that heterodimerize and bind enhancer-box (E-box) elements to promote transcription of *per, cry*, and other clock-controlled genes (Fig. 3(1)) [10, 22, 205, 237, 253, 320]. The CLOCK-BMAL1 heterodimer consists of two basic helix-loop-helix (bHLH) domain protein subunits followed by the PER-ARNT-SIM (PAS) domains [118]. This is a highly conserved family of transcription factors; the bHLH serves as the DNA binding site, while the PAS domains are more variable and confer specific sites for transactivation/repression [151]. Once transcribed, *per* and *cry* are translated and translocated into the nucleus, where they heterodimerize and reenter the nucleus to inhibit their own transcription (Fig. 3(4, 5)) [118]. This transcription, translation, dimerization, and even translocation are

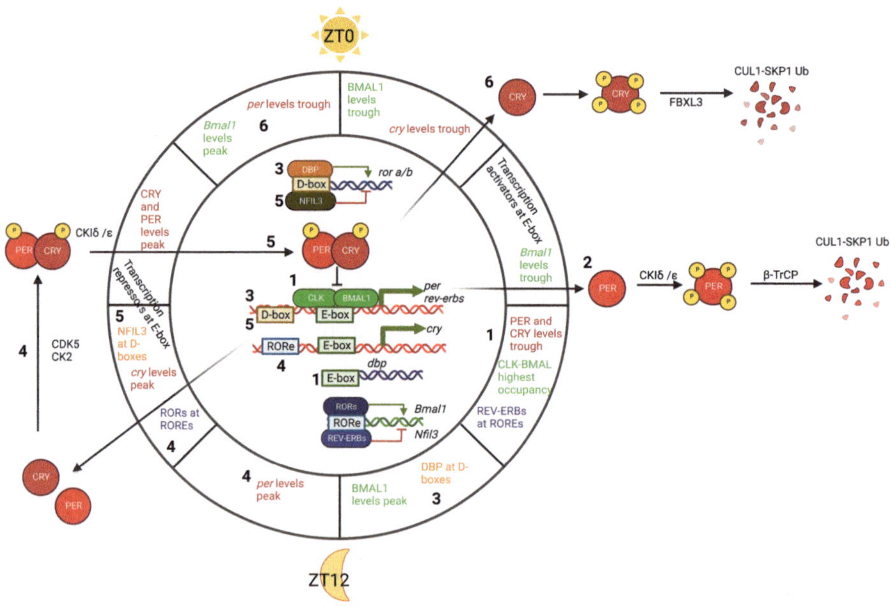

Fig. 3 The mammalian molecular clock cycle in detail. The central circle represents the nucleus, with events depicted around the margin according to clock time, with ZT 0 marked by the sun and ZT 12 by the moon. Sequential events are numbered 1–6: (**1**) CLK and BMAL1 bind E-box elements with maximal occupancy between ZT 6–9 and transcribe *per* throughout the day, as well as *dbp*. (**2**) Because CRY levels are low at earlier time, PER is degraded CUL1-SKP-β-TrCP ubiquitination and degradation pathway. (**3**) DBP shows maximal occupancy at D-boxes between ZT 9–15, promoting transcription of *ror*s to kick-start *cry* transcription and prolong *per* transcription. (**4**) RORs at ROREs together with CLK-BMAL1 kick-start *cry* transcription. CRY in the cytosol can now interact with PER and heterodimerize through the help of CDK5. (**5**) PER-CRY repressive complex translocates into the nucleus, where it inhibits its own transcription, as well as other CLK-BMAL1 transcription. NFIL3 inhibits transcription at D-boxes. (**6**) As *per* levels go down, CRY starts losing its binding partner and gets degraded in the cytosol by CUL1-SKP-FBXL3 degradation. The process continues as *Bmal1* levels build up to restart the loop

tightly regulated through a secondary clock loop as well as many posttranscriptional and translational modifications described in the following sections.

Mammals have paralogs of the core molecular clock components PER, CRY, CLOCK, and BMAL that contribute to robustness of tissue specificity of molecular clock function. Mice have three PER paralogs where PER1 and 2 are present in the brain clock, while PER3 predominantly oscillates in the peripheral tissues. PER2 is the principal component for period determination in the brain clock, while PER1 may regulate PER2 posttranscriptionally [18, 223, 326]. PER3 regulates the period and phase of *per1* expression in the pituitary and lungs and also regulates liver adipogenesis [3, 223]. Interestingly, certain PER3 haplotypes are also associated with delayed sleep phase syndrome [18]. Mice also have two CRY paralogs, with slightly different timing of chromatin occupancy during the circadian cycle [116, 150]. The CLOCK paralog is named NPAS2 and can compensate for CLOCK and maintain circadian rhythms in the SCN, but it is unable to do so in peripheral tissues [65]. Though mammals have a BMAL1 paralog, called BMAL2, the BMAL1 knockout is effectively a double functional knockout due to transcriptional regulation of Bmal2 by BMAL1 [38, 258]. However, constitutive expression of BMAL2 can rescue arrhythmic phenotype in BMAL1 knockout mice [38, 258].

The Secondary Clock Loop (ROR/REV-ERB)

A secondary TTFL works in parallel to the core clock loop to regulate the positive arm of the core TTFL and determine clock phase and amplitude. *Bmal1* mRNA levels are antiphase to *per* mRNA levels; *per* peak after dusk, while *Bmal1* peaks at dawn (Fig. 3(4, 5)) [272]. This suggests that BMAL1 transcription is rhythmically regulated. This regulation occurs through the secondary clock loop, in which CLK and BMAL1 regulate the transcription of retinoic acid-related orphan receptor (ROR) a/b and REV-ERBα/β (Fig. 3(1)) [228, 246]. These two transcription factors compete at REV-ERB/ROR binding elements (ROREs) to repress and activate Bmal1 transcription, respectively, modulating circadian transcription of Bmal1 (Fig. 3(4)). RORa is required for rhythmic Bmal1 expression and consolidation of locomotor activity, while REV-ERBs are not needed for circadian rhythms to function but influence clock output gene cycles, especially in the liver [126, 228, 246]. ROREs are required for phasing the clock. Together with E-boxes, they regulate *cry* expression and cause it to peak a few hours later than *per*, contributing to timing the action of the clock's negative limb (Fig. 3(4)) [288]. These core components of the molecular clock also interact with the clock epigenome and other physiological processes, creating links between these processes and human diseases they cause when dysregulated. Although these TTFLs are interlocked and control much of the transcriptional timing of core clock components, another layer of regulation in the form of posttranslational modifications (PTMs) controls clock period, speed, and turnover [93].

Tissue-Specific Expression of Clock-Controlled Genes

One of the criteria used to define core clock genes was the observation of oscillating mRNA transcripts. Genome-wide association studies show that about half of all mammalian genes display 24-h rhythmic expression in at least one tissue [204, 236, 323]. Of course, not all these genes are "clock" genes. They are instead regulated either directly by the core clock transcription factors or its downstream pathways and are known as clock-controlled genes (CCGs). Indeed, CLK-BMAL1 utilizes E-box element binding sites to directly regulate a fraction of these CCGs, through which REV-ERBs are also transcribed (Fig. 3(4)) [150]. However, a third interlocked loop is also modulated by both CLK-BMAL1 and REV-ERB/ROR that feeds back to regulate RORa/b. D-box protein (DBP) binds D-box elements and modulates gene transcription at the D-box cis-element (Fig. 3(3)). Nuclear factor, interleukin 3 regulated (NFIL3) is D-box's repressor (Fig. 3(5)) [92, 196]. Its transcription is modulated by the REV-ERBs/RORa/b loop (Fig. 3(4)). Together, E-box, ROREs, and D-boxes define a cistrome in which different combinations at promoters of various genes create circadian transcription cycles with tissue-specific phases. These cis-elements have narrow windows of maximum transcription factor occupancy, contributing to the observed "rush hours" of tissue-wide mRNA transcription. CLOCK-BMAL1 occupies E-boxes maximally between ZT 6 and ZT 9, when rhythmic mRNA expression across various tissues peak (Fig. 3(1)) [86]. *Bmal1* mRNA expression peaks at ZT 18–24, when RORa/b occupancy at the ROREs is highest (Fig. 3(6)), while Dbp occupancy between ZT 9 and ZT 16 contributes to RORa/b mRNA peak levels at this time (Fig. 3(3)). Clock-controlled epigenetic regulation also contributes to these occupancy windows, described later. As the clock modulates the timing of transcription across so many biological processes involved in growth, metabolism, immune function, and locomotion, it is not surprising that timing of therapeutics becomes relevant when treating human diseases. A recent study showed that a thousand cycling protein-coding genes across various tissues were therapeutic drug targets or involved in metabolism or transport of drugs [236]. The list of genes the clock regulates is vast and reviewed in many genome-wide association studies. See these for an exhaustive list [204, 236, 323].

Posttranscriptional and Posttranslational Control of Clock Speed

The rhythmic regulation of phosphorylation and ubiquitination states of PER and CRY dictate their cytoplasmic stability, interactions, nuclear localization, and degradation and are critical for maintaining 24-h periodicity of the molecular clock. PER and CRY mRNA transcription happens throughout the day, but their nuclear levels only increase at late night. This is because PER and CRY must heterodimerize to enter the nucleus, but they build up at different speeds in the cytoplasm; with PER being hyperphosphorylated and degraded until there is enough CRY to dimerize and prevent degradation (Fig. 3(2)). Casein kinase I delta and epsilon (CKIδ/ε) is one such kinase that phosphorylates PER marking it for ubiquitination and subsequent degradation [5, 259]. In fact, it was later found that the first circadian mutant in mammals, the *tau* mutant hamster, had a mutation in CKIδ/ε [179]. Interestingly, mutations in CKIδ/ε have been linked to familial advanced sleep phase syndrome [282]. CRY degradation is accomplished via direct, phosphorylation state-dependent interactions between CRY and F-box proteins (Fig. 3(2)) [316]. These competing interactions between phosphorylation and degradation pathways fine-tune the mammalian circadian period.

When PER2 and CRY1 are at optimal levels and are appropriately phosphorylated, they heterodimerize in the cytoplasm in the late night and enter the nucleus (Fig. 3(4, 5)) [197]. Recent evidence was presented that cyclin-dependent kinase 5 (CDK5) modulates this heterodimerization through PER2 phosphorylation [35]. The complete composition of the repressive complex is still unknown; however, studies in mouse liver indicate that it may involve associates of PER1–3, CRY1–2, and CKI δ/ε through regulation of CDK5 [16]. However, this structure likely associates transiently with other tissue-specific components that are dynamically modulated by intrinsic circadian control of each component. Inside the nucleus, the PER-CRY repressive complex represses CLOCK-BMAL1 transcription of *per* and *cry* and other CCGs (Fig. 3(5)).

As described earlier, the positive loop of the core clock TTFL oscillates in its occupation of promoter site. However, the total protein levels of CLK and BMAL1 do not oscillate at the levels necessary to confer this occupational variance at the promoter [35]. Instead, the CLK-BMAL1 complex interacts with a variety of additional proteins that regulate PTMs, altering complex affinity for E-box sites. These interactions and modifications remain incompletely described but include specific phosphorylation events that are necessary for both complex activation and repression [125, 137, 275].

CLOCK and BMAL1's own transcription is regulated by the secondary and tertiary clock loops, described earlier. Once transcribed and translated, CLOCK and BMAL1 must still heterodimerize to enter the nucleus. CDK5-phosphorylated CLK and CK2α-phosphorylated Bmal1 heterodimerize and translocate into the nucleus [275]. Here, they bind E-box elements at the optimal occupancy times and restart the circadian clock loop in the morning.

Other Clocks

Neurospora

Neurospora are used to interrogate the finer aspects of the molecular clock, such as dynamic subcellular localizations of clock components, or a mechanistic understanding of the negative arms "fuzzy" complex [25, 131]. It is an attractive model system to study circadian biology due to its genetic tractability and ease of circadian readout. These filamentous fungi now have whole-genome knock-out libraries, and a bioluminescence reporter for various reporters that can monitor circadian behavior has been established [53, 99]. *Neurospora* also have conserved core molecular clock, where CLK-BMAL are known as WHITE COLLAR-1 (WC-1) and WC-2, while PER-CRY repression function is done by one FREQUENCY (FRQ) protein [190]. It was thought that *Neurospora* growth was not circadian in nature, until "race tubes" were developed that were originally created to determine growth rates of different strains [244]. In these, mutant strains that had circadian growth defects (circadian sporulation) were screened, and *frq* mutant was reported soon after *per* mutant was found in *Drosophila*. Since then, many aspects of the circadian clock components have been found to be conserved in this species. Some striking differences are the fact that both *wc* genes show no cycling and nascent FRQ immediately translocates into the nucleus for its repression activity [317].

Plants

Studying plant clocks could help create more sustainable food production. Many core clock genes of plants were identified in *Arabidopsis*. CCA1 and LHY are MYB-DOMAIN TFs that are transcriptionally activated by TOC1 [248, 270, 295]. They heterodimerize and repress TOC1 transcription in the reminiscent clock TTFL process. Under nonparametric entrainment, light induces transcription of LHY and CCA1. However, the plant clock oscillations do not disappear in constant light and can be fine-tuned by the wavelength and intensity of light [265, 270]. This tuning comes in the form of phase adjustments, where there is a negative correlation between period length and light intensity [317]. Although the core plant clock follows similar arrangement of a core TTFL, the individual components are not homologous. This indicates that the circadian clock's molecular mechanism may have evolved separately at least twice [303]. Studying the circadian biology of plants helps create knowledge about optimal plant growth conditions. For example, a recent study was able to determine minimal effective dosing for a glyphosate-based herbicide based on rhythmic control of glyphosate-responsive genes [27]. These types of studies create a paradigm for agrochemical development and use through the lens of chronobiology, termed "agricultural chronotherapy."

Cyanobacteria

Cyanobacteria do not need a nucleus with transcription-translation feedback loops to persist and keep time. Their clock components are self-sustainable through oscillating phosphorylation that can be reconstituted in a test tube [209]. The clock components consist of the three *kai* (meaning *cycle* in Japanese) genes KaiA, KaiB, and KaiC along with adenosine triphosphate, providing a minimalistic framework for molecular clock studies [152]. Cyanobacteria are entertainable to light and have a core clock; although these aren't truly homologous to any eukaryotic clock component, again providing evidence for independent origins for eukaryotic clocks [317].

Still, the arguments for common ancestor persist with the presence of PAS domains in various clock proteins with the potential to bind cryptochromes [55, 145]. With almost all genes entrainable to various environmental cues, cyanobacteria have been popular organisms to study adaptive significance of circadian programs [46, 132].

Light Entrainment of the Molecular Clock

To entrain to the environmental light cycle, the molecular clock must respond to light input. In mammals, *Per1* and *Per2* are light-inducible [252]. This light induction happens through a highly conserved cAMP response element (CRE) in the *per* promoter region, which is also present in other light-responsive genes [7, 279]. In response to visual light exposure, signals from the retinohypothalamic tract induce phosphorylation of CRE-binding protein (CREB), which translocate into the nucleus and binds these CREs, including at the *per* promotor to induce expression. The phospho-relay pathway and neural signals that entrain the brain clock are described later.

Light exposure at night causes changes in circadian phase, observable at both the behavioral and molecular levels. Light exposure early in the night causes a phase delay (shift of clock timing to later in the 24-h cycle), while light exposure late in the night causes a phase advance (timing shifts earlier) [59]. Light exposure during midnight or during times when light is normally available has no effect on circadian phase. However, when there is prolonged light exposure at night, there is a need to re-entrain to the new light timing. As can be seen with jetlag, this process takes multiple days and can't be explained by the direct regulation of PER by light exposure. Nighttime light exposure induces not only *per1* transcription but also salt-inducible kinase 1 (*sik1*). SIK1 subsequently phosphorylates CREB-regulated transcription coactivator 1 (CRTC1), preventing its interaction with CREB, thus limiting the direct effect of light on *per* transcription [130]. Though knockdown of *sik1* results in rapid molecular and behavioral re-entrainment to jet lag protocols, permitting slower adaptation to new light cycles prevents internal desynchrony.

The Brain Clock

Location

Even as behavioral responses to entrainment cues and core clock genes were being discovered in flies in the 1960s and 1970s, the anatomical location of the clock remained unknown. While we now know that there are molecular clocks across tissues, the central brain clock is necessary for overt behavioral rhythms. Early work in cockroaches found that the clock was in the brain, with bisection of the brain causing a loss of wheel-running locomotor rhythms [235]. Confirming the location of the master pacemaker in the brain, transplantation of brains between *period*-null mutants and donor flies with wild-type or mutant *period* rescued rhythms, with the rescued rhythm displaying the donor period [105]. Around the same time, lesion studies in rats found that ablation of the suprachiasmatic nucleus (SCN) ablates behavioral rhythms in drinking, locomotor activity, and hormone secretion [198, 268]. Rescue studies that grafted wild-type SCN tissue into SCN-ablated hamsters found restoration of locomotor rhythms after transplantation [173]. In animals from flies to humans, the brain clock acts as a network of coupled oscillators to maintain robust environmental entrainment and coordinate output signals that regulate circadian physiological and behavioral rhythms (Fig. 1b). In fact, SCN-specific restoration of *Cry1* in *Cry1/Cry2* double-knockout mice is sufficient to restore circadian control of sleep-wake cycles, suggesting direct regulation of sleep-relevant brain areas by the master SCN clock [189]. In the following sections, we will examine how light can entrains the brain clock (input), how rhythmic oscillation in the SCN is maintained at the network level (oscillator), and diverse SCN output signaling mechanisms that entrain rhythms throughout the body (output). See Table 3 for a list of acronyms for key brain regions and neurotransmitters involved in circadian timekeeping.

Neuroanatomy

A short primer on clock neuroanatomy is necessary to understand the coupled oscillator structure of the brain clock; here we focus on the mammalian SCN, where most of our knowledge comes from studies in rodents. The mouse SCN consists of two nuclei of ~10,000 neurons each, located on each side of the third ventricle, above the optic chiasm (hence the Latin *suprachiasmatic*, meaning “above the chiasm”), though in humans the SCN is much bigger, ~45,000 cells per hemisphere

Table 3 List of acronyms for key brain regions and neurotransmitters involved in circadian timekeeping

Brain Regions	
SCN	Suprachiasmatic nucleus
SPZ	Subparaventricular zone
VLPO	Ventrolateral preoptic area
PVN	Paraventricular nucleus
RHT	Retinohypothalamic tract
Neurotransmitters and neuropeptides	
AVP	Arginine vasopressin
GABA	Gama amino butyric acid
GRP	Gastrin-releasing peptide
NMS	Neuromedin S
PACAP	Pituitary adenylate-cyclase-activating polypeptide
VIP	Vasoactive intestinal peptide

[112]. Each nucleus is further divided into the ventral core and the dorsal shell based on neuropeptide expression (Fig. 1c). The core is defined by expression of the neuropeptides vasoactive intestinal polypeptide (VIP) and gastrin-releasing peptide (GRP), while the shell is defined by expression of arginine vasopressin (AVP). The majority of SCN neurons in both the core and the shell express GABA, though both cell and circuit-level oscillations of the clock TTFL continue when GABA signaling is disrupted [215]. Consequently, the role of SCN GABA signaling in circadian pacemaking remains obscure. Light information is relayed from the eyes to the SCN core via the retinohypothalamic tract (RHT) and then from the core to the shell via VIP signaling. The shell then directs SCN output to other brain regions and provides feedback to the core (Fig. 1c). While the simple division of the SCN into the core and shell is a useful framework for discussion, in fact the SCN circuit organization is more complex, with sub-clusters that form a hierarchical network of differently phased oscillators [182, 200, 222].

Light as the Primary SCN Zeitgeber

Light input resets the molecular clock in the SCN to permit entrainment to the environmental light cycle. All photic input to the SCN in mammals comes from the eyes. The major light input pathway is via the RHT, where retinal ganglion cells make direct monosynaptic contact with retino-recipient neurons in the SCN core. This is a nonimage-forming pathway, with light processed by the melanopsin-producing intrinsically photosensitive retinal ganglion cells [11]. Photic information is also relayed to the SCN via a secondary pathway from the image-forming projection from the eye to the lateral geniculate nucleus of the hypothalamus and thence to the SCN via the geniculo-hypothalamic tract. For simplicity, we will focus on light-induced clock resetting via the RHT.

RHT terminals signal via glutamate and the neuropeptide pituitary adenylate cyclase-activating polypeptide (PACAP). Release of glutamate drives an increase in SCN firing rate and elevation of intracellular calcium levels via activation of AMPA and NMDA receptors and voltage-sensitive calcium channels [54]. PACAP and GABA signaling act both pre- and postsynaptically to regulate the magnitude of the SCN response to light. Increasing intracellular calcium activates multiple signaling pathways that converge to drive phosphorylation of cyclic AMP-responsive element (CRE)-binding protein (CREB) [71, 96]. Phosphorylated CREB enters the nucleus and binds to CRE elements in promotor regions, including the promotor regions of *Per1* and *Per2*, resetting the molecular clock [285]. The peak of SCN firing in both diurnal and nocturnal animals occurs during the subjective day. Light stimulation during the subjective day has little effect on SCN physiology, as the membrane potential is already depolarized, and intracellular calcium levels are already high [310]. Light stimulation at night, when the membrane potential is hyperpolarized and intracellular calcium levels are low results in a larger net change in intracellular calcium and consequent shift in the phase of the molecular clock rhythm.

Interconnectivity Preserves Phase Relationships

While molecular clocks are present across tissues, the SCN is unique because it has topographically organized coupling mechanisms between cells that maintain synchronous molecular clock oscillation even in constant conditions. Within a circadian cycle, PER2 accumulation, as tracked by a PER2::Luciferase reporter, follows a characteristic topographic trajectory starting in the dorsomedial SCN and spreading ventrally and laterally, while deactivation proceeds in the reverse order [188]. The relative phase of PER2 expression in each cell is consistent from cycle to cycle. This stereotyped molecular clock phase dispersal within the SCN provides the opportunity for variously phased output signals to regulate peripheral rhythms. Reciprocal connections within the SCN also allow the network to reinforce individual neuron rhythmicity through rhythmic inputs from other networked cells. This makes the SCN as a whole more resistant to environmental or genetic perturbation than individual neurons.

While dissociated SCN neurons can maintain cell-autonomous molecular clock oscillation and circadian firing rate, they do so with a wide range of periods and consequently quickly fall out of phase with one another. Culturing neurons at high density begins to allow a proportion of neurons to form connections that enable synchronous firing rhythms [296]. Neuronal firing and chemical synapses are required for the tight circadian coupling observed in the SCN. Blocking action potentials with tetrodotoxin for several days allow SCN neurons in slice culture to desynchronize [308]. Analysis of cross-correlations between pairs of neurons within the SCN has identified a correlation between circadian synchrony and excitatory vs. inhibitory synaptic interactions.

The SCN core is necessary for maintaining SCN coupling. When connections between the core and the shell are severed, the core is able to maintain synchrony, while shell neurons continue to oscillate rhythmically but quickly fall into desynchrony [308]. This observation is consistent with anatomical data showing dense projections from the core to the shell, with only sparse feedback from the shell to the core. GABA and the neuropeptide VIP are important for core-to-shell communication. In response to a phase-delaying light pulse, application of a $GABA_A$ receptor antagonist to SCN slices results in a phase shift in the core but a failure in phase-shifting the shell [8]. Loss of GABA signaling from AVP+ neurons alters SCN coupling resulting in dysregulated SCN phase relationships in firing and lengthened behavioral period, without disrupting molecular clock timing [182].

VIP signaling from the core to the shell is also essential in maintaining SCN phase synchrony. VIP acts via its receptor VPAC2 to depolarize SCN neurons by closing potassium channels. Mice lacking either VIP or VPAC2 have weaker circadian behavioral rhythms, and SCN firing and clock gene expression rhythms are dampened due to loss of synchrony. Highlighting the critical role of VIP signaling, restoration of clock function in the VIP-VPAC2 signaling axis alone is sufficient to initiate and sustain circadian locomotor behavior in *Cry*-null mice [222]. Though the effects of VIP and GABA on phase synchrony are the most pronounced, other more sparsely expressed neuropeptides such as GRP and prokineticin2 are necessary for robust timekeeping with and appropriately phased clock output [94, 200]. Recently, the development of a dual-color PER2:luciferase reporter revealed that the presence of a functional clock in AVP+ neurons, but not VIP+ neurons, is essential for SCN network synchrony [256]. This likely reflects the tighter coupling of VIP+ neurons to other neurons with intact clocks.

SCN coupling is important for re-entrainment after a phase-altering light stimulus. After a 6-h phase advance, SCN cells initially show a dispersal in phase, with the core re-entraining to the new phase before the shell and complete resynchronization taking ~8 days. VIP is essential to this process, with mice overexpressing the VPAC2 receptor showing accelerated phase shifts in response to shifts in the lighting schedule.

Output Signals

The tight intercellular coupling in the SCN also allows it to generate more robust and precise rhythmic outputs than single-cell oscillations. While peripheral tissues possess the molecular clock components, rhythms in these tissues gradually dampen out in the absence of signals from the SCN as cells within these tissues become desynchronized. Thus, SCN output signals serve to both coordinate rhythms between tissues and to synchronizing cells within tissues.

Transplantation studies have demonstrated that the SCN mediates rhythmic physiology via both synaptic connections and humoral signals. As described above, complete SCN lesions abolish circadian rhythmicity in behavior and hormone secretion. Transplantation of wild-type SCN tissue into SCN-lesioned or genetically arrhythmic animals is sufficient to restore circadian rhythmicity. Encapsulation of SCN donor grafts in a mesh that prevented axonal outgrowth still rescued locomotor rhythms of genetically arrhythmic hamsters, suggesting that humoral signals are a major output of the SCN [261]. These signals likely include neuropeptides such as vasopressin, which signal to the paraventricular nucleus (PVN) to regulate corticosterone release [135]. However, graft encapsulation does not restore neuroendocrine rhythms [173, 191]. Parabiosis experiments between intact and SCN lesioned hamsters demonstrated that while SCN humoral factors can maintain rhythmic clock gene expression in some tissues, such as the liver and kidney, they are not sufficient to maintain rhythms in other tissues including the heart, spleen, and skeletal muscle [103].

Given the pivotal role of the SCN in orchestrating temporal synchronization across the entire organism, it has surprisingly sparse synaptic outputs. The majority of direct SCN efferents target interneurons of the medial hypothalamus and thalamus, with more scarce connections to endocrine and preautonomic neurons [136]. Multielectrode recordings suggest that while the SCN is essential for the maintenance of rhythmic firing in the majority of thalamic and hypothalamic neurons, there are neurons in these regions that can maintain rhythms in slice culture even after SCN removal [106]. Additional multisynaptic circuits link the SCN to other endocrine control centers such as the pineal gland which regulates release of melatonin, lateral hypothalamic neurons that release orexin, and the pituitary which releases adrenocorticotropic hormone. Efferents from the SCN to neuroendocrine centers allow rhythmic release of hormones that can act broadly across tissues to entrain and synchronize peripheral clocks. The SCN may also release humoral signals directly; candidate factors that may regulate behavioral rhythms include TGFα and prokineticin2 [48, 160]. The SCN provides further indirect regulation of multiple tissues through circadian regulation of factors such as core body temperature, allowing key circadian physiological properties to be regulated in parallel by multiple rhythmic cues.

Circadian Regulation of Sleep

In the two-process model of sleep, sleep timing, duration, and intensity are regulated semi-independently by the circadian clock and by a homeostatic process. Now that we have examined the molecular clock TTFL and the brain clock in the SCN as a master pacemaker, we can mechanistically examine the contribution of clock genes and the brain clock to sleep regulation. The clock likely also regulates sleep indirectly via circadian control of core body temperature. Though mechanistic and anatomical details remain to be investigated, the available data support a model in which the clock and the sleep homeostat semi-independently regulate sleep/wake but also reciprocally regulate one another (Fig. 4).

Sleep Homeostatic Regulation of the Clock

In the two-process model, the clock and the sleep homeostat coordinately regulate sleep but also interact with one another. Sleep deprivation does not reset molecular clocks in the SCN [45, 57, 292]. However, several lines of evidence point to influence of the sleep homeostat on clock output and sensitivity to light. Human sleep studies have shown that as sleep pressure builds, the circadian amplitude of multiple behaviors dampens, including psychomotor vigilance, subjective sleepiness, and certain EEG features of sleep [30, 147, 163]. SCN neuronal activity also changes with sleep state and is reduced with increasing sleep deprivation, which may contribute to dampening of circadian outputs (Deboer et al., [63, 64]). Sleep deprivation also reduces the light sensitivity of the SCN, leading to reduced phase-shifting capacity in animal models and humans [39, 45, 195, 292]. Despite phenomenological evidence for effects of sleep homeostatic processes on circadian processes, mechanistic explanations for this regulation are lacking. Further exploration of how sleep regulatory brain regions influence both SCN and extra-SCN clocks and how potential somnogens may have brain-wide influence on circadian processes would

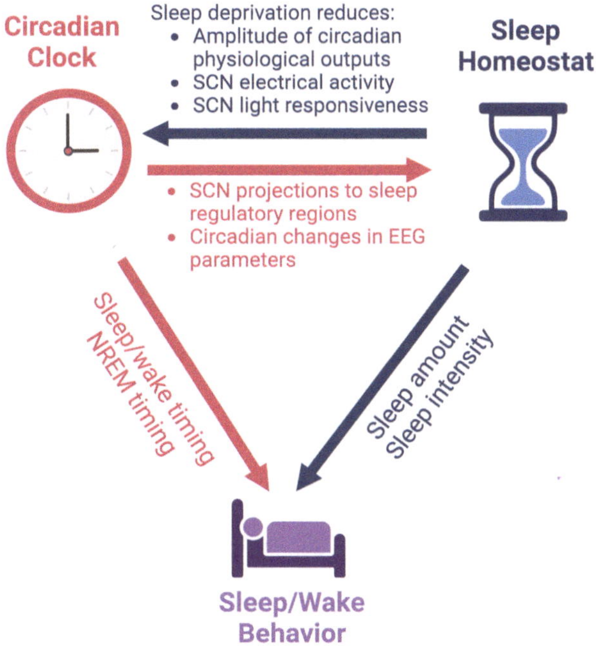

Fig. 4 Sleep is coordinately regulated by circadian and sleep homeostatic mechanisms. Circadian signals initiated in the brain clock (pink) regulate the timing of sleep onset and offset, as well as the timing of NREM sleep. Ablation of the brain clock or molecular clocks results in dysregulation of sleep timing, with limited effects on sleep amount. The sleep homeostatic system tracks waking time to estimate sleep need and regulates total sleep amount and sleep intensity, as measured by slow-wave activity. The circadian clock and sleep homeostat also reciprocally regulate one another, with evidence for clock regulation of the homeostat depicted in pink and for homeostatic regulation of clock output depicted in blue.

provide valuable clues that may elucidate the complex relationship between homeostatic and circadian regulation of sleep.

SCN Regulation of Sleep

SCN lesion studies demonstrate that the SCN is required for mediation of sleep-wake rhythms and regulates sleep distribution around the circadian day with limited, species-specific effects on sleep amount [76, 77, 123, 124, 194, 201, 280, 284, 304]. Sleep deprivation studies show that even during recovery sleep after deprivation, circadian regulation still dictates sleep timing, with Process C opposing recovery sleep during the normal wake phase but promoting sleep during the normal sleep phase [194]. Thus, consistent with the two-process model, SCN regulation of sleep represents Process C, with no consistent effect of SCN ablation on Process

S. This does not mean, however, that the SCN does not directly regulate sleep; it simply does so in a circadian time-dependent manner without regard to sleep pressure. For example, in mice VIP+ neuromedin S+ SCN neurons are normally night-active, and optogenetically manipulating these neurons at night, but not during the day, alters siesta sleep in mice [52]. In humans, forced desynchrony studies demonstrate an effect of circadian rhythms on spontaneous awakening after sleep and on slow wave activity during NREM sleep, again pointing to a role of the clock in timing sleep and even specific sleep stages [69].

The SCN projects to both sleep- and wake-active brain regions. The SCN sends only sparse direct projections to the sleep-promoting ventrolateral preoptic area (VLPO) and to the wake-promoting orexinergic neurons of the lateral hypothalamus [50, 138, 240]. However, the SCN also has indirect contact with sleep regulatory brain regions via its projections to the PVN and subparaventricular zone (SPZ) [180, 216]. In addition to direct projections to sleep regulatory brain regions, the SCN-regulated rhythmic release of melatonin from the pineal gland regulates sleep timing and other aspects of rhythmic physiology [249]. Melatonin release peaks at night in both diurnal and nocturnal animals and has modest, time-dependent sleep-promoting effects in diurnal mammals [36, 238].

How does SCN regulation of sleep differ in diurnal vs. nocturnal animals? On average, SCN peak firing is maximal during the subjective day in both diurnal and nocturnal animals. The timing of the molecular clock TTFL is also the same in both diurnal and nocturnal animals. Consequently, the molecular and brain clocks must regulate sleep and the effect of light stimulation on wakefulness in opposing directions during equivalent phases of the circadian cycle. One candidate site for circadian signal inversion is in the SPZ, which shows opposing cFos rhythms in the diurnal grass rat compared to the nocturnal lab rat [213]. Recent work in the diurnal four-striped grass mouse found changes in ion channel expression and current properties in the SCN itself compared to nocturnal mammals [20]. Further work at the genetic and circuit levels in diurnal animals is required to better understand the differences in circadian regulation of nocturnality vs. diurnality of various behaviors, including sleep.

Clock Genes and Sleep

As clock genes were discovered, whole-organism clock gene knockouts and mutants were analyzed for changes in sleep. While global deletion of core clock genes in rodent models consistently results in loss of sleep-wake rhythms, the effects on sleep itself are varied. Only Bmal1 mutant mice show a change in total sleep time [169]. Several deletion or loss of function mutants affect NREM sleep specifically: *Cry1/Cry2*, *Per3*, and *Npas2* deletion mutants and the *Clock*Δ19 mutant [91, 107, 210, 300]. Other clock gene deletions (*Per1/Per2*, *Per2*, *Timeless*) show no change in sleep architecture, only in timing [156, 166, 260, 306]. For a summary of the effects of clock gene deletions and mutations on sleep in animal models, see Sanchez et al. [243]. It is important to bear in mind that in these global knockout and deletion models, molecular clock function is impaired in all tissues. Circadian oscillations in gene expression and physiology in other sleep-relevant brain regions, and even outside the brain, may play roles in sleep regulation. Thus, the results of global molecular clock manipulations should be interpreted narrowly.

With advances in genetic tools in rodent models, recent studies have examined brain region-specific clock gene deletion to identify how specific clocks in specific SCN subpopulations affect sleep and other circadian behaviors. Consistent with the SCN as the master pacemaker, Bmal1 deletion in the forebrain, broadly in the SCN, or even just in neuromedin S^+ SCN neurons results in loss of circadian locomotor activity rhythms [122, 129, 171]. VIP+ neuron-specific deletion of Bmal1 not only affected rhythmic locomotor behavior but also resulted in increased NREM sleep and disrupted daytime sleep [281]. These SCN-specific deletions help pinpoint subpopulations where clock function contributes to circadian sleep regulation. A converse study restoring the clock only within the SCN of the global *Cry1/Cry2* double-mutant mouse found that the clock in the SCN alone is sufficient to rescue locomotor activity rhythms and NREM sleep changes [189]. However, physiological clocks outside the SCN also contribute to sleep regulation. Deletion of clocks in non-SCN forebrain neurons altered timing of both REM and NREM sleep, while rescue of skeletal muscle clock function in the *Bmal1* global knockout mouse restored total daily sleep time, but not timing [79, 192]. Further mechanistic exploration of the role of specific molecular clock components in SCN subpopulations and sleep regulatory brain regions is needed to untangle how the molecular clockwork coordinates the timing of wake, sleep, and specific sleep states.

Most of what we know about the role of clock genes in sleep comes from animal models; however, several heritable human sleep disorders result from clock gene mutations. Familial advanced sleep phase syndrome (FASPS) causes a very early chronotype, with a ~4 h phase advance in sleep, melatonin, and core body temperature [133]. FASPS is associated with missense mutations in PER2 and CRY2 that result in short period and in TIM which results in normal period but altered light sensitivity [111, 166, 282]. Familial delayed sleep phase syndrome (FDSPS) results in very late chronotypes and is associated with a short-length polymorphism in PER3 and a gain of function mutation in CRY1 [68, 221]. A genome-wide association study to identify genes contributing to variation in sleep timing identified several clock core clock genes and SCN output molecules as well as genes involved in retinal ganglion cell function and metabolism that may contribute to determination of chronotype in humans [134]. These hits provide promising avenues for further mechanistic investigation of the genetics and circuitry of circadian regulation of sleep.

Peripheral Clocks

Peripheral Molecular Clocks

Core molecular clock gene expression has been detected in all mammalian peripheral tissues thus far examined. Even immortalized rat fibroblasts in culture for more than 25 years can display circadian rhythms of gene expression for several days after a serum shock [19]. Under physiological conditions, signals from the SCN provide synchronizing cues for peripheral molecular clocks, sustaining phase synchrony across the body and within tissues. A classic experiment from Yoo et al. demonstrates these principles elegantly by culturing explants of a variety of tissues from PER2::Luciferase mice to track PER2 transcription rhythms [315]. SCN explants can sustain PER2 transcription rhythms for nearly 30 days in culture due to the tight intra-SCN coupling. However, peripheral tissue explants lose phase synchrony of PER2 transcription with the SCN within a few days, with period lengthening in some tissues and shortening in others. In most tissues, whole-explant level rhythms dampen within a week, as cells within the tissue lose phase synchrony with each other, though clocks in individual cells may continue to oscillate. Highlighting the importance of intra-tissue coupling, SCN molecular clock transcription rhythms are quite robust to individual loss of many clock genes, while, for example, in the lung individual deletion of PER1 or CRY1 results in rapid damping of rhythmic clock function [177]. One exception to the observation of molecular clock dampening after loss of SCN input is the liver, in which tissue-level oscillation is sustained, suggesting that some peripheral tissues have intra-organ mechanisms that permit self-sustained rhythms [315]. Recently the cytokine TGF-β was identified as a paracrine signal enabling circadian coupling between cultured fibroblasts [89]. Additional research is needed to uncover organ-specific mechanisms for sustained circadian oscillation.

Entrainment of Peripheral Clocks

Unlike the SCN which is directly entrained exclusively by light, peripheral clocks integrate multiple entrainment cues (Fig. 1b). This includes cues indirectly regulated by the SCN, such as hormonal rhythms and body temperature rhythms, as well as external environmental cues including food, temperature, exercise, circulating metabolites, and more. Under normal physiological conditions, light and peripheral zeitgebers are aligned such that they synergistically sustain synchronous clock timing across tissues. However, circadian misalignment due to mistimed feeding, exercise, etc. can result in clock desynchrony, in which molecular clocks across tissues tell different times. This drives temporal disorganization of gene expression that can impact behavior and increases the risk of metabolic diseases [6]. Here we highlight a few peripheral zeitgebers with relevance to both circadian physiology and sleep regulation.

Temperature

As mentioned earlier, the core body temperature rhythm serves as an entrainment cue for some tissues. The SCN itself is resistant to molecular clock resetting by temperature, though this is again driven by intra-SCN coupling. Decoupling SCN neurons via toxin application or physical several of connections between the core and shell enables temperature to shift clock phase [37]. In other less tightly coupled tissues such as the pituitary and lung, physiological temperature cycles can entrain the molecular clock [37].

Exercise

Exercise can serve as a zeitgeber for skeletal muscle clocks, where resistance exercise can set molecular clock phase in mice and humans [302, 319]. Sleep amount phenotypes found in global *Bmal1* knockout mice can be recapitulated or rescued by skeletal muscle-specific *Bmal1* knockout or restoration [79]. As with temperature, exercise does not affect molecular clock phase in the intact SCN; however, when intra-SCN peptide signaling is impaired, scheduled exercise can promote SCN clock synchrony and rhythmic locomotor behavior. These data suggest there may be mechanistic links between exercise, clocks in the muscle and brain, and sleep that require further investigation [120].

Feeding

Molecular clocks in digestive and metabolic tissues receive cues from the SCN and can also be reset by feeding. Both global and metabolic tissue-specific clock mutant animals show alterations in feeding rhythm and metabolic function pointing to a key role of clocks across tissues in regulating metabolic function. Both *Clock* and *Per2* null mice have disrupted feeding rhythms and are prone to obesity on high-fat diets [286, 311]. While the effect of global *Bmal1* deletion on obesity and metabolic function varies between studies, specific deletion of *Bmal1* in the liver or adipose tissue makes mice prone to diet-induced obesity [108, 168, 220]. In other instances, modulating molecular clock function can improve metabolic phenotypes. Intestinal deletion of *Bmal1* makes mice resistant to diet-induced obesity [318]. Targeting REV-ERB function with a synthetic agonist that resets hypothalamic clocks and metabolic gene expression rhythms in liver, skeletal muscle, and adipose tissue decreased obesity and improved dyslipidemia and hypoglycemia in diet-induced obese mice [264]. Though food only resets peripheral clocks, peripheral feeding cues also feed back to the brain, where the SCN, hypothalamic feeding centers, and wake- and sleep-promoting brain regions are interconnected, providing central coordination of behavior [212]. Because feeding is a potent zeitgeber for many peripheral tissues, time-restricted feeding may be a useful approach to addressing circadian rhythm disorders induced by shift work or travel [297].

The best-studied peripheral clock in mammals is in the liver, due to its unique ability to maintain robust rhythms without SCN inputs and even without clocks anywhere else in the body [100, 158]. Oscillation of the liver transcriptome drives rhythms in many metabolic processes including carbohydrate homeostasis, lipid and bile acid metabolism, and detoxification [234]. The liver clock is food-entrainable, and daytime feeding of normally nocturnal mice reveres the phase of clock gene expression in the liver without affecting the phase of SCN clock gene expression [60, 269]. When mice are fed during the day, signals from the SCN actively oppose food entrainment of the liver clock, and SCN lesioning allows the liver clock to entrain to the feeding schedule even more rapidly [239]. Recent studies suggest there may even be mechanisms for the liver clock to entrain to environmental light via non-SCN-mediated pathways [158, 298]. However, there are also genes that oscillate in the liver even when the liver clock is in active, suggesting that rhythmic transcription of these genes is directly driven by the SCN [157]. Studies of the liver clock have led to new insights into peripheral clock function that can be used to better understand the complex coordination of rhythms in other peripheral tissues.

Clock Epigenetics and Effects on Nuclear Architecture

Clock Epigenetics

Transcriptional control of the molecular clock works within a background of the oscillatory epigenome. The CLK-CYC heterodimer binds to E-boxes rhythmically, and the timing of this binding is associated with positive transcriptional regulation that makes DNA more permissive to binding to TFs and RNAPII [277]. PTMs of histones like acetylation and methylation modulate chromatin dynamics which in turn modulate transcriptional activity. CLOCK itself is a histone acetyltransferase (HAT) and acetylates the K9 and K14 sites on histone 3 (H3K9K14), making chromatin more permissive to transcription [73]. There are additional histone-modifying proteins that directly interact with the core molecular clock complexes, many of these which target H3K4 to modulate transcriptional activity. For example, CREB-binding protein (CBP) is a HAT that directly binds the CLK-BMAL1 complex along with p300 [117, 121, 176]. It is unclear if CBP acts as a transcriptional coactivator or corepressor, and its role may be tissue-specific [117, 121, 176]. The CLOCK-BMAL1 complex additionally interacts with MLL-1, a methyltransferase that serves as a transcriptional coactivator. CLOCK-BMAL1 is also regulated by transcriptional repressors such as JARID1a and EZH2, which modify histones around the E-box elements making it less permissive for transcription factor binding [72, 82, 291].

Epigenetic regulation also allows factors such as metabolic state or hypoxia to regulate circadian gene transcription. For example, SIRT1 is an NAD+ -dependent HDAC that binds dimerized BMAL and targets H3K9K14, making chromatin clock gene regions less permissive for transcription ([208], p. 1). In mammals, SIRT1 also deacetylates PER2, promoting its degradation [17]. As SIRT1 activity is dependent on the NAD/NADH ratio, it is considered an energy state sensor and thought to be the major link between the core molecular clock and metabolism [28]. Another HDAC associated with the clock involved in metabolic pathways is HDAC3, which is recruited by REV-ERBα forms a complex with nuclear receptor corepressor 1 protein and targets H3K9 [228, 313, 322]. Mice lacking the ability to activate HDAC3 have arrhythmic metabolic genes leading to aberrant energy expenditure, and genomic recruitment of HDAC3 is required for normal lipid metabolism in the liver [9, 87]. Oxygen levels and hypoxia can also act as a zeitgeber and reset circadian clocks, where HIF1α seems to modulate core molecular clock transcript levels in response to changes in oxygen level [2]. The histone demethylase KDM5A is an oxygen sensor involved in oxygen-dependent chromatin remodeling [26]. There is speculation that this demethylase may couple oxygen sensing to circadian function, as KDM5A is also involved in histone acetylation at Per2 promoter regions [72, 217].

Circadian 3D Genome Architecture

The three-dimensional organization of the genome within the nucleus regulates chromatin function, and gene positioning within the 3D genome can influence transcription state. Genes that are highly transcribed are usually sequestered more internally than those that are not [254]. The clock genes and associated oscillatory epigenome are no exception [217]. Recent data from *Drosophila* suggests that PERIOD positions the core clock genes *per* and *tim* to the nuclear periphery during the repression phase [305]. In human fibroblasts during interphase, *per* and *clk* genes come closer together during high transcription of *clk* and low transcription of *per* and vice versa and are categorized as topologically associated domains [47]. In fact, it may not just be a pair of genes but a whole interactome of circadian genes that form cyclical functional clusters. Chromatin capture experiments show that the *dbp* gene has circadian genomic interactions that are dependent on clock function in mouse fibroblasts [4]. Chromatin capture in human fibroblasts has identified physical distance between *per2* and *clock* loci during the circadian cycle as predictors of their transcriptional activity [47]. How tissue specificity of circadian 3D architecture may influence transcriptional programs or disease states remains an exciting field of inquiry.

Environmental Effects on the Clock

Aging

Over the course of normal aging in animals from flies to human, the core molecular feedback loop in individual brain clock neurons remains robust, but clock-driven behavior and physiology dampen [113]. This deficit leads to a reduction in clock phase synchrony—different peripheral clocks tell different times—which may exacerbate age-related decline of other physiological systems. The dampening of circadian physiology with age is reflected behaviorally in degradation of sleep-wake patterns [70, 148] as well as in tissue-specific alterations of transcriptional rhythms, affecting organ function [247, 263]. Genetic ablation of the molecular clock in the brain of both rodents and flies leads to reduced lifespan and health span and premature aging-like phenotypes including altered metabolic function, locomotor impairment, and reduction in muscle and subcutaneous fat [153, 164, 165, 278, 307]. Dietary interventions have the most reproducible effects on lifespan and health span across species and affect the molecular circadian clock [1]. Caloric restriction robustly extends lifespan in a clock-dependent manner [41, 139, 164, 165, 289, 299], while time-restricted feeding paradigms show promise in improving metabolic and muscle tissue function with age, likely via interactions with peripheral circadian clocks [257, 278, 293]. Conversely, consumption of high-fat and high-sugar diets leads to reductions in lifespan and disrupts molecular and behavioral circadian rhythms [23, 33, 149, 175, 224, 314]. The tight links between diet, circadian function, and life/health span suggest a diet x clock effect on regulating circadian phase synchrony through unknown mechanisms that merit further investigation to develop of lifestyle and pharmacological interventions to improve lifespan and health span in the aging population.

Circadian Misalignment

In today's world, many people have lifestyles out of sync with the light/dark cycle, which we will group together as circadian misalignment. Nearly everyone in the developed world spends most of the day indoors, exposed to too little bright light for robust clock resetting, and remains exposed to dim light, or blue light from screens, well after sunset. Shift work, travel across time zones, or social jetlag in which people observe different sleep-wake schedules on work days vs. free days contributes to even further decoupling of behavior from the light-dark cycle. Many of these environmental effects on circadian rhythms also result in reduced sleep, and further research is needed to identify contributions of reduced sleep vs. circadian desynchrony to identify optimal interventions to improve health outcomes for shift workers and other vulnerable populations.

Animal models of circadian misalignment include time-restricted or mis-timed feeding experiments as well as jet lag protocols. In jet lag protocols, light cycles are phase-shifted by several hours every few days for the duration of the experiment. Clock gene expression in the SCN core re-entrains to phase-shifting light schedules within a day; however, the SCN shell takes 5–9 days to synchronize with the new phase [206]. This is consistent with re-entrainment of rodent locomotor activity patterns taking about 1 day per hour of phase shift to adapt to a new light schedule [274]. In general, mammals are better able to re-entrain to phase delays (lights turn on later in the new schedule) than to phase advances (lights turn on earlier in the new schedule). This may be due to the ability of SCN *Cry1* expression rhythms to adapt more quickly to phase delays than phase advances [233]. Of course, we can't examine SCN clock gene expression in humans; however, studies have examined the effects of phase-shifting light stimuli on core body temperature, cortisol, melatonin, feeding, and other metrics. As in rodent models, the time needed for re-entrainment of these parameters is a little less than 1 day per hour of phase shifts, with faster re-entrainment to phase delays than phase advances [80, 193]. Additionally, intentional lighting, exercise, feeding schedules, and administration of melatonin or glucocorticoids can hasten re-entrainment in both animal models and humans [14, 40, 51, 81, 143, 187, 241, 287, 309].

Repeated phase-shifting of activity, sometimes termed chronic jet lag, is encountered by people with shift work or extreme social jet lag. In animal models, chronic jet lag is modeled by repeated phase shifts in the light schedule for the duration of experiments. In rodents, chronic jet lag results in metabolic dysregulation [102, 202, 242, 312], impairments in learning and memory [67, 115, 178, 276, 321], faster progression of tumor growth [88, 101, 142, 144, 219], and shortened lifespan [62, 128, 225]. Retrospective epidemiological studies of long-term human shift workers have identified increased risks of cancer, mood disorders, cardiometabolic disorders, and reduced cognitive performance [24, 49, 83, 110, 141, 186, 255, 283].

Chronopharmacology

The presence of molecular clocks and consequent daily transcriptional rhythms in every tissue of an organism results in diurnal physiological variation of most organ systems. Physiological rhythms impose circadian constraints on both pharmacokinetics and pharmacodynamics. Consequently, there are diurnal variations in drug efficacy, as well as side-effect profiles that should be considered in designing treatment regimens. As clinical treatment becomes more personalized, the ability to characterize an individual patient's circadian phase may prove useful in optimizing the timing of drug administration to maximize efficacy and minimize side effects. Chronotherapy can also encompass non-pharmaceutical interventions such as timing phototherapy, meals, or exercise to facilitate circadian synchrony in patients with sleep, mood, or metabolic disorders.

Drug absorption, distribution, metabolism, and elimination are all influenced by circadian rhythms in organ function. For oral drugs, diurnal rhythms in gastric pH and GI tract motility drive diurnal rhythms in drug solubility and absorption [98, 114, 199]. Increased morning blood flow to the gastrointestinal tract also contributes to increase daytime drug distribution [174]. Transporter proteins in the gut show circadian variation inexpression, which can drive circadian rhythms of absorption [13, 267]. The blood-brain barrier also shows clock-regulated rhythms in xenobiotic efflux, indicating that there are optimal times to allow drugs access to the brain—or conversely to prevent off-target access to the brain [56, 229, 324, 325]. Lipophilic drugs show the most pronounced diurnal rhythms of absorptions, with a daytime maximum [271]. The hepatobiliary system shows rhythmic excretion of bile acids, lipids, and xenobiotics with multiple late-day peaks [74, 75, 170, 181]. Renal blood flow and glomerular filtration rate in the kidney also show circadian oscillation contributing to rhythmic drug excretion [155].

Characterizing the effects of circadian rhythms on pharmacodynamics is more complex. In general, drug efficacy is enhanced at the time when the target is most highly expressed. Circadian rhythms in drug targets, as well as in transporters, enzymes, etc. can all interact to affect drug efficacy. A simple illustrative example is the antiangiogenic agent SU1498, a VEGFR-2 tyrosine kinase inhibitor, which is maximally effective in the early light phase when production of VEGF is highest [159, 207]. Pharmacological interventions can also target clock function more directly, as exemplified by the use of melatonin and melatonin receptor agonists, which modify clock output signaling, for the treatment of circadian sleep disorders [97, 172]. Melatonin pathway-targeting drugs must be taken with a few hours of bedtime, at the time when melatonin secretion would naturally be increasing, to effectively resynchronize circadian rhythms and improve sleep disorders.

Predicting the best time for drug delivery must account for both pharmacokinetic and pharmacodynamic rhythms. Studies in animal models are done in laboratory conditions that allow relatively accurate prediction of circadian phase; however, in the real world, identifying patient circadian phase can be difficult. Classically, circadian phase estimates come from serially sampling some tissue over the course of a day, which is not practical in a clinical setting. Recent advances in mathematical modeling from large transcriptomic patient datasets have developed single-timepoint sampling approaches that can facilitate personalized chronopharmacology tailored to a patient's individual circadian phase [34, 167, 301].

References

1. Acosta-Rodríguez VA, Rijo-Ferreira F, Green CB, Takahashi JS. Importance of circadian timing for aging and longevity. Nat Commun. 2021;12:2862. https://doi.org/10.1038/s41467-021-22922-6.
2. Adamovich Y, Ladeuix B, Golik M, Koeners MP, Asher G. Rhythmic oxygen levels reset circadian clocks through HIF1α. Cell Metab. 2017;25:93–101. https://doi.org/10.1016/j.cmet.2016.09.014.

3. Aggarwal A, Costa MJ, Rivero-Gutiérrez B, Ji L, Morgan SL, Feldman BJ. The circadian clock regulates adipogenesis by a Per3 crosstalk pathway to Klf15. Cell Rep. 2017;21:2367–75. https://doi.org/10.1016/j.celrep.2017.11.004.
4. Aguilar-Arnal L, Hakim O, Patel VR, Baldi P, Hager GL, Sassone-Corsi P. Cycles in spatial and temporal chromosomal organization driven by the circadian clock. Nat Struct Mol Biol. 2013;20:1206–13. https://doi.org/10.1038/nsmb.2667.
5. Akashi M, Tsuchiya Y, Yoshino T, Nishida E. Control of intracellular dynamics of mammalian period proteins by casein kinase I ε (CKIε) and CKIδ in cultured cells. Mol Cell Biol. 2002;22:1693–703. https://doi.org/10.1128/MCB.22.6.1693-1703.2002.
6. Albrecht U. Timing to perfection: the biology of central and peripheral circadian clocks. Neuron. 2012;74:246–60. https://doi.org/10.1016/j.neuron.2012.04.006.
7. Albrecht U, Sun ZS, Eichele G, Lee CC. A differential response of two putative mammalian circadian regulators, mper1and mper2, to light. Cell. 1997;91:1055–64. https://doi.org/10.1016/S0092-8674(00)80495-X.
8. Albus H, Vansteensel MJ, Michel S, Block GD, Meijer JH. A GABAergic mechanism is necessary for coupling dissociable ventral and dorsal regional oscillators within the circadian clock. Curr Biol. 2005;15:886–93. https://doi.org/10.1016/j.cub.2005.03.051.
9. Alenghat T, Meyers K, Mullican SE, Leitner K, Adeniji-Adele A, Avila J, Bućan M, Ahima RS, Kaestner KH, Lazar MA. Nuclear receptor corepressor and histone deacetylase 3 govern circadian metabolic physiology. Nature. 2008;456:997–1000. https://doi.org/10.1038/nature07541.
10. Allada R, White NE, So WV, Hall JC, Rosbash M. A mutant *drosophila* homolog of mammalian clock disrupts circadian rhythms and transcription of period and timeless. Cell. 1998;93:791–804. https://doi.org/10.1016/S0092-8674(00)81440-3.
11. Altimus CM, Güler AD, Villa KL, McNeill DS, LeGates TA, Hattar S. Rods-cones and melanopsin detect light and dark to modulate sleep independent of image formation. Proc Natl Acad Sci. 2008;105:19998–20003. https://doi.org/10.1073/pnas.0808312105.
12. Amano H, Fukuda Y, Yokoo T, Yamaoka K. Interleukin-6 level among shift and night workers in Japan: cross-sectional analysis of the J-HOPE study. J Atheroscler Thromb. 2018;25:1206–14. https://doi.org/10.5551/jat.42036.
13. Ando H, Yanagihara H, Sugimoto K, Hayashi Y, Tsuruoka S, Takamura T, Kaneko S, Fujimura A. Daily rhythms of p-glycoprotein expression in mice. Chronobiol Int. 2005;22:655–65. https://doi.org/10.1080/07420520500180231.
14. Ángeles-Castellanos M, Amaya JM, Salgado-Delgado R, Buijs RM, Escobar C. Scheduled food hastens re-entrainment more than melatonin does after a 6-h phase advance of the light-dark cycle in rats. J Biol Rhythm. 2011;26:324–34. https://doi.org/10.1177/0748730411409715.
15. Antoch MP, Song E-J, Chang A-M, Vitaterna MH, Zhao Y, Wilsbacher LD, Sangoram AM, King DP, Pinto LH, Takahashi JS. Functional identification of the mouse circadian clock gene by transgenic BAC rescue. Cell. 1997;89:655–67. https://doi.org/10.1016/S0092-8674(00)80246-9.
16. Aryal RP, Kwak PB, Tamayo AG, Gebert M, Chiu P-L, Walz T, Weitz CJ. Macromolecular assemblies of the mammalian circadian clock. Mol Cell. 2017;67:770–782.e6. https://doi.org/10.1016/j.molcel.2017.07.017.
17. Asher G, Gatfield D, Stratmann M, Reinke H, Dibner C, Kreppel F, Mostoslavsky R, Alt FW, Schibler U. SIRT1 regulates circadian clock gene expression through PER2 deacetylation. Cell. 2008;134:317–28. https://doi.org/10.1016/j.cell.2008.06.050.
18. Bae K, Jin X, Maywood ES, Hastings MH, Reppert SM, Weaver DR. Differential functions of mPer1, mPer2, and mPer3 in the SCN circadian clock. Neuron. 2001;30:525–36. https://doi.org/10.1016/S0896-6273(01)00302-6.
19. Balsalobre A, Damiola F, Schibler U. A serum shock induces circadian gene expression in mammalian tissue culture cells. Cell. 1998;93:929–37. https://doi.org/10.1016/S0092-8674(00)81199-X.

20. Bano-Otalora B, Moye MJ, Brown T, Lucas RJ, Diekman CO, Belle MD. Daily electrical activity in the master circadian clock of a diurnal mammal. eLife. 2021;10:e68179. https://doi.org/10.7554/eLife.68179.
21. Bargiello TA, Young MW. Molecular genetics of a biological clock in *drosophila*. Proc Natl Acad Sci. 1984;81:2142–6. https://doi.org/10.1073/pnas.81.7.2142.
22. Bargiello TA, Jackson FR, Young MW. Restoration of circadian behavioural rhythms by gene transfer in drosophila. Nature. 1984;312:752–4. https://doi.org/10.1038/312752a0.
23. Barnea M, Madar Z, Froy O. High-fat diet delays and fasting advances the circadian expression of adiponectin signaling components in mouse liver. Endocrinology. 2009;150:161–8. https://doi.org/10.1210/en.2008-0944.
24. Baron KG, Reid KJ. Circadian misalignment and health. Int Rev Psychiatry. 2014;26:139–54. https://doi.org/10.3109/09540261.2014.911149.
25. Bartholomai BM, Gladfelter AS, Loros JJ, Dunlap JC. PRD-2 mediates clock-regulated perinuclear localization of clock gene RNAs within the circadian cycle of neurospora. Proc Natl Acad Sci USA. 2022;119:e2203078119. https://doi.org/10.1073/pnas.2203078119.
26. Batie M, Frost J, Frost M, Wilson JW, Schofield P, Rocha S. Hypoxia induces rapid changes to histone methylation and reprograms chromatin. Science. 2019;363:1222–6. https://doi.org/10.1126/science.aau5870.
27. Belbin FE, Hall GJ, Jackson AB, Schanschieff FE, Archibald G, Formstone C, Dodd AN. Plant circadian rhythms regulate the effectiveness of a glyphosate-based herbicide. Nat Commun. 2019;10:3704. https://doi.org/10.1038/s41467-019-11709-5.
28. Bellet MM, Sassone-Corsi P. Mammalian circadian clock and metabolism – the epigenetic link. J Cell Sci. 2010;123:3837–48. https://doi.org/10.1242/jcs.051649.
29. Besedovsky L, Lange T, Haack M. The sleep-immune crosstalk in health and disease. Physiol Rev. 2019;99:1325–80. https://doi.org/10.1152/physrev.00010.2018.
30. Blatter K, Graw P, Münch M, Knoblauch V, Wirz-Justice A, Cajochen C. Gender and age differences in psychomotor vigilance performance under differential sleep pressure conditions. Behav Brain Res. 2006;168:312–7. https://doi.org/10.1016/j.bbr.2005.11.018.
31. Borbély AA, Wirz-Justice A. Sleep, sleep deprivation and depression. A hypothesis derived from a model of sleep regulation. Hum Neurobiol. 1982;1:205–10.
32. Borbély AA, Daan S, Wirz-Justice A, Deboer T. The two-process model of sleep regulation: a reappraisal. J Sleep Res. 2016;25:131–43. https://doi.org/10.1111/jsr.12371.
33. Branecky KL, Niswender KD, Pendergast JS. Disruption of daily rhythms by high-fat diet is reversible. PLoS One. 2015;10:e0137970. https://doi.org/10.1371/journal.pone.0137970.
34. Braun R, Kath WL, Iwanaszko M, Kula-Eversole E, Abbott SM, Reid KJ, Zee PC, Allada R. Universal method for robust detection of circadian state from gene expression. Proc Natl Acad Sci. 2018;115:E9247–56. https://doi.org/10.1073/pnas.1800314115.
35. Brenna A, Albrecht U. Phosphorylation and circadian molecular timing. Front Physiol. 2020;11:11.
36. Brzezinski A, Vangel MG, Wurtman RJ, Norrie G, Zhdanova I, Ben-Shushan A, Ford I. Effects of exogenous melatonin on sleep: a meta-analysis. Sleep Med Rev. 2005;9:41–50. https://doi.org/10.1016/j.smrv.2004.06.004.
37. Buhr ED, Yoo S-H, Takahashi JS. Temperature as a universal resetting cue for mammalian circadian oscillators. Science. 2010;330:379–85. https://doi.org/10.1126/science.1195262.
38. Bunger MK, Wilsbacher LD, Moran SM, Clendenin C, Radcliffe LA, Hogenesch JB, Simon MC, Takahashi JS, Bradfield CA. Mop3 is an essential component of the master circadian pacemaker in mammals. Cell. 2000;103:1009–17. https://doi.org/10.1016/S0092-8674(00)00205-1.
39. Burgess HJ. Partial sleep deprivation reduces phase advances to light in humans. J Biol Rhythm. 2010;25:460–8. https://doi.org/10.1177/0748730410385544.
40. Burke TM, Markwald RR, Chinoy ED, Snider JA, Bessman SC, Jung CM, Wright KP. Combination of light and melatonin time cues for phase advancing the human circadian clock. Sleep. 2013;36:1617–24. https://doi.org/10.5665/sleep.3110.

41. Cabrera D, Young MW, Axelrod S. Time-restricted feeding prolongs lifespan in drosophila in a peripheral clock-dependent manner. bioRxiv. 2020; https://doi.org/10.1101/2020.09.14.296368.
42. Campbell SS, Broughton RJ. Rapid decline in body temperature before sleep: fluffing the physiological pillow? Chronobiol Int. 1994;11:126–31. https://doi.org/10.3109/07420529409055899.
43. Cappuccio FP, D'Elia L, Strazzullo P, Miller MA. Sleep duration and all-cause mortality: a systematic review and meta-analysis of prospective studies. Sleep. 2010;33:585–92. https://doi.org/10.1093/sleep/33.5.585.
44. Carrasco-Benso MP, Rivero-Gutierrez B, Lopez-Minguez J, Anzola A, Diez-Noguera A, Madrid JA, Lujan JA, Martínez-Augustin O, Scheer FAJL, Garaulet M. Human adipose tissue expresses intrinsic circadian rhythm in insulin sensitivity. FASEB J. 2016;30:3117–23. https://doi.org/10.1096/fj.201600269RR.
45. Challet E, Turek FW, Laute M-A, Van Reeth O. Sleep deprivation decreases phase-shift responses of circadian rhythms to light in the mouse: role of serotonergic and metabolic signals. Brain Res. 2001;909:81–91. https://doi.org/10.1016/S0006-8993(01)02625-7.
46. Chavan AG, Swan JA, Heisler J, Sancar C, Ernst DC, Fang M, Palacios JG, Spangler RK, Bagshaw CR, Tripathi S, Crosby P, Golden SS, Partch CL, LiWang A. Reconstitution of an intact clock reveals mechanisms of circadian timekeeping. Science. 2021;374:eabd4453. https://doi.org/10.1126/science.abd4453.
47. Chen H, Chen J, Muir LA, Ronquist S, Meixner W, Ljungman M, Ried T, Smale S, Rajapakse I. Functional organization of the human 4D Nucleome. Proc Natl Acad Sci. 2015;112:8002–7. https://doi.org/10.1073/pnas.1505822112.
48. Cheng MY, Bullock CM, Li C, Lee AG, Bermak JC, Belluzzi J, Weaver DR, Leslie FM, Zhou Q-Y. Prokineticin 2 transmits the behavioural circadian rhythm of the suprachiasmatic nucleus. Nature. 2002;417:405–10. https://doi.org/10.1038/417405a.
49. Cho K, Ennaceur A, Cole JC, Suh CK. Chronic jet lag produces cognitive deficits. J Neurosci Off J Soc Neurosci. 2000;20:RC66.
50. Chou TC, Bjorkum AA, Gaus SE, Lu J, Scammell TE, Saper CB. Afferents to the ventrolateral preoptic nucleus. J Neurosci. 2002;22:977–90. https://doi.org/10.1523/JNEUROSCI.22-03-00977.2002.
51. Christian CA, Harrington ME. Three days of novel wheel access diminishes light-induced phase delays in vivo with no effect on *per1* induction by light. Chronobiol Int. 2002;19:671–82. https://doi.org/10.1081/CBI-120006079.
52. Collins B, Pierre-Ferrer S, Muheim C, Lukacsovich D, Cai Y, Spinnler A, Herrera CG, Wen S, Winterer J, Belle MDC, Piggins HD, Hastings M, Loudon A, Yan J, Földy C, Adamantidis A, Brown SA. Circadian VIPergic neurons of the suprachiasmatic nuclei sculpt the sleep-wake cycle. Neuron. 2020;108:486–499.e5. https://doi.org/10.1016/j.neuron.2020.08.001.
53. Colot HV, Park G, Turner GE, Ringelberg C, Crew CM, Litvinkova L, Weiss RL, Borkovich KA, Dunlap JC. A high-throughput gene knockout procedure for Neurospora reveals functions for multiple transcription factors. Proc Natl Acad Sci USA. 2006;103:10352–7. https://doi.org/10.1073/pnas.0601456103.
54. Colwell CS. Linking neural activity and molecular oscillations in the SCN. Nat Rev Neurosci. 2011;12:553–69. https://doi.org/10.1038/nrn3086.
55. Crosthwaite SK, Dunlap JC, Loros JJ. Neurospora wc-1 and wc-2: transcription, Photoresponses, and the origins of circadian rhythmicity. Science. 1997;276:763–9. https://doi.org/10.1126/science.276.5313.763.
56. Cuddapah VA, Zhang SL, Sehgal A. Regulation of the blood–brain barrier by circadian rhythms and sleep. Trends Neurosci. 2019;42:500–10. https://doi.org/10.1016/j.tins.2019.05.001.
57. Curie T, Maret S, Emmenegger Y, Franken P. *In vivo* imaging of the central and peripheral effects of sleep deprivation and suprachiasmatic nuclei lesion on PERIOD-2 protein in mice. Sleep. 2015;38:1381–94. https://doi.org/10.5665/sleep.4974.

58. Czeisler CA, Weitzman ED, Moore-Ede MC, Zimmerman JC, Knauer RS. Human sleep: its duration and organization depend on its circadian phase. Science. 1980;210:1264–7. https://doi.org/10.1126/science.7434029.
59. Daan S. Colin Pittendrigh, Jurgen Aschoff, and the natural entrainment of circadian systems. J Biol Rhythm. 2000;15:195–207. https://doi.org/10.1177/074873040001500301.
60. Damiola F, Le Minh N, Preitner N, Kornmann B, Fleury-Olela F, Schibler U. Restricted feeding uncouples circadian oscillators in peripheral tissues from the central pacemaker in the suprachiasmatic nucleus. Genes Dev. 2000;14:2950–61. https://doi.org/10.1101/gad.183500.
61. Darlington TK, Wager-Smith K, Ceriani MF, Staknis D, Gekakis N, Steeves TDL, Weitz CJ, Takahashi JS, Kay SA. Closing the circadian loop: CLOCK-induced transcription of its own inhibitors *per* and *tim*. Science. 1998;280:1599–603. https://doi.org/10.1126/science.280.5369.1599.
62. Davidson AJ, Sellix MT, Daniel J, Yamazaki S, Menaker M, Block GD. Chronic jet-lag increases mortality in aged mice. Curr Biol. 2006;16:R914–6. https://doi.org/10.1016/j.cub.2006.09.058.
63. Deboer T, Vansteensel MJ, Détári L, Meijer JH. Sleep states alter activity of suprachiasmatic nucleus neurons. Nat Neurosci. 2003;6:1086–90. https://doi.org/10.1038/nn1122.
64. Deboer T, Détári L, Meijer JH. Long term effects of sleep deprivation on the mammalian circadian pacemaker. Sleep. 2007;30:257–62. https://doi.org/10.1093/sleep/30.3.257.
65. DeBruyne JP, Weaver DR, Reppert SM. CLOCK and NPAS2 have overlapping roles in the suprachiasmatic circadian clock. Nat Neurosci. 2007;10:543–5. https://doi.org/10.1038/nn1884.
66. Depner CM, Stothard ER, Wright KP. Metabolic consequences of sleep and circadian disorders. Curr Diab Rep. 2014;14:507. https://doi.org/10.1007/s11892-014-0507-z.
67. Devan BD, Goad EH, Petri HL, Antoniadis EA, Hong NS, Ko CH, Leblanc L, Lebovic SS, Lo Q, Ralph MR, McDonald RJ. Circadian phase-shifted rats show normal acquisition but impaired long-term retention of place information in the water task. Neurobiol Learn Mem. 2001;75:51–62. https://doi.org/10.1006/nlme.1999.3957.
68. Dijk D-J, Archer SN. PERIOD3, circadian phenotypes, and sleep homeostasis. Sleep Med Rev. 2010;14:151–60. https://doi.org/10.1016/j.smrv.2009.07.002.
69. Dijk D, Czeisler C. Contribution of the circadian pacemaker and the sleep homeostat to sleep propensity, sleep structure, electroencephalographic slow waves, and sleep spindle activity in humans. J Neurosci. 1995;15:3526–38. https://doi.org/10.1523/JNEUROSCI.15-05-03526.1995.
70. Dijk D-J, Duffy JF, Czeisler CA. Contribution of circadian physiology and sleep homeostasis to age-related changes in human sleep. Chronobiol Int. 2000;17:285–311. https://doi.org/10.1081/CBI-100101049.
71. Ding JM, Faiman LE, Hurst WJ, Kuriashkina LR, Gillette MU. Resetting the biological clock: mediation of nocturnal CREB phosphorylation via light, glutamate, and nitric oxide. J Neurosci. 1997;17:667–75. https://doi.org/10.1523/JNEUROSCI.17-02-00667.1997.
72. DiTacchio L, Le HD, Vollmers C, Hatori M, Witcher M, Secombe J, Panda S. Histone lysine demethylase JARID1a activates CLOCK-BMAL1 and influences the circadian clock. Science. 2011;333:1881–5. https://doi.org/10.1126/science.1206022.
73. Doi M, Hirayama J, Sassone-Corsi P. Circadian regulator CLOCK is a histone acetyltransferase. Cell. 2006;125:497–508. https://doi.org/10.1016/j.cell.2006.03.033.
74. Duane WC, Levitt DG, Mueller SM, Behrens JC. Regulation of bile acid synthesis in man. Presence of a diurnal rhythm. J Clin Invest. 1983;72:1930–6. https://doi.org/10.1172/JCI111157.
75. Duez H, van der Veen JN, Duhem C, Pourcet B, Touvier T, Fontaine C, Derudas B, Baugé E, Havinga R, Bloks VW, Wolters H, van der Sluijs FH, Vennström B, Kuipers F, Staels B. Regulation of bile acid synthesis by the nuclear receptor rev-erbα. Gastroenterology. 2008;135:689–698.e5. https://doi.org/10.1053/j.gastro.2008.05.035.

76. Easton A, Meerlo P, Bergmann B, Turek FW. The suprachiasmatic nucleus regulates sleep timing and amount in mice. Sleep. 2004;27:1307–18. https://doi.org/10.1093/sleep/27.7.1307.
77. Edgar D, Dement W, Fuller C. Effect of SCN lesions on sleep in squirrel monkeys: evidence for opponent processes in sleep-wake regulation. J Neurosci. 1993;13:1065–79. https://doi.org/10.1523/JNEUROSCI.13-03-01065.1993.
78. Edgar RS, Stangherlin A, Nagy AD, Nicoll MP, Efstathiou S, O'Neill JS, Reddy AB. Cell autonomous regulation of herpes and influenza virus infection by the circadian clock. Proc Natl Acad Sci. 2016;113:10085–90. https://doi.org/10.1073/pnas.1601895113.
79. Ehlen JC, Brager AJ, Baggs J, Pinckney L, Gray CL, DeBruyne JP, Esser KA, Takahashi JS, Paul KN. *Bmal1* function in skeletal muscle regulates sleep. eLife. 2017;6:e26557. https://doi.org/10.7554/eLife.26557.
80. Elliott AL, Mills JN, Minors DS, Waterhouse JM. The effect of real and simulated time-zone shifts upon the circadian rhythms of body temperature, plasma 11-hydroxycorticosteroids, and renal excretion in human subjects. J Physiol. 1972;221:227–57. https://doi.org/10.1113/jphysiol.1972.sp009750.
81. Escobar C, Espitia-Bautista E, Guzmán-Ruiz MA, Guerrero-Vargas NN, Hernández-Navarrete MÁ, Ángeles-Castellanos M, Morales-Pérez B, Buijs RM. Chocolate for breakfast prevents circadian desynchrony in experimental models of jet-lag and shift-work. Sci Rep. 2020;10:6243. https://doi.org/10.1038/s41598-020-63227-w.
82. Etchegaray J-P, Yang X, DeBruyne JP, Peters AHFM, Weaver DR, Jenuwein T, Reppert SM. The polycomb group protein EZH2 is required for mammalian circadian clock function. J Biol Chem. 2006;281:21209–15. https://doi.org/10.1074/jbc.M603722200.
83. Evans JA, Davidson AJ. Health consequences of circadian disruption in humans and animal models. In: Progress in molecular biology and translational science. Elsevier; 2013. p. 283–323. https://doi.org/10.1016/B978-0-12-396971-2.00010-5.
84. Everson CA. Sustained sleep deprivation impairs host defense. Am J Physiol-Regul Integr Comp Physiol. 1993;265:R1148–54. https://doi.org/10.1152/ajpregu.1993.265.5.R1148.
85. Everson CA, Toth LA. Systemic bacterial invasion induced by sleep deprivation. Am J Physiol-Regul Integr Comp Physiol. 2000;278:R905–16. https://doi.org/10.1152/ajpregu.2000.278.4.R905.
86. Fang B, Everett LJ, Jager J, Briggs E, Armour SM, Feng D, Roy A, Gerhart-Hines Z, Sun Z, Lazar MA. Circadian enhancers coordinate multiple phases of rhythmic gene transcription in vivo. Cell. 2014;159:1140–52. https://doi.org/10.1016/j.cell.2014.10.022.
87. Feng D, Liu T, Sun Z, Bugge A, Mullican SE, Alenghat T, Liu XS, Lazar MA. A circadian rhythm orchestrated by histone deacetylase 3 controls hepatic lipid metabolism. Science. 2011;331:1315–9. https://doi.org/10.1126/science.1198125.
88. Filipski E, Delaunay F, King VM, Wu M-W, Claustrat B, Gréchez-Cassiau A, Guettier C, Hastings MH, Francis L. Effects of chronic jet lag on tumor progression in mice. Cancer Res. 2004;64:7879–85. https://doi.org/10.1158/0008-5472.CAN-04-0674.
89. Finger A-M, Jäschke S, del Olmo M, Hurwitz R, Granada AE, Herzel H, Kramer A. Intercellular coupling between peripheral circadian oscillators by TGF-β signaling. Sci Adv. 2021;7:eabg5174. https://doi.org/10.1126/sciadv.abg5174.
90. Franken P. A role for clock genes in sleep homeostasis. Curr Opin Neurobiol. 2013;23:864–72. https://doi.org/10.1016/j.conb.2013.05.002.
91. Franken P, Dudley CA, Estill SJ, Barakat M, Thomason R, O'Hara BF, McKnight SL. NPAS2 as a transcriptional regulator of non-rapid eye movement sleep: genotype and sex interactions. Proc Natl Acad Sci. 2006;103:7118–23. https://doi.org/10.1073/pnas.0602006103.
92. Gachon F, Fonjallaz P, Damiola F, Gos P, Kodama T, Zakany J, Duboule D, Petit B, Tafti M, Schibler U. The loss of circadian PAR bZip transcription factors results in epilepsy. Genes Dev. 2004;18:1397–412. https://doi.org/10.1101/gad.301404.
93. Gallego M, Virshup DM. Post-translational modifications regulate the ticking of the circadian clock. Nat Rev Mol Cell Biol. 2007;8:139–48. https://doi.org/10.1038/nrm2106.

94. Gamble KL, Allen GC, Zhou T, McMahon DG. Gastrin-releasing peptide mediates light-like resetting of the suprachiasmatic nucleus circadian pacemaker through cAMP response element-binding protein and *per1* activation. J Neurosci. 2007;27:12078–87. https://doi.org/10.1523/JNEUROSCI.1109-07.2007.
95. Gekakis N, Staknis D, Nguyen HB, Davis FC, Wilsbacher LD, King DP, Takahashi JS, Weitz CJ. Role of the CLOCK protein in the mammalian circadian mechanism. Science. 1998;280:1564–9. https://doi.org/10.1126/science.280.5369.1564.
96. Ginty DD, Kornhauser JM, Thompson MA, Bading H, Mayo KE, Takahashi JS, Greenberg ME. Regulation of CREB phosphorylation in the suprachiasmatic nucleus by light and a circadian clock. Science. 1993;260:238–41. https://doi.org/10.1126/science.8097062.
97. Golombek DA, Pandi-Perumal SR, Brown GM, Cardinali DP. Some implications of melatonin use in chronopharmacology of insomnia. Eur J Pharmacol. 2015;762:42–8. https://doi.org/10.1016/j.ejphar.2015.05.032.
98. Goo RH, Moore JG, Greenberg E, Alazraki NP. Circadian variation in gastric emptying of meals in humans. Gastroenterology. 1987;93:515–8. https://doi.org/10.1016/0016-5085(87)90913-9.
99. Gooch VD, Mehra A, Larrondo LF, Fox J, Touroutoutoudis M, Loros JJ, Dunlap JC. Fully codon-optimized luciferase uncovers novel temperature characteristics of the Neurospora clock. Eukaryot Cell. 2008;7:28–37. https://doi.org/10.1128/EC.00257-07.
100. Greco CM, Koronowski KB, Smith JG, Shi J, Kunderfranco P, Carriero R, Chen S, Samad M, Welz P-S, Zinna VM, Mortimer T, Chun SK, Shimaji K, Sato T, Petrus P, Kumar A, Vaca-Dempere M, Deryagin O, Van C, Kuhn JMM, Lutter D, Seldin MM, Masri S, Li W, Baldi P, Dyar KA, Muñoz-Cánoves P, Benitah SA, Sassone-Corsi P. Integration of feeding behavior by the liver circadian clock reveals network dependency of metabolic rhythms. Sci Adv. 2021;7:eabi7828. https://doi.org/10.1126/sciadv.abi7828.
101. Greene MW. Circadian rhythms and tumor growth. Cancer Lett. 2012;318:115–23. https://doi.org/10.1016/j.canlet.2012.01.001.
102. Guerrero-Vargas NN, Espitia-Bautista E, Buijs RM, Escobar C. Shift-work: is time of eating determining metabolic health? Evidence from animal models. Proc Nutr Soc. 2018;77:199–215. https://doi.org/10.1017/S0029665117004128.
103. Guo H, Brewer JM, Champhekar A, Harris RBS, Bittman EL. Differential control of peripheral circadian rhythms by suprachiasmatic-dependent neural signals. Proc Natl Acad Sci. 2005;102:3111–6. https://doi.org/10.1073/pnas.0409734102.
104. Halberg F, Johnson EA, Brown BW, Bittner JJ. Susceptibility rhythm to *E. Coli* endotoxin and bioassay. Exp Biol Med. 1960;103:142–4. https://doi.org/10.3181/00379727-103-25439.
105. Handler AM, Konopka RJ. Transplantation of a circadian pacemaker in drosophila. Nature. 1979;279:236–8. https://doi.org/10.1038/279236a0.
106. Harding C, Bechtold DA, Brown TM. Suprachiasmatic nucleus-dependent and independent outputs driving rhythmic activity in hypothalamic and thalamic neurons. BMC Biol. 2020;18:134. https://doi.org/10.1186/s12915-020-00871-8.
107. Hasan S, Veen DR, Winsky-Sommerer R, Hogben A, Laing EE, Koentgen F, Dijk D, Archer SN. A human sleep homeostasis phenotype in mice expressing a primate-specific *PER3* variable-number tandem-repeat coding-region polymorphism. FASEB J. 2014;28:2441–54. https://doi.org/10.1096/fj.13-240135.
108. Hasan N, Nagata N, Morishige J, Islam MT, Jing Z, Harada K, Mieda M, Ono M, Fujiwara H, Daikoku T, Fujiwara T, Maida Y, Ota T, Shimba S, Kaneko S, Fujimura A, Ando H. Brown adipocyte-specific knockout of Bmal1 causes mild but significant thermogenesis impairment in mice. Mol Metab. 2021;49:101202. https://doi.org/10.1016/j.molmet.2021.101202.
109. Haspel JA, Chettimada S, Shaik RS, Chu J-H, Raby BA, Cernadas M, Carey V, Process V, Hunninghake GM, Ifedigbo E, Lederer JA, Englert J, Pelton A, Coronata A, Fredenburgh LE, Choi AMK. Circadian rhythm reprogramming during lung inflammation. Nat Commun. 2014;5:4753. https://doi.org/10.1038/ncomms5753.

110. Haus EL, Smolensky MH. Shift work and cancer risk: potential mechanistic roles of circadian disruption, light at night, and sleep deprivation. Sleep Med Rev. 2013;17:273–84. https://doi.org/10.1016/j.smrv.2012.08.003.
111. Hirano A, Shi G, Jones CR, Lipzen A, Pennacchio LA, Xu Y, Hallows WC, McMahon T, Yamazaki M, Ptáček LJ, Fu Y-H. A cryptochrome 2 mutation yields advanced sleep phase in humans. eLife. 2016;5:e16695. https://doi.org/10.7554/eLife.16695.
112. Hofman MA, Fliers E, Goudsmit E, Swaab DF. Morphometric analysis of the suprachiasmatic and paraventricular nuclei in the human brain: sex differences and age-dependent changes. J Anat. 1988;160:127–43.
113. Hood S, Amir S. The aging clock: circadian rhythms and later life. J Clin Invest. 2017;127:437–46. https://doi.org/10.1172/JCI90328.
114. Hoogerwerf WA, Shahinian VB, Cornélissen G, Halberg F, Bostwick J, Timm J, Bartell PA, Cassone VM. Rhythmic changes in colonic motility are regulated by *period* genes. Am J Physiol-Gastrointest Liver Physiol. 2010;298:G143–50. https://doi.org/10.1152/ajpgi.00402.2009.
115. Horsey EA, Maletta T, Turner H, Cole C, Lehmann H, Fournier NM. Chronic jet lag simulation decreases hippocampal neurogenesis and enhances depressive behaviors and cognitive deficits in adult male rats. Front Behav Neurosci. 2020;13:272. https://doi.org/10.3389/fnbeh.2019.00272.
116. Horst GTJ van der, Muijtjens M, Kobayashi K, Takano R, Kanno S, Takao M, Wit J de, Verkerk A, Eker APM, Leenen D van, Buijs R, Bootsma D, Hoeijmakers JHJ, Yasui A. Mammalian Cry1 and Cry2 are essential for maintenance of circadian rhythms. Nature. 1999;398:627–30. https://doi.org/10.1038/19323.
117. Hosoda H, Kato K, Asano H, Ito M, Kato H, Iwamoto T, Suzuki A, Masushige S, Kida S. CBP/p300 is a cell type-specific modulator of CLOCK/BMAL1-mediated transcription. Mol Brain. 2009;2:34. https://doi.org/10.1186/1756-6606-2-34.
118. Huang N, Chelliah Y, Shan Y, Taylor CA, Yoo S-H, Partch C, Green CB, Zhang H, Takahashi JS. Crystal structure of the heterodimeric CLOCK:BMAL1 transcriptional activator complex. Science. 2012;337:189–94. https://doi.org/10.1126/science.1222804.
119. Hubbard J, Gent TC, Hoekstra MMB, Emmenegger Y, Mongrain V, Landolt H-P, Adamantidis AR, Franken P. Rapid fast-delta decay following prolonged wakefulness marks a phase of wake-inertia in NREM sleep. Nat Commun. 2020;11:3130. https://doi.org/10.1038/s41467-020-16915-0.
120. Hughes ATL, Samuels RE, Baño-Otálora B, Belle MDC, Wegner S, Guilding C, Northeast RC, Loudon ASI, Gigg J, Piggins HD. Timed daily exercise remodels circadian rhythms in mice. Commun Biol. 2021;4:761. https://doi.org/10.1038/s42003-021-02239-2.
121. Hung H-C, Maurer C, Kay SA, Weber F. Circadian transcription depends on limiting amounts of the transcription co-activator nejire/CBP. J Biol Chem. 2007;282:31349–57. https://doi.org/10.1074/jbc.m702319200.
122. Husse J, Zhou X, Shostak A, Oster H, Eichele G. Synaptotagmin10-Cre, a driver to disrupt clock genes in the SCN. J Biol Rhythm. 2011;26:379–89. https://doi.org/10.1177/0748730411415363.
123. Ibuka N, Kawamura H. Loss of circadian rhythm in sleep-wakefulness cycle in the rat by suprachiasmatic nucleus lesions. Brain Res. 1975;96:76–81. https://doi.org/10.1016/0006-8993(75)90574-0.
124. Ibuka N, Nihonmatsu I, Sekiguchi S. Sleep-wakefulness rhythms in mice after suprachiasmatic nucleus lesions. Waking Sleeping. 1980;4:167–73.
125. Iitaka C, Miyazaki K, Akaike T, Ishida N. A role for glycogen synthase kinase-3beta in the mammalian circadian clock. J Biol Chem. 2005;280:29397–402. https://doi.org/10.1074/jbc.M503526200.
126. Ikeda R, Tsuchiya Y, Koike N, Umemura Y, Inokawa H, Ono R, Inoue M, Sasawaki Y, Grieten T, Okubo N, Ikoma K, Fujiwara H, Kubo T, Yagita K. REV-ERBα and REV-ERBβ function as key factors regulating mammalian circadian output. Sci Rep. 2019;9:10171. https://doi.org/10.1038/s41598-019-46656-0.

127. Ingram LA, Simpson RJ, Malone E, Florida-James GD. Sleep disruption and its effect on lymphocyte redeployment following an acute bout of exercise. Brain Behav Immun. 2015;47:100–8. https://doi.org/10.1016/j.bbi.2014.12.018.
128. Inokawa H, Umemura Y, Shimba A, Kawakami E, Koike N, Tsuchiya Y, Ohashi M, Minami Y, Cui G, Asahi T, Ono R, Sasawaki Y, Konishi E, Yoo S-H, Chen Z, Teramukai S, Ikuta K, Yagita K. Chronic circadian misalignment accelerates immune senescence and abbreviates lifespan in mice. Sci Rep. 2020;10:2569. https://doi.org/10.1038/s41598-020-59541-y.
129. Izumo M, Pejchal M, Schook AC, Lange RP, Walisser JA, Sato TR, Wang X, Bradfield CA, Takahashi JS. Differential effects of light and feeding on circadian organization of peripheral clocks in a forebrain Bmal1 mutant. eLife. 2014;3:e04617. https://doi.org/10.7554/eLife.04617.
130. Jagannath A, Butler R, Godinho SIH, Couch Y, Brown LA, Vasudevan SR, Flanagan KC, Anthony D, Churchill GC, Wood MJA, Steiner G, Ebeling M, Hossbach M, Wettstein JG, Duffield GE, Gatti S, Hankins MW, Foster RG, Peirson SN. The CRTC1-SIK1 pathway regulates entrainment of the circadian clock. Cell. 2013;154:1100–11. https://doi.org/10.1016/j.cell.2013.08.004.
131. Jankowski MS, Griffith D, Shastry DG, Pelham JF, Ginell GM, Thomas J, Karande P, Holehouse AS, Hurley JM. The formation of a fuzzy complex in the negative arm regulates the robustness of the circadian clock (preprint). Mol Biol. 2022; https://doi.org/10.1101/2022.01.04.474980.
132. Johnson CH. Testing the adaptive value of circadian systems. In: Young MW, editor. Methods in enzymology, circadian rhythms. Academic; 2005. p. 818–37. https://doi.org/10.1016/S0076-6879(05)93043-7.
133. Jones CR, Campbell SS, Zone SE, Cooper F, DeSano A, Murphy PJ, Jones B, Czajkowski L, Ptček LJ. Familial advanced sleep-phase syndrome: a short-period circadian rhythm variant in humans. Nat Med. 1999;5:1062–5. https://doi.org/10.1038/12502.
134. Jones SE, Lane JM, Wood AR, van Hees VT, Tyrrell J, Beaumont RN, Jeffries AR, Dashti HS, Hillsdon M, Ruth KS, Tuke MA, Yaghootkar H, Sharp SA, Jie Y, Thompson WD, Harrison JW, Dawes A, Byrne EM, Tiemeier H, Allebrandt KV, Bowden J, Ray DW, Freathy RM, Murray A, Mazzotti DR, Gehrman PR, Lawlor DA, Frayling TM, Rutter MK, Hinds DA, Saxena R, Weedon MN. Genome-wide association analyses of chronotype in 697,828 individuals provides insights into circadian rhythms. Nat Commun. 2019;10:343. https://doi.org/10.1038/s41467-018-08259-7.
135. Jones JR, Chaturvedi S, Granados-Fuentes D, Herzog ED. Circadian neurons in the paraventricular nucleus entrain and sustain daily rhythms in glucocorticoids. Nat Commun. 2021;12:5763. https://doi.org/10.1038/s41467-021-25959-9.
136. Kalsbeek A, Palm IF, La Fleur SE, Scheer FAJL, Perreau-Lenz S, Ruiter M, Kreier F, Cailotto C, Buijs RM. SCN outputs and the hypothalamic balance of life. J Biol Rhythm. 2006;21:458–69. https://doi.org/10.1177/0748730406293854.
137. Kamon S, Harada Y, Sakai M, Todo T, Fukada Y. Serine phosphorylation of mCRY1 and mCRY2 by mitogen-activated protein kinase. Genes Cells Devoted Mol Cell Mech. 2004;9:697. https://doi.org/10.1111/j.1356-9597.2004.00758.x.
138. Kantor S, Mochizuki T, Janisiewicz AM, Clark E, Nishino S, Scammell TE. Orexin neurons are necessary for the circadian control of REM sleep. Sleep. 2009;32:1127–34. https://doi.org/10.1093/sleep/32.9.1127.
139. Katewa SD, Akagi K, Bose N, Rakshit K, Camarella T, Zheng X, Hall D, Davis S, Nelson CS, Brem RB, Ramanathan A, Sehgal A, Giebultowicz JM, Kapahi P. Peripheral circadian clocks mediate dietary restriction-dependent changes in lifespan and fat metabolism in drosophila. Cell Metab. 2016;23:143–54. https://doi.org/10.1016/j.cmet.2015.10.014.
140. Keller M, Mazuch J, Abraham U, Eom GD, Herzog ED, Volk H-D, Kramer A, Maier B. A circadian clock in macrophages controls inflammatory immune responses. Proc Natl Acad Sci. 2009;106:21407–12. https://doi.org/10.1073/pnas.0906361106.

141. Kervezee L, Kosmadopoulos A, Boivin DB. Metabolic and cardiovascular consequences of shift work: the role of circadian disruption and sleep disturbances. Eur J Neurosci. 2020;51:396–412. https://doi.org/10.1111/ejn.14216.
142. Kettner NM, Voicu H, Finegold MJ, Coarfa C, Sreekumar A, Putluri N, Katchy CA, Lee C, Moore DD, Fu L. Circadian homeostasis of liver metabolism suppresses hepatocarcinogenesis. Cancer Cell. 2016;30:909–24. https://doi.org/10.1016/j.ccell.2016.10.007.
143. Kiessling S, Eichele G, Oster H. Adrenal glucocorticoids have a key role in circadian resynchronization in a mouse model of jet lag. J Clin Invest. 2010;120:2600–9. https://doi.org/10.1172/JCI41192.
144. Kiessling S, Beaulieu-Laroche L, Blum ID, Landgraf D, Welsh DK, Storch K-F, Labrecque N, Cermakian N. Enhancing circadian clock function in cancer cells inhibits tumor growth. BMC Biol. 2017;15:13. https://doi.org/10.1186/s12915-017-0349-7.
145. Kim SJ, Chi C, Pattanayak G, Dinner AR, Rust MJ. KidA, a multi-PAS domain protein, tunes the period of the cyanobacterial circadian oscillator. Proc Natl Acad Sci USA. 2022;119:e2202426119. https://doi.org/10.1073/pnas.2202426119.
146. King DP, Zhao Y, Sangoram AM, Wilsbacher LD, Tanaka M, Antoch MP, Steeves TDL, Vitaterna MH, Kornhauser JM, Lowrey PL, Turek FW, Takahashi JS. Positional cloning of the mouse circadian gene. Cell. 1997;89:641–53. https://doi.org/10.1016/S0092-8674(00)80245-7.
147. Knoblauch V. Homeostatic control of slow-wave and spindle frequency activity during human sleep: effect of differential sleep pressure and brain topography. Cereb Cortex. 2002;12:1092–100. https://doi.org/10.1093/cercor/12.10.1092.
148. Koh K, Evans JM, Hendricks JC, Sehgal A. A drosophila model for age-associated changes in sleep: wake cycles. Proc Natl Acad Sci USA. 2006;103:13843–7. https://doi.org/10.1073/pnas.0605903103.
149. Kohsaka A, Laposky AD, Ramsey KM, Estrada C, Joshu C, Kobayashi Y, Turek FW, Bass J. High-fat diet disrupts behavioral and molecular circadian rhythms in mice. Cell Metab. 2007;6:414–21. https://doi.org/10.1016/j.cmet.2007.09.006.
150. Koike N, Yoo S-H, Huang H-C, Kumar V, Lee C, Kim T-K, Takahashi JS. Transcriptional architecture and chromatin landscape of the core circadian clock in mammals. Science. 2012;338:349–54. https://doi.org/10.1126/science.1226339.
151. Kolonko M, Greb-Markiewicz B. bHLH–PAS proteins: their structure and intrinsic disorder. Int J Mol Sci. 2019;20:3653. https://doi.org/10.3390/ijms20153653.
152. Kondo T, Tsinoremas NF, Golden SS, Johnson CH, Kutsuna S, Ishiura M. Circadian clock mutants of cyanobacteria. Science. 1994;266:1233–6. https://doi.org/10.1126/science.7973706.
153. Kondratov RV, Kondratova AA, Gorbacheva VY, Vykhovanets OV, Antoch MP. Early aging and age-related pathologies in mice deficient in BMAL1, the core component of the circadian clock. Genes Dev. 2006;20:1868–73. https://doi.org/10.1101/gad.1432206.
154. Konopka RJ, Benzer S. Clock Mutants of Drosophila melanogaster. Proc Natl Acad Sci U S A. 1971;68:2112–6. https://doi.org/10.1073/pnas.68.9.2112.
155. Koopman MG, Koomen GCM, Krediet RT, de Moor EAM, Hoek FJ, Arisz L. Circadian rhythm of glomerular filtration rate in normal individuals. Clin Sci. 1989;77:105–11. https://doi.org/10.1042/cs0770105.
156. Kopp C, Albrecht U, Zheng B, Tobler I. Homeostatic sleep regulation is preserved in mPer1 and mPer2 mutant mice: sleep homeostasis in mPer1 and mPer2 mutant mice. Eur J Neurosci. 2002;16:1099–106. https://doi.org/10.1046/j.1460-9568.2002.02156.x.
157. Kornmann B, Schaad O, Bujard H, Takahashi JS, Schibler U. System-driven and oscillator-dependent circadian transcription in mice with a conditionally active liver clock. PLoS Biol. 2007;5:e34. https://doi.org/10.1371/journal.pbio.0050034.
158. Koronowski KB, Kinouchi K, Welz P-S, Smith JG, Zinna VM, Shi J, Samad M, Chen S, Magnan CN, Kinchen JM, Li W, Baldi P, Benitah SA, Sassone-Corsi P. Defining the indepen-

dence of the liver circadian clock. Cell. 2019;177:1448–1462.e14. https://doi.org/10.1016/j.cell.2019.04.025.
159. Koyanagi S, Kuramoto Y, Nakagawa H, Aramaki H, Ohdo S, Soeda S, Shimeno H. A molecular mechanism regulating circadian expression of vascular endothelial growth factor in tumor cells. Cancer Res. 2003;63:7277–83.
160. Kramer A, Yang F-C, Snodgrass P, Li X, Scammell TE, Davis FC, Weitz CJ. Regulation of daily locomotor activity and sleep by hypothalamic EGF receptor signaling. Science. 2001;294:2511–5. https://doi.org/10.1126/science.1067716.
161. Krauchi K, Wirz-Justice A. Circadian rhythm of heat production, heart rate, and skin and core temperature under unmasking conditions in men. Am J Physiol Regul Integr Comp Physiol. 1994;267:R819–29. https://doi.org/10.1152/ajpregu.1994.267.3.R819.
162. Kräuchi K, Cajochen C, Werth E, Wirz-Justice A. Functional link between distal vasodilation and sleep-onset latency? Am J Physiol Regul Integr Comp Physiol. 2000;278:R741–8. https://doi.org/10.1152/ajpregu.2000.278.3.R741.
163. Kräuchi K, Knoblauch V, Wirz-Justice A, Cajochen C. Challenging the sleep homeostat does not influence the thermoregulatory system in men: evidence from a nap vs. sleep-deprivation study. Am J Physiol Regul Integr Comp Physiol. 2006;290:R1052–61. https://doi.org/10.1152/ajpregu.00381.2005.
164. Krishnan N, Kretzschmar D, Rakshit K, Chow E, Giebultowicz JM. The circadian clock gene period extends healthspan in aging Drosophila melanogaster. Aging. 2009;1:937–48. https://doi.org/10.18632/aging.100103.
165. Kumar S, Mohan A, Sharma VK. Circadian dysfunction reduces lifespan in Drosophila melanogaster. Chronobiol Int. 2005;22:641–53. https://doi.org/10.1080/07420520500179423.
166. Kurien P, Hsu P-K, Leon J, Wu D, McMahon T, Shi G, Xu Y, Lipzen A, Pennacchio LA, Jones CR, Fu Y-H, Ptáček LJ. TIMELESS mutation alters phase responsiveness and causes advanced sleep phase. Proc Natl Acad Sci. 2019;116:12045–53. https://doi.org/10.1073/pnas.1819110116.
167. Laing EE, Möller-Levet CS, Poh N, Santhi N, Archer SN, Dijk D-J. Blood transcriptome based biomarkers for human circadian phase. eLife. 2017;6:e20214. https://doi.org/10.7554/eLife.20214.
168. Lamia KA, Storch K-F, Weitz CJ. Physiological significance of a peripheral tissue circadian clock. Proc Natl Acad Sci. 2008;105:15172–7. https://doi.org/10.1073/pnas.0806717105.
169. Laposky A, Easton A, Dugovic C, Walisser J, Bradfield C, Turek F. Deletion of the mammalian circadian clock gene *BMAL1/Mop3* alters baseline sleep architecture and the response to sleep deprivation. Sleep. 2005;28:395–410. https://doi.org/10.1093/sleep/28.4.395.
170. Le Martelot G, Claudel T, Gatfield D, Schaad O, Kornmann B, Sasso GL, Moschetta A, Schibler U. REV-ERBα participates in circadian SREBP signaling and bile acid homeostasis. PLoS Biol. 2009;7:e1000181. https://doi.org/10.1371/journal.pbio.1000181.
171. Lee IT, Chang AS, Manandhar M, Shan Y, Fan J, Izumo M, Ikeda Y, Motoike T, Dixon S, Seinfeld JE, Takahashi JS, Yanagisawa M. Neuromedin s-producing neurons act as essential pacemakers in the suprachiasmatic nucleus to couple clock neurons and dictate circadian rhythms. Neuron. 2015;85:1086–102. https://doi.org/10.1016/j.neuron.2015.02.006.
172. Lee Y, Field JM, Sehgal A. Circadian rhythms, disease and chronotherapy. J Biol Rhythm. 2021;36:503–31. https://doi.org/10.1177/07487304211044301.
173. Lehman M, Silver R, Gladstone W, Kahn R, Gibson M, Bittman E. Circadian rhythmicity restored by neural transplant. Immunocytochemical characterization of the graft and its integration with the host brain. J Neurosci. 1987;7:1626–38. https://doi.org/10.1523/JNEUROSCI.07-06-01626.1987.
174. Lemmer B, Nold G. Circadian changes in estimated hepatic blood flow in healthy subjects. Br J Clin Pharmacol. 1991;32:627–9. https://doi.org/10.1111/j.1365-2125.1991.tb03964.x.
175. Liao S, Amcoff M, Nässel DR. Impact of high-fat diet on lifespan, metabolism, fecundity and behavioral senescence in drosophila. Insect Biochem Mol Biol. 2020;133:103495. https://doi.org/10.1016/j.ibmb.2020.103495.

176. Lim C, Lee J, Choi C, Kim J, Doh E, Choe J. Functional role of CREB-binding protein in the circadian clock system of Drosophila melanogaster. Mol Cell Biol. 2007;27:4876–90. https://doi.org/10.1128/MCB.02155-06.
177. Liu AC, Welsh DK, Ko CH, Tran HG, Zhang EE, Priest AA, Buhr ED, Singer O, Meeker K, Verma IM, Doyle FJ, Takahashi JS, Kay SA. Intercellular coupling confers robustness against mutations in the SCN circadian clock network. Cell. 2007;129:605–16. https://doi.org/10.1016/j.cell.2007.02.047.
178. Loh DH, Navarro J, Hagopian A, Wang LM, Deboer T, Colwell CS. Rapid changes in the light/dark cycle disrupt memory of conditioned fear in mice. PLoS One. 2010;5:e12546. https://doi.org/10.1371/journal.pone.0012546.
179. Lowrey PL, Shimomura K, Antoch MP, Yamazaki S, Zemenides PD, Ralph MR, Menaker M, Takahashi JS. Positional Syntenic cloning and functional characterization of the mammalian circadian mutation tau. Science. 2000;288:483–92.
180. Lu J, Zhang Y-H, Chou TC, Gaus SE, Elmquist JK, Shiromani P, Saper CB. Contrasting effects of ibotenate lesions of the paraventricular nucleus and subparaventricular zone on sleep–wake cycle and temperature regulation. J Neurosci. 2001;21:4864–74. https://doi.org/10.1523/JNEUROSCI.21-13-04864.2001.
181. Ma K, Xiao R, Tseng H-T, Shan L, Fu L, Moore DD. Circadian dysregulation disrupts bile acid homeostasis. PLoS One. 2009;4:e6843. https://doi.org/10.1371/journal.pone.0006843.
182. Maejima T, Tsuno Y, Miyazaki S, Tsuneoka Y, Hasegawa E, Islam MT, Enoki R, Nakamura TJ, Mieda M. GABA from vasopressin neurons regulates the time at which suprachiasmatic nucleus molecular clocks enable circadian behavior. Proc Natl Acad Sci. 2021;118:e2010168118. https://doi.org/10.1073/pnas.2010168118.
183. Majde J, Krueger J. Links between the innate immune system and sleep. J Allergy Clin Immunol. 2005;116:1188–98. https://doi.org/10.1016/j.jaci.2005.08.005.
184. Marcheva B, Ramsey KM, Buhr ED, Kobayashi Y, Su H, Ko CH, Ivanova G, Omura C, Mo S, Vitaterna MH, Lopez JP, Philipson LH, Bradfield CA, Crosby SD, JeBailey L, Wang X, Takahashi JS, Bass J. Disruption of the CLOCK components CLOCK and BMAL1 leads to hypoinsulinaemia and diabetes. Nature. 2010;466:627–31. https://doi.org/10.1038/nature09253.
185. Markwald RR, Wright KP. Circadian misalignment and sleep disruption in shift work: implications for fatigue and risk of weight gain and obesity. In: Shiromani P, Horvath T, Redline S, Van Cauter E, editors. Sleep loss and obesity. New York: Springer New York; 2012. p. 101–18. https://doi.org/10.1007/978-1-4614-3492-4_8.
186. Marquié J-C, Tucker P, Folkard S, Gentil C, Ansiau D. Chronic effects of shift work on cognition: findings from the VISAT longitudinal study. Occup Environ Med. 2015;72:258–64. https://doi.org/10.1136/oemed-2013-101993.
187. Marumoto N, Murakami N, Kuroda H, Murakami T. Melatonin accelerates reentrainment of circadian locomotor activity rhythms to new light-dark cycles in the rat. Jpn J Physiol. 1996;46:347–51. https://doi.org/10.2170/jjphysiol.46.347.
188. Maywood ES, Chesham JE, O'Brien JA, Hastings MH. A diversity of paracrine signals sustains molecular circadian cycling in suprachiasmatic nucleus circuits. Proc Natl Acad Sci. 2011;108:14306–11. https://doi.org/10.1073/pnas.1101767108.
189. Maywood ES, Chesham JE, Winsky-Sommerer R, Hastings MH. Restoring the molecular clockwork within the suprachiasmatic hypothalamus of an otherwise clockless mouse enables circadian phasing and stabilization of sleep-wake cycles and reverses memory deficits. J Neurosci. 2021;41:8562–76. https://doi.org/10.1523/JNEUROSCI.3141-20.2021.
190. McClung CR, Fox BA, Dunlap JC. The Neurospora clock gene frequency shares a sequence element with the drosophila clock gene period. Nature. 1989;339:558–62. https://doi.org/10.1038/339558a0.
191. Meyer-Bernstein EL, Jetton AE, Matsumoto S, Markuns JF, Lehman MN, Bittman EL. Effects of suprachiasmatic transplants on circadian rhythms of neuroendocrine function in golden hamsters. Endocrinology. 1999;140:207–18. https://doi.org/10.1210/endo.140.1.6428.

192. Mieda M, Hasegawa E, Kessaris N, Sakurai T. Fine-tuning circadian rhythms: the importance of Bmal1 expression in the ventral forebrain. Front Neurosci. 2017;11:55. https://doi.org/10.3389/fnins.2017.00055.
193. Mills JN, Minors DS, Waterhouse JM. Adaptation to abrupt time shifts of the oscillator(s) controlling human circadian rhythms. J Physiol. 1978;285:455–70. https://doi.org/10.1113/jphysiol.1978.sp012582.
194. Mistlberger RE, Bergmann BM, Waldenar W, Rechtschaffen A. Recovery sleep following sleep deprivation in intact and suprachiasmatic nuclei-lesioned rats. Sleep. 1983;6:217–33. https://doi.org/10.1093/sleep/6.3.217.
195. Mistlberger RE, Landry GJ, Marchant EG. Sleep deprivation can attenuate light-induced phase shifts of circadian rhythms in hamsters. Neurosci Lett. 1997;238:5–8. https://doi.org/10.1016/S0304-3940(97)00815-X.
196. Mitsui S, Yamaguchi S, Matsuo T, Ishida Y, Okamura H. Antagonistic role of E4BP4 and PAR proteins in the circadian oscillatory mechanism. Genes Dev. 2001;15:995–1006. https://doi.org/10.1101/gad.873501.
197. Miyazaki K, Mesaki M, Ishida N. Nuclear entry mechanism of rat PER2 (rPER2): role of rPER2 in nuclear localization of CRY protein. Mol Cell Biol. 2001;21:6651–9. https://doi.org/10.1128/MCB.21.19.6651-6659.2001.
198. Moore RY, Eichler VB. Loss of a circadian adrenal corticosterone rhythm following suprachiasmatic lesions in the rat. Brain Res. 1972;42:201–6. https://doi.org/10.1016/0006-8993(72)90054-6.
199. Moore JG, Englert E. Circadian rhythm of gastric acid secretion in man. Nature. 1970;226:1261–2. https://doi.org/10.1038/2261261a0.
200. Morris EL, Patton AP, Chesham JE, Crisp A, Adamson A, Hastings MH. Single-cell transcriptomics of suprachiasmatic nuclei reveal a Prokineticin-driven circadian network. EMBO J. 2021;40:10.15252/embj.2021108614.
201. Mouret J, Coindet J, Debilly G, Chouvet G. Suprachiasmatic nuclei lesions in the rat: alterations in sleep circadian rhythms. Electroencephalogr Clin Neurophysiol. 1978;45:402–8. https://doi.org/10.1016/0013-4694(78)90191-8.
202. Mukherji A, Kobiita A, Damara M, Misra N, Meziane H, Champy M-F, Chambon P. Shifting eating to the circadian rest phase misaligns the peripheral clocks with the master SCN clock and leads to a metabolic syndrome. Proc Natl Acad Sci. 2015;112:E6691. https://doi.org/10.1073/pnas.1519807112.
203. Mullington JM, Simpson NS, Meier-Ewert HK, Haack M. Sleep loss and inflammation. Best Pract Res Clin Endocrinol Metab. 2010;24:775–84. https://doi.org/10.1016/j.beem.2010.08.014.
204. Mure LS, Le HD, Benegiamo G, Chang MW, Rios L, Jillani N, Ngotho M, Kariuki T, Dkhissi-Benyahya O, Cooper HM, Panda S. Diurnal transcriptome atlas of a primate across major neural and peripheral tissues. Science. 2018;359:eaao0318. https://doi.org/10.1126/science.aao0318.
205. Myers MP, Wager-Smith K, Wesley CS, Young MW, Sehgal A. Positional cloning and sequence analysis of the drosophila clock gene, timeless. Science. 1995;270:805–8. https://doi.org/10.1126/science.270.5237.805.
206. Nagano M, Adachi A, Nakahama K, Nakamura T, Tamada M, Meyer-Bernstein E, Sehgal A, Shigeyoshi Y. An abrupt shift in the day/night cycle causes desynchrony in the mammalian circadian center. J Neurosci. 2003;23:6141–51. https://doi.org/10.1523/JNEUROSCI.23-14-06141.2003.
207. Nakagawa H, Koyanagi S, Takiguchi T, Kuramoto Y, Soeda S, Shimeno H, Higuchi S, Ohdo S. 24-hour oscillation of mouse methionine aminopeptidase2, a regulator of tumor progression, is regulated by clock gene proteins. Cancer Res. 2004;64:8328–33. https://doi.org/10.1158/0008-5472.CAN-04-2122.
208. Nakahata Y, Kaluzova M, Grimaldi B, Sahar S, Hirayama J, Chen D, Guarente LP, Sassone-Corsi P. The NAD+-dependent deacetylase SIRT1 modulates CLOCK-mediated chroma-

tin remodeling and circadian control. Cell. 2008;134:329–40. https://doi.org/10.1016/j.cell.2008.07.002.
209. Nakajima M, Imai K, Ito H, Nishiwaki T, Murayama Y, Iwasaki H, Oyama T, Kondo T. Reconstitution of circadian oscillation of cyanobacterial KaiC phosphorylation in vitro. Science. 2005;308:414–5. https://doi.org/10.1126/science.1108451.
210. Naylor E, Bergmann BM, Krauski K, Zee PC, Takahashi JS, Vitaterna MH, Turek FW. The circadian *clock* mutation alters sleep homeostasis in the mouse. J Neurosci. 2000;20:8138–43. https://doi.org/10.1523/JNEUROSCI.20-21-08138.2000.
211. Nguyen J, Wright KP. Influence of weeks of circadian misalignment on leptin levels. Nat Sci Sleep. 2009;2009:9. https://doi.org/10.2147/NSS.S7624.
212. Northeast RC, Vyazovskiy VV, Bechtold DA. Eat, sleep, repeat: the role of the circadian system in balancing sleep–wake control with metabolic need. Curr Opin Physio. 2020;15:183–91. https://doi.org/10.1016/j.cophys.2020.02.003.
213. Nunez AA, Bult A, McElhinny TL, Smale L. Daily rhythms of Fos expression in hypothalamic targets of the suprachiasmatic nucleus in diurnal and nocturnal rodents. J Biol Rhythm. 1999;14:300–6. https://doi.org/10.1177/074873099129000713.
214. Ogle W. On the diurnal variations in the temperature of the human body in health. St Georges Hosp Rep. 1866;1:221–45.
215. Ono D, Honma K, Yanagawa Y, Yamanaka A, Honma S. Role of GABA in the regulation of the central circadian clock of the suprachiasmatic nucleus. J Physiol Sci. 2018;68:333–43. https://doi.org/10.1007/s12576-018-0604-x.
216. Ono D, Mukai Y, Hung CJ, Chowdhury S, Sugiyama T, Yamanaka A. The mammalian circadian pacemaker regulates wakefulness via CRF neurons in the paraventricular nucleus of the hypothalamus. Sci Adv. 2020;6:eabd0384. https://doi.org/10.1126/sciadv.abd0384.
217. Pacheco-Bernal I, Becerril-Pérez F, Aguilar-Arnal L. Circadian rhythms in the three-dimensional genome: implications of chromatin interactions for cyclic transcription. Clin Epigenetics. 2019;11:79. https://doi.org/10.1186/s13148-019-0677-2.
218. Page AJ, Christie S, Symonds E, Li H. Circadian regulation of appetite and time restricted feeding. Physiol Behav. 2020;220:112873. https://doi.org/10.1016/j.physbeh.2020.112873.
219. Papagiannakopoulos T, Bauer MR, Davidson SM, Heimann M, Subbaraj L, Bhutkar A, Bartlebaugh J, Vander Heiden MG, Jacks T. Circadian rhythm disruption promotes lung tumorigenesis. Cell Metab. 2016;24:324–31. https://doi.org/10.1016/j.cmet.2016.07.001.
220. Paschos GK, Ibrahim S, Song W-L, Kunieda T, Grant G, Reyes TM, Bradfield CA, Vaughan CH, Eiden M, Masoodi M, Griffin JL, Wang F, Lawson JA, FitzGerald GA. Obesity in mice with adipocyte-specific deletion of clock component Arntl. Nat Med. 2012;18:1768–77. https://doi.org/10.1038/nm.2979.
221. Patke A, Murphy PJ, Onat OE, Krieger AC, Özçelik T, Campbell SS, Young MW. Mutation of the human circadian clock gene *CRY1* in familial delayed sleep phase disorder. Cell. 2017;169:203–215.e13. https://doi.org/10.1016/j.cell.2017.03.027.
222. Patton AP, Edwards MD, Smyllie NJ, Hamnett R, Chesham JE, Brancaccio M, Maywood ES, Hastings MH. The VIP-VPAC2 neuropeptidergic axis is a cellular pacemaking hub of the suprachiasmatic nucleus circadian circuit. Nat Commun. 2020;11:3394. https://doi.org/10.1038/s41467-020-17110-x.
223. Pendergast JS, Friday RC, Yamazaki S. Distinct functions of Period2 and Period3 in the mouse circadian system revealed by in vitro analysis. PLoS One. 2010;5:e8552. https://doi.org/10.1371/journal.pone.0008552.
224. Pendergast JS, Branecky KL, Yang W, Ellacott KLJ, Niswender KD, Yamazaki S. High-fat diet acutely affects circadian organisation and eating behavior. Eur J Neurosci. 2013;37:1350–6. https://doi.org/10.1111/ejn.12133.
225. Penev PD, Kolker DE, Zee PC, Turek FW. Chronic circadian desynchronization decreases the survival of animals with cardiomyopathic heart disease. Am J Physiol Heart Circ Physiol. 1998;275:H2334–7. https://doi.org/10.1152/ajpheart.1998.275.6.H2334.

226. Poggiogalle E, Jamshed H, Peterson CM. Circadian regulation of glucose, lipid, and energy metabolism in humans. Metabolism. 2018;84:11–27. https://doi.org/10.1016/j.metabol.2017.11.017.
227. Prather AA, Janicki-Deverts D, Hall MH, Cohen S. Behaviorally assessed sleep and susceptibility to the common cold. Sleep. 2015;38:1353–9. https://doi.org/10.5665/sleep.4968.
228. Preitner N, Damiola F, Luis-Lopez-Molina, Zakany J, Duboule D, Albrecht U, Schibler U. The orphan nuclear receptor REV-ERBα controls circadian transcription within the positive limb of the mammalian circadian oscillator. Cell. 2002;110:251–60. https://doi.org/10.1016/S0092-8674(02)00825-5.
229. Pulido RS, Munji RN, Chan TC, Quirk CR, Weiner GA, Weger BD, Rossi MJ, Elmsaouri S, Malfavon M, Deng A, Profaci CP, Blanchette M, Qian T, Foreman KL, Shusta EV, Gorman MR, Gachon F, Leutgeb S, Daneman R. Neuronal activity regulates blood-brain barrier efflux transport through endothelial circadian genes. Neuron. 2020;108:937–952.e7. https://doi.org/10.1016/j.neuron.2020.09.002.
230. Puttonen S, Viitasalo K, Härmä M. Effect of shiftwork on systemic markers of inflammation. Chronobiol Int. 2011;28:528–35. https://doi.org/10.3109/07420528.2011.580869.
231. Ralph MR, Menaker M. A mutation of the circadian system in Golden hamsters. Science. 1988;241:1225–7. https://doi.org/10.1126/science.3413487.
232. Reddy P, Zehring WA, Wheeler DA, Pirrotta V, Hadfield C, Hall JC, Rosbash M. Molecular analysis of the period locus in Drosophila melanogaster and identification of a transcript involved in biological rhythms. Cell. 1984;38:701–10. https://doi.org/10.1016/0092-8674(84)90265-4.
233. Reddy AB, Field MD, Maywood ES, Hastings MH. Differential resynchronisation of circadian clock gene expression within the suprachiasmatic nuclei of mice subjected to experimental jet lag. J Neurosci. 2002;22:7326–30. https://doi.org/10.1523/JNEUROSCI.22-17-07326.2002.
234. Reinke H, Asher G. Circadian clock control of liver metabolic functions. Gastroenterology. 2016;150:574–80. https://doi.org/10.1053/j.gastro.2015.11.043.
235. Roberts SK. Circadian activity rhythms in cockroaches. III. The role of endocrine and neural factors. J Cell Physiol. 1966;67:473–86. https://doi.org/10.1002/jcp.1040670312.
236. Ruben MD, Wu G, Smith DF, Schmidt RE, Francey LJ, Lee YY, Anafi RC, Hogenesch JB. A database of tissue-specific rhythmically expressed human genes has potential applications in circadian medicine. Sci Transl Med. 2018;10:eaat8806. https://doi.org/10.1126/scitranslmed.aat8806.
237. Rutila JE, Suri V, Le M, So WV, Rosbash M, Hall JC. CYCLE is a second bHLH-PAS clock protein essential for circadian rhythmicity and transcription of drosophila period and timeless. Cell. 1998;93:805–14. https://doi.org/10.1016/S0092-8674(00)81441-5.
238. Sack RL, Hughes RJ, Edgar DM, Lewy AJ. Sleep-promoting effects of melatonin: at what dose, in whom, under what conditions, and by what mechanisms? Sleep. 1997;20:908–15. https://doi.org/10.1093/sleep/20.10.908.
239. Saini C, Liani A, Curie T, Gos P, Kreppel F, Emmenegger Y, Bonacina L, Wolf J-P, Poget Y-A, Franken P, Schibler U. Real-time recording of circadian liver gene expression in freely moving mice reveals the phase-setting behavior of hepatocyte clocks. Genes Dev. 2013;27:1526–36. https://doi.org/10.1101/gad.221374.113.
240. Sakurai T, Nagata R, Yamanaka A, Kawamura H, Tsujino N, Muraki Y, Kageyama H, Kunita S, Takahashi S, Goto K, Koyama Y, Shioda S, Yanagisawa M. Input of orexin/hypocretin neurons revealed by a genetically encoded tracer in mice. Neuron. 2005;46:297–308. https://doi.org/10.1016/j.neuron.2005.03.010.
241. Salgado-Delgado R, Angeles-Castellanos M, Saderi N, Buijs RM, Escobar C. Food intake during the normal activity phase prevents obesity and circadian desynchrony in a rat model of night work. Endocrinology. 2010;151:1019–29. https://doi.org/10.1210/en.2009-0864.
242. Salgado-Delgado RC, Saderi N, Basualdo, M. del C, Guerrero-Vargas NN, Escobar C, Buijs RM. Shift work or food intake during the rest phase promotes metabolic disruption and

desynchrony of liver genes in male rats. PLoS One. 2013;8:e60052. https://doi.org/10.1371/journal.pone.0060052.
243. Sanchez REA, Kalume F, de la Iglesia HO. Sleep timing and the circadian clock in mammals: past, present and the road ahead. Semin Cell Dev Biol. 2022;126:3–14. https://doi.org/10.1016/j.semcdb.2021.05.034.
244. Sargent ML, Briggs WR, Woodward DO. Circadian nature of a rhythm expressed by an invertaseless strain of Neurospora crassa. Plant Physiol. 1966;41:1343–9.
245. Sargent C, Zhou X, Matthews R, Darwent D, Roach G. Daily rhythms of hunger and satiety in healthy men during one week of sleep restriction and circadian misalignment. Int J Environ Res Public Health. 2016;13:170. https://doi.org/10.3390/ijerph13020170.
246. Sato TK, Panda S, Miraglia LJ, Reyes TM, Rudic RD, McNamara P, Naik KA, FitzGerald GA, Kay SA, Hogenesch JB. A functional genomics strategy reveals Rora as a component of the mammalian circadian clock. Neuron. 2004;43:527–37. https://doi.org/10.1016/j.neuron.2004.07.018.
247. Sato S, Solanas G, Peixoto FO, Bee L, Symeonidi A, Schmidt MS, Brenner C, Masri S, Benitah SA, Sassone-Corsi P. Circadian reprogramming in the liver identifies metabolic pathways of aging. Cell. 2017;170:664–677.e11. https://doi.org/10.1016/j.cell.2017.07.042.
248. Schaffer R, Ramsay N, Samach A, Corden S, Putterill J, Carré IA, Coupland G. The late elongated hypocotyl mutation of Arabidopsis disrupts circadian rhythms and the photoperiodic control of flowering. Cell. 1998;93:1219–29. https://doi.org/10.1016/s0092-8674(00)81465-8.
249. Scheer FAJL, Czeisler CA. Melatonin, sleep, and circadian rhythms. Sleep Med Rev. 2005;9:5–9. https://doi.org/10.1016/j.smrv.2004.11.004.
250. Scheer FAJL, Morris CJ, Shea SA. The internal circadian clock increases hunger and appetite in the evening independent of food intake and other behaviors: body clock controls hunger. Obesity. 2013;21:421–3. https://doi.org/10.1002/oby.20351.
251. Scheiermann C, Gibbs J, Ince L, Loudon A. Clocking in to immunity. Nat Rev Immunol. 2018;18:423–37. https://doi.org/10.1038/s41577-018-0008-4.
252. Schwartz WJ, Tavakoli-Nezhad M, Lambert CM, Weaver DR, de la Iglesia HO. Distinct patterns of *period* gene expression in the suprachiasmatic nucleus underlie circadian clock photoentrainment by advances or delays. Proc Natl Acad Sci. 2011;108:17219–24. https://doi.org/10.1073/pnas.1107848108.
253. Sehgal A, Price JL, Man B, Young MW. Loss of circadian behavioral rhythms and per RNA oscillations in the drosophila mutant timeless. Science. 1994;263:1603–6. https://doi.org/10.1126/science.8128246.
254. Shachar S, Misteli T. Causes and consequences of nuclear gene positioning. J Cell Sci. 2017;130:1501–8. https://doi.org/10.1242/jcs.199786.
255. Shan Z, Li Y, Zong G, Guo Y, Li J, Manson JE, Hu FB, Willett WC, Schernhammer ES, Bhupathiraju SN. Rotating night shift work and adherence to unhealthy lifestyle in predicting risk of type 2 diabetes: results from two large US cohorts of female nurses. BMJ. 2018;363:k4641. https://doi.org/10.1136/bmj.k4641.
256. Shan Y, Abel JH, Li Y, Izumo M, Cox KH, Jeong B, Yoo S-H, Olson DP, Doyle FJ, Takahashi JS. Dual-color single-cell imaging of the Suprachiasmatic nucleus reveals a circadian role in network synchrony. Neuron. 2020;108:164–179.e7. https://doi.org/10.1016/j.neuron.2020.07.012.
257. Sherman H, Genzer Y, Cohen R, Chapnik N, Madar Z, Froy O. Timed high-fat diet resets circadian metabolism and prevents obesity. FASEB J. 2012;26:3493–502. https://doi.org/10.1096/fj.12-208868.
258. Shi S, Hida A, McGuinness OP, Wasserman DH, Yamazaki S, Johnson CH. Circadian clock gene BMAL1 is not essential; functional replacement with its paralog, BMAL2. Curr Biol. 2010;20:316–21. https://doi.org/10.1016/j.cub.2009.12.034.
259. Shirogane T, Jin J, Ang XL, Harper JW. SCFbeta-TRCP controls clock-dependent transcription via casein kinase 1-dependent degradation of the mammalian period-1 (Per1) protein. J Biol Chem. 2005;280:26863–72. https://doi.org/10.1074/jbc.M502862200.

260. Shiromani PJ, Xu M, Winston EM, Shiromani SN, Gerashchenko D, Weaver DR. Sleep rhythmicity and homeostasis in mice with targeted disruption of mPeriod genes. Am J Physiol Regul Integr Comp Physiol. 2004;287:R47–57. https://doi.org/10.1152/ajpregu.00138.2004.
261. Silver R, LeSauter J, Tresco PA, Lehman MN. A diffusible coupling signal from the transplanted suprachiasmatic nucleus controlling circadian locomotor rhythms. Nature. 1996;382:810–3. https://doi.org/10.1038/382810a0.
262. Silver AC, Arjona A, Hughes ME, Nitabach MN, Fikrig E. Circadian expression of clock genes in mouse macrophages, dendritic cells, and B cells. Brain Behav Immun. 2012;26:407–13. https://doi.org/10.1016/j.bbi.2011.10.001.
263. Solanas G, Peixoto FO, Perdiguero E, Jardí M, Ruiz-Bonilla V, Datta D, Symeonidi A, Castellanos A, Welz PS, Caballero JM, Sassone-Corsi P, Muñoz-Cánoves P, Benitah SA. Aged stem cells reprogram their daily rhythmic functions to adapt to stress. Cell. 2017;170:678–692.e20. https://doi.org/10.1016/j.cell.2017.07.035.
264. Solt LA, Wang Y, Banerjee S, Hughes T, Kojetin DJ, Lundasen T, Shin Y, Liu J, Cameron MD, Noel R, Yoo S-H, Takahashi JS, Butler AA, Kamenecka TM, Burris TP. Regulation of circadian behaviour and metabolism by synthetic REV-ERB agonists. Nature. 2012;485:62–8. https://doi.org/10.1038/nature11030.
265. Somers DE, Devlin PF, Kay SA. Phytochromes and cryptochromes in the entrainment of the Arabidopsis circadian clock. Science. 1998;282:1488–90. https://doi.org/10.1126/science.282.5393.1488.
266. Spiegel K, Tasali E, Leproult R, Scherberg N, Van Cauter E. Twenty-four-hour profiles of acylated and total ghrelin: relationship with glucose levels and impact of time of day and sleep. J Clin Endocrinol Metab. 2011;96:486–93. https://doi.org/10.1210/jc.2010-1978.
267. Stearns AT, Balakrishnan A, Rhoads DB, Ashley SW, Tavakkolizadeh A. Diurnal rhythmicity in the transcription of jejunal drug transporters. J Pharmacol Sci. 2008;108:144–8. https://doi.org/10.1254/jphs.08100SC.
268. Stephan FK, Zucker I. Circadian rhythms in drinking behavior and locomotor activity of rats are eliminated by hypothalamic lesions. Proc Natl Acad Sci. 1972;69:1583–6. https://doi.org/10.1073/pnas.69.6.1583.
269. Stokkan K-A, Yamazaki S, Tei H, Sakaki Y, Menaker M. Entrainment of the circadian clock in the liver by feeding. Science. 2001;291:490–3. https://doi.org/10.1126/science.291.5503.490.
270. Strayer C, Oyama T, Schultz TF, Raman R, Somers DE, Más P, Panda S, Kreps JA, Kay SA. Cloning of the Arabidopsis clock gene TOC1, an autoregulatory response regulator homolog. Science. 2000;289:768–71. https://doi.org/10.1126/science.289.5480.768.
271. Sukumaran S, Almon RR, DuBois DC, Jusko WJ. Circadian rhythms in gene expression: relationship to physiology, disease, drug disposition and drug action☆. Adv Drug Deliv Rev. 2010;62:904–17. https://doi.org/10.1016/j.addr.2010.05.009.
272. Takahashi JS. Transcriptional architecture of the mammalian circadian clock. Nat Rev Genet. 2017;18:164–79. https://doi.org/10.1038/nrg.2016.150.
273. Takahashi JS. The 50th anniversary of the Konopka and Benzer 1971 paper in PNAS: "Clock Mutants of *Drosophila melanogaster*". Proc Natl Acad Sci. 2021;118:e2110171118. https://doi.org/10.1073/pnas.2110171118.
274. Takamure M, Murakami N, Takahashi K, Kuroda H, Etoh T. Rapid reentrainment of the circadian clock itself, but not the measurable activity rhythms, to a new light-dark cycle in the rat. Physiol Behav. 1991;50:443–9. https://doi.org/10.1016/0031-9384(91)90092-3.
275. Tamaru T, Hattori M, Honda K, Nakahata Y, Sassone-Corsi P, van der Horst GTJ, Ozawa T, Takamatsu K. CRY drives cyclic CK2-mediated BMAL1 phosphorylation to control the mammalian circadian clock. PLoS Biol. 2015;13:e1002293. https://doi.org/10.1371/journal.pbio.1002293.
276. Tapp WN, Holloway FA. Phase shifting circadian rhythms produces retrograde amnesia. Science. 1981;211:1056–8. https://doi.org/10.1126/science.7193351.

277. Taylor P, Hardin PE. Rhythmic E-box binding by CLK-CYC controls daily cycles in per and tim transcription and chromatin modifications. Mol Cell Biol. 2008;28:4642–52. https://doi.org/10.1128/MCB.01612-07.
278. Tevy MF, Giebultowicz J, Pincus Z, Mazzoccoli G, Vinciguerra M. Aging signaling pathways and circadian clock-dependent metabolic derangements. Trends Endocrinol Metab. 2013;24:229–37. https://doi.org/10.1016/j.tem.2012.12.002.
279. Tischkau SA, Mitchell JW, Tyan S-H, Buchanan GF, Gillette MU. Ca^{2+}/cAMP response element-binding protein (CREB)-dependent activation of Per1 is required for light-induced signaling in the suprachiasmatic nucleus circadian clock. J Biol Chem. 2003;278:718–23. https://doi.org/10.1074/jbc.M209241200.
280. Tobler I, Borbély AA, Groos G. The effect of sleep deprivation on sleep in rats with suprachiasmatic lesions. Neurosci Lett. 1983;42:49–54. https://doi.org/10.1016/0304-3940(83)90420-2.
281. Todd WD, Venner A, Anaclet C, Broadhurst RY, De Luca R, Bandaru SS, Issokson L, Hablitz LM, Cravetchi O, Arrigoni E, Campbell JN, Allen CN, Olson DP, Fuller PM. Suprachiasmatic VIP neurons are required for normal circadian rhythmicity and comprised of molecularly distinct subpopulations. Nat Commun. 2020;11:4410. https://doi.org/10.1038/s41467-020-17197-2.
282. Toh KL, Jones CR, He Y, Eide EJ, Hinz WA, Virshup DM, Ptáček LJ, Fu Y-H. An hPer2 phosphorylation site mutation in familial advanced sleep phase syndrome. Science. 2001;291:1040–3. https://doi.org/10.1126/science.1057499.
283. Torquati L, Mielke GI, Brown WJ, Burton NW, Kolbe-Alexander TL. Shift work and poor mental health: a meta-analysis of longitudinal studies. Am J Public Health. 2019;109:e13–20. https://doi.org/10.2105/AJPH.2019.305278.
284. Trachsel L, Edgar DM, Seidel WF, Craig Heller H. Sleep homeostasis in suprachiasmatic nuclei-lesioned rats: effects of sleep deprivation and triazolam administration. Brain Res. 1992;589:253–61. https://doi.org/10.1016/0006-8993(92)91284-L.
285. Travnickova-Bendova Z, Cermakian N, Reppert SM, Sassone-Corsi P. Bimodal regulation of mPeriod promoters by CREB-dependent signaling and CLOCK/BMAL1 activity. Proc Natl Acad Sci. 2002;99:7728–33. https://doi.org/10.1073/pnas.102075599.
286. Turek FW, Joshu C, Kohsaka A, Lin E, Ivanova G, McDearmon E, Laposky A, Losee-Olson S, Easton A, Jensen DR, Eckel RH, Takahashi JS, Bass J. Obesity and metabolic syndrome in circadian *clock* mutant mice. Science. 2005;308:1043–5. https://doi.org/10.1126/science.1108750.
287. Ubaldo-Reyes LM, Buijs RM, Escobar C, Ángeles-Castellanos M. Scheduled meal accelerates entrainment to a 6-h phase advance by shifting central and peripheral oscillations in rats. Eur J Neurosci. 2017;46:1875–86. https://doi.org/10.1111/ejn.13633.
288. Ukai-Tadenuma M, Yamada RG, Xu H, Ripperger JA, Liu AC, Ueda HR. Delay in feedback repression by cryptochrome 1 is required for circadian clock function. Cell. 2011;144:268–81. https://doi.org/10.1016/j.cell.2010.12.019.
289. Ulgherait M, Chen A, McAllister SF, Kim HX, Delventhal R, Wayne CR, Garcia CJ, Recinos Y, Oliva M, Canman JC, Picard M, Owusu-Ansah E, Shirasu-Hiza M. Circadian regulation of mitochondrial uncoupling and lifespan. Nat Commun. 2020;11:1–16. https://doi.org/10.1038/s41467-020-15617-x.
290. Van Cauter E, Blackman JD, Roland D, Spire JP, Refetoff S, Polonsky KS. Modulation of glucose regulation and insulin secretion by circadian rhythmicity and sleep. J Clin Invest. 1991;88:934–42. https://doi.org/10.1172/JCI115396.
291. Van Der Vlag J, Otte AP. Transcriptional repression mediated by the human polycomb-group protein EED involves histone deacetylation. Nat Genet. 1999;23:474–8. https://doi.org/10.1038/70602.
292. van Diepen HC, Lucassen EA, Yasenkov R, Groenen I, Ijzerman AP, Meijer JH, Deboer T. Caffeine increases light responsiveness of the mouse circadian pacemaker. Eur J Neurosci. 2014;40:3504–11. https://doi.org/10.1111/ejn.12715.

293. Villanueva JE, Livelo C, Trujillo AS, Chandran S, Woodworth B, Andrade L, Le HD, Manor U, Panda S, Melkani GC. Time-restricted feeding restores muscle function in drosophila models of obesity and circadian-rhythm disruption. Nat Commun. 2019;10:2700. https://doi.org/10.1038/s41467-019-10563-9.
294. Vitaterna MH, King DP, Chang A-M, Kornhauser JM, Lowrey PL, McDonald JD, Dove WF, Pinto LH, Turek FW, Takahashi JS. Mutagenesis and mapping of a mouse gene, *clock*, essential for circadian behavior. Science. 1994;264:719–25. https://doi.org/10.1126/science.8171325.
295. Wang ZY, Tobin EM. Constitutive expression of the CIRCADIAN CLOCK ASSOCIATED 1 (CCA1) gene disrupts circadian rhythms and suppresses its own expression. Cell. 1998;93:1207–17. https://doi.org/10.1016/s0092-8674(00)81464-6.
296. Webb AB, Angelo N, Huettner JE, Herzog ED. Intrinsic, nondeterministic circadian rhythm generation in identified mammalian neurons. Proc Natl Acad Sci. 2009;106:16493–8. https://doi.org/10.1073/pnas.0902768106.
297. Wehrens SMT, Christou S, Isherwood C, Middleton B, Gibbs MA, Archer SN, Skene DJ, Johnston JD. Meal timing regulates the human circadian system. Curr Biol. 2017;27:1768–1775.e3. https://doi.org/10.1016/j.cub.2017.04.059.
298. Welz P-S, Zinna VM, Symeonidi A, Koronowski KB, Kinouchi K, Smith JG, Guillén IM, Castellanos A, Furrow S, Aragón F, Crainiciuc G, Prats N, Caballero JM, Hidalgo A, Sassone-Corsi P, Benitah SA. BMAL1-driven tissue clocks respond independently to light to maintain homeostasis. Cell. 2019;177:1436–1447.e12. https://doi.org/10.1016/j.cell.2019.05.009.
299. Wilson KA, Beck JN, Nelson CS, Hilsabeck TA, Promislow D, Brem RB, Kapahi P. GWAS for lifespan and decline in climbing ability in flies upon dietary restriction reveal decima as a mediator of insulin-like peptide production. Curr Biol. 2020;30:2749–2760.e3. https://doi.org/10.1016/j.cub.2020.05.020.
300. Wisor JP, O'Hara BF, Terao A, Selby CP, Kilduff TS, Sancar A, Edgar DM, Franken P. A role for cryptochromes in sleep regulation. BMC Neurosci. 2002;3:20. https://doi.org/10.1186/1471-2202-3-20.
301. Wittenbrink N, Ananthasubramaniam B, Münch M, Koller B, Maier B, Weschke C, Bes F, de Zeeuw J, Nowozin C, Wahnschaffe A, Wisniewski S, Zaleska M, Bartok O, Ashwal-Fluss R, Lammert H, Herzel H, Hummel M, Kadener S, Kunz D, Kramer A. High-accuracy determination of internal circadian time from a single blood sample. J Clin Invest. 2018;128:3826–39. https://doi.org/10.1172/JCI120874.
302. Wolff G, Esser KA. Scheduled exercise phase shifts the circadian clock in skeletal muscle. Med Sci Sports Exerc. 2012;44:1663–70. https://doi.org/10.1249/MSS.0b013e318255cf4c.
303. Wulund L, Reddy AB. A brief history of circadian time: the emergence of redox oscillations as a novel component of biological rhythms. Perspect. Sci., proceedings of the Beilstein Bozen symposium 2014 – chemistry and time. 2015;6:27–37. https://doi.org/10.1016/j.pisc.2015.08.002.
304. Wurts SW, Edgar DM. Circadian and homeostatic control of rapid eye movement (REM) sleep: promotion of REM tendency by the suprachiasmatic nucleus. J Neurosci. 2000;20:4300–10. https://doi.org/10.1523/JNEUROSCI.20-11-04300.2000.
305. Xiao Y, Yuan Y, Jimenez M, Soni N, Yadlapalli S. Clock proteins regulate spatiotemporal organization of clock genes to control circadian rhythms. Proc Natl Acad Sci USA. 2021;118:1–10. https://doi.org/10.1073/pnas.2019756118.
306. Xu Y, Toh KL, Jones CR, Shin J-Y, Fu Y-H, Ptáček LJ. Modeling of a human circadian mutation yields insights into clock regulation by PER2. Cell. 2007;128:59–70. https://doi.org/10.1016/j.cell.2006.11.043.
307. Xu K, DiAngelo JR, Hughes ME, Hogenesch JB, Sehgal A. The circadian clock interacts with metabolic physiology to influence reproductive fitness. Cell Metab. 2011;13:639–54. https://doi.org/10.1016/j.cmet.2011.05.001.

308. Yamaguchi S, Isejima H, Matsuo T, Okura R, Yagita K, Kobayashi M, Okamura H. Synchronization of cellular clocks in the suprachiasmatic nucleus. Science. 2003;302:1408–12. https://doi.org/10.1126/science.1089287.
309. Yamanaka Y, Honma S, Honma K. Daily exposure to a running wheel entrains circadian rhythms in mice in parallel with development of an increase in spontaneous movement prior to running-wheel access. Am J Physiol Regul Integr Comp Physiol. 2013;305:R1367–75. https://doi.org/10.1152/ajpregu.00389.2013.
310. Yan L, Silver R. Resetting the brain clock: time course and localization of mPER1 and mPER2 protein expression in suprachiasmatic nuclei during phase shifts: light response of mPER1 and mPER2 in phase shifts. Eur J Neurosci. 2004;19:1105–9. https://doi.org/10.1111/j.1460-9568.2004.03189.x.
311. Yang S, Liu A, Weidenhammer A, Cooksey RC, McClain D, Kim MK, Aguilera G, Abel ED, Chung JH. The role of mPer2 clock gene in glucocorticoid and feeding rhythms. Endocrinology. 2009;150:2153–60. https://doi.org/10.1210/en.2008-0705.
312. Yasumoto Y, Hashimoto C, Nakao R, Yamazaki H, Hiroyama H, Nemoto T, Yamamoto S, Sakurai M, Oike H, Wada N, Yoshida-Noro C, Oishi K. Short-term feeding at the wrong time is sufficient to desynchronize peripheral clocks and induce obesity with hyperphagia, physical inactivity and metabolic disorders in mice. Metabolism. 2016;65:714–27. https://doi.org/10.1016/j.metabol.2016.02.003.
313. Yin L, Lazar MA. The orphan nuclear receptor rev-erbalpha recruits the N-CoR/histone deacetylase 3 corepressor to regulate the circadian Bmal1 gene. Mol Endocrinol Baltim Md. 2005;19:1452–9. https://doi.org/10.1210/me.2005-0057.
314. Yokoyama Y, Nakamura TJ, Yoshimoto K, Ijyuin H, Tachikawa N, Oda H, Shiraishi R, Shinohara K, Kumadaki K, Honda S, Nakamura A, Kitamura N, Tsubota K, Watanabe M. A high-salt/high fat diet alters circadian locomotor activity and glucocorticoid synthesis in mice. PLoS One. 2020;15:e0233386. https://doi.org/10.1371/journal.pone.0233386.
315. Yoo S-H, Yamazaki S, Lowrey PL, Shimomura K, Ko CH, Buhr ED, Siepka SM, Hong H-K, Oh WJ, Yoo OJ, Menaker M, Takahashi JS. PERIOD2::LUCIFERASE real-time reporting of circadian dynamics reveals persistent circadian oscillations in mouse peripheral tissues. Proc Natl Acad Sci. 2004;101:5339–46. https://doi.org/10.1073/pnas.0308709101.
316. Yoo S-H, Mohawk JA, Siepka SM, Shan Y, Huh SK, Hong H-K, Kornblum I, Kumar V, Koike N, Xu M, Nussbaum J, Liu X, Chen Z, Chen ZJ, Green CB, Takahashi JS. Competing E3 ubiquitin ligases govern circadian periodicity by degradation of CRY in nucleus and cytoplasm. Cell. 2013;152:1091–105. https://doi.org/10.1016/j.cell.2013.01.055.
317. Young MW, Kay SA. Time zones: a comparative genetics of circadian clocks. Nat Rev Genet. 2001;2:702–15. https://doi.org/10.1038/35088576.
318. Yu F, Wang Z, Zhang T, Chen X, Xu H, Wang F, Guo L, Chen M, Liu K, Wu B. Deficiency of intestinal Bmal1 prevents obesity induced by high-fat feeding. Nat Commun. 2021;12:5323. https://doi.org/10.1038/s41467-021-25674-5.
319. Zambon AC, McDearmon EL, Salomonis N, Vranizan KM, Johansen KL, Adey D, Takahashi JS, Schambelan M, Conklin BR. Time- and exercise-dependent gene regulation in human skeletal muscle. Genome Biol. 2003;4:R61. https://doi.org/10.1186/gb-2003-4-10-r61.
320. Zehring WA, Wheeler DA, Reddy P, Konopka RJ, Kyriacou CP, Rosbash M, Hall JC. P-element transformation with period locus DNA restores rhythmicity to mutant, arrhythmic *Drosophila melanogaster*. Cell. 1984;39:369–76. https://doi.org/10.1016/0092-8674(84)90015-1.
321. Zelinski EL, Hong NS, McDonald RJ. Persistent impairments in hippocampal function following a brief series of photoperiod shifts in rats. Anim Cogn. 2014;17:127–41. https://doi.org/10.1007/s10071-013-0645-8.
322. Zhang J, Kalkum M, Chait BT, Roeder RG. The N-CoR-HDAC3 nuclear receptor corepressor complex inhibits the JNK pathway through the integral subunit GPS2. Mol Cell. 2002;9:611–23. https://doi.org/10.1016/S1097-2765(02)00468-9.

323. Zhang R, Lahens NF, Ballance HI, Hughes ME, Hogenesch JB. A circadian gene expression atlas in mammals: implications for biology and medicine. Proc Natl Acad Sci USA. 2014;111:16219–24. https://doi.org/10.1073/pnas.1408886111.
324. Zhang SL, Yue Z, Arnold DM, Artiushin G, Sehgal A. A circadian clock in the blood-brain barrier regulates xenobiotic efflux. Cell. 2018;173:130–139.e10. https://doi.org/10.1016/j.cell.2018.02.017.
325. Zhang SL, Lahens NF, Yue Z, Arnold DM, Pakstis PP, Schwarz JE, Sehgal A. A circadian clock regulates efflux by the blood-brain barrier in mice and human cells. Nat Commun. 2021;12:1–12. https://doi.org/10.1038/s41467-020-20795-9.
326. Zheng B, Albrecht U, Kaasik K, Sage M, Lu W, Vaishnav S, Li Q, Sun ZS, Eichele G, Bradley A, Lee CC. Nonredundant roles of the mPer1 and mPer2 genes in the mammalian circadian clock. Cell. 2001;105:683–94. https://doi.org/10.1016/s0092-8674(01)00380-4.

Genetic Markers of Individual Differences in Response to Sleep Loss

Courtney E. Casale and Namni Goel

Abstract This chapter focuses on genotype-driven interindividual differences in phenotypic neurobehavioral responses to different types of sleep loss, including sleep restriction and total sleep deprivation. We highlight the utility of the candidate gene approach to further examine differential neurobehavioral vulnerability to sleep loss in humans, although the genome-wide association approach and other methods are also important for the investigation of biomarkers of such vulnerability. Notable genetic polymorphisms indicative of individual differences to sleep deprivation include those found in core circadian clock genes (*BHLHE41/DEC2* and *PER3*), genes related to synaptic plasticity, cognitive development and functioning (*BDNF* and *COMT*), immune and clearance genes (*AQP4*, *DQB1*0602*, and *TNFα*), adenosine genes (*ADA* and *ADORA2A*), and dopaminergic genes (*DRD2* and *DAT*). In addition to discussing the relevance of these polymorphisms to responses to sleep loss, we review the efficacy of several mitigation strategies and countermeasures for the neurobehavioral deficits induced by insufficient sleep, including banking sleep, recovery sleep, caffeine, and naps. The discovery of reliable genetic markers of differential resilience and vulnerability to sleep deprivation is essential to guide future research in sleep and circadian science regarding predictors, countermeasures, and treatments in applied settings.

Keywords Sleep deprivation · Candidate genes · Polymorphisms · Circadian genes · Interindividual differences · Vulnerability/resilience

---

C. E. Casale · N. Goel (✉)
Biological Rhythms Research Laboratory, Department of Psychiatry and Behavioral Sciences, Rush University Medical Center, Chicago, IL, USA
e-mail: courtney_casale@rush.edu; namni_goel@rush.edu

P. Gehrman et al. (eds.), *Genetics of Sleep and Sleep Disorders*,
https://doi.org/10.1007/978-3-031-62723-1_15

Introduction

The biological clock, located in the suprachiasmatic nuclei (SCN) of the anterior hypothalamus, regulates the timing of sleep onset and offset, as well as waking behavior, promoting circadian rhythmicity in sleepiness and neurobehavioral and cognitive performance [1, 2]. According to the two-process model of sleep regulation, the interaction of Process S (a homeostatic process) and Process C (a circadian process) modulates the timing of sleep and wakefulness, as well as waking neurobehavioral functions [3–7]. The drive for sleep (Process S) increases while awake and decreases during sleep [1]. When the homeostatic drive raises above a certain threshold, sleep onset is induced, and when it falls below a different threshold, wakefulness occurs [1]. The cycle of sleep and wakefulness (Process C) represents the daily oscillations of these thresholds, which promotes wakefulness at certain times throughout the day [8].

Deficits in neurobehavioral functioning caused by sleep loss have been well-established in previous research [4, 9, 10]. The magnitude of such decrements differs between individuals in a stable, phenotypic manner, whereby some individuals are resilient to insufficient sleep (i.e., minimally affected), and others are vulnerable (i.e., greatly affected) [11–22].

This chapter discusses the genetic basis of differential phenotypic responses to insufficient sleep, particularly concerning neurobehavioral resilience and vulnerability. It also examines the utility of several countermeasures for sleep loss-induced decrements in neurobehavioral performance, such as caffeine, naps, banking sleep, and recovery sleep, and includes a discussion of candidate gene studies utilizing caffeine to counteract such decrements. The chapter culminates with a discussion of future directions in this area of research.

Interindividual Differences in Neurobehavioral Responses to Sleep Deprivation

Categorizing Differential Neurobehavioral Resilience and Vulnerability to Sleep Deprivation

A variety of neurobehavioral test metrics have been used to distinguish individuals who are resilient and those who are vulnerable to sleep loss. Commonly used and validated metrics for such classification include behavioral attention tasks such as the Psychomotor Vigilance Test (PVT) [23], working memory tests such as n-back tasks [24] and the Digit Span Task [25], cognitive throughput tasks such as the Digit Symbol Substitution Test [26], measures of subjective sleepiness such as the Karolinska Sleepiness Scale (KSS) [27], and measures of subjective fatigue, vigor, and mood such as those derived from the Profile of Mood States (POMS) [28]. Given the variety of neurobehavioral domains and the general lack of relationships between performance on objective and self-rated metrics [12, 17, 18, 29, 30], the determination of reliable categorization methods for resilience and vulnerability is quite complex. Thus, it is important to use a variety of targeted neurobehavioral tests to investigate individual differences by accurately capturing the complete essence of a cognitive phenotype (i.e., cognitive endophenotyping) [31].

While several categorization methods have been utilized in past research to classify individuals as resilient or vulnerable to sleep loss, the optimal methods by which to do so remain elusive. Approaches involving either raw performance or self-rated scores on neurobehavioral metrics during sleep deprivation [15, 32–35] or involving difference scores accounting for baseline performance [19–21, 36–42] have been used most commonly in past work. Another approach that considers intraindividual variance in performance also has been posited as a potential method to characterize differential resilience and vulnerability to sleep deprivation. Notably, this approach considers time-of-day variation in performance [9, 43–48], as circadian factors may significantly impact neurobehavioral functioning throughout the day. However, future studies utilizing this approach are needed, since very few published studies have categorized differential vulnerability using intraindividual variation in performance [49–51]. Not only have several categorization approaches been used in the past; studies have used a variety of thresholds to divide individuals into resilient and vulnerable groups—a median split [32–34, 36, 39, 40, 52–56], a tertile split [15, 19, 38, 42], a quartile split [35, 57], other numeric splits [37, 58], and ±1 standard deviation of neurobehavioral performance [20, 21]. Interestingly, a recent set of studies revealed that the various approaches and thresholds often do not categorize individuals as resilient or vulnerable in the same manner, either within a neurobehavioral test metric or between metrics [49–51]. Despite this inconsistency and the need for further research into reliable determining factors of neurobehavioral resilience and vulnerability, it remains important to explore potential biological indicators of such characteristics.

Biomarkers and Predictors of Resilience and Vulnerability to Sleep Deprivation

While research to identify reliable predictors of neurobehavioral resilience and vulnerability to sleep loss is ongoing, biomarkers (objective proxies that enable remote detection of biological processes, regardless of their mechanistic role [59]) for differentiating neurobehavioral responses to chronic sleep restriction (SR; several consecutive nights with reduced total sleep time) or total sleep deprivation (TSD; one or more nights without sleep) have been discovered via genetic and molecular omic (e.g., epigenomic, metabolomic, transcriptomic, and proteomic) methods [4]. There is much interest surrounding such biomarkers and other factors, including neurobehavioral performance, cardiovascular variables, hormonal markers, and sleep characteristics, in relation to predicting differential vulnerability to sleep loss [59–62].

Of these, genetic polymorphisms (variants in DNA sequence in a single nucleotide) are perhaps one of the most researched indicators of such individual differences.

Genetic Polymorphisms Associated with Differential Neurobehavioral Responses to Sleep Deprivation

Individuals exhibit differential neurobehavioral resilience and vulnerability to sleep deprivation in a heritable and stable manner. One twin study revealed that, during acute TSD, 56.2% of the total PVT performance variance in monozygotic twins, yet only 14.5% of the variance in dizygotic twins, was attributable to variance between pairs of twins [63]. Findings from this study [63] and others involving unrelated participants [64] demonstrate the substantial impact an individual's genetic profile has on their phenotypic neurobehavioral responses to sleep loss. While neurobehavioral performance outcomes during sleep loss are typically normally distributed [13, 65], thereby suggesting a polygenic nature of such performance [4], previous research using the candidate gene approach has identified several specific genetic polymorphisms related to such differential neurobehavioral responses (see Table 1 for a list of the relevant genes). This investigative approach helps to determine the ways in which certain genetic variants impact neurobehavioral performance during both sleep-deprived and rested conditions. Genome-wide association (GWA) studies are also useful in the search for genetic markers of resilience and vulnerability. Findings from previous GWA studies offer support of robust, phenotypic interindividual differences in sleep parameters, which may be extended to the determination of differential neurobehavioral vulnerability in response to sleep loss (reviewed in [66]). Additionally, it is important to include techniques that involve a perturbation

Table 1 Genes associated with neurobehavioral resilience and vulnerability to sleep loss

Gene	References
ADA	Bachmann et al. [98]; Kuna et al. [63]; Reichert et al. [99]
ADORA2A	Baur et al. [102]; Bodenmann et al. [100]; Elmenhorst et al. [103]; Erblang et al. [81]; Erblang et al. [88]; Rupp et al. [80]
AQP4	Ulv Larsen et al. [92]
BDNF	Bachmann et al. [83]; Grant et al. [82]; Satterfield et al. [84]
BHLHE41/ DEC2	Pellegrino et al. [73]; Webb and Fu [74]
COMT	Bodenmann et al. [90]; Bodenmann et al. [91]; Erblang et al. [81]; Erblang et al. [88]; Goel et al. [89]; Satterfield et al. [87]; Valoman et al. [86]
DAT1	Holst et al. [107]; Satterfield et al. [106]
*DQB1*0602*	Goel et al. [95]
DRD2	Holst et al. [107]; Muck et al. [105]; Whitney et al. [104]
PER3	Erblang et al. [81]; Goel et al. [78]; Groeger et al. [77]; Kuna et al. [63]; Lo et al. [79]; Maire et al. [75]; Rupp et al. [80]; Viola et al. [76]
TNFα	Erblang et al. [81]; Erblang et al. [88]; Satterfield et al. [97]; Skeiky et al. [96]

of the system (e.g., enforced sleep deprivation), theory-driven genotyping, cognitive endophenotyping, and selective sampling when studying differential vulnerability, especially when doing so in relatively small samples [31]. Below, we summarize the genetic factors that explain a portion of the related interindividual variance in neurobehavioral resilience and vulnerability to sleep loss. For reviews of GWA and candidate gene studies related to sleep parameters during normal sleep and rested conditions and related to circadian rhythm sleep disorders, we refer the reader to Goel [66], Yamazaki and Goel [67], Dutta [68], Bolsius et al. [69], and Garfield [70], and for reports detailing other biomarker and omic methods used in the context of sleep loss, we refer the reader to Goel [59], Mullington et al. [71], and Uyhelji et al. [72], as these topics are beyond the scope of this chapter.

Core Circadian Clock Genes

BHLHE41/DEC2

BHLHE41/DEC2, a circadian gene involved in the regulation of sleep quantity, has been associated with differential neurobehavioral vulnerability to both chronic SR and acute TSD [73]. In one twin study, findings revealed an association between the Y362H variant and fewer lapses on the PVT (>500 ms reaction time) during TSD, reduced sleep duration, and less recovery sleep following TSD [73]. However, this variant was not related to non-rapid eye movement sleep duration, as evidenced by a lack of a significant difference between the twins [73]. The presence of the Y362H variant also reduced *BHLHE41/DEC2*'s ability to suppress in vitro NPAS2/BMAL1 and CLOCK/BMAL1 transactivation [73]. Additionally, the P384R mutation of *BHLHE41/DEC2* is involved in familial natural short sleep [74]. This relationship may also contribute to the differential neurobehavioral resilience to sleep deprivation associated with *BHLHE41/DEC2*, though further research is needed.

PER3

A 54-nucleotide variable-number tandem repeat (VNTR), repeating in either four or five units, is located in the coding region of *PER3*, a core circadian clock gene, and is associated with neurobehavioral vulnerability to sleep loss. Studies involving TSD have found that $PER3^{5/5}$ (five-repeat allele) individuals had greater decrements compared to individuals with the $PER3^{4/4}$ (four-repeat allele) genotype in a variety of neurobehavioral domains, including sustained attention [75], cognitive performance [76], and executive function in the early morning [77]. However, neither neurobehavioral vulnerability to acute TSD in twins [63] nor to chronic SR in unrelated individuals [78] were differentially impacted by *PER3* genotype; conversely, two other studies did find differences in neurobehavioral responses to SR associated with this polymorphism [79, 80].

Previous work has also investigated the *PER3* C7827519G single-nucleotide polymorphism (SNP) in relation to differential responses to sleep loss. However, findings from a recent study involving 38 h of TSD revealed neither an association between the *PER3* genotype and neurobehavioral response to TSD nor an interaction between the *PER3* genotype, TSD, and caffeine on PVT performance at any time during the protocol [81].

Synaptic Plasticity, Cognitive Development and Functioning Genes

BDNF

BDNF, the brain-derived neurotrophic factor gene, is critically involved in neuronal development and plasticity, which is important for neurobehavioral functioning. Much research has focused on the Val66Met polymorphism of *BDNF* as a marker of differential neurobehavioral resilience and vulnerability to sleep loss. Several studies have shown that individuals with the *Met* allele had worse performance than *Val/Val* individuals on neurobehavioral metrics during extended wakefulness. In one study, after 20 h of wakefulness, *Met* carriers had poorer response inhibition performance on the Stroop Task, an assessment of cognitive flexibility, than *Val/Val* individuals, suggesting the role of this *BDNF* polymorphism in differential impairment [82]. Similarly, another study found that *Met* carriers had worse performance than *Val/Val* individuals on a verbal two-back working memory task after being awake for 40 h [83]. Yet, this study also reported that the *BDNF* genotype did not influence sustained attention performance or self-rated sleepiness or well-being [83]. Conversely, a study involving simulated night-shift schedules found that *Val/Met* individuals had fewer lapses on the PVT toward the end of night shifts in the second simulation bout as compared to those in the first simulation bout, suggesting that the heterozygote genotype may enable greater adaptation to circadian misalignment [84]. PVT performance of *Val/Val* homozygotes, however, did not differ between night shift bouts [84]. Altogether, these findings support the idea that genotypically regulated differences in BDNF protein expression during sleep loss may be another important factor for determining differential neurobehavioral vulnerability [85].

COMT

COMT, the gene that encodes the catechol-O-methyltransferase protein, breaks down catecholamines including dopamine, epinephrine, and norepinephrine. *COMT* has a functional Val158Met polymorphism, which contributes to differential neurobehavioral responses to sleep loss. Following one night of TSD, *Val* carriers had poorer behavioral attention [86] and adaptive decision-making [87] performance than *Met* carriers. Similarly, a recent study found that *G/G* individuals (homozygous *Val* allele carriers) performed similarly to what the authors predetermined to be the most vulnerable tertile of their sample on the Go/No-Go task, which measures inhibition control, after a night of TSD [88]. However, one study found that this polymorphism did not influence cognitive performance or subjective or physiological sleepiness during chronic SR [89]. Additionally, results from a recent study showed that, after 26 h awake, this *COMT* polymorphism explained 9.1% of the total variance in PVT response speed and 14.4% of the total variance in Go/No-Go commission errors, further suggesting an important association between *COMT* genotype and neurobehavioral responses to sleep loss [88].

Studies have also found that this *COMT* polymorphism modulates the effect of the pharmacological stimulant, modafinil, on several neurobehavioral outcomes, including executive functioning, behavioral attention, and self-reported well-being, though it was not associated with subjective sleepiness [90, 91]. In addition to investigations of the impact of this *COMT* polymorphism on performance impairment during TSD independently, it has been investigated in relation to the mitigatory effect of caffeine on sleep loss-induced performance decrements. Findings from a recent study demonstrated that, while there was no main effect of *COMT* genotype on PVT performance or KSS scores overall, after 20 h of TSD, *Met* allele carriers had poorer PVT performance than *Val/Val* individuals in the placebo condition, though this effect was not observed after 26 h or 32 h of TSD [81]. When the participants were provided caffeine, however, this genotypic performance difference was not observed, suggesting caffeine's utility during sleep loss [81].

Immune and Clearance Genes

AQP4

The astrocytic water channel Aquaporin 4 (AQP4) is involved in the circulation of cerebrospinal fluid (CSF) throughout the brain. Several SNPs in the *AQP4* gene that modulate the expression of AQP4 were recently found in association with differential neurobehavioral responses to sleep deprivation. Results showed that, after 40 h of TSD, individuals with the low-AQP4-expressing HtMi variant haplotype (minor allele carriers) had less of an increase in self-rated sleepiness and less of a reduction in PVT response speed than HtMa haplotype individuals (major allele carriers), suggesting that the HtMi haplotype may be related to greater neurobehavioral resilience to sleep loss [92]. Importantly, these findings must be considered in relation to CSF flow and brain clearance since these processes appear crucial to guarding against the development of neurodegenerative diseases.

*DQB1*0602*

The human leukocyte antigen *DQB1*0602* allele is related to narcolepsy [93, 94] and has been investigated in the context of neurobehavioral vulnerability to sleep loss. *DQB1*0602*-positive individuals reported greater sleepiness than

*DQB1*0602*-negative individuals during five consecutive SR nights and reported greater fatigue and sleepiness during baseline [95]. Similarly, *DQB1*0602*-positive individuals experienced differentially greater increases in fatigue during SR [95]. However, *DQB1*0602* allele status had no impact on cumulative neurobehavioral performance decrements induced by SR, since *DQB1*0602*-positive and *DQB1*0602*-negative groups were comparable in their total decreases in cognitive performance and increases in physiological sleepiness [95].

TNFα

TNFα, the tumor necrosis factor alpha gene, encodes the proinflammatory cytokine *TNFα*. *TNFα* has a G308A polymorphism in its promoter region, which has been studied in relation to differential responses to sleep loss. A recent study found that, after 26 h awake, this *TNFα* polymorphism explained 11.7% of the total variance in PVT lapses, 10.3% of the total variance in PVT response speed, and 8.4% of the total variance in KSS scores, suggesting the impact of this polymorphism on neurobehavioral responses to sleep loss [88]. Notably, two previous studies showed that *A* allele carriers demonstrated greater performance resilience on the PVT during TSD [96, 97]. Results also revealed that *TNFα A* allele carriers exhibited greater sensitivity to the effects of caffeine than *G/G* individuals during TSD [81], although this effect may be associated with the greater performance degradation experienced by *A* allele carriers during TSD, as the *TNFα* genotype did not impact PVT lapses [81]. However, during 48 h of TSD, there was no interaction of *TNFα* genotype and caffeine (either 200 or 300 mg doses) on PVT performance, even though a differential genetic impact on PVT performance variance alone was observed [96].

Adenosine Genes

ADA

Adenosine deaminase (ADA) is an enzyme responsible for breaking down and regulating intra- and extracellular adenosine. The *ADA* gene has a G22A polymorphism that contributes to differential neurobehavioral responses to sleep deprivation. A recent study found that this *ADA* polymorphism explained 14.4% of the total variance in KSS scores after 26 h awake, suggesting its relation to differential resilience and vulnerability [88]. Another TSD study found that participants with the *G/A* genotype performed more poorly on the PVT and also reported greater sleepiness and fatigue than *G/G* individuals [98]. In a twin study, however, PVT performance during 38 h of TSD did not differ by *ADA* genotype [63]. Additionally, when this *ADA* polymorphism was investigated during a protocol involving 40 h of TSD (i.e., high sleep pressure) or 40 h of multiple naps (i.e., low sleep pressure), *G/A* individuals had greater self-reported sleepiness than *G/G* individuals during TSD only [99]. Participants with the *G/A* genotype also had worse self-reported well-being during TSD than during the nap condition and demonstrated poorer working memory performance during TSD [99]. Furthermore, *G/G* individuals exhibited better PVT performance during the nap condition at the end of the night than they did during TSD [99].

ADORA2A

Several polymorphisms of the adenosine A2A receptor (*ADORA2A*) gene have also been associated with differential neurobehavioral vulnerability to sleep loss. One frequently studied SNP is the T1083C polymorphism (also known as rs5751876): one study found that, after 32 h of TSD, individuals carrying the *T* allele had poorer vigilant attention performance than *C/C* individuals, as measured by lapses on the PVT [81]. Findings from another study similarly demonstrated greater PVT performance resilience of *C/T* individuals during chronic SR, as compared with *T/T* individuals [80]. Furthermore, *ADORA2A* contains eight SNPs that create eight distinct haplotypes: findings have shown that carriers of the HT4 haplotype exhibited better PVT performance after a night of TSD compared with individuals without the HT4 haplotype [100]. Another relevant polymorphism is the rs5751862 SNP: a recent study found that carriers of the *G/G* genotype of this *ADORA2A* polymorphism performed similarly to what the authors predetermined to be the most resilient tertile of their sample on PVT lapses and the most resilient tertile of their sample on the Go/No-Go task following a night of TSD [88]. The same study also reported that, after 26 h awake, the rs5751876, rs5751862, rs2236624, and rs3761422 *ADORA2A* polymorphisms were associated with performance on a variety of neurobehavioral metrics: the rs5751876 polymorphism explained 14.1% of the total variance in KSS scores; the rs5751862 polymorphism explained 16.1% of the total variance in PVT lapses, 7.8% of the total variance in PVT response speed, and 12.2% of the variance in Go/No-Go commission errors; the rs2236624 polymorphism explained 6.5% of the total variance in PVT lapses; and the rs3761422 polymorphism explained 9.2% of the total variance in PVT response speed [88]. Associations between the latter polymorphisms and differential vulnerability to sleep loss have been less studied, thus warranting further research.

Furthermore, the *ADORA2A* polymorphisms contribute to an individual's sensitivity to caffeine, thus influencing whether caffeine may be particularly useful in the maintenance of neurobehavioral performance during extended wakefulness. Caffeine and adenosine are structurally similar, making caffeine an effective stimulant during sleep loss [101]. Results from one study showed that caffeinated coffee effectively mitigated SR-induced performance deficits on vigilant attention and executive control tasks, but did not impact subjective sleepiness, in T1083C *ADORA2A C/C* carriers [102]. Furthermore, after 40 h of TSD, caffeine counteracted PVT performance decrements in non-HT4 allele carriers but had no effect on performance in HT4-haplotype individuals [100], though another study involving 38 h of TSD found no impact of the *ADORA2A* genotype on caffeine's ability to mitigate sleep loss-induced PVT performance deficits [81]. Additionally, although A_1 adenosine receptor (A_1AR) availability has been associated with recovery sleep's ability to mitigate neurobehavioral performance decrements following 52 h of TSD, the *ADORA2A* genotype did not moderate this effect [103]. Taken together, these findings are particularly important to the ongoing investigation of whether caffeine consumption may be a more beneficial countermeasure for certain individuals based on their *ADORA2A* genotype.

Dopaminergic Genes

DRD2 and *DAT*

Both separately and in conjunction, the dopamine D2 receptor gene (*DRD2*) C957T polymorphism and the dopamine transporter gene (*DAT1*) 3'-UTR VNTR polymorphism have also been linked to neurobehavioral resilience and vulnerability to sleep loss. Results from one study demonstrated that, during 38 h of TSD, individuals with the *DRD2 C/C* genotype showed considerable resistance to the detrimental effects of sleep loss on a measure of cognitive flexibility, whereas the performance of *T/T* individuals was greatly impacted [104]. However, this study did not reveal a genotypic influence on PVT performance resilience [104]. Another study found that *DRD2 T/T* individuals had greater PVT performance declines with increased time spent on the task during TSD than *C* allele carriers, suggesting that *DRD2* genotype influences the magnitude of the time-on-task (TOT) effect [105]. The TOT effect on PVT performance during TSD was also modulated by the *DAT1* genotype, as 10-repeat allele (*10R*) carriers exhibited less severe TOT decrements compared with nine-repeat allele (*9R*) carriers [106]. Moreover, when examined in combination with each other, possessing the *DAT1-DRD2 10R/10R-C/T* or *9R-C/C* genotypes was related to greater performance resilience on the PVT during TSD, as compared with other genotype combinations [107]. Individuals with the *DAT1-DRD2 9R-C/C* genotype also exhibited the greatest resilience against increases in subjective sleepiness during TSD [107]. Altogether, these results implicate the influence of dopaminergic pathways in differential neurobehavioral responses to sleep loss, as well as demonstrate the importance of considering multiple genetic polymorphisms in combination with each other when determining resilience and vulnerability to sleep deprivation.

The Candidate Gene Approach: Strengths and Weaknesses

Given the existence of robust, phenotypic individual differences in neurobehavioral responses to sleep loss, the candidate gene approach has been particularly useful in investigating the impact of genetic variants on such trait-like differential resilience and vulnerability [4]. This approach enables the exploration of possible associations between specific genetic polymorphisms and phenotypic neurobehavioral responses to sleep loss, especially in a laboratory setting involving relatively small sample sizes. Despite its utility, no studies to date in our field have used the candidate gene approach to explicitly determine causal relationships between genotypes and phenotypic responses. Previous research has demonstrated, however, that several of the aforementioned SNPs, either individually or in combination, reliably contribute to the interindividual variance in neurobehavioral responses to sleep deprivation (reviewed in Satterfield et al. [31]). Importantly, the continued use of the candidate gene approach is critical to the investigation of the influence of specific genes on differential neurobehavioral vulnerability to sleep loss.

Countermeasures for Sleep Deprivation-Induced Neurobehavioral Decrements

Although the neurobehavioral and physiological effects of sleep loss may be quite severe, they often are underestimated by sleep-deprived individuals, which can be especially problematic in applied settings. Ideally, an individual should consistently obtain sufficient sleep aligned with their own circadian rhythms to guard against the diminished neurobehavioral performance and adverse health outcomes associated with sleep deprivation [108], yet work and societal demands often make this difficult to accomplish. Populations such as shift workers, on-call medical personnel, students pulling "all-nighters," transmeridian travelers, and other individuals whose jobs require extended wakefulness are particularly impacted in this regard [9]. Driving impairments and accident risk are also directly increased by sleep deprivation, as sleepiness-related crashes yield similar injury and fatality rates to alcohol-related crashes [109–111] but are often underestimated [9, 112–115]. Therefore, mitigation strategies, including banking sleep, recovery sleep, caffeine, and naps, are necessary to combat the severity of such negative effects. Investigating the efficacy of such countermeasures in relation to the aforementioned candidate genes is critical, as this research may offer more definitive and individualized recommendations as to which mitigation strategies would be most effective based on an individual's genetic make-up.

Impact of Banking Sleep on Neurobehavioral Performance

Previous research has shown that banking sleep (increasing total sleep time to 8–9 h per night for several consecutive nights) protects against various deficits in neurobehavioral performance caused by insufficient sleep, including those on sustained attention tasks [116, 117] and high-cognitive-load decision tasks [118]. Banking sleep may also help effectively manage excessive daytime sleepiness, fatigue, and stress [119, 120], which is advantageous in settings such as military operations [121] and shift work [122], in which optimal neurobehavioral functioning is critical. Studies have demonstrated that the positive effects of banking sleep might be maintained during a recovery sleep opportunity following a period of sleep deprivation [116, 117]. Further investigation regarding this strategy is necessary to definitively establish whether banking sleep can reliably mitigate neurobehavioral performance decrements during sleep deprivation and subsequent recovery periods, particularly concerning differential phenotypic and genotypic characteristics. Likewise, additional work is also necessary to determine whether certain individuals may benefit more from banking sleep than others due to their genetic profile. For example, the *BHLHE41/DEC2* polymorphism has been associated with short sleep [74], suggesting that increasing sleep duration through banking may not be as useful for individuals with the P384R mutation; however, a direct link has yet to be examined.

Impact of Recovery Sleep on Neurobehavioral Performance

Recovery sleep (increased nightly sleep following sleep deprivation) is another mitigation strategy shown to aid in the restoration of neurobehavioral functioning following a period of sleep loss. A variety of neurobehavioral domains may benefit from a recovery sleep opportunity, as studies have reported improvements in cognitive performance; reductions in sleepiness, fatigue, and sleep propensity; improvements in mood; and increases in alertness associated with recovery sleep [123–126]. However, recovery sleep may be less effective in restoring performance on other neurobehavioral metrics, including spatial orientation, vigilance and working memory tasks, inhibition as defined by a pinball task, and self-rated vigor as defined by the POMS, as studies have shown that sleep deprivation-induced performance decrements on such tasks were not fully reversed following recovery sleep [10, 126–128]. Although more research is needed to discover reliable biomarkers of responses to recovery sleep, interindividual genotypic and phenotypic differences may account for some of the discrepancies in previous findings, since differential vulnerability may affect how much recovery sleep is sufficient to return certain aspects of neurobehavioral functioning to their baseline levels [126]. Accordingly, biomarkers and genetic polymorphisms potentially influencing the effectiveness of recovery sleep warrant further investigation.

Impact of Caffeine on Neurobehavioral Performance

Caffeine has also been established as an effective countermeasure for sleep loss-induced neurobehavioral performance deficits [129–135]. The mitigatory effects of acute caffeine consumption (i.e., doses from <80 mg to 600 mg) on sleep deprivation-induced performance declines span across a variety of cognitive domains, including memory, attention, information processing, executive functioning, and driving tasks (reviewed in [136]). As aforementioned, caffeine's mitigatory qualities have also been linked to genetic polymorphisms, specifically the *ADORA2A* [100, 102] and *COMT* [81] genotypes, which further demonstrates caffeine's biological utility. Notably, as the pressure for sleep increases during prolonged wakefulness, caffeine becomes less effective at preventing performance declines [4]. Robust individual differences in response to both sleep deprivation and caffeine also add an additional layer of complexity to the effectiveness of this mitigation strategy [4, 133, 134].

Impact of Napping on Neurobehavioral Performance

Napping during the day also aids in protecting against performance decrements during periods of increased sleepiness and decreased alertness [137–139]. However, it is necessary to consider that although naps are advantageous, periods of rest are usually followed by sleep inertia (a period of diminished performance and increased grogginess) upon awakening [140], which may impact performance variables, such as subjective sleepiness, in a trait-like manner [141]. This may be especially true following longer naps involving slow-wave sleep, yet recent studies have shown mixed findings in this regard [142]. The frequency and timing of naps are also critical factors, as a recent study found that napping often and later in the day may be associated with poorer sleep quality and increased sleep fragmentation [143]. It is also important to consider the potential impacts of daytime napping on cardiometabolic health, given a recent report that suggested genetic links between napping and measures associated with cardiometabolic risk, including blood pressure and waist circumference [144]. Thus, since naps alone are unable to prevent the negative effects of sleep deprivation under all conditions [4, 133, 140] and across all neurobehavioral domains [145, 146], combining caffeine consumption and a short nap may provide maximum protection against sleep loss-induced decrements, though further research into this strategy is needed [140, 147].

Conclusions and Perspectives

Sufficient sleep is necessary for the maintenance of waking neurobehavioral function, yet it is often difficult to achieve due to a number of internal and external factors. Given that decrements in neurobehavioral functioning resulting from sleep loss may negatively impact performance in a variety of real-world settings, it is essential to determine reliable predictors of differential resilience and vulnerability. As discussed, several candidate genes and genetic polymorphisms associated with circadian rhythms, synaptic plasticity, cognitive and immune functioning, and neurotransmitter transmission have been found to be related to neurobehavioral resilience and vulnerability to sleep deprivation [81, 82, 84, 86–88, 92, 96, 104–106] (also see review [4]). Notably, past research has also identified genetic polymorphisms associated with the efficacy of specific countermeasures and mitigation strategies used to combat sleep loss-induced deficits such as caffeine [81, 96, 100, 102] (also see review [4]). Thus, these findings suggest that particular countermeasures may offer greater utility to certain individuals based on their own genetic profile. In addition to the continued search for candidate genes and genetic markers associated with responses to sleep loss, establishing causality between such genes and countermeasures through the candidate gene approach is an essential component of individual differences research.

Additionally, there is a continued need to investigate the genetic determinants of differential resilience and vulnerability to sleep loss in diverse demographic and clinical populations. Studies have found relationships between various sleep characteristics and an individual's ethnicity and/or race [148–151], which are likely driven in part by genetic ancestry and by environmental and societal pressures. Clinical and/or subclinical symptomology and conditions have also been found in association with candidate genes involved in resilience and vulnerability to sleep deprivation, including *ADA*, *ADORA2A*, and *PER3* [152–157] though more research into these genotypic relationships in the context of sleep loss is needed to elucidate the precise connections between such symptomology and neurobehavioral vulnerability. Self-rated personality traits may also contribute to interindividual differences in sleep characteristics [158, 159] and vulnerability to sleep loss [160]; however, such genetic relationships are complex and quite difficult to explore due to the polygenic nature of personality. Thus, in order to create a more generalizable framework of differential neurobehavioral resilience and vulnerability to sleep loss, further research on the genetic underpinnings of responses to sleep deprivation is needed. Importantly, this research will ultimately lead to the development and implementation of individualized mitigation strategies and treatments based on one's own genetic and neurobehavioral performance profiles, which is crucial for optimizing performance in applied settings requiring extended wakefulness.

Author Contributions Writing—original draft preparation, C.E.C. and N.G.; writing—review and editing, C.E.C. and N.G.; funding acquisition, N.G. All authors have read and agreed to the published version of the manuscript.

Funding This manuscript was supported by the National Aeronautics and Space Administration (NASA) grant NNX14AN49G and grant 80NSSC20K0243 (to N.G.) and the National Institutes of Health grant NIH R01DK117488 (to N.G.).

References

1. Goel N, Basner M, Dinges DF. Phenotyping of neurobehavioral vulnerability to circadian phase during sleep loss. Methods Enzymol. 2015;552:285–308. https://doi.org/10.1016/bs.mie.2014.10.024.
2. Achermann P, Dijk DJ, Brunner DP, Borbély AA. A model of human sleep homeostasis based on EEG slow-wave activity; quantitative comparison of data and simulations. Brain Res Bull. 1993;31:97–113. https://doi.org/10.1016/0361-9230(93)90016-5.
3. Borbély AA. A two process model of sleep regulation. Hum Neurobiol. 1982;1:195–204.
4. Goel N. Neurobehavioral effects and biomarkers of sleep loss in healthy adults. Curr Neurol Neurosci Rep. 2017;17:89. https://doi.org/10.1007/s11910-017-0799-x.
5. Goel N, Basner M, Rao H, Dinges DF. Circadian rhythms, sleep deprivation, and human performance. Prog Mol Biol Transl Sci. 2013;119:155–90. https://doi.org/10.1016/B978-0-12-396971-2.00007-5.
6. Khalsa SBS, Jewett ME, Duffy JF, Czeisler CA. The timing of the human circadian clock is accurately represented by the core body temperature rhythm following phase shifts to a three-cycle light stimulus near the critical zone. J Biol Rhythms. 2000;15:524–30. https://doi.org/10.1177/074873040001500609.
7. Van Dongen HPA, Dinges DF. Investigating the interaction between the homeostatic and circadian processes of sleep-wake regulation for the prediction of waking neurobehavioral performance. J Sleep Res. 2003;12:181–7. https://doi.org/10.1046/j.1365-2869.2003.00357.x.
8. Edgar DM, Dement WC, Fuller CA. Effect of SCN lesions on sleep in squirrel monkeys: evidence for opponent processes in sleep-wake regulation. J Neurosci. 1993;13:1065–79. https://doi.org/10.1523/JNEUROSCI.13-03-01065.1993.
9. Goel N, Rao H, Durmer JS, Dinges DF. Neurocognitive consequences of sleep deprivation. Semin Neurol. 2009;29:320–39. https://doi.org/10.1055/s-0029-1237117.
10. Belenky G, Wesensten NJ, Thorne DR, Thomas ML, Sing HC, Redmond DP, Russo MB, Balkin TJ. Patterns of performance degradation and restoration during sleep restriction and subsequent recovery: a sleep dose-response study. J Sleep Res. 2003;12:1–12. https://doi.org/10.1046/j.1365-2869.2003.00337.x.
11. Dijkman M, Sachs N, Levine E, Mallis M, Carlin MM, Gillen KA, Powell JW, Samuel S, Mullington J, Rosekind MR, Dinges DF. Effects of reduced stimulation on neurobehavioral alertness depend on circadian phase during human sleep deprivation. J Sleep Res. 1997;26:265.
12. Van Dongen HP, Maislin G, Mullington JM, Dinges DF. The cumulative cost of additional wakefulness: dose-response effects on neurobehavioral functions and sleep physiology from chronic sleep restriction and total sleep deprivation. Sleep. 2003;26:117–26. https://doi.org/10.1093/sleep/26.2.117.
13. Van Dongen HP, Baynard MD, Maislin G, Dinges DF. Systematic interindividual differences in neurobehavioral impairment from sleep loss: evidence of trait-like differential vulnerability. Sleep. 2004;27:423–33.
14. Van Dongen HP, Belenky G. Individual differences in vulnerability to sleep loss in the work environment. Ind Health. 2009;47:518–26. https://doi.org/10.2486/indhealth.47.518.
15. Chua EC, Yeo SC, Lee IT, Tan LC, Lau P, Cai S, Zhang X, Puvanendran K, Gooley JJ. Sustained attention performance during sleep deprivation associates with instability in behavior and physiologic measures at baseline. Sleep. 2014;37:27–39. https://doi.org/10.5665/sleep.3302.
16. Rupp TL, Wesensten NJ, Balkin TJ. Trait-like vulnerability to total and partial sleep loss. Sleep. 2012;35:1163–72. https://doi.org/10.5665/sleep.2010.
17. Dennis LE, Wohl RJ, Selame LA, Goel N. Healthy adults display long-term trait-like neurobehavioral resilience and vulnerability to sleep loss. Sci Rep. 2017;7:14889. https://doi.org/10.1038/s41598-017-14006-7.

18. Yamazaki EM, Goel N. Robust stability of trait-like vulnerability or resilience to common types of sleep deprivation in a large sample of adults. Sleep. 2021;43:zsz292. https://doi.org/10.1093/sleep/zsz292.
19. Galli O, Jones CW, Larson O, Basner M, Dinges DF. Predictors of interindividual differences in vulnerability to neurobehavioral consequences of chronic partial sleep restriction. Sleep. 2022;45(1):zsab278. https://doi.org/10.1093/sleep/zsab278.
20. Mathew GM, Strayer SM, Bailey DS, Buzzell K, Ness KM, Schade MM, Nahmod NG, Buxton OM, Chang AM. Changes in subjective motivation and effort during sleep restriction moderate interindividual differences in attentional performance in healthy young men. Nat Sci Sleep. 2021;13:1117–36. https://doi.org/10.2147/NSS.S294409.
21. Mathew GM, Strayer SM, Ness KM, Schade MM, Nahmod NG, Buxton OM, Chang AM. Interindividual differences in attentional vulnerability moderate cognitive performance during sleep restriction and subsequent recovery in healthy young men. Sci Rep. 2021;11:19147. https://doi.org/10.1038/s41598-021-95884-w.
22. LaGoy AD, Cashmere JD, Beckner ME, Eagle SR, Sinnott AM, Conkright WR, Miller E, Derrow C, Dretsch MN, Flanagan SD, Nindl BC, Connaboy C, Germain A, Ferrarelli F. A trait of mind: stability and robustness of sleep across sleep opportunity manipulations during simulated military operational stress. Sleep. 2022;45(2):zsab219. https://doi.org/10.1093/sleep/zsab219.
23. Dinges DF, Powell JW. Microcomputer analyses of performance on a portable, simple visual RT task during sustained operations. Behav Res Methods Instrum Comput. 1985;17:652–5. https://doi.org/10.3758/BF03200977.
24. Kirchner WK. Age differences in short-term retention of rapidly changing information. J Exp Psychol. 1958;55:352–8. https://doi.org/10.1037/h0043688.
25. Wechsler D. WAIS-R manual: Wechsler adult intelligence scale-revised. San Antonio: The Psychological Corporation; 1981.
26. Wechsler D. The measurement of adult intelligence. The Williams and Wilkins Company: Baltimore, MD, USA; 1939.
27. Åkerstedt T, Gillberg M. Subjective and objective sleepiness in the active individual. Int J Neurosci. 1990;52:29–37. https://doi.org/10.3109/00207459008994241.
28. McNair DM, Lorr M, Droppleman LF. Manual: Profile of mood states. San Diego: EdITS; 1971.
29. Leproult R, Colecchia EF, Berardi AM, Stickgold R, Kosslyn SM, Van Cauter E. Individual differences in subjective and objective alertness during sleep deprivation are stable and unrelated. Am J Physiol Regul Integr Comp Physiol. 2003;284:R280–90. https://doi.org/10.1152/ajpregu.00197.2002.
30. Tkachenko O, Dinges DF. Interindividual variability in neurobehavioral response to sleep loss: a comprehensive review. Neurosci Biobehav Rev. 2018;89:29–48. https://doi.org/10.1016/j.neubiorev.2018.03.017.
31. Satterfield BC, Stucky B, Landolt HP, Van Dongen HPA. Unraveling the genetic underpinnings of sleep deprivation-induced impairments in human cognition. Prog Brain Res. 2019;246:127–58. https://doi.org/10.1016/bs.pbr.2019.03.026.
32. Caldwell JL, Schroeder VM, Kunkle CL, Stephenson HG. Differential effects of modafinil on performance of low-performing and high-performing individuals during total sleep deprivation. Pharmacol Biochem Behav. 2020;196:172968. https://doi.org/10.1016/j.pbb.2020.172968.
33. Moreno-Villanueva M, von Scheven G, Feiveson A, Bürkle A, Wu H, Goel N. The degree of radiation-induced DNA strand breaks is altered by acute sleep deprivation and psychological stress and is associated with cognitive performance in humans. Sleep. 2018;41:zsy067. https://doi.org/10.1093/sleep/zsy067.
34. Patanaik A, Kwoh CK, Chua EC, Gooley JJ, Chee MW. Classifying vulnerability to sleep deprivation using baseline measures of psychomotor vigilance. Sleep. 2015;38:723–34. https://doi.org/10.5665/sleep.4664.

35. Chua EC, Sullivan JP, Duffy JF, Klerman EB, Lockley SW, Kristal BS, Czeisler CA, Gooley JJ. Classifying attentional vulnerability to total sleep deprivation using baseline features of Psychomotor Vigilance Test performance. Sci Rep. 2019;9:12102. https://doi.org/10.1038/s41598-019-48280-4.
36. Chee MW, Tan JC. Lapsing when sleep deprived: neural activation characteristics of resistant and vulnerable individuals. NeuroImage. 2010;51:835–43. https://doi.org/10.1016/j.neuroimage.2010.02.031.
37. Chee MW, Chuah LY, Venkatraman V, Chan WY, Philip P, Dinges DF. Functional imaging of working memory following normal sleep and after 24 and 35 h of sleep deprivation: correlations of fronto-parietal activation with performance. NeuroImage. 2006;31:419–28. https://doi.org/10.1016/j.neuroimage.2005.12.001.
38. Kong D, Asplund CL, Ling A, Chee MW. Increased automaticity and altered temporal preparation following sleep deprivation. Sleep. 2015;38:1219–27. https://doi.org/10.5665/sleep.4896.
39. Chuah LY, Chong DL, Chen AK, Rekshan WR 3rd, Tan JC, Zheng H, Chee MW. Donepezil improves episodic memory in young individuals vulnerable to the effects of sleep deprivation. Sleep. 2009;32:999–1010. https://doi.org/10.1093/sleep/32.8.999.
40. Xu J, Zhu Y, Fu C, Sun J, Li H, Yang X, Li W, Qin W, Shi D, Tian J. Frontal metabolic activity contributes to individual differences in vulnerability toward total sleep deprivation-induced changes in cognitive function. J Sleep Res. 2016;25:169–80. https://doi.org/10.1111/jsr.12354.
41. Riontino L, Cavallero C. Individual differences in working memory efficiency modulate proactive interference after sleep deprivation. Psychol Res. 2021;85:480–90. https://doi.org/10.1007/s00426-020-01292-6.
42. Patanaik A, Zagorodnov V, Kwoh CK, Chee MWL. Predicting vulnerability to sleep deprivation using diffusion model parameters. J Sleep Res. 2014;23:576–84. https://doi.org/10.1111/jsr.12166.
43. Banks S, Dinges DF. Behavioral and physiological consequences of sleep restriction. J Clin Sleep Med. 2007;3:519–28.
44. Doran SM, Van Dongen HP, Dinges DF. Sustained attention performance during sleep deprivation: evidence of state instability. Arch Ital Biol. 2001;139:253–67.
45. Schmidt C, Collette F, Cajochen C, Peigneux P. A time to think: circadian rhythms in human cognition. Cogn Neuropsychol. 2007;24:755–89. https://doi.org/10.1080/02643290701754158.
46. Blatter K, Cajochen C. Circadian rhythms in cognitive performance: methodological constraints, protocols, theoretical underpinnings. Physiol Behav. 2007;90:196–208. https://doi.org/10.1016/j.physbeh.2006.09.009.
47. Adam M, Rétey JV, Khatami R, Landolt HP. Age-related changes in the time course of vigilant attention during 40 hours without sleep in men. Sleep. 2006;29:55–7. https://doi.org/10.1093/sleep/29.1.55.
48. Rabbitt P, Osman P, Moore B, Stollery B. There are stable individual differences in performance variability, both from moment to moment and from day to day. Q J Exp Psychol A. 2001;54:981–1003. https://doi.org/10.1080/713756013.
49. Brieva TE, Casale CE, Yamazaki EM, Antler CA, Goel N. Cognitive throughput and working memory raw scores consistently differentiate resilient and vulnerable groups to sleep loss. Sleep. 2021;44(12):zsab197. https://doi.org/10.1093/sleep/zsab197.
50. Casale CE, Yamazaki EM, Brieva TE, Antler CA, Goel N. Raw scores on subjective sleepiness, fatigue, and vigor metrics consistently define resilience and vulnerability to sleep loss. Sleep. 2022;45(1):zsab228. https://doi.org/10.1093/sleep/zsab228.
51. Yamazaki EM, Casale CE, Brieva TE, Antler CA, Goel N. Concordance of multiple methods to define resiliency and vulnerability to sleep loss depends on Psychomotor Vigilance Test metric. Sleep. 2022;45(1):zsab249. https://doi.org/10.1093/sleep/zsab249.

52. Michael L, Passmann S, Becker R. Electrodermal lability as an indicator for subjective sleepiness during total sleep deprivation. J Sleep Res. 2012;21:470–8. https://doi.org/10.1111/j.1365-2869.2011.00984.x.
53. Salfi F, Lauriola M, Tempesta D, Calanna P, Socci V, De Gennaro L, Ferrara M. Effects of total and partial sleep deprivation on reflection impulsivity and risk-taking in deliberative decision-making. Nat Sci Sleep. 2020;12:309–24. https://doi.org/10.2147/NSS.S250586.
54. Rocklage M, Williams V, Pacheco J, Schnyer DM. White matter differences predict cognitive vulnerability to sleep deprivation. Sleep. 2009;32:1100–3. https://doi.org/10.1093/sleep/32.8.1100.
55. Yeo BTT, Tandi J, Chee MWL. Functional connectivity during rested wakefulness predicts vulnerability to sleep deprivation. NeuroImage. 2015;111:147–58. https://doi.org/10.1016/j.neuroimage.2015.02.018.
56. Diekelmann S, Born J, Wagner U. Sleep enhances false memories depending on general memory performance. Behav Brain Res. 2010;208:425–9. https://doi.org/10.1016/j.bbr.2009.12.021.
57. St Hilaire MA, Kristal BS, Rahman SA, Sullivan JP, Quackenbush J, Duffy JF, Barger LK, Gooley JJ, Czeisler CA, Lockley SW. Using a single daytime performance test to identify most individuals at high-risk for performance impairment during extended wake. Sci Rep. 2019;9:16681. https://doi.org/10.1038/s41598-019-52930-y.
58. Frey DJ, Badia P, Wright KP Jr. Inter- and intra-individual variability in performance near the circadian nadir during sleep deprivation. J Sleep Res. 2004;13:305–15. https://doi.org/10.1111/j.1365-2869.2004.00429.x.
59. Goel N. "Omics" approaches for sleep and circadian rhythm research: biomarkers for identifying differential vulnerability to sleep loss. Curr Sleep Med Rep. 2015;1:38–46. https://doi.org/10.1007/s40675-014-0003-7.
60. Yamazaki EM, Rosendahl-Garcia KM, Casale CE, MacMullen LE, Ecker AJ, Kirkpatrick JN, Goel N. Left ventricular ejection time measured by echocardiography differentiates neurobehavioral resilience and vulnerability to sleep loss and stress. Front Physiol. 2022;12:795321. https://doi.org/10.3389/fphys.2021.795321.
61. Yamazaki EM, Antler CA, Casale CE, MacMullen LE, Ecker AJ, Goel N. Cortisol and C-reactive protein vary during sleep loss and recovery but are not markers of neurobehavioral resilience. Front Physiol. 2021;12:782860. https://doi.org/10.3389/fphys.2021.782860.
62. Andrillon T, Burns A, Mackay T, Windt J, Tsuchiya N. Predicting lapses of attention with sleep-like slow waves. Nat Commun. 2021;12:3657. https://doi.org/10.1038/s41467-021-23890-7.
63. Kuna ST, Maislin G, Pack FM, Staley B, Hachadoorian R, Coccaro EF, Pack AI. Heritability of performance deficit accumulation during acute sleep deprivation in twins. Sleep. 2012;35:1223–33. https://doi.org/10.5665/sleep.2074.
64. Goel N, Dinges DF. Sleep deprivation: biomarkers for identifying and predicting individual differences in response to sleep loss. In: Thorpy MJ, Billiard M, editors. Sleepiness: Causes, consequences and treatment. Cambridge, UK: Cambridge University Press; 2011. pp. 101–10.
65. Van Dongen HP, Maislin G, Dinges DF. Dealing with inter-individual differences in the temporal dynamics of fatigue and performance: importance and techniques. Aviat Space Environ Med. 2004;75:A147–54.
66. Goel N. Genetic markers of sleep and sleepiness. Sleep Med Clin. 2017;12:289–99. https://doi.org/10.1016/j.jsmc.2017.03.005.
67. Yamazaki EM, Goel N. Genetics of circadian and sleep measures in adults: implications for sleep medicine. Curr Sleep Med Rep. 2020;6:32–45. https://doi.org/10.1007/s40675-020-00165-z.
68. Dutta R. Do genes matter in sleep? – A comprehensive update. J Neurosci Neurol Disord. 2020;4:014–23. https://doi.org/10.29328/journal.jnnd.1001029.
69. Bolsius YG, Zurbriggen MD, Kim JK, Kas MJ, Meerlo P, Aton SJ, Havekes R. The role of clock genes in sleep, stress and memory. Biochem Pharmacol. 2021;191:114493. https://doi.org/10.1016/j.bcp.2021.114493.

70. Garfield V. Sleep duration: a review of genome-wide association studies (GWAS) in adults from 2007 to 2020. Sleep Med Rev. 2021;56:101413. https://doi.org/10.1016/j.smrv.2020.101413.
71. Mullington JM, Abbott SM, Carroll JE, Davis CJ, Dijk DJ, Dinges DF, Gehrman PR, Ginsburg GS, Gozal D, Haack M, Lim DC, Macrea M, Pack AI, Plante DT, Teske JA, Zee PC. Developing biomarker arrays predicting sleep and circadian-coupled risks to health. Sleep. 2016;39:727–36. https://doi.org/10.5665/sleep.5616.
72. Uyhelji HA, Kupfer DM, White VL, Jackson ML, Van Dongen HPA, Burian DM. Exploring gene expression biomarker candidates for neurobehavioral impairment from total sleep deprivation. BMC Genomics. 2018;19:341. https://doi.org/10.1186/s12864-018-4664-3.
73. Pellegrino R, Kavakli IH, Goel N, Cardinale CJ, Dinges DF, Kuna ST, Maislin G, Van Dongen HP, Tufik S, Hogenesch JB, Hakonarson H, Pack AI. A novel BHLHE41 variant is associated with short sleep and resistance to sleep deprivation in humans. Sleep. 2014;37:1327–36. https://doi.org/10.5665/sleep.3924.
74. Webb JM, Fu YH. Recent advances in sleep genetics. Curr Opin Neurobiol. 2021;69:19–24. https://doi.org/10.1016/j.conb.2020.11.012.
75. Maire M, Reichert CF, Gabel V, Viola AU, Strobel W, Krebs J, Landolt HP, Bachmann V, Cajochen C, Schmidt C. Sleep ability mediates individual differences in the vulnerability to sleep loss: evidence from a PER3 polymorphism. Cortex. 2014;52:47–59. https://doi.org/10.1016/j.cortex.2013.11.008.
76. Viola AU, Archer SN, James LM, Groeger JA, Lo JC, Skene DJ, von Schantz M, Dijk DJ. PER3 polymorphism predicts sleep structure and waking performance. Curr Biol. 2007;17:613–8. https://doi.org/10.1016/j.cub.2007.01.073.
77. Groeger JA, Viola AU, Lo JC, von Schantz M, Archer SN, Dijk DJ. Early morning executive functioning during sleep deprivation is compromised by a PERIOD3 polymorphism. Sleep. 2008;31:1159–67.
78. Goel N, Banks S, Mignot E, Dinges DF. PER3 polymorphism predicts cumulative sleep homeostatic but not neurobehavioral changes to chronic partial sleep deprivation. PLoS One. 2009;4:e5874. https://doi.org/10.1371/journal.pone.0005874.
79. Lo JC, Groeger JA, Santhi N, Arbon EL, Lazar AS, Hasan S, von Schantz M, Archer SN, Dijk DJ. Effects of partial and acute total sleep deprivation on performance across cognitive domains, individuals and circadian phase. PLoS One. 2012;7:e45987. https://doi.org/10.1371/journal.pone.0045987.
80. Rupp TL, Wesensten NJ, Newman R, Balkin TJ. PER3 and ADORA2A polymorphisms impact neurobehavioral performance during sleep restriction. J Sleep Res. 2013;22:160–5. https://doi.org/10.1111/j.1365-2869.2012.01062.x.
81. Erblang M, Sauvet F, Drogou C, Quiquempoix M, Van Beers P, Guillard M, Rabat A, Trignol A, Bourrilhon C, Erkel MC, Léger D, Thomas C, Gomez-Merino D, Chennaoui M. Genetic determinants of neurobehavioral responses to caffeine administration during sleep deprivation: a randomized, cross over study (NCT03859882). Genes. 2021;12:555. https://doi.org/10.3390/genes12040555.
82. Grant LK, Cain SW, Chang AM, Saxena R, Czeisler CA, Anderson C. Impaired cognitive flexibility during sleep deprivation among carriers of the Brain Derived Neurotrophic Factor (BDNF) Val66Met allele. Behav Brain Res. 2018;338:51–5. https://doi.org/10.1016/j.bbr.2017.09.025.
83. Bachmann V, Klein C, Bodenmann S, Schäfer N, Berger W, Brugger P, Landolt HP. The BDNF Val66Met polymorphism modulates sleep intensity: EEG frequency- and state-specificity. Sleep. 2012;35:335–44. https://doi.org/10.5665/sleep.1690.
84. Satterfield BC, Savenkova MI, Karatsoreos IN, Jackson ML, Belenky G, Van Dongen HPA. Interleukin-6 (IL-6) response to a simulated night-shift schedule is modulated by brain-derived neurotrophic factor (BDNF) genotype. Chronobiol Int. 2020;37:1452–6. https://doi.org/10.1080/07420528.2020.1803901.

85. Britten RA, Wellman LL, Sanford LD. Progressive increase in the complexity and translatability of rodent testing to assess space-radiation induced cognitive impairment. Neurosci Biobehav Rev. 2021;126:159–74. https://doi.org/10.1016/j.neubiorev.2021.01.027.
86. Valomon A, Holst SC, Borrello A, Weigend S, Müller T, Berger W, Sommerauer M, Baumann CR, Landolt HP. Effects of COMT genotype and tolcapone on lapses of sustained attention after sleep deprivation in healthy young men. Neuropsychopharmacology. 2018;43:1599–607. https://doi.org/10.1038/s41386-018-0018-8.
87. Satterfield BC, Hinson JM, Whitney P, Schmidt MA, Wisor JP, Van Dongen HPA. Catechol-O-methyltransferase (COMT) genotype affects cognitive control during total sleep deprivation. Cortex. 2018;99:179–86. https://doi.org/10.1016/j.cortex.2017.11.012.
88. Erblang M, Drogou C, Gomez-Merino D, Rabat A, Guillard M, Van Beers P, Quiquempoix M, Boland A, Deleuze JF, Olaso R, Derbois C, Prost M, Dorey R, Léger D, Thomas C, Chennaoui M, Sauvet F. Genetics and cognitive vulnerability to sleep deprivation in healthy subjects: interaction of ADORA2A, TNF-α and COMT polymorphisms. Life. 2021;11(10):1110. https://doi.org/10.3390/life11101110.
89. Goel N, Banks S, Lin L, Mignot E, Dinges DF. Catechol-O-methyltransferase Val158Met polymorphism associates with individual differences in sleep physiologic responses to chronic sleep loss. PLoS One. 2011;6:e29283. https://doi.org/10.1371/journal.pone.0029283.
90. Bodenmann S, Xu S, Luhmann UF, Arand M, Berger W, Jung HH, Landolt HP. Pharmacogenetics of modafinil after sleep loss: catechol-O-methyltransferase genotype modulates waking functions but not recovery sleep. Clin Pharmacol Ther. 2009;85:296–304. https://doi.org/10.1038/clpt.2008.222.
91. Bodenmann S, Landolt HP. Effects of modafinil on the sleep EEG depend on Val158Met genotype of COMT. Sleep. 2010;33:1027–35. https://doi.org/10.1093/sleep/33.8.1027.
92. Ulv Larsen SM, Landolt HP, Berger W, Nedergaard M, Knudsen GM, Holst SC. Haplotype of the astrocytic water channel AQP4 is associated with slow wave energy regulation in human NREM sleep. PLoS Biol. 2020;18:e3000623. https://doi.org/10.1371/journal.pbio.3000623.
93. Dauvilliers Y, Tafti M. Molecular genetics and treatment of narcolepsy. Ann Med. 2006;38:252–62. https://doi.org/10.1080/07853890500489700.
94. Mignot E, Young T, Lin L, Finn L. Nocturnal sleep and daytime sleepiness in normal subjects with HLA-DQB1*0602. Sleep. 1999;22:347–52.
95. Goel N, Banks S, Mignot E, Dinges DF. DQB1*0602 predicts interindividual differences in physiologic sleep, sleepiness, and fatigue. Neurology. 2010;75:1509–19. https://doi.org/10.1212/WNL.0b013e3181f9615d.
96. Skeiky L, Brager AJ, Satterfield BC, Petrovick M, Balkin TJ, Capaldi VF, Ratcliffe RH, Van Dongen HPA, Hansen DA. TNFα G308A genotype, resilience to sleep deprivation, and the effect of caffeine on psychomotor vigilance performance in a randomized, double-blind, placebo-controlled, crossover study. Chronobiol Int. 2020;37:1461–4. https://doi.org/10.1080/07420528.2020.1821044.
97. Satterfield BC, Wisor JP, Field SA, Schmidt MA, Van Dongen HPA. TNFα G308A polymorphism is associated with resilience to sleep deprivation-induced psychomotor vigilance performance impairment in healthy young adults. Brain Behav Immun. 2015;47:66–74. https://doi.org/10.1016/j.bbi.2014.12.009.
98. Bachmann V, Klaus F, Bodenmann S, Schäfer N, Brugger P, Huber S, Berger W, Landolt HP. Functional ADA polymorphism increases sleep depth and reduces vigilant attention in humans. Cereb Cortex. 2012;22:962–70. https://doi.org/10.1093/cercor/bhr173.
99. Reichert CF, Maire M, Gabel V, Viola AU, Kolodyazhniy V, Strobel W, Götz T, Bachmann V, Landolt HP, Cajochen C, Schmidt C. Insights into behavioral vulnerability to differential sleep pressure and circadian phase from a functional ADA polymorphism. J Biol Rhythms. 2014;29:119–30. https://doi.org/10.1177/0748730414524898.
100. Bodenmann S, Hohoff C, Freitag C, Deckert J, Rétey JV, Bachmann V, Landolt HP. Polymorphisms of ADORA2A modulate psychomotor vigilance and the effects of caffeine on neurobehavioural performance and sleep EEG after sleep deprivation. Br J Pharmacol. 2012;165:1904–13. https://doi.org/10.1111/j.1476-5381.2011.01689.x.

101. McLellan TM, Caldwell JA, Lieberman HR. A review of caffeine's effects on cognitive, physical and occupational performance. Neurosci Biobehav Rev. 2016;71:294–312. https://doi.org/10.1016/j.neubiorev.2016.09.001.
102. Baur DM, Lange D, Elmenhorst EM, Elmenhorst D, Bauer A, Aeschbach D, Landolt HP. Coffee effectively attenuates impaired attention in ADORA2A C/C-allele carriers during chronic sleep restriction. Prog Neuro-Psychopharmacol Biol Psychiatry. 2021;109:110232. https://doi.org/10.1016/j.pnpbp.2020.110232.
103. Elmenhorst D, Elmenhorst EM, Hennecke E, Kroll T, Matusch A, Aeschbach D, Bauer A. Recovery sleep after extended wakefulness restores elevated A1 adenosine receptor availability in the human brain. Proc Natl Acad Sci U S A. 2017;114:4243–8. https://doi.org/10.1073/pnas.1614677114.
104. Whitney P, Hinson JM, Satterfield BC, Grant DA, Honn KA, Van Dongen HPA. Sleep deprivation diminishes attentional control effectiveness and impairs flexible adaptation to changing conditions. Sci Rep. 2017;7:16020. https://doi.org/10.1038/s41598-017-16165-z.
105. Muck RA, Van Dongen HPA, Schmidt MA, Wisor JP, Layton ME, DePriest DM, Honn KA, Satterfield BC. DRD2 C957T genotype modulates the time-on-task effect during total sleep deprivation. Chronobiol Int. 2020;37:1457–60. https://doi.org/10.1080/07420528.2020.1804925.
106. Satterfield BC, Wisor JP, Schmidt MA, Van Dongen HPA. Time-on-task effect during sleep deprivation in healthy young adults is modulated by dopamine transporter genotype. Sleep. 2017;40:zsx167. https://doi.org/10.1093/sleep/zsx167.
107. Holst SC, Müller T, Valomon A, Seebauer B, Berger W, Landolt HP. Functional polymorphisms in dopaminergic genes modulate neurobehavioral and neurophysiological consequences of sleep deprivation. Sci Rep. 2017;7:45982. https://doi.org/10.1038/srep45982.
108. Watson NF, Badr MS, Belenky G, Bliwise DL, Buxton OM, Buysse D, Dinges DF, Gangwisch J, Grandner MA, Kushida C, Malhotra RK, Martin JL, Patel SR, Quan S, Tasali E, Twery M, Croft JB, Maher E, Barrett JA, Thomas SM, Heald JL. Recommended amount of sleep for a healthy adult: a joint consensus statement of the American Academy of Sleep Medicine and Sleep Research Society. Sleep. 2015;38:843–4. https://doi.org/10.5665/sleep.4716.
109. Dawson D, Reid K. Fatigue, alcohol and performance impairment. Nature. 1997;388:235. https://doi.org/10.1038/40775.
110. Fairclough SH, Graham R. Impairment of driving performance caused by sleep deprivation or alcohol: a comparative study. Hum Factors. 1999;41:118–28. https://doi.org/10.1518/001872099779577336.
111. Williamson AM, Feyer AM. Moderate sleep deprivation produces impairments in cognitive and motor performance equivalent to legally prescribed levels of alcohol intoxication. Occup Environ Med. 2000;57:649–55. https://doi.org/10.1136/oem.57.10.649.
112. Connor J, Whitlock G, Norton R, Jackson R. The role of driver sleepiness in car crashes: a systematic review of epidemiological studies. Accid Anal Prev. 2001;33:31–41. https://doi.org/10.1016/s0001-4575(00)00013-0.
113. Horne J, Reyner L. Vehicle accidents related to sleep: a review. Occup Environ Med. 1999;56:289–94. https://doi.org/10.1136/oem.56.5.289.
114. McCartt AT, Ribner SA, Pack AI, Hammer MC. The scope and nature of the drowsy driving problem in New York State. Accid Anal Prev. 1996;28:511–7. https://doi.org/10.1016/0001-4575(96)00021-8.
115. Philip P, Taillard J, Micoulaud-Franchi JA. Sleep restriction, sleep hygiene, and driving safety: the importance of situational sleepiness. Sleep Med Clin. 2019;14:407–12. https://doi.org/10.1016/j.jsmc.2019.07.002.
116. Arnal PJ, Sauvet F, Leger D, van Beers P, Bayon V, Bougard C, Rabat A, Millet GY, Chennaoui M. Benefits of sleep extension on sustained attention and sleep pressure before and during total sleep deprivation and recovery. Sleep. 2015;38:1935–43. https://doi.org/10.5665/sleep.5244.

117. Rupp TL, Wesensten NJ, Bliese PD, Balkin TJ. Banking sleep: realization of benefits during subsequent sleep restriction and recovery. Sleep. 2009;32:311–21. https://doi.org/10.1093/sleep/32.3.311.
118. Alger SE, Brager AJ, Balkin TJ, Capaldi VF, Simonelli G. Effect of cognitive load and emotional valence of distractors on performance during sleep extension and subsequent sleep deprivation. Sleep. 2020;43:zsaa013. https://doi.org/10.1093/sleep/zsaa013.
119. Mantua J, Skeiky L, Prindle N, Trach S, Doty TJ, Balkin TJ, Brager AJ, Capaldi VF, Simonelli G. Sleep extension reduces fatigue in healthy, normally-sleeping young adults. Sleep Sci. 2019;12:21–7. https://doi.org/10.5935/1984-0063.20190056.
120. Ebben MR. Nonpharmacologic management of excessive daytime sleepiness. Sleep Med Clin. 2017;12:479–87. https://doi.org/10.1016/j.jsmc.2017.03.020.
121. Parker RS, Parker P. The impact of sleep deprivation in military surgical teams: a systematic review. J R Army Med Corps. 2017;163:158–63. https://doi.org/10.1136/jramc-2016-000640.
122. Patterson PD, Ghen JD, Antoon SF, Martin-Gill C, Guyette FX, Weiss PM, Turner RL, Buysse DJ. Does evidence support "banking/extending sleep" by shift workers to mitigate fatigue, and/or to improve health, safety, or performance? A systematic review. Sleep Health. 2019;5:359–69. https://doi.org/10.1016/j.sleh.2019.03.001.
123. Drummond SP, Paulus MP, Tapert SF. Effects of two nights sleep deprivation and two nights recovery sleep on response inhibition. J Sleep Res. 2006;15:261–5. https://doi.org/10.1111/j.1365-2869.2006.00535.x.
124. Kamdar BB, Kaplan KA, Kezirian EJ, Dement WC. The impact of extended sleep on daytime alertness, vigilance, and mood. Sleep Med. 2004;5:441–8. https://doi.org/10.1016/j.sleep.2004.05.003.
125. Mantua J, Brager AJ, Alger SE, Adewale F, Skeiky L, Balkin TJ, Capaldi VF, Simonelli G. Self-reported sleep need, subjective resilience, and cognitive performance following sleep loss and recovery sleep. Psychol Rep. 2021;124:210–26. https://doi.org/10.1177/0033294119899896.
126. Yamazaki EM, Antler CA, Lasek CR, Goel N. Residual, differential neurobehavioral deficits linger after multiple recovery nights following chronic sleep restriction or acute total sleep deprivation. Sleep. 2021;44:zsaa224. https://doi.org/10.1093/sleep/zsaa224.
127. Taub JM, Globus GG, Phoebus E, Drury R. Extended sleep and performance. Nature. 1971;233:142–3. https://doi.org/10.1038/233142a0.
128. Smith MG, Wusk GC, Nasrini J, Baskin P, Dinges DF, Roma PG, Basner M. Effects of six weeks of chronic sleep restriction with weekend recovery on cognitive performance and well-being in high-performing adults. Sleep. 2021;44(8):zsab051. https://doi.org/10.1093/sleep/zsab051.
129. Stepan ME, Altmann EM, Fenn KM. Caffeine selectively mitigates cognitive deficits caused by sleep deprivation. J Exp Psychol Learn Mem Cogn. 2021;47(9):1371–82. https://doi.org/10.1037/xlm0001023.
130. Killgore WDS, Kamimori GH. Multiple caffeine doses maintain vigilance, attention, complex motor sequence expression, and manual dexterity during 77 hours of total sleep deprivation. Neurobiol Sleep Circadian Rhythms. 2020;9:100051. https://doi.org/10.1016/j.nbscr.2020.100051.
131. Hansen DA, Ramakrishnan S, Satterfield BC, Wesensten NJ, Layton ME, Reifman J, Van Dongen HPA. Randomized, double-blind, placebo-controlled, crossover study of the effects of repeated-dose caffeine on neurobehavioral performance during 48 h of total sleep deprivation. Psychopharmacology. 2019;236:1313–22. https://doi.org/10.1007/s00213-018-5140-0.
132. Paech GM, Vedova CD, Pajcin M, Grant C, Kamimori G, Banks S. Caffeine has minimal effects on daytime recovery sleep following severe sleep deprivation. Sleep Biol Rhythms. 2016;14:149–56. https://doi.org/10.1007/s41105-015-0031-9.
133. Spaeth AM, Goel N, Dinges DF. Cumulative neurobehavioral and physiological effects of chronic caffeine intake: individual differences and implications for the use of caffeinated energy products. Nutr Rev. 2014;72:34–47. https://doi.org/10.1111/nure.12151.
134. Wesensten NJ, Balkin TJ, Belenky G. Countermeasures for mitigating fatigue in motor vehicle operators. Rev. Hum. Factors Ergon. 2015;10:115–37. https://doi.org/10.1177/1557234X15574827.

135. Quiquempoix M, Sauvet F, Erblang M, Van Beers P, Guillard M, Drogou C, Trignol A, Vergez A, Léger D, Chennaoui M, Gomez-Merino D, Rabat A. Effects of caffeine intake on cognitive performance related to total sleep deprivation and time on task: a randomized cross-over double-blind study. Nat Sci Sleep. 2021;14:457–73. https://doi.org/10.2147/NSS.S342922.
136. Irwin C, Khalesi S, Desbrow B, McCartney D. Effects of acute caffeine consumption following sleep loss on cognitive, physical, occupational and driving performance: a systematic review and meta-analysis. Neurosci Biobehav Rev. 2020;108:877–88. https://doi.org/10.1016/j.neubiorev.2019.12.008.
137. Faraut B, Andrillon T, Vecchierini MF, Leger D. Napping: a public health issue. From epidemiological to laboratory studies. Sleep Med Rev. 2017;35:85–100. https://doi.org/10.1016/j.smrv.2016.09.002.
138. Farhadian N, Khazaie H, Nami M, Khazaie S. The role of daytime napping in declarative memory performance: a systematic review. Sleep Med. 2021;84:134–41. https://doi.org/10.1016/j.sleep.2021.05.019.
139. Lo JC-Y, Koa TB, Ong JL, Gooley JJ, Chee MWL. Staying vigilant during recurrent sleep restriction: dose-response effects of time-in-bed and benefits of daytime napping. Sleep. 2022;45(4):zsac023. https://doi.org/10.1093/sleep/zsac023.
140. Trotti LM. Waking up is the hardest thing I do all day: sleep inertia and sleep drunkenness. Sleep Med Rev. 2017;35:76–84. https://doi.org/10.1016/j.smrv.2016.08.005.
141. Lundholm KR, Honn KA, Skeiky L, Muck RA, Van Dongen HPA. Trait interindividual differences in the magnitude of subjective sleepiness from sleep inertia. Clocks Sleep. 2021;3(2):298–311. https://doi.org/10.3390/clockssleep3020019.
142. Hilditch CJ, McHill AW. Sleep inertia: current insights. Nat Sci Sleep. 2019;11:155–65. https://doi.org/10.2147/NSS.S188911.
143. Mograss M, Abi-Jaoude J, Frimpong E, Chalati D, Moretto U, Tarelli L, Lim A, Dang-Vu TT. The effects of napping on night-time sleep in healthy young adults. J Sleep Res. 2022;31:e13578. https://doi.org/10.1111/jsr.13578.
144. Dashti HS, Daghlas I, Lane JM, Huang Y, Udler MS, Wang H, Ollila HM, Jones SE, Kim J, Wood AR, 23andMe Research Team, Weedon MN, Aslibekyan S, Garaulet M, Saxena R. Genetic determinants of daytime napping and effects on cardiometabolic health. Nat Commun. 2021;12:900. https://doi.org/10.1038/s41467-020-20585-3.
145. Kurniawan IT, Cousins JN, Chong PL, Chee MW. Procedural performance following sleep deprivation remains impaired despite extended practice and an afternoon nap. Sci Rep. 2016;6:36001. https://doi.org/10.1038/srep36001.
146. Pomares FB, Cross N, Jegou A, Nguyen A, Perrault A, Lee K, Smith D, Aydin U, Grova C, Dang-Vu T. Cognitive performance and brain activation recovery after a nap following total sleep deprivation. Sleep Med. 2019;64:S305. https://doi.org/10.1016/j.sleep.2019.11.856.
147. Centofanti S, Banks S, Coussens S, Gray D, Munro E, Nielsen J, Dorrian J. A pilot study investigating the impact of a caffeine-nap on alertness during a simulated night shift. Chronobiol Int. 2020;37:1469–73. https://doi.org/10.1080/07420528.2020.1804922.
148. Johnson DA, Jackson CL, Williams NJ, Alcántara C. Are sleep patterns influenced by race/ethnicity – a marker of relative advantage or disadvantage? Evidence to date. Nat Sci Sleep. 2019;11:79–95. https://doi.org/10.2147/NSS.S169312.
149. Emmanuel P, von Schantz M. Absence of morningness alleles in non-European populations. Chronobiol Int. 2018;35:1758–61. https://doi.org/10.1080/07420528.2018.1506928.
150. Matthews KA, Hall MH, Lee L, Kravitz HM, Chang Y, Appelhans BM, Swanson LM, Neal-Perry GS, Joffe H. Racial/ethnic disparities in women's sleep duration, continuity, and quality, and their statistical mediators: Study of Women's Health Across the Nation. Sleep. 2019;42:zsz042. https://doi.org/10.1093/sleep/zsz042.
151. Prasad B, Saxena R, Goel N, Patel SR. Genetic ancestry for sleep research: leveraging health inequalities to identify causal genetic variants. Chest. 2018;153:1478–96. https://doi.org/10.1016/j.chest.2018.03.024.

152. Hohoff C, Kroll T, Zhao B, Kerkenberg N, Lang I, Schwarte K, Elmenhorst D, Elmenhorst EM, Aeschbach D, Zhang W, Baune BT, Neumaier B, Bauer A, Deckert J. ADORA2A variation and adenosine A1 receptor availability in the human brain with a focus on anxiety-related brain regions: modulation by ADORA1 variation. Transl Psychiatry. 2020;10:406. https://doi.org/10.1038/s41398-020-01085-w.
153. Magee M, Sletten TL, Murray JM, Gordon CJ, Lovato N, Bartlett DJ, Kennaway DJ, Lockley SW, Lack LC, Grunstein RR, Archer SN, Rajaratnam SMW, Delayed Sleep on Melatonin (DelSoM) Study Group. A PERIOD3 variable number tandem repeat polymorphism modulates melatonin treatment response in delayed sleep-wake phase disorder. J Pineal Res. 2020;69:e12684. https://doi.org/10.1111/jpi.12684.
154. Ozsoy F, Yigit S, Nursal AF, Kulu M, Karakus N. The impact of PER3 VNTR polymorphism on the development of schizophrenia in a Turkish population. Cytol Genet. 2021;55:188–93. https://doi.org/10.3103/S0095452721020109.
155. Weiss C, Woods K, Filipowicz A, Ingram KK. Sleep quality, sleep structure, and PER3 genotype mediate chronotype effects on depressive symptoms in young adults. Front Psychol. 2020;11:2028. https://doi.org/10.3389/fpsyg.2020.02028.
156. Tartar JL, Hiffernan FS, Freitas KE, Fins AI, Banks JB. A functional adenosine deaminase polymorphism associates with evening melatonin levels and sleep quality. J Circadian Rhythms. 2021;19:5. https://doi.org/10.5334/jcr.209.
157. Yegin Z, Sarisoy G, Aral AE, Koc H. For whom the circadian clock ticks? Investigation of PERIOD and CLOCK gene variants in bipolar disorder. Chronobiol Int. 2021;38(8):1109–19. https://doi.org/10.1080/07420528.2021.1917594.
158. Križan Z, Hisler G. Personality and sleep: neuroticism and conscientiousness predict behaviourally recorded sleep years later. Eur J Personal. 2019;33:133–53. https://doi.org/10.1002/per.2191.
159. Križan Z, Hisler G, Krueger RF, McGue M. Why is personality tied to sleep quality? A biometric analysis of twins. J Res Pers. 2021;90:104048. https://doi.org/10.1016/j.jrp.2020.104048.
160. Killgore WD, Richards JM, Killgore DB, Kamimori GH, Balkin TJ. The trait of introversion-extraversion predicts vulnerability to sleep deprivation. J Sleep Res. 2007;16:354–63. https://doi.org/10.1111/j.1365-2869.2007.00611.x.

Sleep and Genetic Neurodevelopmental Disorders

Anne Marie Morse and Eric Strong

Abstract Sleep is a brain-derived homeostatic process. Neurodevelopmental milestones typically parallel the natural ontogeny of sleep observed in humans. In fact, an early indicator of neurodivergence is commonly abnormal sleep, which may manifest as reduced duration, impaired quality, or inappropriate timing of sleep. Typically, sleep evolves with predictable changes in duration of sleep required in 24 h, progression toward consolidation of nocturnal sleep, entrainment of circadian rhythms, and maturation of ultradian patterns, as a function of age and brain development. Genetically based neurodevelopmental disorders (NDDs) are providing initial insights to further understanding the relationship between sleep and neurodevelopmental trajectories and how genetics guide these neurodivergent pathways. Recognition and intervention of sleep-wake disorders may benefit developmental trajectories, especially in these high-risk populations.

Keywords Sleep · Neurodevelopment · Neurodivergent · Sleep-wake disorders · Autism spectrum disorder

Introduction

Sleep-wake patterns are a result of a complex process integrating input from the circadian system (process C) and sleep-wake homeostasis (process S). The circadian system is an innate genetically derived biologic rhythm that follows an approximate 24-h cycle. The suprachiasmatic nuclei (SCN), located in the ventral hypothalamus, respond to light exposure to entrain the circadian pattern. Additional exogenous factors (known as zeitgebers), in addition to light, such as drugs (prescription or illicit), temperature, stress, or even eating times can influence the

A. M. Morse (✉) · E. Strong
Geisinger Commonwealth School of Medicine, Janet Weis Children's Hospital, Danville, PA, USA
e-mail: amorse@geisinger.edu; estrong@geisinger.edu

P. Gehrman et al. (eds.), *Genetics of Sleep and Sleep Disorders*,
https://doi.org/10.1007/978-3-031-62723-1_16

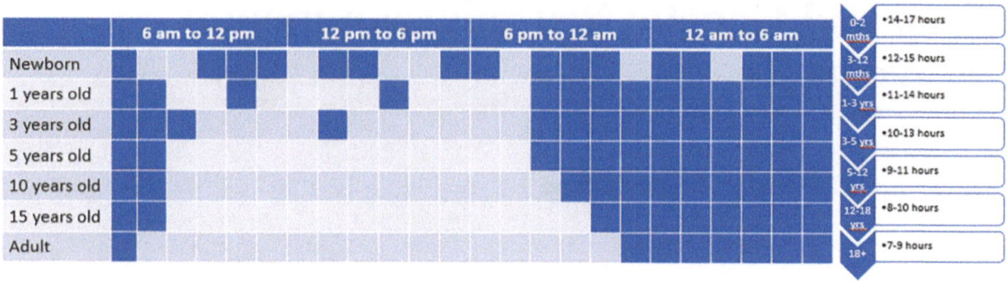

Fig. 1 Sleep hours and consolidation by age

expression of the circadian rhythm. The circadian system influences more than the sleep-wake cycle, as it has been found to be intrinsic to multiple bodily functions including body temperature cycle and hormone production and secretion, as well may other peak times of function such as blood pressure peak and nadir [1–4].

The natural ontogeny of sleep correlates well with the progression of active neuromaturation. Sleep evolves with predictable changes in sleep duration, consolidation of nocturnal sleep, and changes in circadian and ultradian patterns, as a function of age and neurodevelopment (Fig. 1) (1–3). This intimate relationship is well illustrated when examining the dysfunction present in both sleep and developmental trajectories for specific genetic etiologies responsible for developmental brain dysfunction. This relationship becomes increasingly relevant recognizing sleep problems that affect approximately 25% of all children [5] but is observed in 50–95% of individuals with neurodevelopmental disorders (NDDs) [6]. When examining different genetic etiologies of neurodevelopmental disorders, more specific sleep-related phenotypes and risk have been described.

Autism Spectrum Disorder

Autism spectrum disorder (ASD) is common, affecting about 1 in 40 children [7]. ASD is a neurodevelopmental disorder characterized by deficits in social communication and the presence of restricted interests and repetitive behaviors [8]. Sleep problems are frequently reported in ASD [9–11] irrespective of presence or absence of identified underlying genetic etiology. In fact, as many as 80% of patients with ASD will experience sleep issues of some kind, with insomnia and circadian rhythm sleep-wake disorders being the most reported. Additionally, polysomnography (PSG) studies have shown that patients with ASD exhibit a reduction of REM sleep, NREM2 sleep, and slow-wave sleep spindles and an increase in NREM1 sleep [12].

Genetic mutations in circadian or clock genes are more common in patients with ASD and have been suggested as a significant driver in the sleep-wake dysfunction observed [13] but also a potential etiology of the ASD as well [14]. Additional factors influencing circadian patterns observed in ASD include genetic abnormalities associated with melatonin synthesis and function [15]. Abnormalities within the

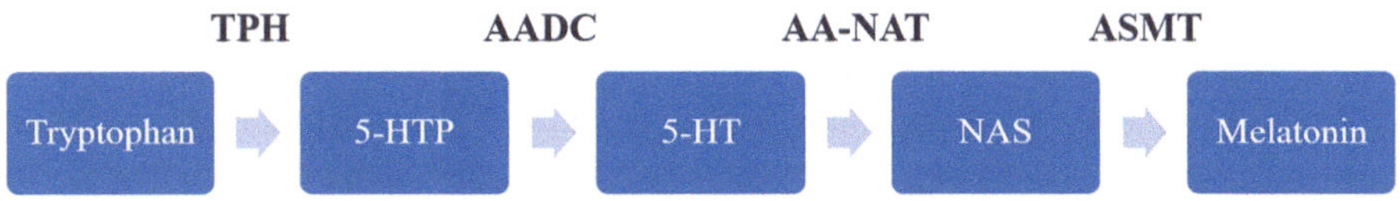

Fig. 2 Melatonin Synthesis

serotonin-NAS-melatonin pathway have been identified in ASD. This pathway begins with tryptophan being converted to serotonin and subsequently processed to melatonin (Fig. 2). L-tryptophan is an essential amino acid, meaning that it is not an amino acid that the body creates itself and must be taken by food or supplement. High levels of serotonin and a melatonin deficiency are observed in individuals with ASD due to disruption in the serotonin-NAS-melatonin pathway [16]. Similar abnormalities are also observed in first-degree relatives of individuals with ASD [16]. Genetic variability within the acetylserotonin O-methyltransferase (ASMT) gene has been implicated as a cause of variable melatonin levels in individuals with ASD. Specifically, two polymorphisms of the ASMT gene located in the promoter are frequently identified and associated with reduced melatonin levels [17]. Other evidence suggests reduced arylalkylamine I-acetyltransferase (AA-NAT) and hydroxyindole-O-methyltransferase enzyme activity as a reason for reduced melatonin levels [18]. For this reason, melatonin or melatonin receptor agonists are commonly used in the setting of sleep initiation difficulties or circadian disruption in individuals with ASD. Despite the identification of melatonin synthesis and function abnormalities in ASD, melatonin itself is not considered as a marker of ASD clinically. However, several sleep-related proteins being evaluated as a potential biomarker for ASD include N-acetylserotonin (NAS), 6-sulphatoxymelatonin (aMT6s), serotonin, 14–3-3 protein, and miR-451 [19].

Angelman Syndrome

Angelman syndrome (AS) is a rare neurodevelopmental disease with a prevalence of 1 in 10,000–24,000 births [20] and is caused by the loss of function of the maternal copy of ubiquitin–protein ligase E3A (UBE3A) at the chromosome 15q11–13 [21]. AS is typically characterized by global developmental delay, severe intellectual disability, lack of language, happy temperament, ataxia, and intractable epilepsy [22]. There are four molecular mechanisms of etiology: maternal deletion of chromosome 15q11–q13, paternal uniparental disomy of chromosome 15q11–q13, imprinting defects, and maternally inherited UBE3A mutations [23–26]. The phenotype of an individual with AS does appear to be influenced by molecular mechanism of etiology for the condition [22]. Irrespective of molecular mechanism of etiology, sleep problems are common. Interestingly, however, the highest prevalence of sleep issues occurs when there is either uniparental disomy or a maternally inherited UBE3A mutation, even though UPD and imprinting are associated with a less severe phenotype overall [22].

Individuals with AS experience a broad range of sleep difficulties including reduced total sleep time (TST), sleep initiation and maintenance difficulties, and premature arousal. A 2018 meta-analysis identified frequent nighttime arousals during sleep and daytime sleepiness as the primary difficulties identified by caregivers of patients with AS [27]. In addition to these clinically observed changes, there are also electrophysiologic changes observed. Delta power, or 2–4 Hz frequency waveforms, has been found to be increased during wake and all stages of non-REM sleep in AS and may serve as a potentially reliable biomarker for AS [28]. In fact, the strength of delta rhythms has a strong correlation with severity of epilepsy, as well as cognitive, motor, and communication deficits [29]. The results of poor-quality sleep and reduced REM sleep have both been shown to negatively impact behavior and seizures [30].

Down Syndrome

Down syndrome (DS) is the most common genetic neurodevelopmental disorder occurring in 1 in 700 live births [31]. The genetic basis of DS is most commonly due to an extra chromosome 21, but about 5% are due to 1 copy of chromosome 21 translocation and in minority of cases due to a mosaicism of trisomy [32]. DS is typically characterized by upslanted palpebral fissures, flattened nasal bridge, clinodactyly of the fifth finger, hypotonia, nuchal folds, and a single palmar crease [31]. There is a significant variability observed in intellectual disability and developmental outcomes in DS.

Typically, medical attention has been focused on the high prevalence of OSA occurring in individuals with DS [33]. In fact, it is recommended that evaluation for OSA with clinical screening questions and nocturnal polysomnography occurs at least once between 0 and 4 years old and then yearly clinical screening questions thereafter [34]. The increased risk for OSA in DS is multifactorial. Midfacial hypoplasia, macroglossia, hypotonia, thick necks, elevated BMI, and hypothyroidism can enhance likelihood of OSA [35–41]. However, sleep difficulties extend past sleep-related breathing disorders and include difficulties with settling and resistance going to sleep, sleep-related anxiety, bruxism (teeth-grinding), fragmented nocturnal sleep, nocturnal enuresis, restlessness, and sleep-related movement disorders. As a result it is not uncommon to encounter excessive daytime sleepiness (EDS) [42].

Electrophysiologic characterization of sleep in DS via PSG demonstrates reduced time spent in REM sleep in adults with DS [43]. These findings were also identified in children with DS where there was reduced time spent in REM sleep, as well as a reduced sleep efficiency and an increased percentage of time spent in SWS [44]. However, in children under 2 years old with DS, no differences in ultradian pattern were observed [45]. The differences observed in sleep architecture did not seem to be due to comorbid OSA as the difference identified pre-adenotonsillectomy for OSA remained post-procedure [46].

DS is also associated with an increased risk of early-onset Alzheimer disease (AD) due to an additional copy of the gene that carries the amyloid precursor protein (APP) [47]. In fact, the prevalence of progressive cognitive impairment in adults with DS has been suggested to be between 23% and 88% when evaluating individuals with DS who are 40 and older, with increasing risk corresponding to more advanced chronologic age [48]. This particular genetic susceptibility is likely exacerbated by sleep-wake disorder comorbidities. As it is known that not only are sleep disturbances a common feature of AD, but it may predict the development of dementia and beta-amyloid pathology [49]. There is evidence to suggest that the sleep-wake cycle control beta-amyloid levels and sleep pathology can exacerbate beta-amyloid pathology and even neuronal injury independent of beta-amyloid [49].

Fragile X Syndrome

Fragile X syndrome (FXS) is one of the most common genetic neurodevelopmental disorders encountered with the prevalence being about 1 in 7000 males and 1 in 11,000 females [50–52]. It is an X-linked disorder that occurs because of the trinucleotide CGG expansion in the fragile X mental retardation (FMR1) gene [50]. Males are typically more affected than females [50]. There are physical features that are typically characteristic in FXS which include hypotonia, a prominent forehead, an elongated face, high arched palate, protruding ears, and in males macroorchidism [53]. Clinical features can include mild to severe ID, self-injurious behaviors, epilepsy, anxiety, depression, language delays, and sleep problems [54]. It should be noted that as many as 60% of males with FXS and 20% of females with FXS will also be diagnosed with ASD [55].

Sleep problems can occur in up to 50% of individuals with FXS [56–58]. Reduced total sleep time related to sleep initiation and maintenance difficulties are common [56, 57, 59]. Polysomnographic studies also demonstrate similar findings with prolonged SOL, increased WASO, increased N1 sleep, and features of disturbed nocturnal sleep [59–61]. Patients with FXS are at increased risk for sleep-disordered breathing with about 7% of individuals having OSA based on data collected for the Fragile X Clinical and Research Consortium Database [62]. Significantly, there is also an increased risk for carriers of the pre-mutation who have fragile X ataxia tremor syndrome, where the risk of OSA is about 3.5 times higher than controls [63].

Rett Syndrome

Rett syndrome (RTT) is a rare genetic disorder, with an incidence of 1/10,000–1/15,000, that manifests as developmental regression primarily afflicting females in early childhood [64–67]. The regression observed typically is characterized as expressive language regression and loss of purposeful hand use, with development of stereotypical writhing hand movements, followed by ongoing motoric impairment with loss of ambulation [68–70]. Pathogenic variants resulting in loss of function in the X-linked gene methyl–CpG-binding protein 2 (MECP2) are the most commonly encountered genetic etiology [68]. Alternative genetic etiologies resulting in a Rett-like disorder occur much more rarely and include pathogenic changes in the genes cyclin-dependent kinase like 5 (CDKL5) and forkhead box G1 (FOXG1). It has been suggested that the three genes share a common pathway that impacts gene transcription modulation resulting in improper brain growth and maturation [71–73].

A recent systematic review and meta-analysis of sleep problems in individuals with Rett syndrome demonstrate a lifelong presence of sleep initiation and maintenance difficulties with additional daytime finding of excessive sleepiness [74]. Sleep disorders have been reported to be highly prevalent in RTT, impacting up to about 80% of individuals [6, 75–77]. In fact, problematic sleep is considered one of the supportive diagnostic criteria for RTT [68]. This systematic review and meta-analysis did confirm that the rates of sleep disorders were higher than those of a neurotypical individual, establishing a generic prevalence rate of 54.1% and found this increased risk existed across the lifespan [74]. Sleep disorders are predominantly characterized in individuals with RTT due to a MECP2 abnormality, as this is the most commonly encountered etiology. In these individuals sleep abnormalities are present across the lifespan but appear to differ based on age. In children between 0 and 18 years old, nocturnal laughing and sleep fragmentation due to recurrent wakings are the more common problems encountered, where EDS is the more likely issue in adults with RTT.

Polysomnographic studies of sleep in RTT have led to inconsistent findings which include reduced TST, impaired sleep efficiency, increased arousals, and reduced REM sleep with increased SWS [6, 74, 76–78]. In CDKL5-related RTT, TST tended to be longer, and there is evidence to suggest that the ultradian pattern also has reduced REM, but N1 was more increased rather than SWS [75]. Daytime respiratory abnormalities due to autonomic dysfunction are well described as part of the RTT phenotype and are commonly described as hyperventilation alternating with prolonged apneic episodes [79]. Sleep-disordered breathing, with elevation in both obstructive and central events, has also been found in individuals with RTT [78]. Those with MECP2, who have epilepsy and scoliosis, without prior history of adenotonsillectomy tend to stratify as higher risk for more severe sleep-disordered breathing [78].

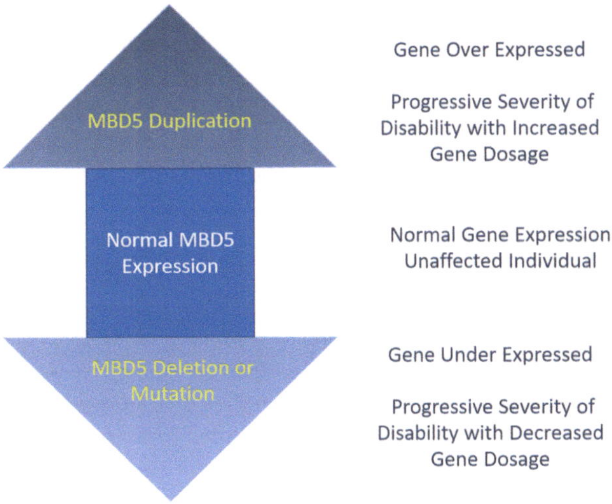

Fig. 3 MBD5 gene dosage and phenotype

MBD5-Associated Neurodevelopmental Disorder (MAND)

Methyl-CpG-binding domain 5 (MBD5)-associated neurodevelopmental disorders (MAND) are rare disorders frequently characterized by features of intellectual disability, epilepsy, speech impairment, ASD, and broad-based ataxic gait [80]. MAND is a catchment term used to describe a group of disorders that impact the function of MBD5 (e.g., 2q23.1 deletion syndrome, 2q23.1 duplication syndrome, and MBD5 variants) [81]. Although MBD5 is a dose-sensitive gene (Fig. 3) [82], sleep disturbances are a common phenotypic feature irrespective of molecular basis of MAND [81].

It has been suggested that the sleep disturbances encountered in MAND are similar to those seen in SMS, AS, and FXS [81]. Features of dysfunctional sleep include sleep fragmentation due to frequent nocturnal awakenings, night terrors, and earlier-than-desired waking in the morning [81]. EDS is common [81]. MBD5 regulates RAI1 mRNA expression [81, 82], which contributes to the observed phenotypic similarity to SMS. Studies evaluating polysomnography in MAND are lacking. Therefore, it is unclear whether there are ultradian differences observed in MAND.

Pitt–Hopkins Syndrome (PTHS)

Pitt–Hopkins syndrome (PTHS) is a rare neurodevelopmental disorder that bears some similarity in phenotype to RTT as it is characterized by moderate to severe ID, episodic hyperventilation, intractable epilepsy, and developmental delay with autistic features. PTHS has distinctive physical features that include microcephaly, a large beaked nose, a wide mouth, cup-shaped ears with broad helices, and clubbed fingertips [83, 84]. PTHS is caused by heterozygous pathogenic variants in the transcription factor 4 (TCF4) gene on chromosome 18q21.1 [85]. The protein encoded by the TCF4 gene resembles the inhibitor of DNA-binding proteins. These proteins act as a negative regulator of the feedback loop that generates intracellular circadian rhythms [86]. It is unclear if this is a contributing pathologic factor to sleep problems encountered in PTHS.

The literature characterizing sleep problems in individuals with PTHS is extremely limited. Reports available suggest increased nocturnal sleep time when compared to both neurotypical peers and some other selected neurodevelopmental disorders (SMS and MAND) [87]. When comparing against SMS and MAND, daytime napping, nocturnal wakings, and bedtime resistance were less common to occur in PTHS [87]. There are no available studies evaluating polysomnography or ultradian features of PTHS. Although abnormalities of respiration in these patients are well-documented, their episodic hyperventilation and apnea due to not seem to occur in sleep [88].

Smith–Magenis Syndrome (SMS)

SMS is a rare neurodevelopmental disorder characterized by intellectual disability, developmental delays, and other congenital anomalies due to haploinsufficiency of RAI1 caused by either a pathogenic variant or microdeletion in retinoic acid-induced 1 gene (RAI1), which is located on the short arm of chromosome 17 [89]. RAI1 regulates transcription with effect on downstream gene expression [90]. The majority of patients with SMS are the result of microdeletions in the RAI1 gene (90%) with the remainder representing heterozygous pathogenic variants of RAI1 (10%) [91]. SMS is suggested to affect about 1 in 25,000 individuals globally; however, it is thought to be underdiagnosed with a more accurate prevalence being closer to 1 in 15,000 individuals.

Sleep abnormalities in SMS are classically characterized by an inversion in circadian rhythm (Fig. 4) resulting in individuals with SMS experiencing EDS, daytime napping, sleep attacks, and behavioral outbursts [92, 93]. Interestingly sleep observed in SMS evolves over time. Infants' families will frequently describe them as the perfect baby and great sleeper, even though objective data has demonstrated reduced TST [94]. Early childhood sleep is typically characterized by a myriad of sleep issues including EDS, disturbed nocturnal sleep, earlier-than-desired waking times, snoring, and bedwetting [95]. In older children, adolescents, and adults with the SMS features of circadian abnormalities become more obvious with difficulty settling for nighttime sleep, reduced nocturnal TST, EDS, daytime napping, sleep attacks, and difficulty with waking from sleep [96]. The circadian dysfunction observed is related to abnormal expression of various clock genes (i.e., CLOCK, PER1, PER2, BMAL1) due to changes in the transcriptional regulator, RAI1 [97].

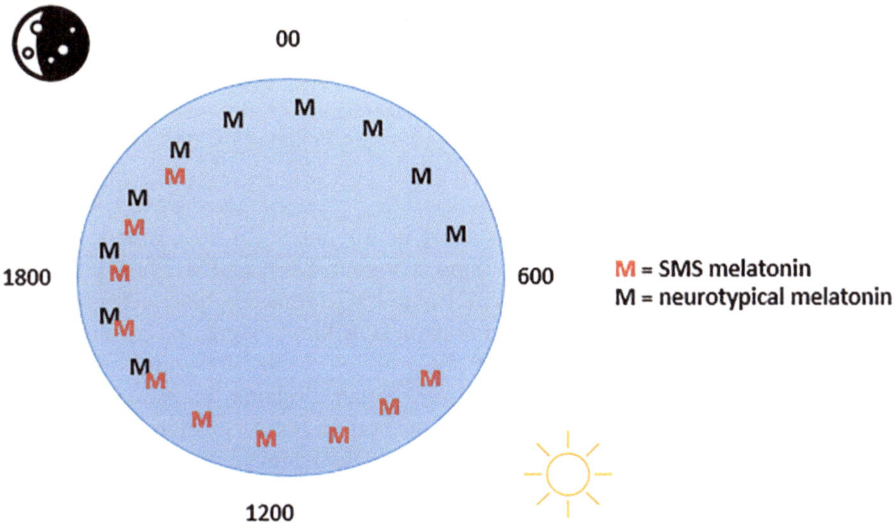

Fig. 4 Circadian pattern of melatonin production in Smith–Magenis syndrome and neurotypical individuals

Williams Syndrome (WS)

Williams syndrome (WS) is a neurodevelopmental disorder with a prevalence of about 1 in 7500 [98] that has multi-organ involvement including neurological, neurocognitive, cardiovascular, musculoskeletal, and endocrine abnormalities [99]. WS results from a microdeletion in the long arm of chromosome 7 (7q11.23). Typically, it involves deletion 1.5–1.7 megabases across 25–30 genes [100]. WS is typically characterized by mild-to-moderate intellectual disability, poor visual-motor integration, and ADHD but uniquely tends to have better-than-expected short-term memory, language, and social skills [101]. Individuals with WS have characteristic facial features with a broad forehead, periorbital fullness, sunken nasal bridge, wide mouth with a prominent lower lip, and small spaced-out teeth [102]. Additional physical findings include growth retardation, congenital heart disease (typically supravalvular aortic stenosis), premature aging, connective tissue weakness, and renal [102].

Sleep difficulties are widely reported in individuals with WS [103–105]. Difficulties settling, sleep anxiety, prolonged sleep onset latencies, and disturbed nocturnal sleep are frequent. Studies evaluating polysomnographic features in WS have demonstrated a reduced sleep efficiency with increased wake after sleep onset and an increase in SWS [104]. Additional findings on polysomnography demonstrate increase likelihood of sleep-disordered breathing and respiratory-related arousals, both of which may also contribute to impaired sleep efficiency observed [104]. It is unclear if individuals with WS are higher risk for elevated periodic limb movements, as there is conflicting data regarding these findings [104]. Patients with

WS have also been shown to have higher levels of cortisol and lower levels of melatonin in evening hours when compared to controls [106].

Tuberous Sclerosis Complex (TSC)

Tuberous sclerosis complex (TSC) is a rare neurogenetic disorder genetic disorder that is due to mutations in the tumor suppressor gene TSC1 or TSC2 [107]. It affects about 1 in 6000 individuals [107]. TSC1 and TSC2 genes regulate cell growth, as they encode a protein complex that inhibits mTOR kinase signaling by inactivating the Rheb GTPase [108]. As a result of TSC1 or TSC2 mutations, uncontrolled cell growth produces tumors across multiple organ systems in individuals with TSC. Almost all patients have central nervous system (CNS) involvement, resulting in a spectrum of neurologic and psychiatric manifestations including seizures, cognitive impairment, ASD, ADHD, depression, and anxiety [107]. TSC-associated neuropsychiatric disorders (TAND) is the term commonly applied to encompass these manifestations [109]. One of the most encountered elements of TAND is sleep problems, impacting 49–73% of individuals with TSC [110–114].

Sleep problems commonly encountered in TSC include difficulty settling for sleep, disturbed nocturnal sleep, and impaired daytime wakefulness and behavior [115]. Studies evaluating polysomnography demonstrate reduced total sleep time, impaired sleep efficiency, increased awakenings, and reduced time spent in REM sleep [110]. The frequency and severity of sleep disorders in TSC are so prominent that a recent meta-analysis concluded that insomnia and excessive daytime sleepiness should be considered as part of a priority assessment in TSC [116].

Treatment Options for NDD-Related Sleep Difficulties

Despite the prevalence of sleep disorders and difficulties in patients with NDDs, head-to-head comparisons of pharmacotherapy options are quite limited. As a result, the medical management of these patients is typically driven by consideration of factors such as age, side-effect profile, and comorbidities. Consideration of a patient's specific challenges surrounding sleep should be taken into account as well. The young child with ASD who has difficulty initiating sleep due to severe hyperactivity and impulsivity would likely benefit from a very different medication regimen when compared to an adolescent with FXS exhibiting self-injurious behavior and frequent nighttime awakenings. Selection of a medication regimen should be a deliberate process that occurs after or in concert with optimization of sleep hygiene routines. Most of the medications outlined below are commonly used for NDD-related sleep difficulties. It is by no means intended to be an exhaustive listing but instead a framework through which to consider one's options in addressing these complex patients.

Melatonin: Melatonin is a biogenic amine with hypnotic properties. Exogenous melatonin is typically the first-line pharmacotherapy option for addressing sleep difficulties in patients with NDDs. A 2019 systematic review identified 19 randomized controlled trials examining the efficacy of melatonin for treating sleep disorders in children. Melatonin significantly improved sleep latency, sleep duration, and wake time after sleep onset. Improvement in sleep was greatest in children with NDDs [117]. Its efficacy and tolerability in specific conditions such as ASD, SMS, FXS, and WS have also been well-established [118–121]. The timing and dosing of melatonin administration is dependent on intended effect, chronobiologic versus hypnotic [122]. Chronobiologic effect means that the goal of treatment is to shift the circadian rhythm, whereas hypnotic effect is with the intention to sedate. Chronobiologic dosing is typically 0.25 mg–3 mg. Two specific disorders where this may be helpful include individuals with a delayed sleep phase syndrome (DSPS) or those with an advanced sleep phase syndrome (ASPS). DSPS is when an individual's circadian timing (endogenous melatonin production) is shifted later than socially and biologically expected by at least 2 hours. To establish a phase advance in individuals with DSPS, there should be melatonin administration in the afternoon to early evening and light exposure on first awakening [123]. On the other hand, ASPS is a circadian shift that moves the bedtime to early evening and wake time to 3 am–5 am. To establish a phase delay in individuals with advanced sleep phase syndrome, there should be melatonin administration in the morning and light exposure in the evening [123]. Typical starting dose of melatonin is 1 mg, and this dose can be titrated as needed to a maximum of 10 mg nightly [122, 124]. It is important to note that melatonin is not regulated by the FDA, and the ingredients of formulations being sold can vary widely. Melatonin is well tolerated by most patients, and the most reported side effects are daytime drowsiness, vivid dreams, or nightmares [125].

Alpha-2 adrenergic receptor agonists: Clonidine and guanfacine are alpha-2 adrenergic receptor agonists, medication which functions by decreasing sympathetic tone [126]. In patients with sleep disorders associated with ASD, clonidine has been found to significantly decrease the number of nighttime awakenings and improvements in initiating sleep. Clonidine is typically initiated at 0.025 mg per day and can be increased as needed to a maximum dose of 0.4 mg per day. Although an effective option—especially for patients with comorbid hyperactivity—monitoring for side effects such as daytime sedation, postural hypotension, bradycardia, and dizziness is necessary [127]. Guanfacine is typically started at 0.5 mg per day, and it can be increased to a maximum dose of 4 mg per day. Commonly reported side effects include nausea, constipation, and bradycardia [128]. Guanfacine has been shown to be helpful for symptoms of hyperactivity and repetitive behaviors in patients with NDDs, but it has not been shown to have any significant effect on sleep difficulties [129].

Antipsychotics: Risperidone and aripiprazole are second-generation antipsychotics that block D2 dopamine receptors and inhibit reuptake of serotonin and norepinephrine. Risperidone is typically dosed between 0.5 mg and 3 mg per day. Risperidone has been proposed to be effective in treating behavioral insomnia, especially in patients who also struggle with aggression, agitation, and self-injurious behaviors [130]. Aripiprazole is dosed between 2 mg and 15 mg per day, and its benefit in the literature has been limited to managing behavioral issues in patients with ASD [131]. Both atypical antipsychotics are associated with adverse effects such as weight gain, sedation, hyperlipidemia, and extrapyramidal symptoms; however, the risk of hyperprolactinemia is significantly higher with risperidone use than aripiprazole [132]. In general, antipsychotics are not recommended by either the American Academy of Sleep Medicine or the American Academy of Neurology for treatment of sleep difficulties. Due to the risk of serious side effects and the need for routine lab monitoring, the use of antipsychotics should be limited to cases in which they are primarily being prescribed for behavioral issues, and these behavioral issues pose a significant risk to the safety of patients or caregivers.

Antidepressants: Mirtazapine is an α2 adrenergic antagonist in the central nervous system and is an antagonist of 5-HT2 and 5-HT3 serotonin receptors. It has been investigated as an option for AS-related sleep disorders given the association between UBE3A and serotonin-mediated pathways that modulate the circadian rhythm and sleep. Studies examining its efficacy have shown some benefit, particularly in a small cohort of AS patients who had been refractory to measures such as melatonin. However, mirtazapine was associated with weight gain and anticholinergic side effects that limited its tolerability in these studies as well [133]. These findings were consistent with previous RCTs that showed benefit in healthy controls [134, 135].

Trazodone is an antidepressant that functions via serotonin antagonism and antihistaminergic activity. It is often used to treat insomnia in pediatric patients of various developmental backgrounds [126]. There is a small amount data showing that patients with AS benefitted from trazodone use for sleep problems, but this has not been studied on a larger scale in patients with other NDDs [136]. Important side effects to be aware of before prescribing trazodone include QTc prolongation, cardiac arrhythmia, priapism, and daytime sedation. Despite the lack of data and side-effect profile, trazodone is a reasonable option for patients whose sleep problems have not responded to previous treatments if they are not at an already increased risk of cardiac arrhythmia or QTc prolongation based on their other medications or medical history. For the sake of this text, it is important to highlight the fact that patients with Rett syndrome are at an increased risk of sudden cardiac death and cardiac arrhythmia, so trazodone should be avoided in this population [137].

Melatonin agonists: Ramelteon and tasimelteon are both MT1 and MT2 receptor agonists [138, 139]. These medications are not FDA-approved for use in children in general, except for tasimelteon which has a specific indication for SMS who are 3 years and older. Prior to this approval, off-label use has shown increased sleep quality and TST in the SMS population [140, 141]. Similarly, case reports have suggested benefits in pediatric patients with ASD-related sleep issues treated with ramelteon [142].

Conclusions

Sleep is a centrally derived homeostatic process. It is a critical component for normal growth and development and contributes substantially to optimal social, cognitive, and neuropsychological function. Sleep disturbances in individuals with genetic neurodevelopmental disorders are pervasive and appear to represent a continuum of the abnormalities observed in overall neurologic development. There are even sleep phenotypes that are unique to the underlying genetic etiology for the developmental brain dysfunction, while some sleep features are overlapping across neurodevelopmental disorders.

The sleep disturbances encountered in individuals with neurodevelopmental disorders are frequently disruptive not only for the individual but also for their family or caregivers. In addition, in childhood this disruption can lead to compromised growth and development because of inadequate TST, EDS, and impaired attention and participation for school, therapeutic services, and social engagement. Accurate and timely diagnosis of sleep dysfunction in individuals with neurodevelopmental disorders results in the opportunity to apply appropriate therapeutic interventions to optimize sleep. In addition to improved quality, duration, and timing of sleep, improvement in development, cognitive, behavioral, and social function is observed. Ongoing research to improve the understanding of the complex interplay between sleep and developmental trajecetories is necessary as growing evidence suggests the possibility of sleep related biomarkers for diagnosis, treatment targets and modification in neurodivergent developmental outcomes may exist.

References

1. Bathory E, Tomopoulos S. Sleep regulation, physiology and development, sleep duration and patterns, and sleep hygiene in infants, toddlers, and preschool-age children. Curr Probl Pediatr Adolesc Health Care. 2017;47(2):29–42.
2. Borb AA, Achermann P. Sleep homeostasis and models of sleep regulation. J Biol Rhythm. 1999;14(6):559–70.
3. Daan S, Beersma DG, Borbely AA. Timing of human sleep: recovery process gated by a circadian pacemaker. Am J Phys. 1984;246(2 Pt 2):R161–83. https://doi.org/10.1152/ajpregu.1984.246.2.R161.
4. Herman JH. Chronobiology of sleep in children. In: Principles and practice of pediatric sleep medicine. Elsevier; 2005. p. 85–99.
5. Owens J. Classification and epidemiology of childhood sleep disorders. Sleep Med Clin. 2007;2(3):353–61.
6. Boban S, Leonard H, Wong K, Wilson A, Downs J. Sleep disturbances in rett syndrome: impact and management including use of sleep hygiene practices. Am J Med Genet A. 2018;176(7):1569–77.
7. Maenner MJ, Shaw KA, Baio J, et al. Prevalence of autism spectrum disorder among children aged 8 years – autism and developmental disabilities monitoring network, 11 sites, United States, 2016. MMWR Surveill Summ. 2020;69(4):1–12. https://doi.org/10.15585/mmwr.ss6904a1.

8. American Psychiatric Association. Diagnostic and statistical manual of mental disorders. American Psychiatric Association; 2013.
9. Glickman G. Circadian rhythms and sleep in children with autism. Neurosci Biobehav Rev. 2010;34(5):755–68.
10. Kotagal S, Broomall E. Sleep in children with autism spectrum disorder. Pediatr Neurol. 2012;47(4):242–51.
11. Karaivazoglou K, Assimakopoulos K. Circadian dysregulation in young children with autism spectrum disorder. J Sleep Med Disord. 2018;5(2):1092.
12. Ballester P, Richdale A, Baker E, Peiro A. Sleep in autism: a biomolecular approach to aetiology and treatment. Sleep Med Rev. 2020;54:101357.
13. Bourgeron T. The possible interplay of synaptic and clock genes in autism spectrum disorders. Cold Spring Harb Symp Quant Biol. 2007;72:645–54. https://doi.org/10.1101/sqb.2007.72.020.
14. Siepka SM, Yoo SH, Park J, Lee C, Takahashi JS. Genetics and neurobiology of circadian clocks in mammals. Cold Spring Harb Symp Quant Biol. 2007;72:251–9. https://doi.org/10.1101/sqb.2007.72.052.
15. Tordjman S, Anderson GM, Pichard N, Charbuy H, Touitou Y. Nocturnal excretion of 6-sulphatoxymelatonin in children and adolescents with autistic disorder. Biol Psychiatry. 2005;57(2):134–8.
16. Pagan C, Delorme R, Callebert J, et al. The serotonin-N-acetylserotonin–melatonin pathway as a biomarker for autism spectrum disorders. Transl Psychiatry. 2014;4(11):e479.
17. Melke J, Goubran Botros H, Chaste P, et al. Abnormal melatonin synthesis in autism spectrum disorders. Mol Psychiatry. 2008;13(1):90–8.
18. Pagan C, Goubran-Botros H, Delorme R, et al. Disruption of melatonin synthesis is associated with impaired 14-3-3 and miR-451 levels in patients with autism spectrum disorders. Sci Rep. 2017;7(1):1–11.
19. Wu Z, Zou J, Wang Q, et al. Autism spectrum disorder (ASD): disturbance of the melatonin system and its implications. Biomed Pharmacother. 2020;130:110496.
20. Keute M, Miller MT, Krishnan ML, et al. Angelman syndrome genotypes manifest varying degrees of clinical severity and developmental impairment. Mol Psychiatry. 2021;26(7):3625–33.
21. Williams CA, Driscoll DJ, Dagli AI. Clinical and genetic aspects of angelman syndrome. Genet Med. 2010;12(7):385–95.
22. Yang L, Shu X, Mao S, Wang Y, Du X, Zou C. Genotype–phenotype correlations in angelman syndrome. Genes. 2021;12(7):987.
23. Nicholls RD. Genomic imprinting and uniparental disomy in angelman and Prader-Willi syndromes: a review. Am J Med Genet. 1993;46(1):16–25.
24. Kaplan LC, Wharton R, Elias E, et al. Clinical heterogeneity associated with deletions in the long arm of chromosome 15: report of 3 new cases and their possible genetic significance. Am J Med Genet. 1987;28(1):45–53.
25. Buiting K, Saitoh S, Gross S, et al. Inherited microdeletions in the angelman and Prader–Willi syndromes define an imprinting centre on human chromosome 15. Nat Genet. 1995;9(4):395–400.
26. Kishino T, Lalande M, Wagstaff J. UBE3A/E6-AP mutations cause angelman syndrome. Nat Genet. 1997;15(1):70–3.
27. Spruyt K, Braam W, Curfs LM. Sleep in angelman syndrome: A review of evidence. Sleep Med Rev. 2018;37:69–84.
28. Levin Y, Hosamane NS, McNair TE, et al. Evaluation of electroencephalography biomarkers for angelman syndrome during overnight sleep. Autism Research. 2022;15:1031.
29. Hipp JF, Frohlich J, Keute M, Tan W, Bird LM. Electrophysiological abnormalities in angelman syndrome correlate with symptom severity. Biol Psychiatry Glob Open Sci. 2021;1(3):201–9.

30. Dosier LBM, Vaughn BV, Fan Z. Sleep disorders in childhood neurogenetic disorders. Children. 2017;4(9):82.
31. Bull MJ. Down syndrome. N Engl J Med. 2020;382(24):2344–52.
32. Hassold T, Sherman S. Down syndrome: genetic recombination and the origin of the extra chromosome 21. Clin Genet. 2000;57(2):95–100.
33. Lee CF, Lee CH, Hsueh WY, Lin MT, Kang KT. Prevalence of obstructive sleep apnea in children with down syndrome: a meta-analysis. J Clin Sleep Med. 2018;14(5):867–75. https://doi.org/10.5664/jcsm.7126.
34. Bull MJ. Committee on Genetics. Health supervision for children with down syndrome. Pediatrics. 2011;128(2):393–406.
35. Uong EC, McDonough JM, Tayag-Kier CE, et al. Magnetic resonance imaging of the upper airway in children with down syndrome. Am J Respir Crit Care Med. 2001;163(3):731–6.
36. Fink GB, Madaus WK, Walker GF. A quantitative study of the face in down's syndrome. Am J Orthod. 1975;67(5):540–53.
37. Strome M. Obstructive sleep apnea in down syndrome children: a surgical approach. Laryngoscope. 1986;96(12):1340–2.
38. Donnelly LF, Shott SR, LaRose CR, Chini BA, Amin RS. Causes of persistent obstructive sleep apnea despite previous tonsillectomy and adenoidectomy in children with down syndrome as depicted on static and dynamic cine MRI. Am J Roentgenol. 2004;183(1):175–81.
39. Prasher V. Overweight and obesity amongst down's syndrome adults. J Intellect Disabil Res. 1995;39(5):437–41.
40. Harris N, Rosenberg A, Jangda S, O'Brien K, Gallagher ML. Prevalence of obesity in international special olympic athletes as determined by body mass index. J Am Diet Assoc. 2003;103(2):235–7.
41. Rajagopal KR, Abbrecht PH, Derderian SS, et al. Obstructive sleep apnea in hypothyroidism. Ann Intern Med. 1984;101(4):491–4.
42. Stores R, Stores G, Buckley S. The pattern of sleep problems in children with down's syndrome and other intellectual disabilities. J Appl Res Intellect Disabil. 1996;9(2):145–59.
43. Diomedi M, Curatolo P, Scalise A, Placidi F, Caretto F, Gigli GL. Sleep abnormalities in mentally retarded autistic subjects: Down's syndrome with mental retardation and normal subjects. Brain Dev. 1999;21(8):548–53.
44. Levanon A, Tarasiuk A, Tal A. Sleep characteristics in children with down syndrome. J Pediatr. 1999;134(6):755–60.
45. Nisbet LC, Phillips NN, Hoban TF, O'Brien LM. Characterization of a sleep architectural phenotype in children with down syndrome. Sleep Breathing. 2015;19(3):1065–71.
46. Mims M, Thottam PJ, Kitsko D, Shaffer A, Choi S. Characterization of sleep architecture in down syndrome patients pre and post airway surgery. Cureus. 2017;9(1):e983.
47. Head E, Helman AM, Powell D, Schmitt FA. Down syndrome, beta-amyloid and neuroimaging. Free Radic Biol Med. 2018;114:102–9.
48. Ballard C, Mobley W, Hardy J, Williams G, Corbett A. Dementia in down's syndrome. Lancet Neurol. 2016;15(6):622–36.
49. Musiek ES, Xiong DD, Holtzman DM. Sleep, circadian rhythms, and the pathogenesis of alzheimer disease. Exp Mol Med. 2015;47(3):e148.
50. Hunter J, Berry-Kravis E, Hipp H, Todd P. FMR1 disorders. 1998 June 16 [updated 2019 nov 21]. GeneReviews®[Internet]. University of Washington, Seattle. 2021.
51. Hersh JH, Saul RA. Committee on Genetics. Health supervision for children with fragile X syndrome. Pediatrics. 2011;127(5):994–1006.
52. McDuffie A, Thurman AJ, Hagerman RJ, Abbeduto L. Symptoms of autism in males with fragile X syndrome: a comparison to nonsyndromic ASD using current ADI-R scores. J Autism Dev Disord. 2015;45(7):1925–37.
53. Salcedo-Arellano MJ, Dufour B, McLennan Y, Martinez-Cerdeno V, Hagerman R. Fragile X syndrome and associated disorders: clinical aspects and pathology. Neurobiol Dis. 2020;136:104740.

54. Bailey DB Jr, Raspa M, Olmsted M, Holiday DB. Co-occurring conditions associated with FMR1 gene variations: findings from a national parent survey. Am J Med Genet A. 2008;146(16):2060–9.
55. Budimirovic D, Haas-Givler B, Blitz R, et al. Consensus of the fragile X clinical and research consortium on clinical practices autism spectrum disorder in fragile X syndrome. Walnut Creek: National Fragile X Foundation Website; 2014.
56. Kronk R, Dahl R, Noll R. Caregiver reports of sleep problems on a convenience sample of children with fragile X syndrome. Am J Intellect Dev Disabil. 2009;114(6):383–92.
57. Kronk R, Bishop EE, Raspa M, Bickel JO, Mandel DA, Bailey DB Jr. Prevalence, nature, and correlates of sleep problems among children with fragile X syndrome based on a large scale parent survey. Sleep. 2010;33(5):679–87.
58. Symons FJ, Byiers BJ, Raspa M, Bishop E, Bailey DB Jr. Self-injurious behavior and fragile X syndrome: findings from the national fragile X survey. Am J Intellect Dev Disabil. 2010;115(6):473–81.
59. Carotenuto M, Roccella M, Pisani F, et al. Polysomnographic findings in fragile X syndrome children with EEG abnormalities. Behav Neurol. 2019;2019:1.
60. Ferri R, Del Gracco S, Elia M, Musumeci SA, Pettinato S. Heart rate variability and autonomic function during sleep in fragile X syndrome. Am J Med Genet. 1999;83(4):296–7.
61. Miano S, Bruni O, Elia M, et al. Sleep phenotypes of intellectual disability: a polysomnographic evaluation in subjects with down syndrome and fragile-X syndrome. Clin Neurophysiol. 2008;119(6):1242–7.
62. Kidd SA, Lachiewicz A, Barbouth D, et al. Fragile X syndrome: a review of associated medical problems. Pediatrics. 2014;134(5):995–1005.
63. Hamlin A, Liu Y, Nguyen DV, Tassone F, Zhang L, Hagerman RJ. Sleep apnea in fragile X premutation carriers with and without FXTAS. Am J Med Genet B Neuropsychiatr Genet. 2011;156(8):923–8.
64. Bienvenu T, Philippe C, De Roux N, et al. The incidence of rett syndrome in France. Pediatr Neurol. 2006;34(5):372–5.
65. Hagberg B, Aicardi J, Dias K, Ramos O. A progressive syndrome of autism, dementia, ataxia, and loss of purposeful hand use in girls: Rett's syndrome: report of 35 cases. Ann Neurol. 1983;14(4):471–9.
66. Laurvick CL, De Klerk N, Bower C, et al. Rett syndrome in Australia: a review of the epidemiology. J Pediatr. 2006;148(3):347–52.
67. Thomas GH. High male: female ratio of germ-line mutations: an alternative explanation for postulated gestational lethality in males in X-linked dominant disorders. Am J Hum Genet. 1996;58(6):1364–8.
68. Neul JL, Kaufmann WE, Glaze DG, et al. Rett syndrome: revised diagnostic criteria and nomenclature. Ann Neurol. 2010;68(6):944–50.
69. Kyle SM, Vashi N, Justice MJ. Rett syndrome: a neurological disorder with metabolic components. Open Biol. 2018;8(2):170216.
70. Percy AK, Zoghbi HY, Glaze DG. Rett syndrome: discrimination of typical and variant forms. Brain Dev. 1987;9(5):458–61.
71. Cobolli Gigli C, Scaramuzza L, De Simone M, et al. Lack of methyl-CpG binding protein 2 (MeCP2) affects cell fate refinement during embryonic cortical development. Cereb Cortex. 2018;28(5):1846–56.
72. Mari F, Azimonti S, Bertani I, et al. CDKL5 belongs to the same molecular pathway of MeCP2 and it is responsible for the early-onset seizure variant of rett syndrome. Hum Mol Genet. 2005;14(14):1935–46.
73. Roche-Martínez A, Gerotina E, Armstrong-Morón J, Sans-Capdevila O, Pineda M. FOXG1, un nuevo gen responsable de la forma congénita del síndrome de rett. Rev Neurol. 2011;52(10):597–602.
74. Spruyt K. Sleep problems in individuals with rett syndrome: a systematic review and meta-analysis. Sleep Epidemiol. 2022;2:100027.

75. Downs J, Leonard H, Boban S, et al. Sleep problems are pervasive in rett syndrome and the CDKL5 disorder. J Intellect Disabil Res. 2016;60:779.
76. Young D, Nagarajan L, de Klerk N, Jacoby P, Ellaway C, Leonard H. Sleep problems in rett syndrome. Brain Dev. 2007;29(10):609–16.
77. Wong K, Leonard H, Jacoby P, Ellaway C, Downs J. The trajectories of sleep disturbances in rett syndrome. J Sleep Res. 2015;24(2):223–33.
78. Zhang X, Spruyt K. Literature cases summarized based on their polysomnographic findings in rett syndrome. Int J Environ Res Public Health. 2022;19(6):3422.
79. Southall DP, Kerr AM, Tirosh E, Amos P, Lang MH, Stephenson JB. Hyperventilation in the awake state: potentially treatable component of rett syndrome. Arch Dis Child. 1988;63(9):1039–48. https://doi.org/10.1136/adc.63.9.1039.
80. Talkowski ME, Mullegama SV, Rosenfeld JA, et al. Assessment of 2q23. 1 microdeletion syndrome implicates MBD5 as a single causal locus of intellectual disability, epilepsy, and autism spectrum disorder. Am J Hum Genet. 2011;89(4):551–63.
81. Mullegama SV, Elsea SH. Clinical and molecular aspects of MBD5-associated neurodevelopmental disorder (MAND). Eur J Hum Genet. 2016;24(9):1235–43.
82. Mullegama SV, Pugliesi L, Burns B, et al. MBD5 haploinsufficiency is associated with sleep disturbance and disrupts circadian pathways common to Smith–Magenis and fragile X syndromes. Eur J Hum Genet. 2015;23(6):781–9.
83. Pitt D, Hopkins I. A syndrome of mental retardation, wide mouth and intermittent overbreathing. J Paediatr Child Health. 1978;14(3):182–4.
84. Peippo MM, Simola KO, Valanne LK, et al. Pitt–Hopkins syndrome in two patients and further definition of the phenotype. Clin Dysmorphol. 2006;15(2):47–54.
85. Amiel J, Rio M, De Pontual L, et al. Mutations in TCF4, encoding a class I basic helix-loop-helix transcription factor, are responsible for pitt-hopkins syndrome, a severe epileptic encephalopathy associated with autonomic dysfunction. Am J Hum Genet. 2007;80(5):988–93.
86. Duffield GE, Watson NP, Mantani A, et al. A role for Id2 in regulating photic entrainment of the mammalian circadian system. Curr Biol. 2009;19(4):297–304.
87. Gandhi A, Zhou D, Alaimo J, Chon E, Fountain MD, Elsea SH. Composite sleep problems observed across Smith–Magenis syndrome, MBD5-associated neurodevelopmental disorder, Pitt–Hopkins syndrome, and ASD. J Autism Dev Disord. 2021;51(6):1852–65.
88. Sweetser DA, Elsharkawi I, Yonker L, Steeves M, Parkin K, Thibert R. Pitt-hopkins syndrome. In: Adam MP, Everman DB, Mirzaa GM, et al., editors. GeneReviews((R)). Seattle: University of Washington; 1993. GeneReviews is a registered trademark of the University of Washington, Seattle. All rights reserved; NBK100240 [bookaccession].
89. Smith AC, McGavran L, Robinson J, et al. Interstitial deletion of (17)(p11. 2p11. 2) in nine patients. Am J Med Genet. 1986;24(3):393–414.
90. Kaplan KA, Elsea SH, Potocki L. Management of sleep disturbances associated with smith-magenis syndrome. CNS Drugs. 2020;34(7):723–30.
91. Edelman E, Girirajan S, Finucane B, et al. Gender, genotype, and phenotype differences in Smith–Magenis syndrome: a meta-analysis of 105 cases. Clin Genet. 2007;71(6):540–50.
92. De Leersnyder H, de Blois M, Claustrat B, et al. Inversion of the circadian rhythm of melatonin in the smith-magenis syndrome. J Pediatr. 2001;139(1):111–6.
93. Smith AC, Dykens E, Greenberg F. Sleep disturbance in Smith-Magenis syndrome (del 17 p11. 2). Am J Med Genet. 1998;81(2):186–91.
94. Duncan W, Gropman A, Morse R, Krasnewich D, Smith A. Good babies sleeping poorly: insufficient sleep in infants with smith-magenis syndrome (SMS). Am J Hum Genet. 2003;73(5):323.
95. Smith AC, Morse RS, Introne W, Duncan WC Jr. Twenty-four-hour motor activity and body temperature patterns suggest altered central circadian timekeeping in Smith–Magenis syndrome, a neurodevelopmental disorder. Am J Med Genet A. 2019;179(2):224–36.
96. Gropman AL, Elsea S, Duncan WC Jr, Smith AC. New developments in smith-magenis syndrome (del 17p11. 2). Curr Opin Neurol. 2007;20(2):125–34.

97. Williams SR, Zies D, Mullegama SV, Grotewiel MS, Elsea SH. Smith-magenis syndrome results in disruption of CLOCK gene transcription and reveals an integral role for RAI1 in the maintenance of circadian rhythmicity. Am J Hum Genet. 2012;90(6):941–9.
98. Strømme P, Bjømstad PG, Ramstad K. Prevalence estimation of williams syndrome. J Child Neurol. 2002;17(4):269–71.
99. Ewart AK, Morris CA, Atkinson D, et al. Hemizygosity at the elastin locus in a developmental disorder, williams syndrome. Nat Genet. 1993;5(1):11–6.
100. Van Hagen J, van Der Geest J, van der Giessen RS, et al. Contribution of CYLN2 and GTF2IRD1 to neurological and cognitive symptoms in williams syndrome. Neurobiol Dis. 2007;26(1):112–24.
101. Bellugi U, Lichtenberger L, Jones W, Lai Z, St. George M. I. the neurocognitive profile of williams syndrome: a complex pattern of strengths and weaknesses. J Cogn Neurosci. 2000;12(Supplement 1):7–29.
102. Burn J. Williams syndrome. J Med Genet. 1986;23(5):389–95. https://doi.org/10.1136/jmg.23.5.389.
103. Annaz D, Hill CM, Ashworth A, Holley S, Karmiloff-Smith A. Characterisation of sleep problems in children with williams syndrome. Res Dev Disabil. 2011;32(1):164–9.
104. Mason TB, Arens R, Sharman J, et al. Sleep in children with williams syndrome. Sleep Med. 2011;12(9):892–7.
105. Goldman S, Malow B, Newman K, Roof E, Dykens E. Sleep patterns and daytime sleepiness in adolescents and young adults with williams syndrome. J Intellect Disabil Res. 2009;53(2):182–8.
106. Sniecinska-Cooper AM, Iles RK, Butler SA, Jones H, Bayford R, Dimitriou D. Abnormal secretion of melatonin and cortisol in relation to sleep disturbances in children with williams syndrome. Sleep Med. 2015;16(1):94–100.
107. Roach ES, Sparagana SP. Diagnosis of tuberous sclerosis complex. J Child Neurol. 2004;19(9):643–9.
108. Manning BD, Tee AR, Logsdon MN, Blenis J, Cantley LC. Identification of the tuberous sclerosis complex-2 tumor suppressor gene product tuberin as a target of the phosphoinositide 3-kinase/akt pathway. Mol Cell. 2002;10(1):151–62.
109. Northrup H, Aronow ME, Bebin EM, et al. Updated international tuberous sclerosis complex diagnostic criteria and surveillance and management recommendations. Pediatr Neurol. 2021;123:50–66.
110. Bruni O, Cortesi F, Giannotti F, Curatolo P. Sleep disorders in tuberous sclerosis: A polysomnographic study. Brain Dev. 1995;17(1):52–6.
111. Hunt A. Development, behaviour and seizures in 300 cases of tuberous sclerosis. J Intellect Disabil Res. 1993;37(1):41–51.
112. Hunt A, Stores G. Sleep disorder and epilepsy in children with tuberous sclerosis: A questionnaire-based study. Dev Med Child Neurol. 1994;36(2):108–15.
113. De Vries PJ, Belousova E, Benedik MP, et al. TSC-associated neuropsychiatric disorders (TAND): findings from the TOSCA natural history study. Orphanet J Rare Dis. 2018;13(1):1–13.
114. Trickett J, Heald M, Oliver C, Richards C. A cross-syndrome cohort comparison of sleep disturbance in children with smith-magenis syndrome, angelman syndrome, autism spectrum disorder and tuberous sclerosis complex. J Neurodev Disord. 2018;10(1):1–14.
115. Moavero R, Voci A, La Briola F, et al. Sleep disorders and neuropsychiatric disorders in a pediatric sample of tuberous sclerosis complex: a questionnaire-based study. Sleep Med. 2022;89:65–70.
116. Agar G, Brown C, Sutherland D, Coulborn S, Oliver C, Richards C. Sleep disorders in rare genetic syndromes: a meta-analysis of prevalence and profile. Mol Autism. 2021;12(1):1–17.
117. McDonagh MS, Holmes R, Hsu F. Pharmacologic treatments for sleep disorders in children: a systematic review. J Child Neurol. 2019;34(5):237–47.

118. Gringras P, Nir T, Breddy J, Frydman-Marom A, Findling RL. Efficacy and safety of pediatric prolonged-release melatonin for insomnia in children with autism spectrum disorder. J Am Acad Child Adolesc Psychiatry. 2017;56(11):948–57. e4
119. Maras A, Schroder CM, Malow BA, et al. Long-term efficacy and safety of pediatric prolonged-release melatonin for insomnia in children with autism spectrum disorder. J Child Adolesc Psychopharmacol. 2018;28(10):699–710.
120. Martens MA, Seyfer DL, Andridge RR, Coury DL. Use and effectiveness of sleep medications by parent report in individuals with williams syndrome. J Dev Behav Pediatr. 2017;38(9):765–71.
121. Wirojanan J, Jacquemont S, Diaz R, et al. The efficacy of melatonin for sleep problems in children with autism, fragile X syndrome, or autism and fragile X syndrome. J Clin Sleep Med. 2009;5(2):145–50.
122. Hardeland R. Divergent importance of chronobiological considerations in high-and low-dose melatonin therapies. Diseases. 2021;9(1):18.
123. Lewy AJ. Melatonin and human chronobiology. Cold Spring Harb Symp Quant Biol. 2007;72:623–36. https://doi.org/10.1101/sqb.2007.72.055.
124. Williams Buckley A, Hirtz D, Oskoui M, et al. Practice guideline: treatment for insomnia and disrupted sleep behavior in children and adolescents with autism spectrum disorder: report of the guideline development, dissemination, and implementation subcommittee of the american academy of neurology. Neurology. 2020;94(9):392–404. https://doi.org/10.1212/WNL.0000000000009033.
125. Foley HM, Steel AE. Adverse events associated with oral administration of melatonin: a critical systematic review of clinical evidence. Complement Ther Med. 2019;42:65–81.
126. Owens JA, Rosen CL, Mindell JA, Kirchner HL. Use of pharmacotherapy for insomnia in child psychiatry practice: a national survey. Sleep Med. 2010;11(7):692–700.
127. Ming X, Gordon E, Kang N, Wagner GC. Use of clonidine in children with autism spectrum disorders. Brain Dev. 2008;30(7):454–60.
128. Scahill L, Aman MG, McDougle CJ, et al. A prospective open trial of guanfacine in children with pervasive developmental disorders. Eur J Child Adolesc Psychiatry. 2006;16(5):589–98.
129. Politte LC, Scahill L, Figueroa J, McCracken JT, King B, McDougle CJ. A randomized, placebo-controlled trial of extended-release guanfacine in children with autism spectrum disorder and ADHD symptoms: an analysis of secondary outcome measures. Neuropsychopharmacology. 2018;43(8):1772–8.
130. Grigg-Damberger M, Ralls F. Treatment strategies for complex behavioral insomnia in children with neurodevelopmental disorders. Curr Opin Pulm Med. 2013;19(6):616–25.
131. Williamson E, Sathe NA, Andrews JC, et al. NBK448262 [bookaccession]; 2017.
132. Balijepalli C, Druyts E, Zoratti MJ, et al. Change in prolactin levels in pediatric patients given antipsychotics for schizophrenia and schizophrenia spectrum disorders: a network meta-analysis. Schizophr Res Treat. 2018;2018:1.
133. Hanzlik E, Klinger SA, Carson R, Duis J. Mirtazapine for sleep disturbances in angelman syndrome: a retrospective chart review of 8 pediatric cases. J Clin Sleep Med. 2020;16(4):591–5.
134. Aslan S, Isik E, Cosar B. The effects of mirtazapine on sleep: a placebo controlled, double-blind study in young healthy volunteers. Sleep. 2002;25(6):666–8.
135. Karsten J, Hagenauw LA, Kamphuis J, Lancel M. Low doses of mirtazapine or quetiapine for transient insomnia: a randomised, double-blind, cross-over, placebo-controlled trial. J Psychopharmacol. 2017;31(3):327–37.
136. Pereira JA, Ravichandran CT, Mullett J, McDougle CJ, Keary CJ. Characterization of sleep habits and medication outcomes for sleep disturbance in children and adults with angelman syndrome. Am J Med Genet A. 2020;182(8):1913–22.
137. Acampa M, Guideri F. Cardiac disease and rett syndrome. Arch Dis Child. 2006;91(5):440–3. https://doi.org/10.1136/adc.2005.090290.
138. Hatta K, Kishi Y, Wada K, et al. Preventive effects of ramelteon on delirium: a randomized placebo-controlled trial. JAMA Psychiatry. 2014;71(4):397–403.

139. Rajaratnam SM, Polymeropoulos MH, Fisher DM, et al. Melatonin agonist tasimelteon (VEC-162) for transient insomnia after sleep-time shift: two randomised controlled multi-centre trials. Lancet. 2009;373(9662):482–91.
140. Baek W, Elsea S. Smith-magenis syndrome treated with ramelteon and amphetamine-dextroamphetamine: Case report and review of the literature. J Genet Disor Genet Rep 5. 2016;4:2.
141. Hull J, Polymeropoulos C, Cho Y, Xiao C, Polymeropoulos M. Tasimelteon improves sleep quality and behavior in individuals with smith-magenis syndrome (SMS) in an open-label study. Sleep Med. 2017;40:e139.
142. Stigler KA, Posey DJ, McDougle CJ. Ramelteon for insomnia in two youths with autistic disorder. J Child Adolesc Psychopharmacol. 2006;16(5):631–6.

Genetic Studies of Sleep in Children

Jonathan A. Mitchell and Christopher M. Cielo

Abstract Sleep is critical for optimal development in childhood. There are multiple sleep health traits that have been shown to have a degree of heritability, indicating that genetic factors do contribute to sleep health traits, such as sleep duration, in childhood. However, the specific genes that regulate sleep health in childhood are largely unknown. In the future, it is likely that larger-scale studies with standardized sleep health phenotyping will facilitate the discovery novel pediatric sleep health loci through genome-wide association studies, and candidate gene studies will additionally contribute to our understanding of how specific genes impact sleep in childhood, especially at key developmental stages. In regard to clinical sleep disorders, specific genetic mutations are known to contribute to the risk of specific sleep disorders. For example, genetic variants are known to increase the risk of sleep-disordered breathing, and obstructive sleep apnea is common among children with Trisomy 21. Mutations in genes comprising the molecular clock have been observed in families with circadian rhythm sleep-wake disorders; for example, mutations in *CRY1* and *PER3* that extend the molecular clock period have been detected in families with advanced sleep-phased syndrome. Current and future findings on the genetics of sleep in childhood will help identify those most at risk for problematic sleep and guide prevention and treatment strategies to improve sleep and therefore health and development in early life.

J. A. Mitchell (✉)
Depratment of Pediatrics, Perelman School of Medicine, University of Pennsylvania, Philadelphia, PA, USA

Division of Gastroenterology, Hepatology, and Nutrition, Children's Hospital of Philadelphia, Philadelphia, PA, USA
e-mail: mitchellj2@chop.edu

C. M. Cielo
Depratment of Pediatrics, Perelman School of Medicine, University of Pennsylvania, Philadelphia, PA, USA

Division of Pulmonary and Sleep Medicine, Children's Hospital of Philadelphia, Philadelphia, PA, USA

P. Gehrman et al. (eds.), *Genetics of Sleep and Sleep Disorders*,
https://doi.org/10.1007/978-3-031-62723-1_17

Keywords Sleep · Twins · GWAS · Genes

Introduction

The American Academy of Sleep Medicine recommends that preschoolers (3–5 years old), children (6–12 years old) and adolescents (13–18 years old) sleep for 10–13 h, 9–12 h, and 8–10 h per day, respectively [1]. Similar recommendations are in place for children and adolescents worldwide. Yet, high proportions of children do not achieve these recommendations. Data from the US-based Youth Risk Behavior Surveillance Survey revealed that most middle school students (53%) and high school students (65%) do not meet these recommendations [2–6]. These data are aligned with international estimates where an estimated 52% of adolescents aged 14–17 years sleep insufficiently [7]. Further, clinical sleep disorders are common among children and adolescents [8]. For example, an estimated 11% of children have insomnia; 7–16% of adolescents have a circadian rhythm sleep-wake disorder; 1–4% of children have obstructive sleep apnea, with the prevalence higher among children with obesity (10–25%); and 30% of children have excessive daytime sleepiness [8].

The high prevalence of suboptimal sleep health and clinical sleep disorders in children has major health and developmental implications. In childhood, obtaining optimal sleep has been associated with less risky behavior [9–15]. Sleep has also been associated with more optimal growth trajectories in childhood [16] and has been implicated in the prevention of childhood obesity and related metabolic health problems [17–26]. Further, it is known that sleep is associated with traits needed to achieve academic potential (e.g., memory) [9–15] and to develop prosocial behaviors (e.g., emotional regulation) [27–30]. There is therefore a need to study determinants of sleep in childhood to better direct prevention and treatment efforts for impaired sleep.

Sleep health traits and clinical sleep disorders are determined by both environmental and genetic factors. The objective of this chapter is to summarize genetic contributions to sleep health traits and clinical sleep disorders in childhood.

Sleep Health Traits

In 2014, a sleep health framework was proposed to quantify how well an individual or population is sleeping across five dimensions: duration, continuity, timing, alertness, and satisfaction [31]. Formally, sleep health is defined as "a multidimensional pattern of sleep-wakefulness, adapted to individual, social, and environmental demands, that promotes physical and mental well-being. Good sleep health is characterized by subjective satisfaction, appropriate timing, adequate duration, high efficiency, and sustained alertness during waking hours [31]." The capacity of an individual to achieve optimal sleep health may be due in part to their genetics. The first section of this chapter will consider the genetics of sleep health traits in children.

Twin Studies, Developmental Programming, and Sleep Health Traits

A key step in the process of identifying genetic determinants of a trait or disease is to first estimate the heritability (i.e., the proportion of the variation in the trait or disease that is due to genetic factors). This can be achieved through twin studies. Importantly, twin studies do not reveal which genes regulate a trait or disease but indicate that genetic discovery efforts are worth pursuing if a relatively high degree of heritability is established. Twin studies compare the similarity of a trait or disease between identical (monozygotic) twins, who share 100% of their genetic material, to that of fraternal (dizygotic) twins, who share on average 50% of their genetic material. If a trait or disease is highly heritable, then a greater similarity for a trait/disease would be observed between identical twins compared to fraternal twins.

Twin studies have reported a range of heritability estimates for sleep health traits in childhood, including sleep duration, sleep timing, insomnia symptoms, and generic sleep quality scores. A meta-analysis of twin studies revealed that the heritability of sleep duration was 17–20% among preschool-aged children, 41–52% among children, and 69% among adolescents [32]. While different methods were used to measure sleep duration across studies and age groups, these data suggest that sleep duration is under genetic regulation in childhood and that the genetic influence increases as children age into adolescence [32]. Indeed, a twin study that used actigraphy to estimate sleep duration and timing reported that 43–62% of the variance in adolescent sleep-wake behaviors was due to genetics factors, with the heritability being greater among older adolescents (16–17 years old) versus younger adolescents (9–14 years old) [33]. In regard to insomnia, one twin study observed that the heritability was 30% when using a combination of self-reported insomnia symptoms from children (aged 8–16 years) and caregivers [34], whereas the heritability of insomnia was reported to be 14–38% when using clinician ratings of insomnia symptoms among children aged 8–18 years [35]. Using a composite score from self-reported data, a twin study reported that the heritability of a sleep quality score was 30% among children aged 10–17 years [36]. Further, adolescent twin studies that used the PSQI report heritability between 33% and 41% for sleep quality [37].

Heritability can also be estimated from common single-nucleotide polymorphisms (SNPs). SNP heritability measures the proportion of variance in a trait or disease that is explained by all measured SNPs [38]. For childhood sleep duration, the Early Genetics and Life-course Epidemiology Consortium used data from five studies of children aged 2–11 years and reported that sleep duration has a SNP heritability of 14% (95% CI: 5–23%) [39].

In further support that sleep is regulated by genetic factors, there are predictable changes in sleep that occur during development across different species. Sleep is highly conserved throughout the animal kingdom [40], and it is consistently observed that animals exhibit longer sleep during early development, indicating a crucial role for sleep in young animals that is genetically programmed [41]. For example, sleep duration typically declines in mice as they age from adolescence to adulthood; in zebrafish, sleep duration decreases as they age from larvae to adults; and in fruit flies, sleep duration is typically highest during the first few days after they hatch and gradually decreases as they age [41]. In humans, it is well established that there are predictable developmental sleep patterns in infancy, childhood, and adolescence. Under the two-process model for sleep regulation, the homeostatic process concerns the accumulation and dissipation of pressure to sleep, whereas the circadian process describes the circadian rhythmicity that promotes alertness during the day and sleep at night [42]. In human infants, the homeostatic process is dominant as indicated by longer sleep duration and daytime napping. Conversely, during puberty in adolescence, the circadian process becomes dominant, as indicated by shorter sleep duration and advanced sleep phase [43]. These predictable developmental patterns suggest that there is likely inherent genetically programmed physiology that contributes to our sleep patterns.

Genome-Wide Association Studies and Sleep Health Traits

Genome-wide association studies (GWAS) can facilitate the discovery of genetic loci implicated in polygenic traits and diseases. In brief, a GWAS investigates common genetic variant data across the entire genome, from a large number of individuals to determine if an allele for a given variant is associated with a particular trait or disease. This approach yields genomic signals that can provide important insights into the genetic basis of complex diseases and traits, and it has been used to identify genetic variants associated with a wide range of conditions, including sleep traits.

GWAS have discovered over 300 genomic signals for a variety of sleep health traits in adults. The majority of discoveries stem from analyses of UK Biobank and 23andMe data, involving sample sizes that far exceed 100,000 participants [44–50]. Seventy-eight signals for self-reported adult *sleep duration* have been discovered [44], with a further 10 signals discovered for actigraphy-estimated sleep duration [47]. The strongest signal resides at the *PAX8* locus [44]. Bioinformatic follow-up analyses provided consistent evidence that genes near the GWAS signals for sleep duration were enriched for expression in the brain, including numerous areas of cortex [44]. Fifty-seven signals have been GWAS-implicated with self-reported *difficulty falling asleep* (an indicator of insomnia) in a UK Biobank study [51], and the addition of 23andMe data has revealed 202 genomic loci associated with insomnia [50]. Furthermore, LD score regression analyses found a genetic correlation between insomnia and sleep duration [51]. GWAS signals for *self-reported chronotype* (i.e., sleep timing) traits total 351 [45, 47], and 6 loci were GWAS-implicated with actigraphy-estimated sleep timing trait (least active 5-h period) [47]. The *RGS16* locus was the strongest signal discovered for chronotype [45].

The majority of sleep-related GWAS efforts have been pursued using sleep health traits measured in adults. As such, pediatric sleep traits have been underserved by genetic discovery efforts to date. Indeed, only two pediatric sleep GWAS have been completed, yielding just a single locus for sleep duration. This discovery was achieved by the Early Genetics and Life-course Epidemiology Consortium, leveraging data from five studies of children aged 2–11 years [39]. The limited number of signals discovered is likely the consequence of the relatively small GWAS sample size of $N = 10{,}544$ that is far below that of the adult GWAS efforts.

Future efforts are needed to develop pediatric Biobank resources that include sleep health traits and genome-wide genotyped data. Ideally, this future data resource would have standardized self-reported sleep health trait data captured in sufficient numbers of children across key developmental periods. The participants contributing data would be diverse in relation to factors such as ancestry, geographic location, age, and sex. Further, sleep health data estimated using standardized actigraphy protocols would provide an ecologically valid assessment of sleep health beyond self-reported data. This resource would allow for fully powered GWAS for genetic signal discovery that is warranted based on twin studies and adult GWAS findings.

Candidate Genes and Sleep Health Traits

In the absence of large sample sizes, candidate gene studies offer an opportunity to gain insight into the genetic regulation of sleep health in childhood. The majority of candidate gene studies for childhood sleep health have focused on variants in genes that are components of the molecular circadian clock. The molecular circadian clock regulates the timing of biological processes and behaviors, including sleep-wake cycles, through a negative feedback loop that creates oscillations in gene expression over each 24-h period [52]. During the daytime, transcription factors encoded by *CLOCK* and *BMAL1/NPAS2* form a heterodimeric protein complex that activates the transcription of *CRY1–3*, *PER1–3*, and *NR1D1* genes; in turn, as the day progresses, the *CRY*, *PER*, and *NR1D1* transcription factors accumulate and suppress the activity of *CLOCK* and *BMAL1* [52]. Molecular clocks are present in essentially all tissues, with the central clock located in the suprachiasmatic nucleus (SCN) of the hypothalamus that can synchronize with molecular clocks located in peripheral tissues [52].

Common variants in the promotor and intron of *NR1D1* have been associated with self-reported sleep duration in adolescents of European ancestry [53];

specifically, among male adolescents AA rs2071570 carriers vs. C/A or CC rs2071570 carriers were observed to have 19-min short sleep duration [53]. In contrast, a study of New Zealand children of European ancestry aged 6 years reported no associations between common *CLOCK* variants and actigraphy-assessed sleep duration; furthermore, in this study no associations were observed between variants in/near *NPAS2*, *BMAL1*, or *TIMELESS* and actigraphy-assessed sleep duration [54]. Similarly, a study involving children with an average age of 8 years from Brazil reported no associations between a *CLOCK* variant and parent-reported sleep timing and sleep duration [55]. And a study of American adolescents of European and non-European ancestry aged 11 years reported no associations between a *PER2* variant and actigraphy-assessed sleep timing (specifically sleep midpoint) [56].

Another candidate gene study approach would be to test if GWAS-implicated variants for adult sleep health traits adults associate with sleep health traits in childhood. This approach would only require testing for 100 s of candidate variants already established to regulate sleep in adults through past GWAS. This has been done in part; in 2019 a GWAS was performed among adults that discovered 77 loci implicated in sleep duration [44]; as part of their follow-up analyses, the investigators leveraged the Early Genetics and Life-course Epidemiology Consortium GWAS summary statistics and found that 45 out of the 77 signals were directionally consistent (i.e., the effect allele point estimates, indication of short or longer sleep duration, were consistent in adults and children) [44]. Relatedly, polygenic risk scores could be calculated using adult GWAS-implicated sleep health variants to test if the polygenic scores associate with sleep health traits in children. While such study designs do not facilitate the discovery of novel genetic loci that regulate pediatric sleep health specifically, they do inform if there is any overlap in genetic variants that regulate sleep health across the lifespan.

Follow-up LD score regression analyses of the Early Genetics and Life-course Epidemiology Consortium GWAS revealed a genetic correction between common variants implicated in type 2 diabetes and pediatric sleep duration [39]. Therefore, GWAS-implicated variants with type 2 diabetes are another set of candidates that could be tested in the context of association with sleep health traits in childhood. It would also be worth considering which GWAS-implicated pubertal onset variants associate with sleep health traits in childhood. We know that adolescents tend to experience a sleep phase delay that coincides with the onset and advancement through pubertal stages. Hundreds of genetic loci have been implicated in the onset of puberty using traits such as age at menarche (female sex), age at voice break (male sex), and age at peak height velocity. It is possible that a proportion of these variants specifically impact the changes in sleep timing that coincides with puberty.

Future candidate gene studies would benefit from having pediatric Biobank resources so that standardized sleep health traits and standardized genotyping and selection of variants in/near candidate genes will be performed. Further, incorporating actigraphy-assessed sleep health traits into candidate gene studies would further our knowledge of the genetic regulation of pediatric sleep.

Genetics of Clinical Sleep Disorders in Children

Clinical sleep disorders in children span the spectrum from sleep-disordered breathing, hypersomnia, insomnia, parasomnias, sleep-related movement disorders, and others, yet only a small portion of these conditions or phenotypes have been linked with an identifiable genetic etiology. Linking pediatric sleep disorders with specific genetic findings is critical, as these conditions are often lifelong, and sensitive and specific diagnostic testing is needed for early identification to prevent long-term sequelae.

Sleep-Disordered Breathing

Sleep-disordered breathing encompasses a spectrum of conditions that result in alternations in respiratory physiology during sleep, including airway collapse and alterations in central respiratory drive. While some clinical sleep disorders seen in children, such as obstructive sleep apnea, may also be seen in adulthood, the pathophysiology may be very different. In other cases, more rare conditions such as congenital central hypoventilation syndrome, may have an onset that is unique to pediatrics.

Obstructive Sleep Apnea

Obstructive sleep apnea (OSA) is characterized by recurrent episodes of airway obstruction during sleep, which can result in sleep disruption and gas exchange abnormality [57] and affects 1–5% of children obstructive sleep apnea (OSA), and is characterized by recurrent episodes of airway obstruction during sleep, which can result in sleep disruption and gas exchange abnormality [57] and affects 1–5% of children [58, 59]. Unlike adults, the primary structural contributor to OSA in children is enlargement of the tonsils and adenoids [60, 61]. Other risk factors include hypotonia, prematurity, obesity, and craniofacial differences such as mandibular hypoplasia [62]. Other risk factors include hypotonia, prematurity, obesity, and craniofacial differences such as mandibular hypoplasia [62].

In addition, there is emerging evidence that familial factors influence the development of OSA in children [63, 64]. There have been few studies examining the genetic etiologies of OSA in children. A recent GWAS study of 1486 children with OSA and 6961 controls, including both those with European and African ancestry, identified a number of loci associated with OSA risk [65]. A locus at 18p11.32 was associated with OSA in the entire cohort, and others at 1p36.22 and 15q26.1 were associated with OSA in European American and African Americans, respectively. Notably, several SNPs identified in these regions have previously been associated with OSA or related clinical symptoms in adult cohorts including the UK Biobank. They lead SNP for the 18p11.32 region, which has previously been associated with throat pain, respiratory failure, and insomnia. In addition, SNPs identified in both the European-American and African-American cohorts have been associated with OSA in adults [66]. Additional studies are needed to validate these findings, particularly across a variety of ethnic backgrounds and ages in the general pediatric population.

In addition to otherwise healthy children, there have been multiple genetic conditions that have been strongly associated with obstructive sleep apnea in children, many of which result in phenotypes that include risk factors such as structural craniofacial narrowing or hypotonia [67]. Identification of known and novel genetic conditions associated with OSA can be critically important in providing the proper management. More than half of children with Trisomy 21 are diagnosed with obstructive sleep apnea [68]. Common features of Trisomy 21 include a flattened nasal bridge, hypotonia, and relative macroglossia, and children with Down syndrome have higher rates of laryngomalacia and subglottic stenosis [69]. The association between Down syndrome and OSA in childhood is so significant that clinical practice guidelines recommend evaluation of all children with this condition by age 4 [70]. Identification of known and novel genetic conditions associated with OSA can be critically important in providing the proper management. More than half of children with Trisomy 21 are diagnosed with obstructive sleep apnea [68]. Common features of Trisomy 21 include a flattened nasal bridge, hypotonia, and relative macroglossia, and children with Down syndrome have higher rates of laryngomalacia and subglottic stenosis [69]. The association between Down syndrome and OSA in childhood is so significant that clinical practice guidelines recommend evaluation of all children with this condition are evaluated for OSA by age 4 [70].

During infancy, genetic conditions resulting in mandibular hypoplasia often lead to very severe OSA as the primary presenting feature [71]. These underlying syndromes, which may also affect other organ systems, come from a variety of loci which often have consequences on embryogenesis [72]. Stickler syndrome, perhaps the most common condition with an associated genetic etiology, is associated with the COL2A1 gene on chromosome 12q13, while Treacher Collins syndrome is mapped to chromosome 5q31-q34. Some rare forms of congenital micrognathia such as Nager and Rodriguez syndromes are inherited as autosomal dominant traits from heterozygous mutations of the SF3B4 gene on chromosome 1q12-q21 and Miller syndrome, which is an autosomal recessive trait caused by compound heterozygous mutations to chromosome 16q22.2. Taken together, large quantities of pathogenic genes associated with syndromic micrognathia have been identified, associated with cellular processing, cell metabolism, cartilage and bone development, and neuromuscular function [73].

There are many other conditions that have a genetic diagnosis and are strongly associated with obstructive sleep apnea during childhood. Beckwith-Wiedemann syndrome is a disorder associated with overgrowth due to changes to chromosome 11p15.5 and frequently results in obstructive sleep apnea due to macroglossia, particularly during infancy [74]. Several autosomal-dominant conditions associated with the FGFR2 gene on chromosome 10q26, such as Apert syndrome, Pfeiffer syndrome, and Crouzon syndrome result in premature fusion of the cranial sutures, often causing severe obstructive sleep apnea due to midface hypoplasia in the first months of life [75]. Mucopolysaccharidoses are a group of disorders including Hunter syndrome and Hurler syndrome associated with deficiency in glycosaminoglycan enzyme metabolism [76]. Soft tissue and skeletal malformations particularly affecting the face and airways are common in these conditions, resulting in severe OSA, sometimes associated with early death. Previously difficult to manage, newer therapies including stem cell transplantation, enzyme replacement therapy, and gene therapy may result in improvement in OSA as well as other associated comorbidities. There are many other conditions that have a genetic diagnosis and are strongly associated with obstructive sleep apnea during childhood. Beckwith-Wiedemann syndrome is a disorder associated with overgrowth due to changes to chromosome 11p15.5 and frequently results in obstructive sleep apnea due to macroglossia, particularly during infancy [74]. Several autosomal-dominant conditions associated with the FGFR2 gene on chromosome 10q26, such as Apert syndrome, Pfeiffer syndrome, and Crouzon syndrome result in premature fusion of the cranial sutures, often causing severe obstructive sleep apnea due to midface hypoplasia in the first months of life [75]. Mucopolysaccharidoses are a group of disorders including Hunter syndrome and Hurler syndrome associated with deficiency in glycosaminoglycan enzyme metabolism [76]. Soft tissue and skeletal malformations particularly affecting the face and airways are common in these conditions, resulting in severe OSA, sometimes associated with early death. Previously difficult to manage, newer therapies including stem cell transplantation, enzyme replacement therapy, and gene therapy may result in improvement in OSA as well as other associated comorbidities [77]. Better identification of conditions causing OSA during the newborn period through newborn screening and genetic testing can ensure that these patients get the appropriate treatment to reduce consequences of untreated OSA.

Several conditions with an identified genetic etiology are associated with both central and obstructive sleep apnea in children. Achondroplasia is characterized by short stature with shortening of the arms and legs as well as distinctive craniofacial features, caused by gain-of-function mutation in the FGR3 gene. Children with achondroplasia are at risk for both obstructive sleep apnea due to midface retrusion and central sleep apnea due to cervicomedullary compression [78]. Prader-Willi syndrome is caused by a lack of the paternal contribution of genes in the p region of chromosome 15, which are typically maternally imprinted, and presents during the newborn period with hypotonia, poor feeding, and failure to thrive, followed by rapid weight gain and hyperphagia which results in rapid weight gain in early childhood, as well as intellectual disability and growth hormone deficiency. Due to factors that may include craniofacial differences, low tone, and obesity, children with Prader-Willi syndrome are also at risk for both central and obstructive sleep apnea, and monitoring is particularly important in those being treated with growth hormone [79, 80].

Congenital Central Hypoventilation Syndrome

Genetic conditions play a critical role in the development of sleep-related hypoventilation in children. First described in 1970, congenital central hypoventilation syndrome (CCHS) is a rare disorder characterized primarily by alveolar hypoventilation as well as multisystem autonomic nervous system dysregulation including attenuated heart rate response to activity, gastrointestinal dysmotility and Hirschsprung disease, pupillary abnormalities, and predisposition to neural crest tumors [81]. Clinically, CCHS is typically identified in the newborn period, but milder phenotypes may not present until later in childhood or even adulthood [82].

CCHS results from mutations of the paired-like homeobox 2B (PHOX2B) gene at the p12 region on chromosome 4 [83]. Mutations are felt to occur as a result of unequal sister chromatid exchange during spermatogenesis [84]. PHOX2B encodes a highly conserved homeodomain transcription factor known to play a key role in the development of autonomic nervous system reflex circuits. In exon 4, PHOX2B contains a repeat sequence of 20 alanines (designated as 20/20). In approximately 90% of CCHS cases, the etiology of CCHS is an amino acid expansion of a polyalanine tract in exon 3. These repeat expansion mutations (PARMs) range from 24 to 33, with a greater number of repeats resulting in a more severe phenotype with an earlier presentation. Most cases will be heterozygous for in-frame PARM coding for 25–27 alanines (designated 20/25 through 20/27), with 28 or more being less common and 24 or 25 expansions less likely to be diagnosed due to milder symptoms. In addition to PARMs, the remaining 10% of patients with CCHS result from non-polyalanine repeat mutations (NPARMs) which are typically frameshift mutations in exon 3. While 90% of these mutations are de novo, PHOX2B gene mutations have an autosomal dominant fashion. Familial recurrence of CCHS may also occur through somatic or germline mosaicism [85]. Some of these milder cases present as late-onset CCHS with prolonged recovery following a surgical procedure or may be identified through the diagnosis of a family member such as a child with more severe phenotype. Familial recurrence of CCHS may also occur through somatic or germline mosaicism [85]. Some of these milder cases present as late-onset CCHS with prolonged recovery following a surgical procedure or may be identified through the diagnosis of a family member such as a child with more severe phenotype [86]. As asymptomatic hypoventilation has been identified in multiple parents of children diagnosed with CCHS, consideration should be made to screen for PHOX2B mutations and clinical signs such as alveolar hypoventilation in first-degree family members of patients diagnosed with CCHS [87].

In mice that have a PHOX2B 20/27 defect, the retrotrapezoid nucleus is absent, and the animals have a similar phenotype to humans, including hypoventilation and central apnea resulting in early death [88], providing a good animal model for better understanding the physiology and assessing future therapeutic targets for CCHS. Identification of the PHOX2B gene has allowed for earlier identification of CCHS, resulting in earlier diagnosis and reducing treatment delays. As a result, children identified with CCHS can now expect to lead much healthier lives with the proper comprehensive management [81].

Rapid-onset obesity with hypothalamic dysregulation, hypoventilation, and autonomic dysregulation (ROHHAD) is a rare condition that similarly results in alveolar hypoventilation and autonomic dysfunction but presents after the neonatal period and also is characterized by rapid weight gain and endocrinologic disorders and neuroendocrine tumors [82, 89], in addition to sleep-related hypoventilation and respiratory complications often associated with obesity, including obstructive sleep apnea and restrictive lung disease. In contrast to CCHS, patients with ROHHAD do not have mutations in PHOX2B, and to date studies have not identified a candidate genetic etiology. The etiology of ROHHAD may be epigenetic, transcriptional, or posttranscriptional misregulation, potentially related to dysfunction of neural crest-derived cells and tissues [90].

Hypersomnolence

Hypersomnolence is a common sleep-related symptom in childhood, particularly during adolescence. The etiology of excessive sleepiness is broad and includes insufficient sleep, underlying medical or psychiatric conditions, medication, or sleep disorders such as OSA, restless leg syndrome, or narcolepsy. The association with a known genetic diagnosis is variable and depends on the underlying etiology.

Narcolepsy

Narcolepsy is a rare, chronic condition that clinically presents with irrepressible sleepiness, sleep paralysis, and hypnogogic hallucinations, sometimes associated with cataplexy. Approximately one-third of patients with narcolepsy develop symptoms before 15 years of age, and approximately 5% develop in children under 5 years. Genetic factors likely play an important role in the development of narcolepsy. Narcolepsy type 1, which includes cataplexy, is characterized by loss of orexin-secreting neurons in the hypothalamus and has a strong association with human leukocyte antigen DQB1*0602 and DQBI*0101, which are present in 95% of those with cataplexy and 96% of those with orexin deficiency. Most cases of narcolepsy are sporadic, and there is no clear pattern of inheritance. As only 1–2% of first-degree relatives of people with narcolepsy and 25–31% of monozygotic twins are affected, there are likely additional epigenetic factors that determine whether the condition manifests [91]. However, some family members of affected individuals may manifest some symptoms without meeting the criteria for diagnosis, suggesting incomplete penetrance. While in most cases, an associated gene responsible for narcolepsy cannot be identified, a genome-wide linkage search did identify a locus on chromosome 4p13-q21 in a cohort of Japanese families [92]. In other studies, mutation associations were found in genes encoding orexin and myelin oligodendrocyte glycoprotein [91]. However, some family members of affected individuals may manifest some symptoms without meeting the criteria for diagnosis, suggesting incomplete penetrance. While in most cases, an associated gene responsible for narcolepsy cannot be identified, a genome-wide linkage search did identify a locus on chromosome 4p13-q21 in a cohort of Japanese families [92]. In other studies, mutations associations were found in genes encoding orexin and myelin oligodendrocyte glycoprotein [93, 94]. In children with narcolepsy, diagnosis is often delayed, which can lead to problems in school and socially. Having better testing for youth, including genetic testing, with suspected narcolepsy both with and without cataplexy would allow for earlier treatment and anticipatory guidance.

Kleine-Levin Syndrome

In addition, Kleine-Levin syndrome, a rare disorder characterized by recurrent episodes of excessive sleepiness lasting days to weeks associated with cognitive or behavior problems, separated by weeks or months of being asymptomatic. Typically presenting during the second decade of life and more common in males, Kleine-Levin syndrome has a prevalence of 2–10 per million, and the diagnosis is typically made based on the clinical symptoms and exclusion of other diagnoses like narcolepsy. While most cases are sporadic, familial cases account for approximately 8% of patients with Kleine-Levin syndrome and concordant monozygotic twins have been reported [95, 96]. A case-control GWAS study that included 673 cases of Kleine-Levin syndrome found an association with the *TRANK1* gene locus as well as circadian regulation genes [97]. However, there have also been associations with birth trauma, illness, and alcohol consumption, suggesting environmental triggers or a potential immune-modulated etiology.

Neurogenetic Conditions Associated with Hypersomnolence

Several neurogenetic conditions have been associated with hypersomnolence, including Moebius syndrome, Fragile X syndrome, Williams syndrome, and Smith-Magenis syndrome, all with distinct chromosomal abnormalities. Some neurogenetic conditions, including Prader-Willi syndrome and Niemann-Pick type C may also present with cataplexy-like symptoms seen in narcolepsy.

Circadian Rhythm Sleep-Wake Disorders

Delayed sleep phase syndrome (DSPS) is a sleep disorder characterized by a persistent delayed timing of sleep onset and awakening, which is significantly later than desired or socially acceptable. In contrast, advanced sleep phase syndrome (ASPS) is a sleep disorder characterized by a persistent advanced pattern of falling asleep and waking up much earlier than desired or socially acceptable. Both DSPS and ASPS are circadian rhythm sleep disorders, as they involve a misalignment between an individual's circadian rhythm and the external demands of their daily schedule. These circadian rhythm disorders are diagnosed through the use of a sleep diary or actigraphy, and polysomnography may be performed to rule out other sleep disorders. DSPS is especially common among adolescents; an estimated 7–16% of adolescents have DSPS [8].

The transcriptional products from the negative feedback loop components of the molecular clock—*CRYs* and *PERs*—repress *CLOCK* and *BMAL1*. A gain-of-function mutation in exon 11 of *CRY1* lengthens the period of the circadian molecular clock and was detected in an adult with DSPS; further, the same mutation was detected in related family members across generations who also had delayed sleep timing [98]. Mutations in the promotor of *PER3* have been identified in DSPS patients aged 28 ± 15 years [99]; the TA2G haplotype in the *PER3* promotor was two- to threefold more common among DSPS patients compared to controls and was associated with increased *PER3* expression [99]. Genetic mutations in molecular clock genes have also been linked to familial ASPS. In one family with ASPS, a mutation in *PER2* was detected that accelerates the molecular clock period [100], and mutations in *PER3* were detected in other families with ASPS [101]. Further, a missense mutation in *CRY2* that accelerates the molecular clock period was detected in a family with ASPS [102]. A *TIMELESS* mutation was detected in a family with ASPS, and through a mutant mouse model and cell lines, it was demonstrated that this mutation destabilizes *CRYs* and *PERs* shortening the circadian clock period [103]. Overall, these data suggest genetic mutations that extend or accelerate the period of the molecular clock may partly determine DSPS and ASPS, respectively, but pediatric-specific studies are needed.

Parasomnias

Parasomnias are a group of episodic behaviors that disrupt sleep and are experienced in some form by most children [104]. Some parasomnias such as nightmares and sleep paralysis occur primarily during REM sleep, and others such as sleep terrors and sleepwalking are more common during NREM sleep. Parasomnias are generally diagnosed clinically and most resolve without requiring treatment. In general, there are few studies evaluating the etiology of parasomnias in children, and most have no known genetic association, although a family history of parasomnias is common. Parasomnias are a group of episodic behaviors that disrupt sleep and are experienced in some form by most children [104]. Some parasomnias such as nightmares and sleep paralysis occur primarily during REM sleep, and others such as sleep terrors and sleepwalking are more common during NREM sleep. Parasomnias are generally diagnosed clinically and most resolve without requiring treatment. In general, there are few studies evaluating the etiology of parasomnias in children, and most have no known genetic association, although a family history of parasomnias is common [105, 106]. An association with HLA DQB1*0501 has been found in children with sleepwalking and other parasomnias [107]. In children with sleep enuresis, both an autosomal dominant and recessive inheritance, associated with loci on multiple chromosomes, have been described [108]. Likely for most parasomnias in children, there is a shared interaction between genetic and environmental factors, and additional studies are needed to better characterize the potential genetic contribution and the relationships between different parasomnias in children and adults.

Sleep-Related Movement Disorders

Sleep-related movement disorders include conditions characterized by simple, often stereotyped movements that fragment sleep or prevent the onset of sleep in children. Similar to parasomnias, most sleep-related movement disorders are diagnosed clinically, and many require no specific treatment. One of the most common sleep-related movement disorders to come to clinical attention in children is restless leg syndrome (RLS), a sensorimotor condition that includes the urge to move the legs and accompanying unpleasant sensations occurring in the evening and during rest, relieved with movement, and causing impairment in sleep or daytime activity. While RLS is likely underdiagnosed in children, nearly half of adults with the condition report that symptoms began during childhood, and there is a strong familial association for children who are diagnosed [109, 110]. Restless leg syndrome is felt to be associated with iron deficiency, and children can be treated with ferrous sulfate when serum ferritin levels are below 50 mcg/dL. There have been several GWAS studies in restless leg syndrome, identifying loci, including MEIS1, which encodes for transcription factors involved in hematopoiesis and limb development [111].

Periodic limb movement disorder is a related condition found in patients who have recurrent rhythmic limb movements during sleep that result in disrupted sleep and/or daytime symptoms. Analyses of families whose children had RLS found concordance of elevated periodic limb movements during sleep, but linkage and genotype analyses have thus far failed to identify an associated genetic locus [112]. Bruxism, characterized by repetitive clenching or grinding of the teeth, typically during NREM sleep, is most common during childhood. Bruxism is sometimes associated with anxiety, sleep disorders such as OSA, as well as medications such as selective serotonin reuptake inhibitors. Birth cohort studies have shown a strong association within families as well as environmental effects, and case-control studies have identified polymorphisms within the HTR2A gene, associated with a serotonin receptor, that may be genetic contributors to bruxism [113, 114]. The genetic contribution for other sleep-related movement disorders in children is poorly understood, but often there are strong associations within families. As many therapies for RLS, PLMD, and associated disorders like restless sleep disorder have significant side effects, additional studies are needed to better understand the relationship between these conditions to specific genes and iron metabolism to support earlier diagnosis and better therapeutics.

Overall Summary

Sleep health traits and clinical sleep disorders are explained in part by genetic factors. With improved sleep health trait phenotyping in larger samples of children, this will facilitate the discovery of novel pediatric sleep health loci through genome-wide association studies. In addition, candidate gene studies will continue to provide opportunities to understand how specific genes impact sleep in childhood, especially at key developmental stages. In regard to clinical sleep disorders, specific genetic mutations are known to contribute to the risk of specific sleep disorders. For example, genetic variants are known to increase the risk of sleep-disordered breathing, and obstructive sleep apnea is common among children with Trisomy 21. Mutations in genes comprising the molecular clock have been observed in families with circadian rhythm sleep-wake disorders; specifically, mutations in *CRY1* and *PER3* that extend the molecular clock period have been detected in families with advanced sleep-phased syndrome. Current and future findings on the genetics of sleep in childhood will help identify those most at risk for problematic sleep and guide prevention and treatment strategies to improve sleep and therefore health and development in early life.

References

1. Paruthi S, Brooks LJ, D'Ambrosio C, et al. Consensus statement of the American Academy of Sleep Medicine on the recommended amount of sleep for healthy children: methodology and discussion. J Clin Sleep Med. 2016;12(11):1549–61. https://doi.org/10.5664/jcsm.6288.
2. Owens J, Wang G, Lewin D, Skora E, Baylor A. Association between short sleep duration and risk behavior factors in middle school students. Sleep. 2017;40(1). https://doi.org/10.1093/sleep/zsw004.
3. Mike TB, Shaw DS, Forbes EE, Sitnick SL, Hasler BP. The hazards of bad sleep-sleep duration and quality as predictors of adolescent alcohol and cannabis use. Drug Alcohol Depend. 2016;168:335–9. https://doi.org/10.1016/j.drugalcdep.2016.08.009.
4. Wheaton AG, Olsen EO, Miller GF, Croft JB. Sleep duration and injury-related risk behaviors among high school students – United States, 2007–2013. MMWR Morb Mortal Wkly Rep. 2016;65(13):337–41. https://doi.org/10.15585/mmwr.mm6513a1.

5. Sivertsen B, Skogen JC, Jakobsen R, Hysing M. Sleep and use of alcohol and drug in adolescence. A large population-based study of Norwegian adolescents aged 16 to 19 years. Drug Alcohol Depend. 2015;149:180–6. https://doi.org/10.1016/j.drugalcdep.2015.01.045.
6. Weaver MD, Barger LK, Malone SK, Anderson LS, Klerman EB. Dose-dependent associations between sleep duration and unsafe behaviors among US high school students. JAMA Pediatr. 2018;172(12):1187–9. https://doi.org/10.1001/jamapediatrics.2018.2777.
7. Kocevska D, Lysen TS, Dotinga A, et al. Sleep characteristics across the lifespan in 1.1 million people from The Netherlands, United Kingdom and United States: a systematic review and meta-analysis. *Nat.* Hum Behav. 2021;5(1):113–22. https://doi.org/10.1038/s41562-020-00965-x.
8. Krishna J, Kalra M, McQuillan ME. Sleep disorders in childhood. Pediatr Rev. 2023;44(4):189–202. https://doi.org/10.1542/pir.2022-005521.
9. Tonetti L, Fabbri M, Filardi M, Martoni M, Natale V. Effects of sleep timing, sleep quality and sleep duration on school achievement in adolescents. Sleep Med. 2015;16(8):936–40. https://doi.org/10.1016/j.sleep.2015.03.026.
10. Hysing M, Haugland S, Stormark KM, Boe T, Sivertsen B. Sleep and school attendance in adolescence: results from a large population-based study. Scand J Public Health. 2015;43(1):2–9. https://doi.org/10.1177/1403494814556647.
11. Asarnow LD, McGlinchey E, Harvey AG. The effects of bedtime and sleep duration on academic and emotional outcomes in a nationally representative sample of adolescents. J Adolesc Health. 2014;54(3):350–6. https://doi.org/10.1016/j.jadohealth.2013.09.004.
12. Gruber R, Michaelsen S, Bergmame L, et al. Short sleep duration is associated with teacher-reported inattention and cognitive problems in healthy school-aged children. Nat Sci Sleep. 2012;4:33–40. https://doi.org/10.2147/NSS.S24607.
13. Gruber R, Laviolette R, Deluca P, Monson E, Cornish K, Carrier J. Short sleep duration is associated with poor performance on IQ measures in healthy school-age children. Sleep Med. 2010;11(3):289–94. https://doi.org/10.1016/j.sleep.2009.09.007.
14. Garner AA, Hansen A, Baxley C, Becker SP, Sidol CA, Beebe DW. Effect of sleep extension on sluggish cognitive tempo symptoms and driving behavior in adolescents with chronic short sleep. Sleep Med. 2017;30:93–6. https://doi.org/10.1016/j.sleep.2016.09.003.
15. Cousins JN, Wong KF, Chee MWL. Multi-night sleep restriction impairs long-term retention of factual knowledge in adolescents. J Adolesc Health. 2019;65:549. https://doi.org/10.1016/j.jadohealth.2019.04.030.
16. Mitchell JA, Rodriguez D, Schmitz KH, Audrain-McGovern J. Sleep duration and adolescent obesity. Pediatrics. 2013;131(5):e1428–34. https://doi.org/10.1542/peds.2012-2368.
17. Santos ESG, Souza OF. Association of sleep duration and blood pressure in adolescents: a multicenter study. Am J Hypertens. 2019;33:77. https://doi.org/10.1093/ajh/hpz129.
18. Patel NJ, Savin KL, Kahanda SN, et al. Sleep habits in adolescents with type 1 diabetes: Variability in sleep duration linked with glycemic control. Pediatr Diabetes. 2018;19:1100. https://doi.org/10.1111/pedi.12689.
19. Hall MH, Lee L, Matthews KA. Sleep duration during the school week is associated with C-reactive protein risk groups in healthy adolescents. Sleep Med. 2015;16(1):73–8. https://doi.org/10.1016/j.sleep.2014.10.005.
20. Meininger JC, Gallagher MR, Eissa MA, Nguyen TQ, Chan W. Sleep duration and its association with ambulatory blood pressure in a school-based, diverse sample of adolescents. Am J Hypertens. 2014;27(7):948–55. https://doi.org/10.1093/ajh/hpt297.
21. Au CT, Ho CK, Wing YK, Lam HS, Li AM. Acute and chronic effects of sleep duration on blood pressure. Pediatrics. 2014;133(1):e64–72. https://doi.org/10.1542/peds.2013-1379.
22. Martinez-Gomez D, Eisenmann JC, Gomez-Martinez S, et al. Sleep duration and emerging cardiometabolic risk markers in adolescents. The AFINOS study. Sleep Med. 2011;12(10):997–1002. https://doi.org/10.1016/j.sleep.2011.05.009.

23. Koren D, Levitt Katz LE, Brar PC, Gallagher PR, Berkowitz RI, Brooks LJ. Sleep architecture and glucose and insulin homeostasis in obese adolescents. Diabetes Care. 2011;34(11):2442–7. https://doi.org/10.2337/dc11-1093.
24. Kong AP, Wing YK, Choi KC, et al. Associations of sleep duration with obesity and serum lipid profile in children and adolescents. Sleep Med. 2011;12(7):659–65. https://doi.org/10.1016/j.sleep.2010.12.015.
25. Javaheri S, Storfer-Isser A, Rosen CL, Redline S. Sleep quality and elevated blood pressure in adolescents. Circulation. 2008;118(10):1034–40. https://doi.org/10.1161/CIRCULATIONAHA.108.766410.
26. Hitze B, Bosy-Westphal A, Bielfeldt F, et al. Determinants and impact of sleep duration in children and adolescents: data of the Kiel Obesity Prevention Study. Eur J Clin Nutr. 2009;63(6):739–46. https://doi.org/10.1038/ejcn.2008.41.
27. Cappuccio FP, Taggart FM, Kandala NB, et al. Meta-analysis of short sleep duration and obesity in children and adults. Sleep. 2008;31(5):619–26.
28. Miller MA, Kruisbrink M, Wallace J, Ji C, Cappuccio FP. Sleep duration and incidence of obesity in infants, children, and adolescents: a systematic review and meta-analysis of prospective studies. Sleep. 2018;41(4). https://doi.org/10.1093/sleep/zsy018.
29. Hart CN, Carskadon MA, Considine RV, et al. Changes in children's sleep duration on food intake, weight, and leptin. Pediatrics. 2013;132(6):e1473–80. https://doi.org/10.1542/peds.2013-1274.
30. Beebe DW, Simon S, Summer S, Hemmer S, Strotman D, Dolan LM. Dietary intake following experimentally restricted sleep in adolescents. Sleep. 2013;36(6):827–34. https://doi.org/10.5665/sleep.2704.
31. Buysse DJ. Sleep health: can we define it? Does it matter? Sleep. 2014;37(1):9–17. https://doi.org/10.5665/sleep.3298.
32. Kocevska D, Barclay NL, Bramer WM, Gehrman PR, Van Someren EJW. Heritability of sleep duration and quality: a systematic review and meta-analysis. Sleep Med Rev. 2021;59:101448. https://doi.org/10.1016/j.smrv.2021.101448.
33. O'Callaghan VS, Hansell NK, Guo W, et al. Genetic and environmental influences on sleep-wake behaviors in adolescence. Sleep Adv. 2021;2(1):zpab018. https://doi.org/10.1093/sleepadvances/zpab018.
34. Gehrman PR, Meltzer LJ, Moore M, et al. Heritability of insomnia symptoms in youth and their relationship to depression and anxiety. Sleep. 2011;34(12):1641–6. https://doi.org/10.5665/sleep.1424.
35. Barclay NL, Kocevska D, Bramer WM, Van Someren EJW, Gehrman P. The heritability of insomnia: A meta-analysis of twin studies. Genes Brain Behav. 2021;20(4):e12717. https://doi.org/10.1111/gbb.12717.
36. Moore M, Slane J, Mindell JA, Burt SA, Klump KL. Genetic and environmental influences on sleep problems: a study of preadolescent and adolescent twins. Child Care Health Dev. 2011;37(5):638–41. https://doi.org/10.1111/j.1365-2214.2011.01230.x.
37. Lewis KJS, Gregory AM. Heritability of sleep and its disorders in childhood and adolescence. Curr Sleep Med Rep. 2021;7(4):155–66. https://doi.org/10.1007/s40675-021-00216-z.
38. Zhu H, Zhou X. Statistical methods for SNP heritability estimation and partition: a review. Comput Struct Biotechnol J. 2020;18:1557–68. https://doi.org/10.1016/j.csbj.2020.06.011.
39. Marinelli M, Pappa I, Bustamante M, et al. Heritability and genome-wide association analyses of sleep duration in children: the EAGLE consortium. Sleep. 2016;39(10):1859–69. https://doi.org/10.5665/sleep.6170.
40. Anafi RC, Kayser MS, Raizen DM. Exploring phylogeny to find the function of sleep. Nat Rev Neurosci. 2019;20(2):109–16. https://doi.org/10.1038/s41583-018-0098-9.
41. Kayser MS, Biron D. Sleep and development in genetically tractable model organisms. Genetics. 2016;203(1):21–33. https://doi.org/10.1534/genetics.116.189589.
42. Borbely AA, Daan S, Wirz-Justice A, Deboer T. The two-process model of sleep regulation: a reappraisal. J Sleep Res. 2016;25(2):131–43. https://doi.org/10.1111/jsr.12371.

43. Jenni OG, Achermann P, Carskadon MA. Homeostatic sleep regulation in adolescents. Sleep. 2005;28(11):1446–54. https://doi.org/10.1093/sleep/28.11.1446.
44. Dashti HS, Jones SE, Wood AR, et al. Genome-wide association study identifies genetic loci for self-reported habitual sleep duration supported by accelerometer-derived estimates. Nat Commun. 2019;10(1):1100. https://doi.org/10.1038/s41467-019-08917-4.
45. Jones SE, Lane JM, Wood AR, et al. Genome-wide association analyses of chronotype in 697,828 individuals provides insights into circadian rhythms. Nat Commun. 2019;10(1):343. https://doi.org/10.1038/s41467-018-08259-7.
46. Jones SE, Tyrrell J, Wood AR, et al. Genome-wide association analyses in 128,266 individuals identifies new morningness and sleep duration loci. PLoS Genet. 2016;12(8):e1006125. https://doi.org/10.1371/journal.pgen.1006125.
47. Jones SE, van Hees VT, Mazzotti DR, et al. Genetic studies of accelerometer-based sleep measures yield new insights into human sleep behaviour. Nat Commun. 2019;10(1):1585. https://doi.org/10.1038/s41467-019-09576-1.
48. Lane JM, Liang J, Vlasac I, et al. Genome-wide association analyses of sleep disturbance traits identify new loci and highlight shared genetics with neuropsychiatric and metabolic traits. Nat Genet. 2017;49(2):274–81. https://doi.org/10.1038/ng.3749.
49. Lane JM, Vlasac I, Anderson SG, et al. Genome-wide association analysis identifies novel loci for chronotype in 100,420 individuals from the UK Biobank. Nat Commun. 2016;7:10889. https://doi.org/10.1038/ncomms10889.
50. Jansen PR, Watanabe K, Stringer S, et al. Genome-wide analysis of insomnia in 1,331,010 individuals identifies new risk loci and functional pathways. Nat Genet. 2019;51(3):394–403. https://doi.org/10.1038/s41588-018-0333-3.
51. Lane JM, Jones SE, Dashti HS, et al. Biological and clinical insights from genetics of insomnia symptoms. Nat Genet. 2019;51(3):387–93. https://doi.org/10.1038/s41588-019-0361-7.
52. Bass J, Lazar MA. Circadian time signatures of fitness and disease. Science. 2016;354(6315):994–9. https://doi.org/10.1126/science.aah4965.
53. Nascimento Ferreira MV, Goumidi L, Carvalho HB, et al. Associations between REV-ERBalpha, sleep duration and body mass index in European adolescents. Sleep Med. 2018;46:56–60. https://doi.org/10.1016/j.sleep.2018.01.014.
54. Krishnan M, Shelling AN, Wall CR, et al. Gene-by-environment interactions of the CLOCK, PEMT, and GHRELIN loci with average sleep duration in relation to obesity traits using a cohort of 643 New Zealand European children. Sleep Med. 2017;37:19–26. https://doi.org/10.1016/j.sleep.2017.05.017.
55. Giovaninni NP, Fuly JT, Moraes LI, et al. Study of the association between 3111T/C polymorphism of the CLOCK gene and the presence of overweight in schoolchildren. J Pediatr. 2014;90(5):500–5. https://doi.org/10.1016/j.jped.2014.01.011.
56. Forbes EE, Dahl RE, Almeida JR, et al. PER2 rs2304672 polymorphism moderates circadian-relevant reward circuitry activity in adolescents. Biol Psychiatry. 2012;71(5):451–7. https://doi.org/10.1016/j.biopsych.2011.10.012.
57. Guilleminault C, Lee JH, Chan A. Pediatric obstructive sleep apnea syndrome. Arch Pediatr Adolesc Med. 2005;159(8):775–85. https://doi.org/10.1001/archpedi.159.8.775.
58. Bixler EO, Vgontzas AN, Lin HM, et al. Sleep disordered breathing in children in a general population sample: prevalence and risk factors. Sleep. 2009;32(6):731–6.
59. Li AM, So HK, Au CT, et al. Epidemiology of obstructive sleep apnoea syndrome in Chinese children: a two-phase community study. Thorax. 2010;65(11):991–997. 65/11/991 [pii]. https://doi.org/10.1136/thx.2010.134858.
60. Schwab RJ, Kim C, Bagchi S, et al. Understanding the anatomic basis for obstructive sleep apnea syndrome in adolescents. Am J Respir Crit Care Med. 2015;191(11):1295–309. https://doi.org/10.1164/rccm.201501-0169OC.
61. Marcus CL, Moore RH, Rosen CL, et al. A randomized trial of adenotonsillectomy for childhood sleep apnea. N Engl J Med. 2013;368(25):2366–76. https://doi.org/10.1056/NEJMoa1215881.

62. Marcus CL, Brooks LJ, Draper KA, et al. Diagnosis and management of childhood obstructive sleep apnea syndrome. Pediatrics. 2012;130(3):e714–e755. peds.2012–1672 [pii]. https://doi.org/10.1542/peds.2012-1672.
63. Xiao L, Su S, Liang J, Jiang Y, Shu Y, Ding L. Analysis of the risk factors associated with obstructive sleep apnea syndrome in chinese children. Front Pediatr. 2022;10:900216. https://doi.org/10.3389/fped.2022.900216.
64. Redline S, Tishler PV, Tosteson TD, et al. The familial aggregation of obstructive sleep apnea. Am J Respir Crit Care Med. 1995;151(3 Pt 1):682–7. https://doi.org/10.1164/ajrccm/151.3_Pt_1.682.
65. Quinlan CM, Chang X, March M, et al. Identification of novel loci in obstructive sleep apnea in European American and African American children. Sleep. 2022;47:zsac182. https://doi.org/10.1093/sleep/zsac182.
66. Zhou W, Nielsen JB, Fritsche LG, et al. Efficiently controlling for case-control imbalance and sample relatedness in large-scale genetic association studies. Nat Genet. 2018;50(9):1335–41. https://doi.org/10.1038/s41588-018-0184-y.
67. Cielo CM, Konstantinopoulou S, Hoque R. OSAS in specific pediatric populations. Curr Probl Pediatr Adolesc Health Care. 2016;46(1):11–8. https://doi.org/10.1016/j.cppeds.2015.10.008.
68. de Miguel-Diez J, Villa-Asensi JR, Alvarez-Sala JL. Prevalence of sleep-disordered breathing in children with Down syndrome: polygraphic findings in 108 children. Sleep. 2003;26(8):1006–9. https://doi.org/10.1093/sleep/26.8.1006.
69. Bertrand P, Navarro H, Caussade S, Holmgren N, Sanchez I. Airway anomalies in children with down syndrome: endoscopic findings. Pediatr Pulmonol. 2003;36(2):137–41. https://doi.org/10.1002/ppul.10332.
70. Bull MJ, Committee on G. Health supervision for children with down syndrome. Pediatrics. 2011;128(2):393–406. https://doi.org/10.1542/peds.2011-1605.
71. Logjes RJH, MacLean JE, de Cort NW, et al. Objective measurements for upper airway obstruction in infants with Robin sequence: what are we measuring? A systematic review. J Clin Sleep Med. 2021;17(8):1717–29. https://doi.org/10.5664/jcsm.9394.
72. Breugem CC, Evans KN, Poets CF, et al. Best practices for the diagnosis and evaluation of infants with robin sequence: a clinical consensus report. JAMA Pediatr. 2016;170(9):894–902. https://doi.org/10.1001/jamapediatrics.2016.0796.
73. Chen Q, Zhao Y, Qian Y, Lu C, Shen G, Dai J. A genetic-phenotypic classification for syndromic micrognathia. J Hum Genet. 2019;64(9):875–83. https://doi.org/10.1038/s10038-019-0630-4.
74. Cielo CM, Duffy KA, Taylor JA, Marcus CL, Kalish JM. Obstructive sleep apnea in children with Beckwith-Wiedemann syndrome. J Clin Sleep Med. 2019;15(3):375–81. https://doi.org/10.5664/jcsm.7656.
75. Resnick CM, Middleton JK, Calabrese CE, Ganjawalla K, Padwa BL. Retropalatal cross-sectional area is predictive of obstructive sleep apnea in patients with syndromic craniosynostosis. Cleft Palate Craniofac J. 2020;57(5):560–5. https://doi.org/10.1177/1055665619882571.
76. Cimaz R, La Torre F. Mucopolysaccharidoses. Curr Rheumatol Rep. 2014;16(1):389. https://doi.org/10.1007/s11926-013-0389-0.
77. Koehne T, Muller-Stover S, Kohn A, et al. Obstructive sleep apnea and craniofacial appearance in MPS type I-Hurler children after hematopoietic stem cell transplantation. Sleep Breath. 2019;23(4):1315–21. https://doi.org/10.1007/s11325-019-01900-9.
78. Afsharpaiman S, Saburi A, Waters KA. Respiratory difficulties and breathing disorders in achondroplasia. Paediatr Respir Rev. 2013;S1526–0542(13)00020–1 [pii]. https://doi.org/10.1016/j.prrv.2013.02.009.
79. Cohen M, Hamilton J, Narang I. Clinically important age-related differences in sleep related disordered breathing in infants and children with Prader-Willi Syndrome. PLoS One. 2014;9(6):e101012. https://doi.org/10.1371/journal.pone.0101012. PONE-D-13-50273 [pii]

80. Aurora RN, Zak RS, Karippot A, et al. Practice parameters for the respiratory indications for polysomnography in children. Sleep. 2011;34(3):379–88.
81. Weese-Mayer DE, Berry-Kravis EM, Ceccherini I, Keens TG, Loghmanee DA, Trang H. An official ATS clinical policy statement: congenital central hypoventilation syndrome: genetic basis, diagnosis, and management. Am J Respir Crit Care Med. 2010;181(6):626–44. 181/6/626 [pii]. https://doi.org/10.1164/rccm.200807-1069ST.
82. Katz ES, McGrath S, Marcus CL. Late-onset central hypoventilation with hypothalamic dysfunction: a distinct clinical syndrome. Pediatr Pulmonol. 2000;29(1):62–68. https://doi.org/10.1002/(sici)1099-0496(200001)29:1<62::aid-ppul10>3.0.co;2-m [pii].
83. Amiel J, Laudier B, Attie-Bitach T, et al. Polyalanine expansion and frameshift mutations of the paired-like homeobox gene PHOX2B in congenital central hypoventilation syndrome. Nat Genet. 2003;33(4):459–61. https://doi.org/10.1038/ng1130.
84. Arai H, Otagiri T, Sasaki A, et al. De novo polyalanine expansion of PHOX2B in congenital central hypoventilation syndrome: unequal sister chromatid exchange during paternal gametogenesis. J Hum Genet. 2007;52(11):921–5. https://doi.org/10.1007/s10038-007-0197-3.
85. Rand CM, Yu M, Jennings LJ, et al. Germline mosaicism of PHOX2B mutation accounts for familial recurrence of congenital central hypoventilation syndrome (CCHS). Am J Med Genet A. 2012;158A(9):2297–301. https://doi.org/10.1002/ajmg.a.35499.
86. Kasi AS, Kun SS, Keens TG, Perez IA. Adult with PHOX2B mutation and late-onset congenital central hypoventilation syndrome. J Clin Sleep Med. 2018;14(12):2079–81. https://doi.org/10.5664/jcsm.7542.
87. Janssen H, Vulto-van Silfhout AT, Jongmans MCJ, van der Hout AH, Overeem S. Severe positional central sleep apnea in an asymptomatic adult with a PHOX2B frameshift mutation. J Clin Sleep Med. 2018;14(8):1427–30. https://doi.org/10.5664/jcsm.7290.
88. Amiel J, Dubreuil V, Ramanantsoa N, et al. PHOX2B in respiratory control: lessons from congenital central hypoventilation syndrome and its mouse models. Respir Physiol Neurobiol. 2009;168(1–2):125–32. https://doi.org/10.1016/j.resp.2009.03.005.
89. Ize-Ludlow D, Gray JA, Sperling MA, et al. Rapid-onset obesity with hypothalamic dysfunction, hypoventilation, and autonomic dysregulation presenting in childhood. Pediatrics. 2007;120(1):e179–88. 120/1/e179 [pii]. https://doi.org/10.1542/peds.2006-3324.
90. Barclay SF, Rand CM, Nguyen L, et al. ROHHAD and Prader-Willi syndrome (PWS): clinical and genetic comparison. Orphanet J Rare Dis. 2018;13(1):124. https://doi.org/10.1186/s13023-018-0860-0.
91. Mignot E. Genetic and familial aspects of narcolepsy. Neurology. 1998;50(2 Suppl 1):S16–22. https://doi.org/10.1212/wnl.50.2_suppl_1.s16.
92. Nakayama J, Miura M, Honda M, Miki T, Honda Y, Arinami T. Linkage of human narcolepsy with HLA association to chromosome 4p13-q21. Genomics. 2000;65(1):84–6. https://doi.org/10.1006/geno.2000.6143.
93. Peyron C, Faraco J, Rogers W, et al. A mutation in a case of early onset narcolepsy and a generalized absence of hypocretin peptides in human narcoleptic brains. Nat Med. 2000;6(9):991–7. https://doi.org/10.1038/79690.
94. Hor H, Bartesaghi L, Kutalik Z, et al. A missense mutation in myelin oligodendrocyte glycoprotein as a cause of familial narcolepsy with cataplexy. Am J Hum Genet. 2011;89(3):474–9. https://doi.org/10.1016/j.ajhg.2011.08.007.
95. Dauvilliers Y, Mayer G, Lecendreux M, et al. Kleine-Levin syndrome: an autoimmune hypothesis based on clinical and genetic analyses. Neurology. 2002;59(11):1739–45. https://doi.org/10.1212/01.wnl.0000036605.89977.d0.
96. Peraita-Adrados R, Vicario JL, Tafti M, Garcia de Leon M, Billiard M. Monozygotic twins affected with Kleine-Levin syndrome. Sleep. 2012;35(5):595–6. https://doi.org/10.5665/sleep.1808.
97. Ambati A, Hillary R, Leu-Semenescu S, et al. Kleine-Levin syndrome is associated with birth difficulties and genetic variants in the TRANK1 gene loci. Proc Natl Acad Sci USA. 2021;118(12):e2005753118. https://doi.org/10.1073/pnas.2005753118.

98. Patke A, Murphy PJ, Onat OE, et al. Mutation of the human circadian clock gene CRY1 in familial delayed sleep phase disorder. Cell. 2017;169(2):203–215 e13. https://doi.org/10.1016/j.cell.2017.03.027.
99. Archer SN, Carpen JD, Gibson M, et al. Polymorphism in the PER3 promoter associates with diurnal preference and delayed sleep phase disorder. Sleep. 2010;33(5):695–701. https://doi.org/10.1093/sleep/33.5.695.
100. Toh KL, Jones CR, He Y, et al. An hPer2 phosphorylation site mutation in familial advanced sleep phase syndrome. Science. 2001;291(5506):1040–3. https://doi.org/10.1126/science.1057499.
101. Zhang L, Hirano A, Hsu PK, et al. A PERIOD3 variant causes a circadian phenotype and is associated with a seasonal mood trait. Proc Natl Acad Sci USA. 2016;113(11):E1536–44. https://doi.org/10.1073/pnas.1600039113.
102. Hirano A, Shi G, Jones CR, et al. A cryptochrome 2 mutation yields advanced sleep phase in humans. elife. 2016;5:e16695. https://doi.org/10.7554/eLife.16695.
103. Kurien P, Hsu PK, Leon J, et al. TIMELESS mutation alters phase responsiveness and causes advanced sleep phase. Proc Natl Acad Sci USA. 2019, 116(24):12045–53. https://doi.org/10.1073/pnas.1819110116.
104. Petit D, Touchette E, Tremblay RE, Boivin M, Montplaisir J. Dyssomnias and parasomnias in early childhood. Pediatrics. 2007;119(5):e1016–25. https://doi.org/10.1542/peds.2006-2132.
105. Hublin C, Kaprio J. Genetic aspects and genetic epidemiology of parasomnias. Sleep Med Rev. 2003;7(5):413–21. https://doi.org/10.1053/smrv.2001.0247.
106. Nguyen BH, Perusse D, Paquet J, et al. Sleep terrors in children: a prospective study of twins. Pediatrics. 2008;122(6):e1164–7. https://doi.org/10.1542/peds.2008-1303.
107. Heidbreder A, Frauscher B, Mitterling T, et al. Not only sleepwalking but NREM parasomnia irrespective of the type is associated with HLA DQB1*05:01. J Clin Sleep Med. 2016;12(4):565–70. https://doi.org/10.5664/jcsm.5692.
108. Fatouh AA, Motawie AA, Abd Al-Aziz AM, et al. Anti-diuretic hormone and genetic study in primary nocturnal enuresis. J Pediatr Urol. 2013;9(6 Pt A):831–7. https://doi.org/10.1016/j.jpurol.2012.11.009.
109. Picchietti D, Allen RP, Walters AS, Davidson JE, Myers A, Ferini-Strambi L. Restless legs syndrome: prevalence and impact in children and adolescents – the Peds REST study. Pediatrics. 2007;120(2):253–66. https://doi.org/10.1542/peds.2006-2767.
110. Xiong L, Jang K, Montplaisir J, et al. Canadian restless legs syndrome twin study. Neurology. 2007;68(19):1631–3. https://doi.org/10.1212/01.wnl.0000261016.90374.fd.
111. Winkelmann J, Schormair B, Lichtner P, et al. Genome-wide association study of restless legs syndrome identifies common variants in three genomic regions. Nat Genet. 2007;39(8):1000–6. https://doi.org/10.1038/ng2099.
112. Muhle H, Neumann A, Lohmann-Hedrich K, et al. Childhood-onset restless legs syndrome: clinical and genetic features of 22 families. Mov Disord. 2008;23(8):1113–21.; quiz 1203. https://doi.org/10.1002/mds.22016.
113. Abe Y, Suganuma T, Ishii M, et al. Association of genetic, psychological and behavioral factors with sleep bruxism in a Japanese population. J Sleep Res. 2012;21(3):289–96. https://doi.org/10.1111/j.1365-2869.2011.00961.x.
114. Rintakoski K, Hublin C, Lobbezoo F, Rose RJ, Kaprio J. Genetic factors account for half of the phenotypic variance in liability to sleep-related bruxism in young adults: a nationwide Finnish twin cohort study. Twin Res Hum Genet. 2012;15(6):714–9. https://doi.org/10.1017/thg.2012.54.

Multi-omics as a Strategy for Understanding Sleep's Role in Memory

Sara J. Aton

Abstract For at least a century (Jenkins and Dallenbach, Am J Psychol 35:605–12, 1924), behavioral data from both human subjects and animal models have supported the idea that sleep in the hours immediately following learning is essential for storing new information in the brain (Aton et al., Encyclopedia of life science. John Wiley and Sons, Ltd, Chichester, 2009b; Rasch and Born, Physiol Rev 93:681–766, 2013; Walker and Stickgold, Neuron 44:121–33, 2004). However, our understanding of the neural mechanisms underlying sleep's effect on memory consolidation has been slowed by the sheer number of neurophysiological variables that are capable of affecting this process and the fact that these variables all change simultaneously when the brain transitions from wake to NREM sleep and from NREM to REM (Puentes-Mestril and Aton, Front Neural Circuits 11, 2017; Puentes-Mestril et al., Sleep 42. pii: zsz095, 2019). In response to this complexity, a discovery-based experimental strategy has been adopted, in which a range of omics technologies have been used to quantify neurobiological responses to sleep deprivation (SD). This chapter provides an overview of advances to our understanding of an important biological issue—the effect sleep has on brain function—driven by omics-based approaches to this question.

Keywords Sleep function · mRNA · Protein · Intracellular trafficking · Cytoskeleton

For at least a century [27], behavioral data from both human subjects and animal models have supported the idea that sleep in the hours immediately following learning is essential for storing new information in the brain [2, 43, 50]. However, our understanding of the neural mechanisms underlying sleep's effect on memory consolidation has been slowed by the sheer number of neurophysiological variables that

---

S. J. Aton (✉)
Department of Molecular, Cellular, and Developmental Biology, University of Michigan, Ann Arbor, MI, USA
e-mail: saton@umich.edu

P. Gehrman et al. (eds.), *Genetics of Sleep and Sleep Disorders*,
https://doi.org/10.1007/978-3-031-62723-1_18

are capable of affecting this process and the fact that these variables all change simultaneously when the brain transitions from wake to NREM sleep and from NREM to REM [40, 41]. In response to this complexity, a discovery-based experimental strategy has been adopted, in which a range of omics technologies have been used to quantify neurobiological responses to sleep deprivation (SD). This chapter provides an overview of advances to our understanding of an important biological issue—the effect sleep has on brain function—driven by omics-based approaches to this question.

Investigating Sleep Function in the Age of Omics

From the earliest days of mRNA quantification, differential display, in situ hybridization, and microarrays were used to quantify transcript changes in the rat brain following periods of experimental sleep deprivation (SD) [8, 9]. These studies provided the first connection between sleep and synaptic plasticity, by demonstrating that after a period of SD, activity-regulated immediate early gene (IEG) expression (including levels of *Cfos*, *Arc*, and *Egr1*) was elevated throughout the brain. These studies further demonstrated that locus coeruleus (LC) noradrenergic signaling was upstream of these changes—6-OHDA or DSP-4 neurotoxic lesions of the LC led to disruption of SD-driven transcript changes [9, 10]. Critically, another finding from these early studies was that in addition to these transcription factor IEGs, transcripts encoding components of ER stress response (e.g., *BIP, Hsp70*) and mitochondrial energetic pathways (e.g., *COX1*) were also more abundant in the brain after SD.

These findings and further characterization of SD-driven transcript changes in the early 2000s using commercial microarrays led to the development of the highly influential synaptic homeostasis hypothesis [11, 46]. These studies demonstrated that in the neocortex, a diverse set of transcripts encoding IEGs, glutamate receptor subunits, *BDNF*, *Homer/Vesl*, and *PSD95*—all of which are mobilized in the context of synaptic plasticity—are upregulated during SD. The synaptic homeostasis hypothesis proposed that a major function of sleep in the mammalian brain is the widespread reduction of excitatory synaptic strength, which serves to offset net strengthening of these synapses across waking experiences. More recently, biochemical data (i.e., increased expression of AMPA receptor subunit GluR1 and phosphorylated CAMK2 in cortical synaptoneurosomes after a period of SD) [49] and anatomical findings (i.e., an increase in synaptic density on dendrites of neocortical pyramidal cells, using serial electron microscopy) [12] have supported the idea that there are differences in neocortical synapses' density and strength between sleeping and sleep-deprived animals.

Theoretical benefits of this process include space savings (due to a reduction in either synaptic numbers or size) and improved signal-to-noise ratio (due to elimination of less efficacious synapses in neural circuits). The "homeostatic" aspect of the hypothesis, as originally posited, that was both intuitively satisfying and intriguing is that it linked synaptic downscaling across periods of sleep to slow wave activity—a biomarker of sleep homeostasis. The hypothesis stipulated that synapses are progressively strengthened across extended wake (i.e., SD), which in turn leads to greater synchrony of activity across the thalamocortical network—and thus greater slow wave activity as the brain transitions to NREM sleep. This enhancement of slow wave activity then dissipates as synaptic strength across the network renormalizes—with reductions in synaptic strength being driven by the high level of firing synchrony during slow wave oscillations.

An important corollary and implication of the synaptic homeostasis hypothesis is that improvements in cognition after sleep, including sleep-dependent memory consolidation, are mediated (at least in part) by synaptic downscaling. This aspect of the hypothesis has (as yet) not been supported by the majority of studies from animals during sleep-dependent memory consolidation. With some exceptions [55], these studies have largely demonstrated evidence of synaptic strengthening occurring in a sleep-dependent manner in the hours following learning [1, 24, 48, 54].

Another important aspect of the hypothesis, related to its origins, is that the same transcriptomic studies that provided inspiration for it also contained additional revelations—and these have equally biologically important implications. For example, early transcriptomic datasets which showed that SD caused increases in IEG and *Glua1* expression in the neocortex also showed increases in *CHOP*, *BiP*, *Cdk110*, and *Hsp70*. This suggests that in addition to affecting pathways that mediate synaptic plasticity, SD has a range of effects on neuronal cell biology—including ER and mitochondrial functions and overall regulation of transcription. These effects of SD were subsequently confirmed by results a carefully designed microarray study comparing effects of different durations of ad lib sleep vs. SD on neocortical and hypothalamic transcripts [31]. In that study, using an unbiased screen of microarray data, the major conclusion was that sleep upregulates macromolecule biosynthesis pathways—including protein, cholesterol, and lipid biogenesis—and downregulates transcription and RNA processing. SD, on the other hand, downregulates cholesterol and lipid metabolism. These data emphasize that the majority of transcripts affected by sleep and SD in the brain are involved in basic cell biology of neurons—an idea which has been supported by recent studies finding changes in chromatin and organelle structure in neurons after a period of SD [51]. These changes may in turn (albeit indirectly) affect neuronal information processing.

What Cellular Changes Are Really Sleep-Dependent?: Caveats to Interpretation

Two broad strategies have been used to determine how sleep affects neurobiological function—experimental SD, to determine effects of losing sleep, and time-of-day studies, to determine processes that cycle together with the daily rhythm in sleep and wake. Both methods have contributed to omics-based approaches to our understanding of sleep function in the brain. However, both also have caveats that should be considered when interpreting these datasets.

As discussed in greater depth elsewhere [23], a major caveat for interpreting neurobiological data from SD experiments is that experimental strategies for implementation of SD vary from lab to lab. For example, some studies assessing SD effects on the brain have used so-called gentle handling, which typically involves cage tapping or shaking, and, if necessary, disturbance of the nesting material, to enforce wakefulness. A motivation behind this approach is applying the minimal stimulus necessary to sustain wake without providing additional, novel sensory stimuli. Other studies (including those providing data to support the synaptic homeostasis hypothesis) have opted for presentation of novel caging, switching bedding with that of a conspecific, treats, manipulanda, and running wheels. This approach has the advantage of promoting more spontaneous wakefulness, with less observation required. On the other hand, however, this approach also provides the sort of novel stimuli that are known to evoke plasticity on their own. Thus, the extent to which SD-evoked cellular responses in the neocortex are induced by novel stimuli vs. wakefulness per se remains unclear, and ultimately, confirmation of these findings using alternate SD methods is needed.

Another method has been used in the past to quantify neuronal transcript and protein changes as a function of sleep vs. wake, which does not rely on experimental SD (thus, circumventing the caveats described above). This strategy uses tissue sampling at different time of day, taking advantage of rodents' daily rhythms of sleep and wake. For example, Diering et al. sampled synaptosomes at ZT4 vs. ZT16 to identify differences in protein abundance and protein phosphorylation levels. The authors found that between these two timepoints, levels of multiple kinases, GluA1 and GluA2, and phosphorylated GluA1 increased in the postsynaptic density—all of which indicated higher overall excitatory synaptic strength during the dark phase [17]. While a reasonable interpretation is that this change was driven by changes due to sleep/wake cycles, an unaddressed issue is the extent to which the circadian clock mediates these changes.

The respective roles of circadian clock vs. sleep-wake regulation of these mechanisms were addressed by two recent studies which carried out both synaptosomal and whole-forebrain transcriptomics and proteomics and synaptosomal phosphoproteomics in the context of either ad lib sleep, or SD at different times of day [7, 35]. Using this strategy, the authors were able to assess which cycling transcripts and proteins were driven by rhythms in localization (i.e., cycling was present in synaptosomes but not whole-forebrain homogenate) or by sleep/wake cycles (i.e., cycling was present unless sleep-wake cycles were disrupted). The authors found that many more transcripts' abundance was rhythmic in the synaptosomal fraction (more than 2000) than across the forebrain as a whole (only 155)—suggesting that localization of these transcripts at synapses may be mediated by a daily rhythm. Of these cycling synaptic transcripts, nearly three quarters were no longer cycling when sleep-wake cycles were disrupted experimentally. In other words, rhythms of most synaptic transcripts' abundance are controlled by sleep and wake. Given this finding, it is not surprising that rhythms of protein abundance at synapses were also dependent on sleep-wake cycles, with cyclic abundance disrupted by SD [35]. Moreover, the authors found that a large number of synaptic proteins (more than 2000) undergo rhythmic phosphorylation, and this rhythm is driven by sleep-wake cycles [7].

A final, important caveat to interpretation for both experimental SD or time-of-day omics studies is that it cannot be assumed that cellular changes are uniform across the brain. In large part, these studies have used "bulk" methods, where starting materials include large volumes of brain tissue—e.g., whole forebrain, neocortex, hypothalamus, or cerebellum, or subcellular fractions thereof. This means that transcript changes quantified from these datasets reflect summed changes across multiple brain regions and cell types. Until recently, an unstated assumption is that neurobiological processes affected by sleep vs. wake were similar across the brain, but as described in greater detail below, this is proving to be inaccurate.

New Techniques Leading to New Neurobiological Insights

More recent studies have used recently developed genetic tools and approaches to provide a more granular view of the cellular effects of SD in the brain. These include RNAseq, which allows for quantification of splice variants of differentially spliced genes, as well as nonprotein-coding RNAs, and translating ribosome affinity purification (TRAP), which allows quantification of transcripts associated with ribosomal subunits, including those being actively transcribed. These studies have provided new insights into the biological processes affected by SD.

Recent studies have used RNAseq to quantify transcripts altered as a function of SD in the whole forebrain [35], neocortex [5, 18], and hippocampus [14, 21]. One example of data captured by RNAseq, but not identified in prior studies, is the finding that in the hippocampus, the abundance of several long noncoding RNAs (lncRNAs) and other nonprotein-coding RNA types is affected by SD [14, 21]. SD also selectively alters levels for specific slice variants of *Homer1* and *Fosb* (i.e., *Homer1a* and *ΔFosb*) [14]. Intriguingly, while these specific splice variants are transcribed in an activity-dependent manner to promote plasticity and are selectively upregulated after learning in the hippocampus, learning-driven increases in these splice isoforms are occluded by subsequent SD. This is one plausible mechanism by which post-learning sleep disruption could impair memory storage in the hippocampus [14]. This interpretation would be compatible with the synaptic homeostasis hypothesis described above and perhaps would be the first illustration of this mechanism as a contributing factor in SD-driven cognitive disruption. However, these same studies provided additional insights that are less compatible with the hypothesis and suggest other contributing mechanisms, as described below.

TRAP, which involves immune purification of ribosomes and associated transcripts from brain or tissue homogenate, has further refined our understanding of the effects of sleep disruption on the brain and how sleep interacts with prior learning. Using cell type-specific expression of affinity-tagged ribosomal proteins (e.g., eGFP-L10a or HA-Rpl22 [RiboTag]), this allows cell type-specific characterization of the effects of an intervention such as SD. The first sleep-related studies carried out using TRAP (with subsequent microarray analysis on affinity-purified mRNAs) focused on effects of sleep loss on astrocytes [4] and oligodendrocytes [3]. Among forebrain astrocytes, transcripts encoding proteins involved in cellular proliferation, cytoskeletal regulation, and progenitor development were associated with ribosomes at higher levels after sleep, while transcripts involved in cellular metabolism, endocytosis, and nucleotide binding were increased after SD [4]. Among forebrain oligodendrocytes, transcripts encoding proteins involved in lipid metabolism, cellular proliferation, and axon guidance were present on ribosomes at higher levels after sleep, while transcripts involved in apoptosis, cell differentiation, and chromatin remodeling were more abundant after SD [3].

More recently, TRAP has been used to profile ribosome-associated transcripts in different neuron subtypes. Our lab recently characterized SD-driven changes in ribosome-associated transcripts in pyramidal cells vs. parvalbumin-expressing (PV+) interneurons, in the neocortex vs. hippocampus, to test whether the effects of SD across cell types and brain structures were similar or different [42]. These studies, which combined TRAP following varying durations of sleep vs. SD with qPCR for activity-regulated genes involved in synaptic plasticity (including various IEGs) and core clock genes, demonstrated that neuronal responses to SD are more diverse than previously known. For example, while neocortical neurons of both types showed progressive increases in IEG and plasticity-related transcripts on ribosomes as SD progressed over 6 h, hippocampal neurons underwent far fewer changes in these transcripts with SD. In both brain structures, PV+ interneurons had blunted IEG and plasticity-related transcript responses to SD when compared with neighboring Camk2a + pyramidal neurons, even after 6 h of SD. SD-driven changes in core clock protein-encoding transcripts similarly varied as a function of brain structure and cell type. Using in situ hybridization for these same transcripts, we found that changes to expression of IEGs and *Pvalb* itself varied greatly after SD—in the neocortex by cell layer and in the hippocampus by subregion (e.g., CA1, CA3, or DG) [42]. Taken together, these findings present a picture where cellular responses to SD that are far more diverse than previously imagined.

TRAP combined with RNAseq (TRAPseq) has also been used as a strategy to determine whether the set transcripts (including splice variants) associated with ribosomes after SD differ from those present in bulk tissue from specific brain structures. For example, Lyons et al. recently found that transcript profiles on ribosomes in Camk2a + principal neurons in the hippocampus after SD differed from profiles (based on microarray from a previous study) of bulk hippocampal tissue. SD also appeared to cause selective enrichment/de-enrichment of certain splice variants from differentially spliced genes at the ribosome [30]. Intriguingly while pathway analysis of microarray data suggested effects of SD on transcription regulatory proteins, analysis of transcripts associated with ribosomes from excitatory neurons showed no evidence of these pathways being regulated. At the same time, transcripts encoding proteins that suppress translation are downregulated in microarray data after SD, which in theory could promote overall translation rates through derepression. However, again, no change was observed at the level of excitatory neurons' ribosomes. Together, these data suggest that either (1) the processes of gene transcription and protein translation are differentially regulated in the context of SD (leading to differential effects on transcripts vs. proteins in different cellular pathways) or (2) the cellular pathways affected (at both the transcriptional and translational levels) by SD in different cell populations in the hippocampus are distinct.

Assessing Neurobiological Mechanisms Engaged by Sleep During Active Memory Consolidation

How do these cellular mechanisms come into play in the context of memory storage, within neurons activated by learning in structures like the hippocampus? More recently, our lab used TRAPseq to characterize transcripts from different cell populations within the hippocampus and from ribosomes in different subcellular compartments within those cell populations [14]. These distinct pools of RNA were taken from the hippocampi from the same individual animals for direct comparison. To separate these populations of transcripts, hippocampi were taken from mice expressing HA-Rpl22 in Camk2a + neurons. As part of the RNA isolation process, samples were taken from both the cytosol of lysed cells (the component of the tissue typically used for RNAseq, qPCR, and microarray studies) and also from the resuspended plasma and organelle membrane-containing "pellet," which is usually generated (and then discarded) after tissue homogenization and centrifugation. Analysis of transcripts associated with the two fractions showed that they were clearly segregated based on the cellular functions of the proteins they encoded—cytosolic transcripts encoded proteins with functions in the cytoplasm and nucleus, while those in the membrane fraction encoded proteins with functions in the plasma membrane, ER and Golgi, and at synapses. These two sample fractions were then subdivided further, into a fraction sampled from the entire hippocampus (including all neurons and non-neuron cell types), a fraction affinity purified using an antibody against HA (i.e., including transcripts on ribosomes in excitatory hippocampal neurons), and a fraction affinity purified using an antibody against a highly phosphorylated ribosomal subunit S6. Because phosphorylation of S6 at these sites is mediated by numerous kinases that are activated in response to neuronal activity (including PKA, PKC, ERK, and S6K downstream of mTOR), this fraction represents transcripts on ribosomes from the most active neurons within the hippocampal circuit [29, 39]. These fractions were profiled from mice that either were kept in their home cage (a negative control for learning) or underwent single-trial contextual fear conditioning at ZT0. Mice in these two groups then either were allowed ad lib sleep in their home cage over the next 6 h or underwent 6 h of home cage SD (a manipulation which has previously been shown to disrupt consolidation of contextual fear memory) [22, 38].

This study design, followed by RNAseq of the resulting transcript pools, led to several surprising findings. Most surprisingly, we found that across all three cell population pools (input from the entire hippocampus, anti-HA TRAP representing excitatory neurons, and anti-pS6 TRAP representing highly active neurons), a large number of cytosolic transcripts (several hundred) were altered as a function of SD, but only a small handful of cytosolic transcripts were altered by learning—regardless of subsequent sleep or SD. These cytosolic transcripts included many identified in prior transcriptomic studies of SD effects on the brain, including IEGs such as *Arc* and *Cfos* [14]. In stark contrast, while SD alone significantly altered roughly 100–200 transcripts in the membrane fraction of each of the three cell population pools, these changes were dwarfed by the vastly larger number of transcripts (into the thousands) that were altered by prior learning. This finding has critical implications for our understanding of which neurobiological processes altered by sleep vs. SD could be important mediators of sleep function with respect to memory. Because most transcriptomic profiling of brain regions after sleep vs. SD has used the cytosolic fraction (not the membrane fraction) as the starting material, it is likely that SD-driven changes in neurobiological processes occurring at cell and organelle membranes, which are most relevant to memory storage, have been overlooked.

To better understand the biological processes that are affected at the interface of learning and subsequent sleep, we compared how transcripts on membrane-associated ribosomes in neurons were differentially affected when learning was followed by sleep (leading to successful memory consolidation) vs. SD (where consolidation is disrupted). When mice were sleep-deprived after learning, we found that (in both Camk2a + neurons and pS6+, highly active neurons) transcripts involved in cell stress pathway components (ER stress, unfolded protein response, and autophagy) are increased. This finding is consistent with prior transcriptomic studies of the effects of SD. Intriguingly, transcripts associated with glycolysis are also selectively altered on membrane-bound ribosomes when learning is followed by SD [14]. This pathway is upregulated in neurons, with enzymes localizing selectively around the synaptic compartment, as a neuronal response to metabolic stress [16, 26]. Moreover, in the most active neurons, post-learning SD led to increased expression of transcripts involved in GABAergic and GABA synapse neuromodulatory (endocannabinoid and nNOS) signaling pathways, suggesting enhanced inhibitory neurotransmission in the hippocampus with SD. In contrast, we found that when learning is followed by sleep, pS6+ neurons' ribosomes showed increased transcripts encoding actin cytoskeleton signaling components (a pathway essential for hippocampal LTP and memory consolidation) [24] and protein synthesis machinery (which also plays an essential role in sleep-dependent memory) [47]. It is tempting to speculate that synaptic remodeling and biogenesis would preferentially occur in the most active neurons in the network (i.e., those "engram neurons" activated during learning) to promote successful memory storage. Together, these data suggest a number of processes, from the subcellular to the hippocampal network level, which could contribute to disruption of memory storage by SD, and facilitation by post-learning sleep.

Another recent study from our lab took advantage of anti-pS6 TRAP followed by RNAseq as a method to identify neuron populations that are selectively activated in the hippocampus as a function of sleep vs. sleep deprivation [15]. In this study, expression levels for modules of co-regulated transcripts were quantified as a function of prior sleep time, across both freely sleeping and SD mice, using weighted gene co-expression network analysis (WGCNA). Using this technique, we identified modules of transcripts present on pS6+ ribosomes at levels that were inversely proportional to prior sleep time (i.e., the transcripts were upregulated as a function of wake time). Transcripts within these modules were then used for cell type-specific expression analysis (CSEA), to identify sets of transcripts co-expressed selectively in genetically defined cell types, based on prior transcriptomic profiling studies. This analysis revealed, somewhat unsurprisingly, that wake-active neuromodulatory inputs (of orexinergic and cholinergic neuron groups) were among the cell types activated in the hippocampus as a function of SD. However, it also identified somatostatin and neuropeptide Y-expressing (SST+ and NPY+) inhibitory interneurons among the population of selectively wake-active hippocampal neurons. We verified the selective activation of this population in the DG hilus (which contains many SST+/NPY+ interneurons), along with inhibition of surrounding excitatory DG granule cells, after SD using immunohistochemistry for IEGs (see also [13]). This network-level change after SD validated our unbiased TRAPseq findings from pS6+ hippocampal neurons [14] and suggested a clear mechanism by which SD could disrupt hippocampal memory storage. To test this, we used chemogenetics to selectively activate or inhibit DG SST+ interneurons in the hours following contextual fear conditioning, in an effort to mimic or block DG network-level changes caused by SD. We found that activation of the SST+ population decreased activity in neighboring DG granule cells and blocked fear memory consolidation, while inhibition of SST+ interneurons increased activity in neighboring granule cells and improved memory consolidation. Because SST+ interneurons are known to be highly active in response to cholinergic signaling [34], we speculated that SST+ interneuron activation due to SD might be driven by increased cholinergic input. We tested this using chemogenetics to selectively activate or inhibit medial septal cholinergic neurons, which project selectively to the hippocampus. We found that (similar to manipulations of SST+ interneurons), increasing cholinergic input to the hippocampus after learning reduced granule cell activation and blocked memory consolidation, while inhibition of cholinergic input increased granule cells' activity and improved consolidation [15]. Taken together, our initial unbiased approach led to a series of findings suggesting that SD interferes with memory via selective, acetylcholine-mediated activation of an inhibitory "gate" at the level of the hippocampus, which prevents memory processing within the DG [44].

Will Newer Technologies Lead to Newer Revelations?

Recent work has used RNAseq and TRAPseq to identify novel, testable, hypothetical mechanisms by which sleep promotes memory storage in the brain, and SD disrupts memory mechanisms. The technical advances of the past two decades have greatly expanded our view of the neurobiological processes affected by sleep and SD. With new technologies on the horizon, it is likely that our mechanistic understanding of sleep's role in memory will change dramatically in the coming years. These techniques are already being applied to investigate how sleep and SD affect brain function. For example, single-cell RNAseq and proteomics have recently been applied to assess cell type-specific SD effects in the neocortex, hypothalamus, and brainstem [28]—a strategy that will provide valuable insights into how different brain cell types (e.g., astrocytes and neurons) interact in the context of SD. Another recently developed technique, assay for transposase-accessible chromatin using sequencing (ATACseq), has been used to map changes to the landscape of chromatin across the genome as a function of sleep vs. sleep loss [25]. This work demonstrated that as little as 3 h of SD can dramatically change the accessibility of chromatin, particularly at intergenic sites distal to coding sequences (rather than more proximal promoter regions). This suggests that brief SD can change the binding of enhancers and repressors to DNA, which may in turn have longer-lasting implications for coordinated changes in gene expression. These changes are likely to affect memory processing, as sleep-dependent histone acetylation and methylation have recently been implicated in memory consolidation [19, 32]. Yet another evolving technology, spatial transcriptomics, has recently been used by our lab to address the question of how sleep and SD impact the transcriptome of different regions of the hippocampus in the context of fear memory consolidation [52]. Data from this study suggest that SD effects on hippocampal structures—even analogous structures like the granule and pyramidal cell body layers of DG, CA1, and CA3—are quite distinct from one another, leading to effects on different cellular pathways in the hours following learning. These recently developed techniques (and others still under development) will refine our understanding of how learning and sleep intersect at the cellular and network levels.

Future Directions: Linking Genetic Changes to Physiological Changes in the Brain During Sleep

Despite decades of study, the mechanistic link between the cell biological changes occurring in the brain during sleep and other aspects of sleep physiology (e.g., thalamocortical and hippocampal network oscillations, changes in neuromodulation) remains to be clarified [40, 41]. It is tempting to speculate regarding the role of network oscillations, which have been directly implicated in consolidation of new memories [6, 36–38, 45, 53] and sleep-dependent cortical plasticity [20, 33], in regulating intracellular (e.g., transcription and translation) pathways. Future studies will be needed to clarify which aspects of sleep physiology are not only drivers of memory storage but also drivers of the molecular events that accompany sleep and wake in the context of cognition.

References

1. Aton SJ, Seibt J, Dumoulin M, Jha SK, Steinmetz N, et al. Mechanisms of sleep-dependent consolidation of cortical plasticity. Neuron. 2009a;61:454–66.
2. Aton SJ, Seibt J, Frank MG. Sleep and memory. In: Encyclopedia of life science. Chichester: John Wiley and Sons, Ltd; 2009b.
3. Bellesi M, Pfister-Genskow M, Maret S, Keles S, Tononi G, Cirelli C. Effects of sleep and wake on oligodendrocytes and their precursors. J Neurosci. 2013;33:14288–300.
4. Bellesi M, de Vivo L, Tononi G, Cirelli C. Effects of sleep and wake on astrocytes: clues from molecular and ultrastructural studies. BMC Biol. 2015;13:66.
5. Bjorness TE, Kulkarni A, Rybalchenko V, Suzuki A, Bridges C, et al. An essential role for MEF2C in the cortical response to loss of sleep in mice. elife. 2020;9:e58331.
6. Boyce R, Glasgow SD, Williams S, Adamantidis A. Causal evidence for the role of REM sleep theta rhythm in contextual memory consolidation. Science. 2016;352:812–6.
7. Bruning F, Noya SB, Bange T, Koutsouli S, Rudolph JD, et al. Sleep-wake cycles drive daily dynamics of synaptic phosphorylation. Science. 2019;366:eaav3617.
8. Cirelli C, Tononi G. Differences in gene expression between sleep and waking as revealed by mRNA differential display. Brain Res Mol Brain Res. 1998;56:293–305.
9. Cirelli C, Tononi G. Differential expression of plasticity-related genes in waking and sleep and their regulation by the noradrenergic system. JNeurosci. 2000;20:9187–94.
10. Cirelli C, Tononi G. Locus Ceruleus control of state-dependent gene expression. J Neurosci. 2004;24:5410–9.
11. Cirelli C, Gutierrez CM, Tononi G. Extensive and divergent effects of sleep and wakefulness on brain gene expression. Neuron. 2004;41:35–43.
12. de Vivo L, Bellesi M, Marshall W, Bushong EA, Ellisman H, et al. Ultrastructural evidence for synaptic scaling across the wake/sleep cycle. Science. 2017;355:507–10.
13. Delorme JE, Kodoth V, Aton SJ. Sleep loss disrupts Arc expression in dentate gyrus neurons. Neurobiol Learn Mem. 2019;160:73–82.
14. Delorme J, Wang L, Kodoth V, Wang Y, Ma J, et al. Hippocampal neurons' cytosolic and membrane-bound ribosomal transcript profiles are differentially regulated by learning and subsequent sleep. Proc Natl Acad Sci USA. 2021a;118:e2108534118.
15. Delorme J, Wang L, Roig Kuhn F, Kodoth V, Ma J, et al. Sleep loss drives acetylcholine- and somatostatin interneuron-mediated gating of hippocampal activity, to inhibit memory consolidation. Proc Natl Acad Sci USA. 2021b;118:e2019318118.
16. Diaz-Garcia CM, Mongeon R, Lahmann C, Koveal D, Zucker H, Yellen G. Neuronal stimulation triggers neuronal glycolysis and not lactate uptake. Cell Metab. 2017;26:361–74.
17. Diering GH, Nirujogi RS, Roth RH, Worley PF, Pandey A, Huganir RL. Homer1a drives homeostatic scaling-down of excitatory synapses during sleep. Science. 2017;355:511–5.
18. Diessler S, Jan M, Emmenegger Y, Guex N, Middleton B, et al. A systems genetics resource and analysis of sleep regulation in the mouse. PLoS Biol. 2018;16:e2005750.
19. Duan R, Liu X, Wang T, Wu L, Gao X, Zhang Z. Histone acetylation regulation in sleep deprivation-induced spatial memory impairment. Neurochem Res. 2016;41:2223–32.
20. Durkin J, Suresh AK, Colbath J, Broussard C, Wu J, et al. Cortically coordinated NREM thalamocortical oscillations play an essential, instructive role in visual system plasticity. Proceedings National Academy of Sciences. 2017;114:10485–90.

21. Gaine ME, Bahl E, Chatterjee S, Michaelson JJ, Abel T, Lyons LC. Altered hippocampal transcriptome dynamics following sleep deprivation. Mol Brain. 2021;14:125.
22. Graves LA, Heller EA, Pack AI, Abel T. Sleep deprivation selectively impairs memory consolidation for contextual fear conditioning. Learn Mem. 2003;10:168–76.
23. Havekes R, Aton SJ. Impacts of sleep loss versus waking experience on brain plasticity: parallel or orthogonal? Trends in Neuroscience. 2020;43:385–93.
24. Havekes R, Park AJ, Tudor JC, Luczak VG, Hansen RT, et al. Sleep deprivation causes memory deficits by negatively impacting neuronal connectivity in hippocampal area CA1. elife. 2016;5. pii: e13424.
25. Hor CN, Yeung J, Jan M, Emmenegger Y, Hubbard J, et al. Sleep-wake-driven and circadian contributions to daily rhythms in gene expression and chromatin accessibility in the murine cortex. Proc Natl Acad Sci USA. 2019;116:25773–83.
26. Jang S, Nelson JC, Bend EG, Rodriguez-Laureano L, Tueros FG, et al. Glycolytic enzymes localize to synapses under energy stress to support synaptic function. Neuron. 2016;90:278–91.
27. Jenkins JG, Dallenbach KM. Obliviscence during sleep and waking. Am J Psychol. 1924;35:605–12.
28. Jha PK, Valekunja UK, Ray S, Nollet M, Reddy AB. Single-cell transcriptomics and cell-specific proteomics reveals molecular signatures of sleep. Communications Biology. 2022;5:846.
29. Knight ZA, Tan K, Birsoy K, Schmidt S, Garrison JL, et al. Molecular profiling of activated neurons by phosphorylated ribosome capture. Cell. 2012;151:1126–37.
30. Lyons LC, Chatterjee S, Vanrobaeys Y, Gaine ME, Abel T. Translational changes induced by acute sleep deprivation uncovered by TRAP-Seq. Mol Brain. 2020;13:165.
31. Mackiewicz M, Shockley KR, Romer MA, Galante RJ, Zimmerman JE, et al. Macromolecule biosynthesis – a key function of sleep. Physiol Genomics. 2007;31:441–57.
32. Massart R, Freyburger M, Suderman M, Paquet J, El Helou J, et al. The genome-wide landscape of DNA methylation and hydroxymethylation in response to sleep deprivation impacts on synaptic plasticity genes. Transl Psychiatry. 2014;4:e347.
33. Miyamoto D, Hirai D, Fung CC, Inutsuka A, Odagawa M, et al. Top-down cortical input during NREM sleep consolidates perceptual memory. Science. 2016;352:1315–8.
34. Munoz W, Tremblay R, Levenstein D, Rudy B. Layer-specific modulation of neocortical dendritic inhibition during active wakefulness. Science. 2017;355:954–9.
35. Noya SB, Colamea D, Bruning F, Spinnler A, Mircsof D, et al. The forebrain synaptic transcriptome is organized by clocks but its proteome is driven by sleep. Science. 2019;366:eaav2642.
36. Ognjanovski N, Maruyama D, Lashner N, Zochowski M, Aton SJ. CA1 hippocampal network activity changes during sleep-dependent memory consolidation. Front Syst Neurosci. 2014;8:61.
37. Ognjanovski N, Schaeffer S, Mofakham S, Wu J, Maruyama D, et al. Parvalbumin-expressing interneurons coordinate hippocampal network dynamics required for memory consolidation. Nat Commun. 2017;8:15039.
38. Ognjanovski N, Broussard C, Zochowski M, Aton SJ. Hippocampal network oscillations rescue memory consolidation deficits caused by sleep loss. Cereb Cortex. 2018;28:3711–23.
39. Pirbhoy PS, Farris S, Steward O. Synaptic activation of ribosomal protein S6 phosphorylation occurs locally in activated dendritic domains. Learn Mem. 2016;23:255–69.
40. Puentes-Mestril C, Aton SJ. Linking network activity to synaptic plasticity during sleep: hypotheses and recent data. Frontiers in Neural Circuits. 2017;11 https://doi.org/10.3389/fncir.2017.00061.
41. Puentes-Mestril C, Roach J, Niethard N, Zochowski M, Aton SJ. How rhythms of the sleeping brain tune memory and synaptic plasticity. Sleep. 2019;42. pii: zsz095.
42. Puentes-Mestril C, Delorme J, Wang L, Donnelly M, Popke D, et al. Sleep loss drives brain region- and cell type-specific alterations in ribosome-associated transcripts involved in synaptic plasticity and cellular timekeeping. J Neurosci. 2021;41:5386–98.
43. Rasch B, Born J. About sleep's role in memory. Physiol Rev. 2013;93:681–766.

44. Raven F, Aton S. The Engram's dark horse: how interneurons regulate state-dependent memory processing and plasticity. Front Neural Circuits. 2021;15:750541.
45. Skilling QM, Eniwaye B, Clawson BC, Shaver J, Ognjanovski N, et al. Acetylcholine-gated current translates wake neuronal firing rate information into a spike timing-based code in non-REM sleep, stabilizing neural network dynamics during memory consolidation. PLoS Comput Biol. 2021;17:e1009424.
46. Tononi G, Cirelli C. Sleep and synaptic homeostasis: a hypothesis. Brain Res Bull. 2003;62:143–50.
47. Tudor JC, Davis EJ, Peixoto L, Wimmer ME, van Tilborg E, et al. Sleep deprivation impairs memory by attenuating mTORC1-dependent protein synthesis. Sci Signal. 2016;9:ra41.
48. Vecsey CG, Baillie GS, Jaganath D, Havekes R, Daniels A, et al. Sleep deprivation impairs cAMP signalling in the hippocampus. Nature. 2009;461:1122–5.
49. Vyazovskiy VV, Cirelli C, Pfister-Genskow M, Faraguna U, Tononi G. Molecular and electrophysiological evidence for net synaptic potentiation in wake and depression in sleep. Nat Neurosci. 2008;11:200–8.
50. Walker MP, Stickgold R. Sleep-dependent learning and memory consolidation. Neuron. 2004;44:121–33.
51. Wang L, Aton SJ. Perspective - ultrastructural analyses reflect the effects of sleep and sleep loss on neuronal cell biology. Sleep. 2022;45:zsac047.
52. Wang L, Park L, Wu W, King D, Vega Medina A, Raven F, Martinez JD, Ensing A, McDonald K, Yang Z, Jiang S, Aton SJ. Sleep-dependent engram reactivation during hippocampal memory consolidation associated with subregion-specific biosynthetic changes. iScience. 2024;27(4):109408. https://doi.org/10.1016/j.isci.2024.109408
53. Xia F, Richards BA, Tran MM, Josselyn SA, Takehara-Nishiuchi K, Frankland PW. Parvalbumin-positive interneurons mediate neocortical-hippocampal interactions that are necessary for memory consolidation. elife. 2017;6:e27868.
54. Yang G, Lai CS, Cichon J, Ma L, Li W, Gan WB. Sleep promotes branch-specific formation of dendritic spines after learning. Science. 2014;344:1173–8.
55. Zhou Y, Lai CSW, Bai Y, Li W, Zhou R, et al. REM sleep promotes experience-dependent dendritic spine elimination in the mouse cortex. Nat Commun. 2020;11:4819.

Index

P. Gehrman et al. (eds.), *Genetics of Sleep and Sleep Disorders*,
https://doi.org/10.1007/978-3-031-62723-1